Springer Handbook of Enzymes Volume 22

Dietmar Schomburg and
Ida Schomburg (Eds.)

Springer Handbook of Enzymes

Volume 22
Class 1 · Oxidoreductases VII
EC 1.4

coedited by Antje Chang

Second Edition

 Springer

Professor DIETMAR SCHOMBURG
e-mail: d.schomburg@uni-koeln.de

Dr. IDA SCHOMBURG
e-mail: i.schomburg@uni-koeln.de

Dr. ANTJE CHANG
e-mail: a.chang@uni-koeln.de

University to Cologne
Institute for Biochemistry
Zülpicher Strasse 47
50674 Cologne
Germany

Library of Congress Control Number: 2005921724

ISBN 3-540-23848-4   2nd Edition Springer Berlin Heidelberg New York

The first edition was published as Volume 6 (ISBN 3-540-56435-7) of the "Enzyme Handbook".

Springer is a part of Springer Science+Business Media
springeronline.com
© Springer-Verlag Berlin Heidelberg 2005
Printed in The Netherlands

Cover design: Erich Kirchner, Heidelberg
Typesetting: medionet AG, Berlin

Printed on acid-free paper   2/3141/xv-5 4 3 2 1 0

Attention all Users
of the "Springer Handbook of Enzymes"

Information on this handbook can be found on the internet at
http://www.springeronline.com
choosing "Chemistry" and then "Reference Works".

A complete list of all enzyme entries either as an alphabetical Name Index or as the EC-Number Index is available at the above mentioned URL. You can download and print them free of charge.

A complete list of all synonyms (> 25,000 entries) used for the enzymes is available in print form (ISBN 3-540-41830-X).

Save 15 %

We recommend a standing order for the series to ensure you automatically receive all volumes and all supplements and save 15 % on the list price.

Preface

Today, as the full information about the genome is becoming available for a rapidly increasing number of organisms and transcriptome and proteome analyses are beginning to provide us with a much wider image of protein regulation and function, it is obvious that there are limitations to our ability to access functional data for the gene products – the proteins and, in particular, for enzymes. Those data are inherently very difficult to collect, interpret and standardize as they are widely distributed among journals from different fields and are often subject to experimental conditions. Nevertheless a systematic collection is essential for our interpretation of genome information and more so for applications of this knowledge in the fields of medicine, agriculture, etc. Progress on enzyme immobilisation, enzyme production, enzyme inhibition, coenzyme regeneration and enzyme engineering has opened up fascinating new fields for the potential application of enzymes in a wide range of different areas.

The development of the enzyme data information system BRENDA was started in 1987 at the German National Research Centre for Biotechnology in Braunschweig (GBF) and is now continuing at the University at Cologne, Institute of Biochemistry. The present book "Springer Handbook of Enzymes" represents the printed version of this data bank. The information system has been developed into a full metabolic database.

The enzymes in this Handbook are arranged according to the Enzyme Commission list of enzymes. Some 3,700 "different" enzymes are covered. Frequently enzymes with very different properties are included under the same EC-number. Although we intend to give a representative overview on the characteristics and variability of each enzyme, the Handbook is not a compendium. The reader will have to go to the primary literature for more detailed information. Naturally it is not possible to cover all the numerous literature references for each enzyme (for some enzymes up to 40,000) if the data representation is to be concise as is intended.

It should be mentioned here that the data have been extracted from the literature and critically evaluated by qualified scientists. On the other hand, the original authors' nomenclature for enzyme forms and subunits is retained. In order to keep the tables concise, redundant information is avoided as far as possible (e.g. if K_m values are measured in the presence of an obvious cosubstrate, only the name of the cosubstrate is given in parentheses as a commentary without reference to its specific role).

The authors are grateful to the following biologists and chemists for invaluable help in the compilation of data: Cornelia Munaretto and Dr. Antje Chang.

Cologne
Spring 2005

Dietmar Schomburg, Ida Schomburg

List of Abbreviations

A	adenine
Ac	acetyl
ADP	adenosine 5'-diphosphate
Ala	alanine
All	allose
Alt	altrose
AMP	adenosine 5'-monophosphate
Ara	arabinose
Arg	arginine
Asn	asparagine
Asp	aspartic acid
ATP	adenosine 5'-triphosphate
Bicine	N,N'-bis(2-hydroxyethyl)glycine
C	cytosine
cal	calorie
CDP	cytidine 5'-diphosphate
CDTA	trans-1,2-diaminocyclohexane-N,N,N,N-tetraacetic acid
CMP	cytidine 5'-monophosphate
CoA	coenzyme A
CTP	cytidine 5'-triphosphate
Cys	cysteine
d	deoxy-
D-	(and L-) prefixes indicating configuration
DFP	diisopropyl fluorophosphate
DNA	deoxyribonucleic acid
DPN	diphosphopyridinium nucleotide (now NAD^+)
DTNB	5,5'-dithiobis(2-nitrobenzoate)
DTT	dithiothreitol (i.e. Cleland's reagent)
EC	number of enzyme in Enzyme Commission's system
E. coli	Escherichia coli
EDTA	ethylene diaminetetraacetate
EGTA	ethylene glycol bis(-aminoethyl ether) tetraacetate
ER	endoplasmic reticulum
Et	ethyl
EXAFS	extended X-ray absorption fine structure
FAD	flavin-adenine dinucleotide
FMN	flavin mononucleotide (riboflavin 5'-monophosphate)
Fru	fructose
Fuc	fucose
G	guanine
Gal	galactose

GDP	guanosine 5'-diphosphate
Glc	glucose
GlcN	glucosamine
GlcNAc	N-acetylglucosamine
Gln	glutamine
Glu	glutamic acid
Gly	glycine
GMP	guanosine 5'-monophosphate
GSH	glutathione
GSSG	oxidized glutathione
GTP	guanosine 5'-triphosphate
Gul	gulose
h	hour
H4	tetrahydro
HEPES	4-(2-hydroxyethyl)-1-piperazineethane sulfonic acid
His	histidine
HPLC	high performance liquid chromatography
Hyl	hydroxylysine
Hyp	hydroxyproline
IAA	iodoacetamide
Ig	immunoglobulin
Ile	isoleucine
Ido	idose
IDP	inosine 5'-diphosphate
IMP	inosine 5'-monophosphate
ITP	inosine 5'-triphosphate
K_m	Michaelis constant
L-	(and D-) prefixes indicating configuration
Leu	leucine
Lys	lysine
Lyx	lyxose
M	mol/l
mM	millimol/l
m-	meta-
Man	mannose
MES	2-(N-morpholino)ethane sulfonate
Met	methionine
min	minute
MOPS	3-(N-morpholino)propane sulfonate
Mur	muramic acid
MW	molecular weight
NAD^+	nicotinamide-adenine dinucleotide
NADH	reduced NAD
$NADP^+$	NAD phosphate
NADPH	reduced NADP
NAD(P)H	indicates either NADH or NADPH
NBS	N-bromosuccinimide

NDP	nucleoside 5'-diphosphate
NEM	N-ethylmaleimide
Neu	neuraminic acid
NMN	nicotinamide mononucleotide
NMP	nucleoside 5'-monophosphate
NTP	nucleoside 5'-triphosphate
o-	*ortho-*
Orn	ornithine
p-	*para-*
PBS	phosphate-buffered saline
PCMB	*p*-chloromercuribenzoate
PEP	phosphoenolpyruvate
pH	$-\log10[H^+]$
Ph	phenyl
Phe	phenylalanine
PHMB	*p*-hydroxymercuribenzoate
PIXE	proton-induced X-ray emission
PMSF	phenylmethane-sulfonylfluoride
p-NPP	*p*-nitrophenyl phosphate
Pro	proline
Q_{10}	factor for the change in reaction rate for a 10 °C temperature increase
Rha	rhamnose
Rib	ribose
RNA	ribonucleic acid
mRNA	messenger RNA
rRNA	ribosomal RNA
tRNA	transfer RNA
Sar	N-methylglycine (sarcosine)
SDS-PAGE	sodium dodecyl sulfate polyacrylamide gel electrophoresis
Ser	serine
T	thymine
t_H	time for half-completion of reaction
Tal	talose
TDP	thymidine 5'-diphosphate
TEA	triethanolamine
Thr	threonine
TLCK	N^α-*p*-tosyl-L-lysine chloromethyl ketone
T_m	melting temperature
TMP	thymidine 5'-monophosphate
Tos-	tosyl-(*p*-toluenesulfonyl-)
TPN	triphosphopyridinium nucleotide (now NADP$^+$)
Tris	tris(hydroxymethyl)-aminomethane
Trp	tryptophan
TTP	thymidine 5'-triphosphate
Tyr	tyrosine
U	uridinc
U/mg	μmol/(mg*min)

UDP	uridine 5'-diphosphate
UMP	uridine 5'-monophosphate
UTP	uridine 5'-triphosphate
Val	valine
Xaa	symbol for an amino acid of unknown constitution in peptide formula
XAS	X-ray absorption spectroscopy
Xyl	xylose

List of Deleted and Transferred Enzymes

Since its foundation in 1956 the Nomenclature Committee of the International Union of Biochemistry and Molecular Biology (NC-IUBMB) has continually revised and updated the list of enzymes. Entries for new enzymes have been added, others have been deleted completely, or transferred to another EC number in the original class or to different EC classes, catalyzing other types of chemical reactions. The old numbers have not been allotted to new enzymes; instead the place has been left vacant or cross-references given to the changes in nomenclature.

Deleted and Transferred Enzymes
For EC class 1.4 these changes are:

Recommended name	Old EC number	Alteration
D-proline reductase	1.4.1.6	deleted, included in EC 1.4.4.1
tyramine oxidase	1.4.3.9	deleted, included in EC 1.4.3.4
D-proline reductase (dithiol)	1.4.4.1	transferred to EC 1.21.4.1

Index of Recommended Enzyme Names

Description of Data Fields

All information except the nomenclature of the enzymes (which is based on the recommendations of the Nomenclature Committee of IUBMB (International Union of Biochemistry and Molecular Biology) and IUPAC (International Union of Pure and Applied Chemistry) is extracted from original literature (or reviews for very well characterized enzymes). The quality and reliability of the data depends on the method of determination, and for older literature on the techniques available at that time. This is especially true for the fields Molecular Weight and Subunits.

The general structure of the fields is: **Information – Organism – Commentary – Literature**

The information can be found in the form of numerical values (temperature, pH, K_m etc.) or as text (cofactors, inhibitors etc.).

Sometimes data are classified as Additional Information. Here you may find data that cannot be recalculated to the units required for a field or also general information being valid for all values. For example, for Inhibitors, Additional Information may contain a list of compounds that are not inhibitory.

The detailed structure and contents of each field is described below. If one of these fields is missing for a particular enzyme, this means that for this field, no data are available.

1 Nomenclature

EC number
The number is as given by the IUBMB, classes of enzymes and subclasses defined according to the reaction catalyzed.

Systematic name
This is the name as given by the IUBMB/IUPAC Nomenclature Committee

Recommended name
This is the name as given by the IUBMB/IUPAC Nomenclature Committee

Synonyms
Synonyms which are found in other databases or in the literature, abbreviations, names of commercially available products. If identical names are frequently used for different enzymes, these will be mentioned here, cross references are given. If another EC number has been included in this entry, it is mentioned here.

CAS registry number

The majority of enzymes have a single chemical abstract (CAS) number. Some have no number at all, some have two or more numbers. Sometimes two enzymes share a common number. When this occurs, it is mentioned in the commentary.

2 Source Organism

For listing organisms their systematic name is preferred. If these are not mentioned in the literature, the names from the respective literature are used. For example if an enzyme from yeast is described without being specified further, yeast will be the entry. This field defines the code numbers for the organisms in which the enzyme with the respective EC number is found. These code numbers (form <_>) are displayed together with each entry in all fields of Brenda where organism-specific information is given.

3 Reaction and Specificity

Catalyzed reaction

The reaction as defined by the IUBMB. The commentary gives information on the mechanism, the stereochemistry, or on thermodynamic data of the reaction.

Reaction type

According to the enzyme class a type can be attributed. These can be oxidation, reduction, elimination, addition, or a name (e.g. Knorr reaction)

Natural substrates and products

These are substrates and products which are metabolized in vivo. A natural substrate is only given if it is mentioned in the literature. The commentary gives information on the pathways for which this enzyme is important. If the enzyme is induced by a specific compound or growth conditions, this will be included in the commentary. In *Additional information* you will find comments on the metabolic role, sometimes only assumptions can be found in the references or the natural substrates are unknown.

In the listings, each natural substrate (indicated by a bold **S**) is followed by its respective product (indicated by a bold **P**). Products are given with organisms and references included only if the respective authors were able to demonstrate the formation of the specific product. If only the disappearance of the substrate was observed, the product is included without organisms of references. In cases with unclear product formation only a ? as a dummy is given.

Substrates and products

All natural or synthetic substrates are listed (not in stoichiometric quantities). The commentary gives information on the reversibility of the reaction, on isomers accepted as substrates and it compares the efficiency of substrates. If a specific substrate is accepted by only one of several isozymes, this will be stated here.

The field *Additional Information* summarizes compounds that are not accepted as substrates or general comments which are valid for all substrates.

In the listings, each substrate (indicated by a bold **S**) is followed by its respective product (indicated by a bold **P**). Products are given with organisms and references included if the respective authors demonstrated the formation of the specific product. If only the disappearance of the substrate was observed, the product will be included without organisms or references. In cases with unclear product formation only a ? as a dummy is given.

Inhibitors

Compounds found to be inhibitory are listed. The commentary may explain experimental conditions, the concentration yielding a specific degree of inhibition or the inhibition constant. If a substance is activating at a specific concentration but inhibiting at a higher or lower value, the commentary will explain this.

Cofactors, prosthetic groups

This field contains cofactors which participate in the reaction but are not bound to the enzyme, and prosthetic groups being tightly bound. The commentary explains the function or, if known, the stereochemistry, or whether the cofactor can be replaced by a similar compound with higher or lower efficiency.

Activating Compounds

This field lists compounds with a positive effect on the activity. The enzyme may be inactive in the absence of certain compounds or may require activating molecules like sulfhydryl compounds, chelating agents, or lipids. If a substance is activating at a specific concentration but inhibiting at a higher or lower value, the commentary will explain this.

Metals, ions

This field lists all metals or ions that have activating effects. The commentary explains the role each of the cited metal has, being either bound e.g. as Fe-S centers or being required in solution. If an ion plays a dual role, activating at a certain concentration but inhibiting at a higher or lower concentration, this will be given in the commentary.

Turnover number (min^{-1})

The k_{cat} is given in the unit min^{-1}. The commentary lists the names of the substrates, sometimes with information on the reaction conditions or the type of reaction if the enzyme is capable of catalyzing different reactions with a single substrate. For cases where it is impossible to give the turnover number in the defined unit (e.g., substrates without a defined molecular weight, or an undefined amount of protein) this is summarized in Additional Information.

Specific activity (U/mg)

The unit is micromol/minute/milligram of protein. The commentary may contain information on specific assay conditions or if another than the natural substrate was used in the assay. Entries in Additional Information are included if the units of the activity are missing in the literature or are not calculable to the obligatory unit. Information on literature with a detailed description of the assay method may also be found.

K_m-Value (mM)

The unit is mM. Each value is connected to a substrate name. The commentary gives, if available, information on specific reaction condition, isozymes or presence of activators. The references for values which cannot be expressed in mM (e.g. for macromolecular, not precisely defined substrates) are given in Additional Information. In this field we also cite literature with detailed kinetic analyses.

K_i-Value (mM)

The unit of the inhibition constant is mM. Each value is connected to an inhibitor name. The commentary gives, if available, the type of inhibition (e.g. competitive, non-competitive) and the reaction conditions (pH-value and the temperature). Values which cannot be expressed in the requested unit and references for detailed inhibition studies are summerized under Additional information.

pH-Optimum

The value is given to one decimal place. The commentary may contain information on specific assay conditions, such as temperature, presence of activators or if this optimum is valid for only one of several isozymes. If the enzyme has a second optimum, this will be mentioned here.

pH-Range

Mostly given as a range e.g. 4.0–7.0 with an added commentary explaining the activity in this range. Sometimes, not a range but a single value indicating the upper or lower limit of enzyme activity is given. In this case, the commentary is obligatory.

Temperature optimum (°C)

Sometimes, if no temperature optimum is found in the literature, the temperature of the assay is given instead. This is always mentioned in the commentary.

Temperature range (°C)

This is the range over which the enzyme is active. The commentary may give the percentage of activity at the outer limits. Also commentaries on specific assay conditions, additives etc.

4 Enzyme Structure

Molecular weight

This field gives the molecular weight of the holoenzyme. For monomeric enzymes it is identical to the value given for subunits. As the accuracy depends on the method of determination this is given in the commentary if provided in the literature. Some enzymes are only active as multienzyme complexes for which the names and/or EC numbers of all participating enzymes are given in the commentary.

Subunits

The tertiary structure of the active species is described. The enzyme can be active as a monomer a dimer, trimer and so on. The stoichiometry of subunit composition is given. Some enzymes can be active in more than one state of complexation with differing effectivities. The analytical method is included.

Posttranslational modifications

The main entries in this field may be proteolytic modification, or side-chain modification, or no modification. The commentary will give details of the modifications e.g.:
-   proteolytic modification <1> (<1>, propeptide Name) [1];
-   side-chain modification <2> (<2>, N-glycosylated, 12% mannose) [2];
-   no modification [3]

5 Isolation / Preparation / Mutation / Application

Source / tissue

For multicellular organisms, the tissue used for isolation of the enzyme or the tissue in which the enzyme is present is given. Cell-lines may also be a source of enzymes.

Localization

The subcellular localization is described. Typical entries are: cytoplasm, nucleus, extracellular, membrane.

Purification

The field consists of an organism and a reference. Only references with a detailed description of the purification procedure are cited.

Renaturation

Commentary on denaturant or renaturation procedure.

Crystallization

The literature is cited which describes the procedure of crystallization, or the X-ray structure.

Cloning

Lists of organisms and references, sometimes a commentary about expression or gene structure.

Engineering

The properties of modified proteins are described.

Application

Actual or possible applications in the fields of pharmacology, medicine, synthesis, analysis, agriculture, nutrition are described.

6 Stability

pH-Stability

This field can either give a range in which the enzyme is stable or a single value. In the latter case the commentary is obligatory and explains the conditions and stability at this value.

Temperature stability

This field can either give a range in which the enzyme is stable or a single value. In the latter case the commentary is obligatory and explains the conditions and stability at this value.

Oxidation stability

Stability in the presence of oxidizing agents, e.g. O_2, H_2O_2, especially important for enzymes which are only active under anaerobic conditions.

Organic solvent stability

The stability in the presence of organic solvents is described.

General stability information

This field summarizes general information on stability, e.g., increased stability of immobilized enzymes, stabilization by SH-reagents, detergents, glycerol or albumins etc.

Storage stability

Storage conditions and reported stability or loss of activity during storage.

References

Authors, Title, Journal, Volume, Pages, Year.

1 Nomenclature

EC number
1.4.1.1

Systematic name
L-alanine:NAD$^+$ oxidoreductase (deaminating)

Recommended name
alanine dehydrogenase

Synonyms
(S)alanine dehydrogenase
(S)alanine:NAD oxidoreductase
40 kDa antigen
AlaDH
L-alanine dehydrogenase
NAD-dependent alanine dehydrogenase
NAD-linked alanine dehydrogenase
NADH-dependent alanine dehydrogenase
TB43
alanine oxidoreductase
α-alanine dehydrogenase
dehydrogenase, alanine

CAS registry number
9029-06-5

2 Source Organism

<1> *Geobacillus stearothermophilus* (gene cloned and expressed in Escherichia coli [1]) [1, 8, 19, 58, 64]
<2> *Bacillus sphaericus* (gene cloned and expressed in Escherichia coli [8]) [2, 8, 9, 19, 21, 37, 64]
<3> *Acholeplasma laidlawii* (mycoplasma [3]) [3]
<4> *Bacillus megaterium* [4]
<5> *Frankia sp.* [5]
<6> *Cunninghamella elegans* [6]
<7> *Saccharopolyspora erythrea* [7]
<8> *Rhodobacter capsulatus* [10]
<9> *Streptomyces aureofaciens* [11, 13, 64]

<10> *Streptomyces fradiae* [12]
<11> *Bacillus subtilis* [14, 20, 21, 24, 25, 27, 28, 33, 40, 45, 50, 52, 55, 58, 62, 63, 64]
<12> *Propionibacterium freudenreichii* (subsp. shermanii) [15]
<13> *Pseudomonas sp.* (strain MA [16,22]; strain AM1 [22]) [16, 22, 64]
<14> *Bacillus cereus* (strain T [45]) [17, 30, 45, 64]
<15> *Desulfotomaculum ruminis* [18]
<16> *Desulfovibrio sp.* (strain 20020 and strain 20028 [18]) [18]
<17> *Bacillus sp.* (screening of thermostable alanine dehydrogenase in thermophilic Bacillus strains [19]) [19]
<18> *Streptomyces phaeochromogenes* [21]
<19> *Bacillus flavothermus* [23]
<20> *Euglena gracilis* (Z [29]) [29]
<21> *Rhizobium japonicum* [31]
<22> *Rhodopseudomonas capsulata* [32]
<23> *Thermus thermophilus* [33, 64]
<24> *Halobacterium salinarum* [34, 36, 41, 45, 64]
<25> *Streptomyces clavuligerus* [21, 35]
<26> *Halobacterium cutirubrum* [38, 45, 64]
<27> *Anabaena cylindrica* [31, 39, 64]
<28> *Bacillus licheniformis* [42, 64]
<29> *Streptomyces hygroscopicus* [21, 43]
<30> *Bacillus thuringiensis* [44]
<31> *Halobacterium halobium* [45]
<32> *Desulfovibrio desulfuricans* [26]
<33> *Bacillus coagulans* [19]
<34> *Bacillus japonicum* [64]
<35> *Streptomyces griseoluteus* [21]
<36> *Streptomyces ruber* [21]
<37> *Streptomyces albolongus* [21]
<38> *Streptomyces lydicus* [21]
<39> *Streptomyces caespitosus* [21]
<40> *Streptomyces coelicolor* [21]
<41> *Streptomyces melanosporea* [21]
<42> *Streptomyces olivochromogenes* [21]
<43> *Streptomyces albus* [21]
<44> *Streptomyces griseus* [21]
<45> *Streptomyces roseochromogenes* [21]
<46> *Streptomyces roseus* [21]
<47> *Streptomyces bobiliae* [21]
<48> *Streptomyces flavus* [21]
<49> *Streptomyces aureus* [21]
<50> *Bacillus caldolyticus* [23]
<51> *Phormidium lapideum* [28, 49, 54]
<52> *Bilophila wadsworthia* [46]
<53> *Mycobacterium tuberculosis* [47, 53, 64]
<54> *Rhizobium leguminosarum* (bv. viciae strain 3841 [48]) [48]

<55> *Bacterium sp.* (psychrophile strain PA-43, isolated from a sea urchin [49])
     [49]
<56> *Vibrio proteolyticus* [49, 58]
<57> *Shewanella sp.* (strain Ac10 [58]) [49, 58]
<58> *Rhizobium sp.* [51]
<59> *Mycobacterium smegmatis* [56, 59, 61]
<60> *Archaeoglobus fulgidus* (glycine and alanine dehydrogenase activities are
     catalyzed by the same proteon in Mycobacterium smegamatis [57]) [57]
<61> *Carnobacterium sp.* (strain St2 [58]) [58]
<62> *Enterobacter aerogenes* [60]

3 Reaction and Specificity

Catalyzed reaction
   L-alanine + H_2O + NAD^+ = pyruvate + NH_3 + NADH + H^+ (<1>, reductive
   amination: sequential mechanism with partially random binding [1]; <2>,
   reductive amination proceeds through a sequential mechanism containing
   partially random binding. NADH binds first to the enzyme, and then pyruvate and ammonia bind in a random fashion [2]; <11>, predominantly ordered kinetic mechanism in which NAD^+ adds before L-Ala, and ammonia,
   pyruvate, and NADH are released in that order [24]; <11>, mechanism [25];
   <2>, sequential ordered ternary-binary mechanism, the enzyme is A-stereospecific, the pro-R hydrogen at C-4 of the reduced nicotinamide ring of
   NADH is exclusively transferred to pyruvate [37]; <58>, the kinetic mechanism at pH 8.5 is determined to be ter-bi Theorell-Chance. In the amination
   direction, the substrates add in the order: NADH, NH_4^+, pyruvate, with NH_4^+
   binding in rapid-equilibrium. In the reverse direction, NAD adds first followed by L-Ala [51]; <53>, oxidative deamination proceeds by an random-
   ordered mechanism [53]; <62>, sequential ordered binary-ternary mechanism. NAD^+ binds first to the enzyme, followed by L-Ala. The products are
   released in the order: NH_4^+, pyruvate and NADH [60])

Reaction type
   oxidation
   oxidative deamination
   redox reaction
   reduction
   reductive amination

Natural substrates and products
  S L-Ala + H_2O + NAD^+ <1, 5, 7, 8, 9, 11, 12, 13, 21, 22, 53, 59> (<5,7>, NH_4^+
   assimilating enzyme [5,7]; <7>, enzyme formation is effectively induced
   by Ala [7]; <8>, alternative route for ammonia assimilation when glutamine synthetase is inactivated [10]; <9>, physiological role in nitrogen
   metabolism in Streptomycetes [11]; <12>, key enzyme in the metabolism
   of alanine, Ala is onyl fermented after lactate exhaustion and then at a
   slow rate [15]; <13>, role in NH_3 assimilation and alanine catabolism

[16]; <21>, enzyme is involved in primary ammonium assimilation and the rapid formation of key metabolites pyruvate and NH_3 from L-Ala and also in cell differentiation [31]; <22>, no function in NH_4^+-assimilation but rather is required to supply the cells with the appropriate quantities of organic carbon when the organism grows at the expense of L-Ala as C-source and N-source [32]; <53>, since the physiological environment of the organism has a neutral pH, it can be assumed that the enzyme catalyzes exclusively the formation of L-Ala [53]; <59>, enzyme is required for utilization of alanine as a sole nitrogen source, enzyme is required for optimal growth under anaerobic conditions [56]; <11>, the physiological role is to catabolize L-Ala to pyruvate and NH_3, inducible by L-Ala, D-Ala and 11 other D-amino acids. The enzyme catabolizes L-Ala, and thereby limits the amount of L-Ala available to alanine racemase for the synthesis of D-Ala [63]) (Reversibility: ? <1, 5, 7, 8, 11, 12, 13, 21, 22, 53, 59> [1, 5, 7, 10, 11, 15, 16, 31, 32, 53, 56, 63]) [1, 5, 7, 10, 11, 15, 16, 31, 32, 53, 56, 63]

P pyruvate + NH_3 + NADH

S pyruvate + NH_3 + NADH <54> (<54>, primary route for alanine synthesis in isolated bacteroids, alanine synthesis and secretion contributes to the efficiency of N_2-fixation and therefore biomass accumulation [48]) (Reversibility: ? <54> [48]) [48]

P L-Ala + NAD^+ + H_2O

S Additional information <7, 11, 20, 29, 52, 59> (<7>, important role in biosynthesis of erythromycin [7]; <20>, circadian oscillations in alanine dehydrogenase [29]; <29>, enzyme activity during production phase of the antibiotic A 6599, enzyme induction by an excess of alanine or ammonia [43]; <52>, enzyme is involved in taurine metabolism [46]; <59>, strong induction immediately after deflection from aerobic growth suggests that alanine dehydrogenase may be required for the adaption from aerobic growth to anaerobic dormacy, the induction of alanine dehydrogenase may also support the maintenance of the NAD^+ pool when oxygen as the terminal electron acceptor becomes limiting [61]; <11>, the enzyme is required for normal sporulation [62]) [7, 29, 43, 46, 61, 62]

P ?

Substrates and products

S 2-oxo-3-methylbutanoate + NH_3 + NADH <2> (Reversibility: ? <2> [37]) [37]

P L-Val + NAD^+ + H_2O

S 2-oxobutanoate + NH_3 + NADH <1, 2, 10, 18, 24> (<2>, 62% of the activity with pyruvate [2]; <10>, 15% of maximal activity [12]; <18>, 15.8% of the activity with pyruvate [21]; <25>, no activity [35]) (Reversibility: r <1, 2, 24> [1, 3, 36, 37]; ? <2, 10, 18> [2, 12, 21]) [1, 2, 3, 12, 21, 36, 37]

P 2-aminobutanoate + H_2O + NAD^+

S 2-oxobutanoate + NH_3 + NADH <25, 51> (<25>, 3.7% of the activity with pyruvate [35]; <51>, 4.1% of the activity with pyruvate [28]) (Reversibility: ? <25, 51> [28, 35]) [28, 35]

P 2-aminobutanoate + H_2O + NAD^+

S 2-oxoglutarate + NH_3 + NADH <4> (<4>, 11.7% of the activity with pyruvate [4]; <26>, no activity [38]) (Reversibility: ? <4> [4]) [4]

P L-Gln + NAD^+ + H_2O

S 3-bromopyruvate + NH_3 + NADH <4> (<4>, 5.3% of the activity with pyruvate [4]) (Reversibility: ? <4> [4]) [4]

P 2-amino-3-bromopropanoate + NAD^+ + H_2O

S 3-fluoropyruvate + NH_3 + NADH <2> (<2>, 97% of the activity with pyruvate [2]) (Reversibility: ? <2> [2, 9]) [2, 9]

P 3-fluoro-L-alanine + H_2O + NAD^+ <2> [9]

S 3-hydroxypyruvate + NH_3 + NADH <2, 18, 25> (<25>, 6.9% of the activity with pyruvate [35]; <2>, 123% of the activity with pyruvate [2]; <18>, 32.3% of the activity with pyruvate [21]) (Reversibility: ? <2, 18, 25> [2, 21, 35]) [2, 21, 35]

P L-Ser + H_2O + NAD^+

S 3-hydroxypyruvate + NH_3 + NADH <51> (<51>, 15.4% of the activity with pyruvate [28]) (Reversibility: ? <51> [28]) [28]

P Ser + H_2O + NAD^+

S 4-methyl-2-oxopentanoate + NH_3 + NADH <4, 26> (<4>, 1.6% of the activity with pyruvate [4]) (Reversibility: ? <4, 26> [4, 38]) [4, 38]

P 2-amino-4-methylpentanoate + NAD^+ + H_2O

S D-Ala + H_2O + NAD^+ <4, 52> (<4>, 32.3% of the activity with L-Ala [4]; <52>, 3.5% of the activity with L-Ala [46]) (Reversibility: ? <4, 52> [4, 46]) [4, 46]

P pyruvate + NH_3 + NADH

S D-Ser + H_2O + NAD^+ <4, 52> (<4>, 1.3% of the activity with L-Ser [4]; <52>, 4.4% of the activity with L-Ala [46]; <52>, in the reverse reaction the enzyme reacts at 9.3% of the activity with pyruvate [46]) (Reversibility: r <52> [46]; ? <4> [4]) [4, 46]

P 3-hydroxypyruvate + NH_3 + NADH <52> [46]

S Gly + H_2O + NAD^+ <59> (Reversibility: r <59> [59]) [59]

P glyoxylate + NH_3 + NADH <59> [59]

S L-2-aminobutanoate + H_2O + NAD^+ <1, 2, 11, 18, 24, 26, 52, 53, 62> (<1>, 7.5% of the activity with L-Ala [1]; <2>, 2.4% of the activity with L-Ala [2]; <18>, 1.6% of the activity with L-Ala [21]; <1>, in the reverse reaction 2-oxobutanoate reacts with 79% of the activity with pyruvate [1]; <11>, 13.3% of the activity with L-Ala [20]; <52>, 10.4% of the activity with L-Ala [46]; <52>, in the reverse reaction the enzyme reacts at 15.3% of the activity with pyruvate [46]; <53>, 0.7% of the activity with L-Ala [53]; <53>, in the reverse reaction 2-oxobutanoate reacts with 1.2% of the activity with pyruvate [53]; <62>, 8.7% of the activity with L-Ala [60]; <62>, in the reverse reaction 2-oxobutanoate reacts with 2.0% of the activity with pyruvate [60]) (Reversibility: r <1, 2, 11, 24, 26, 52, 53, 62> [1, 4, 20, 36, 37, 38, 46, 53, 60]; ? <11, 2> [2, 21]) [1, 2, 4, 20, 21, 36, 37, 38, 46, 53, 60]

P 2-oxobutanoate + NH_3 + NADH <1, 2, 11, 24, 26> [1, 4, 20, 36, 37, 38]

S L-Ala + H_2O + 1,N^6-etheno-NAD^+ <9, 10> (<9>, 45% of the activity with
NAD^+ [11]; <10>, 59% of the activity with NAD^+ [12]) (Reversibility: ?
<9, 10> [11, 12]) [11, 12]

P pyruvate + NH_3 + 1,N^6-etheno-NADH

S L-Ala + H_2O + 3-acetylpyridine-NAD^+ <9, 10, 62> (<9>, as active as
NAD^+ [11]; <10>, 62% of the activity with NAD^+ [12]; <62>, 43% of the
activity with NAD^+ [60]) (Reversibility: ? <9, 10, 62> [11, 12, 60]) [11, 12,
60]

P pyruvate + H_2O + 3-acetylpyridine-NADH

S L-Ala + H_2O + 3-pyridinealdehyde-NAD^+ <9> (<9>, 5.0% of the activity
with NAD^+ [11]) (Reversibility: ? <9> [11]) [11]

P ?

S L-Ala + H_2O + NAD^+ <1-62> (<6>, absolutely specific for L-Ala in oxidative deamination [6]; <2>, the reaction rate for deamination of pyruvate
at the optimum is about 2.8 times higher than that for amination of L-Ala
at the optimum [2]; <13>, rate of the deaminating reaction is only 2.2% of
the aminating reaction [16]; <9,23,27,53>, highly specific for L-Ala [64])
(Reversibility: r <1-62> [1-64]) [1-64]

P pyruvate + NH_3 + NADH <1-62> [1-64]

S L-Ala + H_2O + $NADP^+$ <53, 62> (<53>, 2.5% of the activity with NAD^+
[53]; <53>, in the reverse direction NADPH reacts at 1.6% of the activity
with NADH [53]; <62>, 7% of the activity with NAD^+ [60]) (Reversibility:
r <53> [53]; ? <62> [60]) [53, 60]

P pyruvate + NH_3 + NADPH

S L-Ala + H_2O + deamino-NAD^+ <9, 10, 62> (<9>, 90.2% of the activity
with NAD^+ [11]; <10>, 77% of the activity with NAD^+ [12]; <62>, 4.4%
of the activity with NAD^+ [60]) (Reversibility: ? <9, 10, 62> [11, 12, 60])
[11, 12, 60]

P pyruvate + NH_3 + deamino-NADH

S L-Ala + H_2O + nicotinamide guanine dinucleotide <9, 10, 62> (<9>,
59.9% of the activity with NAD^+ [11]; <10>, 71% of the activity with
NAD^+ [12]; <62>, 69.7% of the activity with NAD^+ [60]) (Reversibility: ?
<9, 10, 62> [11, 12, 60]) [11, 12, 60]

P ?

S L-Ile + H_2O + NAD^+ <11, 23, 52> (<11>, 5% of the activity with L-Ala
[33]; <23>, 0.1% of the activity with L-Ala [33]; <11>, 5% of the activity
with L-Ala [20]) (Reversibility: ? <11, 23, 52> [20, 33, 46]) [20, 33, 46]

P 2-oxo-3-methylpentanoate + NH_3 + NADH

S L-Ser + H_2O + NAD^+ <1, 2, 4, 11, 18, 23, 62> (<1>, 3.5% of the activity
with L-Ala [1]; <4>, 7.9% of the activity with L-Ala [4]; <1>, in the reverse reaction 3-hydroxypyruvate reacts with 0.4% of the activity with
pyruvate [1]; <62>, 4.1% of the activity with L-Ala [60]; <62>, in the
reverse reaction 3-hydroxypyruvate reacts with 14.2% of the activity with
pyruvate [60]; <18>, 2.4% of the activity with L-Ala [21]; <25,53>, no
activity [35,53]) (Reversibility: r <1, 2, 62> [1, 37, 60]; ? <4, 11> [4, 20,
21, 33]) [1, 4, 20, 21, 33, 37, 60]

P 3-hydroxypyruvate + NH_3 + NADH <1, 2> [1, 37]

S L-Val + H_2O + NAD^+ <2, 11, 23, 52, 62> (<11>, 9% of the activity with L-Ala [33]; <23>, 0.2% of the activity with L-Ala [33]; <11>, 9% of the activity with L-Ala [20]; <52>, 4.8% of the activity with L-Ala [46]; <62>, 1.5% of the activity with L-Ala [60]) (Reversibility: ? <2, 11, 23, 52, 62> [20, 33, 37, 46, 60]) [20, 33, 37, 46, 60]

P 3-methyl-2-oxobutanoate + NH_3 + NADH

S L-norvaline + H_2O + NAD^+ <1, 2, 11, 27, 52, 62> (<1>, 1.5% of the activity with L-Ala [1]; <1>, in the reverse reaction 2-oxopentanoate reacts with 6.6% of the activity of pyruvate [1]; <1>, in the reverse reaction 2-oxopentanoate reacts with 0.12% of the activity of pyruvate [1]; <52>, at 4.6% of the activity with L-Ala [46]) (Reversibility: r <1, 2, 11, 62> [1, 20, 37, 60]; ? <27, 52> [39, 46]) [1, 20, 37, 39, 46, 60]

P 2-oxopentanoate + NH_3 + NADH <1, 2, 11> [1, 20, 37]

S glyoxylate + NH_3 + NADH <2, 4, 18, 24, 25, 26, 51, 52> (<4>, 6.2% of the activity with pyruvate [4]; <18>, 13.3% of the activity with pyruvate [21]; <25>, 5.6% of the activity with pyruvate [35]; <51>, 1.2% of the activity with pyruvate [28]) (Reversibility: ? <2, 4, 18, 24, 25, 26, 51, 52> [4, 21, 28, 35, 36, 37, 38, 46]) [4, 21, 28, 35, 36, 37, 38, 46]

P aminoacetate + NAD^+ + H_2O

S hydroxypyruvate + NH_3 + NADH <11, 24> (Reversibility: ? <11, 24> [20, 36]) [20, 36]

P L-Ser + NAD^+ + H_2O

S oxaloacetate + NH_3 + NADH <2, 10, 18, 21, 24, 26, 27, 52> (<2>, 99% of the activity with pyruvate [2]; <10>, 94% of maximal activity [12]; <18>, 43.1% of the activity with pyruvate [21]) (Reversibility: ? <2, 10, 18, 21, 24, 26, 27, 52> [2, 12, 21, 31, 36, 38, 39, 46]) [2, 12, 21, 31, 36, 38, 39, 46]

P L-Asp + H_2O + NAD^+

S Additional information <11, 52> (<11>, A-specific enzyme with regard to stereochemistry of the hydrogen transfer to NAD^+ [27]; <52>, no activity with taurine [46]) [27, 46]

P ?

Inhibitors

2,4,6-trinitrobenzenesulfonic acid <11> (<11>, inactivation follows pseudo first-order kinetics with a 1:1 stoichiometric ratio between the reagent and the enzyme subunit. Partial protection by each of the substrates, NADH or pyruvate. Complete protection only in presence of the ternary complex enzyme-NADH-pyruvate [52]) [52]

2-oxobutanoate <25> (<25>, weak [35]) [35]

3-(2-pyridyldithio)propionate <11> (<11>, inactivation follows pseudo first-order kinetics with a 1:1 stoichiometric ratio between the reagent and the enzyme subunit. Partial protection by each of the substrates, NADH or pyruvate. Complete protection only in presence of the ternary complex enzyme-NADH-pyruvate [52]) [52]

3-bromopyruvate <13> [16]

3-hydroxypyruvate <25> (<25>, competitive with respect to pyruvate [35]) [35]

5'-(p-(fluorosulfonyl)-benzoyl)adenosine <11> (<11>, inactivation follows pseudo-first-order kinetics, complete inactivation of the enzyme can not be obtained even at high reagent concentration [52]) [52]

5,5'-dithiobis(2-nitrobenzoate) <13> [16]

AgNO$_3$ <11> (<11>, 0.1 mM, 57% inhibition [20]) [20]

CaCl$_2$ <11> (<11>, 10 mM, 22% inhibition [20]) [20]

Cd^{2+} <26, 51> (<26>, 10 mM; 36% inhibition of reductive amination [38]) [28, 38]

Cibacron blue <14, 51> [28, 17]

Co^{2+} <26> (<26>, 10 mM, 60% inhibition of reductive amination [38]) [38]

CoCl$_2$ <4, 11> (<11>, 0.1 mM, 53% inhibition [20]) [4, 20]

Cu^{2+} <2, 18, 25, 26, 51> (<26>, 10 mM, 60% inhibition of reductive amination [38]) [21, 28, 35, 37, 38]

CuCl$_2$ <2, 62> (<62>, 1 mM, 52% inhibition [60]) [2, 60]

CuSO$_4$ <4, 10, 11, 53> (<11>, 10 mM, 36% inhibition [20]; <53>, 50.7% inhibition by 1 mM, 96.3% inhibition by 20 mM, EDTA in a 10fold molar excess restores activity almost completely, recombinant enzyme [53]) [4, 12, 20, 53]

D-2-aminobutanoate <26> [38]

D-Ala <10, 21, 23, 25, 26, 28> (<25>, aminating reaction, competitive with respect to NADH, noncompetitive with respect to NH$_4^+$ and pyruvate [35]) [12, 31, 33, 35, 38, 42]

D-Cys <23> [33]

EDTA <6> [6]

F$^-$ <24> [36]

Fe^{2+} <25, 26> (<26>, 10 mM, 85% inhibition of reductive amination [38]) [35, 38]

Fe^{3+} <26> (<26>, 10 mM, 85% inhibition of reductive amination [38]) [38]

GTP <30> (<30>, slight [44]) [44]

Hg^{2+} <25, 51> [28, 35]

HgCl$_2$ <2, 4, 9, 10, 11, 13, 62> (<11>, 0.003 mM, complete inhibition [20]; <62>, 1 mM, 38% inhibition [60]) [2, 4, 11, 12, 16, 20, 37, 60]

I$^-$ <24> [36]

L-Ala <8-10, 13, 21, 22, 25, 27, 28, 51, 58, 62> (<9,25>, reductive amination of pyruvate [11,18,35,64]; <8>, 5 mM, 34% inhibition [10]; <22>, above 15 mM [32]; <58>, substrate inhibition [51]; <62>, uncompetitive inhibition with respect to NADH [60]) [10-12, 16, 28, 31, 32, 35, 39, 42, 51, 60, 64]

L-Asp <22> [32]

L-Cys <22, 23, 27> [32, 33, 39]

L-Gly <21, 27> [31, 39]

L-Phe <51> (<51>, reductive amination of pyruvate, 10 mM, 28% inhibition [28]) [28]

L-Ser <8-10, 13, 21, 22, 25, 27, 51> (<8,9,21,25>, inhibition of aminating activity [10,11,31,35]; <51>, reductive amination of pyruvate [28]; <8>, 10 mM, 26% inhibition [10]) [10-12, 16, 28, 31, 32, 35, 39]

L-Thr <21, 22> (<21>, inhibition of aminating activity [31]) [31, 32]

L-Thr <51> (<51>, reductive amination of pyruvate, 10 mM, 37% inhibition [28]) [28]

L-Trp <51> (<51>, reductive amination of pyruvate, 10 mM, 29% inhibition [28]) [28]

L-glutamic acid <22> [32]

MgCl$_2$ <10> [12]

Mn^{2+} <26> (<26>, 10 mM, 60% inhibition of reductive amination [38]) [38]

MnCl$_2$ <11> (<11>, 10 mM, 12% inhibition [20]) [20]

MnSO$_4$ <11> (<11>, 10 mM, 50% inhibition [20]) [20]

NAD$^+$ <13, 62> (<13>, 0.1 mM, 35% inhibition [16]; <62>, product inhibition [60]) [16, 60]

NADH <2, 12, 58> (<2,58>, product inhibition [2,51]) [2, 15, 51]

NADPH <25> (<25>, inhibition of reductive amination [35]) [35]

NH$_4^+$ <2, 12, 25, 28, 58> (<2>, product inhibition [2]; <25>, inhibition of aminating activity [35]) [2, 15, 35, 42, 51]

Ni^{2+} <26> (<26>, 10 mM, 60% inhibition of reductive amination [38]) [38]

PCMB <2, 9-11, 18, 27, 32, 51> (<26>, reversed by L-Cys [45]; <11>, 0.3 mM; 52% inhibition [20]) [2, 11, 12, 20, 21, 26, 28, 37, 39, 45]

Pb^{2+} <2, 18> [21, 37]

Zn^{2+} <6, 11, 18, 25, 27, 51, 53> (<11>, 10 mM, 62% inhibition [20]; <53>, 26.5% inhibition by 1 mM ZnCl$_2$, 90.1% inhibition by 20 mM ZnCl$_2$, EDTA in a 10fold molar excess restores activity almost completely, recombinant enzyme [53]; <11>, allosteric competitive inhibitor, reversible, inducing conformational change through the intersubunit interaction, positive cooperative binding of the substrate in presence of Zn^{2+} [55]) [6, 20, 21, 28, 35, 39, 40, 53, 55]

borate <27> [39]

glycine <27> [39]

glyoxylate <25> (<25>, weak [35]) [35]

hydroxypyruvate <13, 25> [16, 35]

iodoacetate <6> [6]

mercurials <2, 11, 27, 53> [64]

methylglyoxal <58> [51]

monoiodoacetate <18> [21]

oxaloacetate <22> [32]

oxamate <58> [51]

p-hydroxymercuribenzoate <8> (<8>, no effect on aminating activity, inhibition of deaminating activity [10]) [10]

perchlorate <24> [36]

phosphoenolpyruvate <22> [32]

propionate <58> [51]

pyridoxal 5'-phosphate <51> [28]

pyruvate <2, 4, 8, 12, 22, 58> (<4,58>, substrate inhibition [4,51]; <2,8,12>, product inhibition [2,10,15]; <12>, inhibition at pH 7.5 greater than at pH 9.5 [15]; <22>, above 2 mM [32]) [2, 4, 10, 15, 32, 51]

thiocyanate <24> [36]

Cofactors/prosthetic groups

1,N^6-etheno-NAD$^+$ <9, 10> (<9>, 45% of the activity with NAD$^+$ [11]; <10>, 59% of the activity with NAD$^+$ [12]) [11, 12]

3-acetylpyridine-NAD$^+$ <9, 10, 62> (<9>, as active as NAD$^+$ [11]; <10>, 62% of the activity with NAD$^+$ [12]; <62>, 43% of the activity with NAD$^+$ [60]) [11, 12, 60]

3-pyridinealdehyde-NAD$^+$ <9> (<9>, 5.0% of the activity with NAD$^+$ [11]) [11]

NAD$^+$ <1-62> (<1,8>, specific for [1,10]) [1-64]

NADH <1-62> (<1>, specific for [1]) [1-64]

NADP$^+$ <9, 18, 53, 62> (<9>, 15% of the activity with NAD$^+$ [13]; <18>, 0.6% of the activity with NAD$^+$ [21]; <25>, 8% of the activity with NAD$^+$ [35]; <62>, 7% of the activity with NAD$^+$ [60]; <1,2,4,21,25,30>, no activity [1,4,20,31,35,37,42,44]) [13, 21, 53, 60]

NADPH <8, 18, 53> (<8>, aminating activity absolutely specific for [10]; <18>, 3.5% of the activity with NADH [21]; <1,4,13,21,25>, no activity [1,4,16,31,35]) [10, 21, 53]

deamino-NAD$^+$ <9, 10, 62> (<9>, 90.2% of the activity with NAD$^+$ [11]; <10>, 77% of the activity with NAD$^+$ [12]; <62>, 81% of the activity with NAD$^+$ [60]) [11, 12, 60]

nicotinamide guanine dinucleotide <9, 10, 62> (<9>, 59.9% of the activity with NAD$^+$ [11]; <10>, 71% of the activity with NAD$^+$ [12]) [11, 12, 60]

Activating compounds

1,10-phenanthroline <18> (<18>, activation [21]) [21]

2,2'-dipyridyl <18> (<18>, activation [21]) [21]

2-mercaptoethanol <8, 18, 51> (<8,18,51>, activation [10,21,28]) [10, 21, 28]

Cys <8> (<8>, activates [10]) [10]

EDTA <18> (<18>, activation [21]) [21]

L-Arg <22> (<22>, activation [32]) [32]

diethyldithiocarbamate <18> (<18>, activation [21]) [21]

dithioerythritol <8> (<8>, activates [10]) [10]

dithiothreitol <8, 18> (<8,18>, activation [10,21]) [10, 21]

Metals, ions

Ca^{2+} <6> (<6>, activation [6]) [6]

CdCl$_2$ <24> (<24>, stimulates [36]) [36]

Co^{2+} <6> (<6>, activation [6]) [6]

Cs$^+$ <24, 26> (<24>, CsCl, weak stimulation [36]; <26>, reductive amination is activated equally well by K$^+$, Na$^+$ or NH$_4^+$ and to a lesser extent with Cs$^+$ or Li$^+$ [38]) [36, 38]

Fe^{2+} <6> (<6>, activation [6]) [6]

KCl <24, 26> (<24>, KCl stimulates [36]; <26>, reductive amination is activated equally well by K$^+$, Na$^+$ or NH$_4^+$ and to a lesser extent with Cs$^+$ or Li$^+$, absolute requirement for K$^+$ in oxidative deamination, completely inactive with Na$^+$ or NH$_4^+$ [38]) [36, 38]

Li$^+$ <24> (<24>, LiCl, weak stimulation [36]; <26>, reductive amination is activated equally well by K$^+$, Na$^+$ or NH$_4^+$ and to a lesser extent with Cs$^+$ or Li$^+$ [38]) [36]

Mg^{2+} <6> (<6>, activation [6]) [6]

Mn^{2+} <6> (<6>, activation [6]) [6]

NaCl <24, 26, 31> (<24>, NaCl, active only in presence of high concentrations [36]; <26>, reductive amination is activated equally well by K$^+$, Na$^+$ or NH$_4^+$ and to a lesser extent with Cs$^+$ or Li$^+$ [38]; <24,26>, optimal concentration: 4.3 M [45]; <31>, optimal concentration: 4.0 M [45]) [36, 38, 45]

RbCl <24> (<24>, stimulates [36]) [36]

Additional information <24, 26> (<26>, absolute requirement for high ionic strength for optimum activity [38]; <24>, highest activation by Cl$^-$, much smaller effect occurs in presence of Br$^-$, nitrate, or sulfate [36]) [36, 38]

Turnover number (min^{-1})
Additional information <11, 57, 60, 61> (<11>, amination: maximal turnover is about 650000 per min, deamination: maximal turnover is 54000 per min [20]) [20, 57, 58]

Specific activity (U/mg)
0.564 <32> [26]
11.2 <4> [4]
13.4 <2> [2]
17.1 <12> [15]
19 <28> [42]
22.8 <18> [21]
30.4 <21> [31]
38 <27> [39]
40 <57> [58]
60 <60> (<60>, 25°C [57]) [57]
75 <1> [1]
92 <62> [60]
129 <13> [16]
157 <2> [37]
168 <26> [38]
178 <24> [36]
183 <51> [28]
203 <60> (<60>, 82°C [57]) [57]
Additional information <4, 6, 9-11, 52> (<11>, enzyme assay [20]) [4, 6, 10, 12, 13, 20, 46]

K$_m$-Value (mM)
0.01 <2> (NADH) [9]
0.014 <2> (L-norvaline) [37]
0.018 <2> (L-Ala) [37]
0.02 <2> (L-Val) [37]
0.02 <51> (NADH) [28]
0.022 <58> (NADH) [51]

0.023 <11> (NADH) [20, 33, 64]
0.029 <9> (NADH) [13, 64]
0.031 <24> (NADH) [64]
0.035 <11, 23> (NADH) [33, 64]
0.036 <18> (NAD$^+$, <18>, pH 10.0 [21]) [21]
0.037 <14> (NADH, <64>, soluble enzyme [30]) [30, 64]
0.039 <2> (L-Ser) [37]
0.04 <51> (NAD$^+$) [28]
0.04 <52> (NADH) [46]
0.044 <21> (NADH) [31]
0.047 <18> (NADH) [21]
0.05 <10> (NADH) [12]
0.055 <32> (NADH) [26]
0.059 <13> (NADH) [64]
0.067 <22> (NADH) [32, 60]
0.077 <1> (NADH) [1]
0.086 <34> (NADH) [64]
0.09 <4> (NADH) [4]
0.093 <2> (NADH) [9]
0.098 <58> (NADH) [51]
0.0982 <53> (NADH) [53]
0.1 <2> (NADH) [2]
0.11 <9> (NAD$^+$) [13, 64]
0.11 <27> (pyruvate) [39, 64]
0.12 <23> (NAD$^+$) [33, 64]
0.13 <8> (pyruvate) [10]
0.14 <22> (NAD$^+$) [32]
0.14 <25> (NADH) [35]
0.15 <8> (NAD$^+$) [10]
0.15 <52> (NAD$^+$) [46]
0.16 <62> (NAD$^+$) [60]
0.18 <23> (L-Ala) [64]
0.18 <10, 11, 14> (NAD$^+$, <14>, soluble enzyme [30]) [12, 20, 30, 33, 58, 64]
0.2 <34, 61> (NAD$^+$) [58, 64]
0.2 <26> (NADH) [38, 64]
0.21 <56> (NAD$^+$) [58]
0.22 <62> (pyruvate) [60]
0.23 <10> (pyruvate) [12]
0.24 <57> (NAD$^+$) [58]
0.25 <8> (NADH) [10]
0.26 <2> (NAD$^+$) [2]
0.29 <18> (pyruvate) [21]
0.31 <53> (NAD$^+$) [53]
0.32 <14> (NADH, <14>, Sepharose-bound enzyme [30]) [30]
0.33 <2> (L-2-aminobutanoate) [37]
0.33 <51> (pyruvate) [28]
0.36 <11> (NAD$^+$) [40]

0.37 <58> (L-Ala) [51]
0.37 <22> (pyruvate) [32]
0.4 <27> (L-Ala) [39]
0.43 <58> (pyruvate) [51, 64]
0.45 <22> (L-Ala) [32]
0.47 <62> (L-Ala) [60]
0.48 <14> (pyruvate, <14>, soluble enzyme [30]) [30, 64]
0.49 <34> (pyruvate) [64]
0.5 <26> (NAD^+) [38, 64]
0.5 <2, 25> (pyruvate) [2, 35]
0.52 <14> (pyruvate, <14>, Sepharose-bound enzyme [30]) [30]
0.53 <11> (pyruvate) [20]
0.54 <11> (pyruvate) [20, 33, 64]
0.55 <4> (pyruvate) [4]
0.56 <9> (pyruvate) [13, 64]
0.68 <21> (pyruvate) [31]
0.75 <23> (pyruvate) [33, 64]
0.8 <26> (pyruvate) [38, 64]
0.82 <26> (NH_4^+) [64]
0.95 <24> (pyruvate) [64]
1 <34> (L-Ala) [64]
1.1 <25, 52> (pyruvate) [35, 46]
1.18 <14> (NAD^+, <14>, Sepharose-bound enzyme [30]) [30]
1.25 <8> (L-alanine) [10]
1.45 <53> (pyruvate) [53]
1.5 <24> (NAD^+) [64]
1.6 <52> (L-Ala) [46]
1.61 <2> (3-fluoropyruvate) [9]
1.67 <1> (NAD^+) [1]
1.7 <11> (L-Ala) [33]
1.7 <2> (pyruvate) [37]
1.73 <11> (L-Ala) [20, 58, 64]
1.9 <18> (L-Ala, <18>, pH 8.0 [21]) [21]
2.4 <10> (2-oxobutanoate) [12]
2.5 <11> (hydroxypyruvate) [20]
2.5 <10> (oxalacetate) [12]
3 <27> (oxaloacetate) [39]
3.6 <18> (L-Ala, <18>, pH 9.0 [21]) [21]
3.8 <61> (L-Ala) [58]
4.2 <23> (L-Ala) [33]
4.3 <13> (pyruvate) [64]
5 <9, 51> (L-Ala) [13, 28, 64]
5 <1, 32> (pyruvate) [1, 26]
5.3 <24> (L-Ala) [64]
5.4 <11> (L-Ala) [40]
5.5 <58> (NH_4^+) [51]
6.67 <9> (NH_4^+) [13, 64]

7 <11> (2-oxohexanoate) [20]
7 <26> (L-Ala) [38, 64]
7.1 <18> (L-Ala, <18>, pH 10.0 [21]) [21]
7.6 <57> (L-Ala) [58]
8 <27> (NH_4^+) [64]
8.9 <34> (NH_4^+) [64]
9.1 <25> (L-Ala) [35]
10 <10> (L-Ala) [12]
10.5 <2> (L-alanine) [2]
11 <2> (2-oxo-3-methylbutanoate) [37]
11 <2> (2-oxobutanoate) [37]
11.3 <14> (L-Ala, <14>, Sepharose-bound enzyme [30]) [30]
11.6 <10> (NH_4^+) [12]
12 <11> (3-methyl-2-oxobutanoate) [20]
12 <2> (glyoxylate) [37]
12.5 <14> (L-Ala, <14>, soluble enzyme [30]) [30, 64]
13.3 <1> (L-alanine) [1]
13.8 <53> (L-Ala) [53]
16 <8> (NH_4^+) [10]
16 <11> (glyoxalate) [20]
20 <25> (NH_4^+) [35]
22 <14> (NH_4^+, <14>, Sepharose-bound enzyme [30]) [30]
23 <11> (2-oxobutanoate) [20]
23 <2> (2-oxopentanoate) [37]
24 <32> (NH_4^+) [26]
26 <13> (NH_4^+) [64]
28 <22> (NH_4^+) [32]
28.2 <2> (NH_3) [9]
30 <56> (L-Ala) [58]
30 <14> (NH_4^+, <14>, soluble enzyme [30]) [30, 64]
31 <52> (NH_4^+) [46]
33 <11> (2-oxopentanoate) [20]
35.4 <53> (NH_4^+) [53]
38 <2, 11> (NH_4^+) [2, 20, 33, 64]
45 <11> (4-methyl-2-oxopentanoate) [20]
57 <11> (L-Ser) [20]
59 <23> (NH_3) [33, 64]
61 <18> (NH_4^+) [21]
62 <11> (L-norvalione) [20]
66.7 <62> (NH_4^+) [60]
80 <2> (3-hydroxypyruvate) [37]
111 <51> (NH_4^+) [28]
150 <11> (L-Val) [20]
176 <11> (L-2-aminobutanoate) [20]
235 <4> (NH_4^+) [4]
250 <11> (L-Ile) [20]
500 <24> (NH_4^+) [64]

1060 <2> (NH_3) [9]

Additional information <6, 12, 18, 27> (<27>, K_m-value for NH_4^+ varies from 5 mM to 133 mM depending on the pH, being lowest at high pH levels, pH 8.7 or above [39]) [6, 15, 21, 39]

K_i-Value (mM)

0.092 <58> (pyruvate) [51]

0.64 <11> (D-Cys) [33]

0.9 <11> (D-Cys) [33]

1 <58> (L-Ala) [51]

1.6 <25> (L-Ala, <25>, with respect to NADH [35]) [35]

2 <25> (L-Ala, <25>, with respect to NH_4^+ [35]) [35]

3 <25> (L-Ala, <25>, with respect to pyruvate [35]) [35]

3.7 <58> (NH_4^+) [51]

5 <11> (D-Ala) [33]

20 <11> (D-Ala) [33]

Additional information <58> [51]

pH-Optimum

7-7.5 <53> (<53>, reductive amination [53]) [53]

7.8 <2> (<2>, reaction with 3-fluoropyruvate, NADH and NH_4^+ [9]) [9]

7.9-9.1 <28> [42]

8 <4, 6, 18, 23, 27> (<4,6,18,23,27>, reductive amination [4,6,21,39,64]) [4, 6, 21, 33, 39, 64]

8.2 <2, 8> (<2,8>, amination of pyruvate [2,10]) [2, 10]

8.4 <51> (<51>, reductive amination [28]) [28]

8.4-8.6 <25> (<25>, pyruvate amination [35]) [35]

8.5 <9, 10, 11, 23, 34, 55> (<9,11,34>, reductive amination of pyruvate [33,45,64]; <9,10>, reductive amination [12, 13]; <55>, NADH oxidation [49]) [12, 13, 33, 45, 49, 64]

8.5-9 <14> (<14>, reductive amination [17]) [17]

8.6 <21> (<21>, reductive amination [31]) [31]

8.7 <62> (<62>, amination of pyruvate [60]) [60]

8.8-9 <11> (<11>, amination [20]) [20]

9 <2, 13, 20, 22, 24, 26, 31, 52, 55> (<13,24,26>, both directions [36,38,64]; <2,13,22>, reductive amination [16, 29, 32, 37]; <52>, reductive amination of pyruvate [46]; <55>, reductive amination of pyruvate [49]) [16, 29, 32, 36, 37, 38, 45, 46, 49]

9-9.5 <14> (<14>, amination reaction, soluble and Sepharose-bound enzyme [30]) [30]

9-11.5 <52> (<52>, oxidative deamination of L-Ala [46]) [46]

9.2 <51> (<51>, oxidative deamination [28]) [28]

9.2-9.6 <12> [15]

9.3 <53> (<53>, reductive amination [64]) [64]

9.5 <1> (<1>, reductive amination of pyruvate [1]) [1]

9.6 <27> (<27>, oxidative deamination [39,64]) [39, 64]

9.8 <22, 32, 53> (<22,32,53>, oxidative deamination [26,32,64]) [26, 32, 64]

10 <6, 9, 10, 18, 34, 55> (<6,9,10,18>, oxidative deamination [6,12,13,21]; <55>, NAD$^+$ reduction [49]) [6, 12, 13, 21, 49, 64]

10-10.5 <2, 9, 11, 21> (<2,9,11,21,34>, oxidative deamination [20,31,37,64]) [20, 31, 37, 64]

10-11 <53> (<53>, oxidative deamination [53]) [53]

10.1 <11> [40]

10.5 <2, 4, 8, 23, 55> (<2,4,8,23>, deamination of L-Ala [4,10, 33,37]; <23,55>, oxidative deamination of L-Ala [49,64]) [4, 10, 33, 37, 49, 64]

10.5-11 <14> (<14>, oxidative deamination [17]) [17, 30]

10.7-11.2 <1> (<1>, oxidative deamination [1]) [1]

10.9 <62> (<62>, deamination of L-Ala [60]) [60]

pH-Range

6-8 <53> (<53>, pH 6.0: about 30% of maximal activity, pH 8.5: about 25% of maximal activity, reductive amination [53]) [53]

6.5-9 <6> (<6>, pH 6.5: about 30% of activity maximum, pH 9.0: about 55% of maximal activity, reductive amination [6]) [6]

6.8-8.8 <27> (<27>, pH 6.8: about 50% of maximal activity, pH 8.8: about 60% of maximal activity, reductive amination [39]) [39]

7-10 <2> (<2>, pH 7.0: about 50% of maximal activity, pH 10.0: about 50% of maximal activity, oxidative deamination of L-Ala [2]) [2]

7-10.8 <12> (<12>, pH 7.0: 13% of maximal activity, pH 10.8: 52% of maximal activity [15]) [15]

7.2-9.2 <51> (<51>, pH 7.2: about 40% of maximal activity, pH 9.2: 40% of maximal activity, reductive amination of pyruvate [28]) [28]

7.2-10 <11> (<11>, pH 7.2: about 10% of activity maximum, pH 10.0: about 30% of activity maximum, reductive amination [20]) [20]

7.8-9.4 <20> (<20>, pH 7.8: about 50% of activity maximum, pH 9.4: about 90% of activity maximum, reductive amination [29]) [29]

7.8-10 <51> (<51>, pH 7.8: about 50% of maximal activity, pH 10.0: about 65% of maximal activity, oxidative deamination of L-Ala [28]) [28]

7.8-10.2 <1> (<1>, pH 7.8: 54% of activity maximum, pH 10.2: 72% of activity maximum, reductive amination of pyruvate [1]) [1]

8-10.5 <6> (<6>, pH 8.0: about 60% of activity maximum, pH 10.5: about 80% of activity maximum, oxidative deamination [6]) [6]

8-11 <2, 11> (<2>, pH 8.0: about 50% of maximal activity, pH 11.0: about 95% of maximal activity, reductive amination of pyruvate [2]; <11>, pH 8: about 35% of activity maximum, pH 11: about 75% of activity maximum, oxidative deamination [20]) [2, 20]

8.5-10.6 <27> (<27>, pH 8.5: about 60% of maximal activity, pH 10.6: about 40% of maximal activity, oxidative deamination [39]) [39]

8.5-11.5 <53> (<53>, pH 8.5: about 35% of maximal activity, pH 11.5: about 20% of maximal activity, oxidative deamination [53]) [53]

9.3-11.8 <1> (<1>, pH 9.3: about 55% of activity maximum, pH 11.8: about 63% of activity maximum, oxidative deamination [1]) [1]

Temperature optimum (°C)

27 <8> (<8>, deaminating activity [10]) [10]

30 <8> (<8>, aminating activity [10]) [10]

40 <6, 24> (<24>, presence of NaNO$_3$ [41]) [6, 41]

45 <24> (<24>, presence of 3.5 M NaNO$_3$ [36]) [36]

45-50 <55> (<55>, NAD$^+$ reduction [49]) [49]

50-55 <52> (<52>, oxidative deamination [46]) [46]

52 <4> [4]

55-60 <52> (<52>, reductive amination [46]) [46]

58 <19> (<19>, enzyme prepared from cells grown at 34°C [23]) [23]

60 <10, 18, 19, 24> (<19>, enzyme prepared from cells grown at 43°C [23]; <24>, presence of NaCl [41]) [12, 21, 23, 41]

60-65 <26> (<26>, reductive amination, in presence of KCl [38]) [38]

60-70 <51> (<51>, reductive amination [28]) [28]

62 <19> (<19>, enzyme prepared from Bacillus flavothermus cells grown at 52°C or at 70°C [23]) [23]

65 <24, 26> (<24>, in presence of 3.5 M NaCl [36]; <26>, 3.5 NaCl [45]) [36, 45]

65-70 <26> (<26>, reductive amination in presence of NaCl [38]) [38]

70 <2, 24> (<24>, in presence of KCl [41]; <24>, 3.5 M KCl [36]; <2>, increase of activity up to [37]) [36, 37, 41]

75 <9> [13]

Temperature range (°C)

40-65 <26, 55> (<26>, 40°C: about 50% of activity maximum, 65°C: optimum, sharp decrease of activity above, no activity at 75°C [38]; <55>, 40°C: 70% of maximal activity, 65°C: about 50% of maximal activity [49]) [38, 49]

73-75 <19> (<19>, not active above, temperature maximum depends on growth temperature of cells [23]) [23]

4 Enzyme Structure

Molecular weight

58000 <24> (<24>, non-denaturing PAGE [36]) [36]

60000 <24> (<24>, gel filtration [36]) [36]

72500 <26> [38]

92000 <25> (<25>, sucrose density gradient centrifugation [35]) [35]

168000 <21> (<21>, gel filtration [31]) [31]

190000 <34> [64]

198000 <9> (<9>, analytical ultracentrifugation [11]) [11]

205000-210000 <10> (<10>, gel filtration, equilibrium ultracentrifugation [12]) [12]

217000 <13> (<13>, gel filtration [16]) [16]

220000-230000 <11> (<11>, equilibrium sedimentation [20]) [20]

230000 <2> (<2>, gel filtration [2,37]) [2, 37]

231000 <2> (<2>, equilibrium sedimentation [37]) [37]

235000 <2> (<2>, recombinant enzyme, gel filtration [8]) [8]
240000 <1, 18, 51> (<1>, recombinant enzyme, equilibrium sedimentation, gel filtration [1]; <18,51>, gel filtration [21,28]) [1, 21, 28]
240000-241000 <8> (<8>, gel filtration [10]) [10]
245000 <62> (<62>, gel filtration [60]) [60]
246000 <8> (<8>, non-denaturing PAGE [10]) [10]
255000 <14> (<14>, non-denaturing PAGE [17]) [17]
260000 <55> (<55>, gel filtration [49]) [49]
270000 <27> (<27>, gel filtration [39]) [39]
273000 <52> (<52>, gel filtration [46]) [46]
280000 <11> (<11>, gel filtration [40]) [40]
290000 <23> (<23>, equilibrium sedimentation [33]) [33]
395000 <9> (<9>, gel filtration [13]) [13]

Subunits

hexamer <1, 2, 8, 11, 14, 18, 23, 26, 27, 51, 52, 55, 57, 62> (<2>, 6 * 38000, SDS-PAGE [2,38,40]; <18>, 6 * 39000, SDS-PAGE [21]; <2>, 6 * 39465, calculation from nucleotide sequence [8]; <2>, 6 * 39500, recombinant enzyme, SDS-PAGE [8]; <62>, 6 * 39807, calculation from nucleotide sequence [60]; <52>, 6 * 39900, calculation from nucleotide sequence [46]; <1>, 6 * 40000, recombinant enzyme, SDS-PAGE [1]; <62>, 6 * 40000, SDS-PAGE [60]; <51>, 6 * 41000, SDS-PAGE [28]; <8,14,52>, 6 * 42000, SDS-PAGE [10,17,46]; <55>, 6 * 42300, SDS-PAGE [49]; <27,57>, 6 * 43000, SDS-PAGE [39,58]; <23>, 6 * 48000, SDS-PAGE [33]) [1, 2, 8, 10, 17, 21, 28, 33, 38-40, 46, 49, 58, 60]
monomer <24> (<24>, 1 * 58000-60000, SDS-PAGE [36]) [36]
octamer <9> (<9>, 8 * 48000, SDS-PAGE [13]) [13]
tetramer <10, 13, 21> (<10>, 4 * 51000, SDS-PAGE [12]; <13>, 4 * 53000, SDS-PAGE [16]; <21>, 4 * 42000, SDS-PAGE [31]) [12, 16, 31]

5 Isolation/Preparation/Mutation/Application

Source/tissue

bacteroid <21, 54> [31, 48]
heterocyst <27> [39]
hypha <5> (<5>, restricted to vegetative hyphae, vesicles lack complete pathway for assimilating ammonia beyond the glutamine stage [5]) [5]
mycelium <6, 29> [6, 43]
spore <11> (<11>, resting [40]) [40]
vegetative <21, 27> (<21>, free living [31]) [31, 39]

Localization

intracellular <59> [59]
membrane <16> (<16>, enzyme is partly associated with the membrane fraction [18]) [18]
soluble <15> [18]

Purification

<1> (gene cloned and expressed in Escherichia coli [1]) [1]
<2> (recombinant enzyme [8]) [2, 8, 37]
<4> [4]
<6> (partial [6]) [6]
<8> [10]
<9> [13]
<10> [12]
<11> [20, 40]
<12> [15]
<13> [16]
<14> (partial [30]) [17, 30]
<18> [21]
<20> [29]
<21> [31]
<22> [32]
<23> [33]
<24> [36]
<25> [35]
<26> (partial [38]) [38]
<27> [39]
<28> [42]
<32> [26]
<51> [28, 54]
<52> [46]
<53> (recombinant enzyme [47]) [47]
<55> [49]
<60> [57]
<62> [60]

Renaturation

<14> (the ability of denatured immobilized subunits to pick up subunits from solution proves their capacity to fold back to the native conformation after urea treatment [17]) [17]
<24> (partial renaturation of the enzyme denatured by 6 M urea is linked to a corresponding reappearance of α-helix and of β-structure [34]) [34]

Crystallization

<2> [37]
<18> [21]
<23> [33]
<51> (hanging-drop method [54]) [54]

Cloning

<1> (cloned and expressed in Escherichia coli [1]) [1]
<2> (cloned and expressed in Escherichia coli [8]) [8]
<53> (host vector system for high-level expression in Escherichia coli [47]) [47]

<57> (expression in Escherichia coli [58]) [58]
<59> [56]
<61> (expression in Escherichia coli [58]) [58]
<62> (expression in Escherichia coli JM109 [60]) [60]

Application

analysis <11> (<11>, determination of L-Ala [50]) [50]
synthesis <2, 60> (<2>, continuous production of 3-fluoro-L-alanine with
alanine dehydrogenase [9]; <60>, batch synthesis of L-Ala at room temperature via reductive amination of pyruvate [57]) [9, 57]

6 Stability

pH-Stability

2.3-10.6 <2> (<2>, 37°C, 10 min, stable [2]) [2]
3 <11> (<11>, rapid and complete inactivation below [20]) [20]
5 <11> (<11>, stable above [20]) [20]
6-9 <14, 51> (<14>, 25°C, 24 h, concentration 0.001 mg/ml, stable [30];
<51>, 50°C, 5 min, stable [28]) [28, 30]
6-10 <1, 2> (<2>, 50°C, 5 min, stable [37]; <1>, 55°C, 10 min, stable [1]) [1,
37]
6-10.5 <62> (<62>, 30°C, 10 min, most stable in the pH-range [60]) [60]
7-10 <18> (<18>, 20°C, 3 h, stable [21]) [21]

Temperature stability

20 <11> (<11>, 10 min, stable [40]) [40]
22 <60> (<60>, 139 h, no loss of activity [57]) [57]
30 <4, 62> (<4>, pH 7.5, half-life: 22 days [4]; <62>, pH 6.0-10.5, 10 min,
stable [60]) [4, 60]
37 <2, 14, 55> (<2>, pH 2.3-10.6, 10 min, stable [2]; <14>, 28 d, immobilized
enzyme retains 85% of its activity in 0.1 M phosphate buffer, pH 8 [30];
<55>, 90 min, stable [49]) [2, 30, 49]
41 <61> (<61>, 30 min, 50% loss of activity [58]) [58]
42 <55> (<55>, 90 min, stable [49]) [49]
45 <18> (<18>, 20 min, pH 7.0, stable [21]) [21]
50 <2, 6, 8, 18, 40, 51, 55, 62> (<2>, pH 6.0-10.0, 5 min, stable [37]; <6>, 35%
loss of activity after 10 min, 65% loss of activity after 20 min [6]; <8>, 50%
loss of aminating activity after 20 min, 50% loss of deaminating activity after
27 min [10]; <18>, 20 min, pH 7.0, 29% loss of activity [21]; <51>, 5 min, pH
6.0-9.0, stable [28]; <55>, half-life: 42 min [49]; <62>, pH 7.4, 10 min, stable
up to [60]) [6, 10, 21, 28, 37, 49, 60]
55 <1, 18, 26> (<1>, pH 6.0-11.0, 10 min, stable [1]; <26>, pH 8.5, 50 min,
40% loss of activity [38]; <18>, 20 min, pH 7.0, 50% loss of activity [21]) [1,
21, 38]
56 <55> (<55>, half-life: 2.25 min [49]) [49]
59 <14, 57> (<14>, 10 min, 50% loss of activity of soluble enzyme [30];
<57>, 30 min, 50% loss of activity [58]) [30, 58]

60 <2, 18, 26, 51, 52, 53, 55> (<18>, 20 min, pH 7.0, 70% loss of activity [21]; <26>, pH 8.5, 10 min, about 45% loss of activity [38]; <2>, pH 7.2, 10 mM potassium phosphate buffer, 5 min, stable up to [37]; <51>, 10 min, in presence of 0.3 mM NADH, 30% loss of activity [28]; <52>, inactivation after about 1 min [46]; <55>, half-life: 15 s [49]; <53>, 4 h, 25% loss of activity, recombinant enzyme [53]) [21, 28, 37, 38, 46, 49, 53]
63 <11, 56> (<11>, melting temperature [33]; <56>, 30 min, 50% loss of activity [58]) [33, 58]
64 <11> (<11>, 30 min, 50% loss of activity [58]) [58]
65 <2, 14, 52> (<2>, 5 min, 50% loss of activity [8]; <14>, 10 min, complete loss of activity, soluble enzyme [30]; <52>, inactivation after about 10 s [46]) [8, 30, 46]
65.7 <53> (<53>, 5 min, 50% loss of activity, recombinant enzyme [53]) [53]
67 <14> (<14>, 10 min, 50% loss of activity, immobilized enzyme [30]) [30]
70 <19, 26, 50> (<19>, 5 min, stable in presence of substrates, 20% loss of activity in absence of substrates [23]; <50> 5 min, 13% loss of activity in absence of substrates, 2% loss of activity in presence of substrates [23]; <26>, pH 8.5, 10 min, 90% loss of activity [38]) [23, 38]
75 <1, 2, 14, 17> (<1>, 30 min, stable, [1]; <2>, 60 min, pH 5.5-9.5, stable [2]; <1,2,17,33>, 5 min, stable up to [19]; <14>, 10 min, complete loss of activity, immobilized enzyme [30]) [1, 2, 19, 30]
80 <2> (<2>, 20 min, stable [2]) [2]
81 <1> (<1>, 30 min, 50% loss of activity [58]) [58]
83 <1, 2, 17, 33> (<1,2,17,33>, 5 min, 50% loss of activity [19]) [19]
85 <1> (<1>, 5 min, substantial loss of activity [1]; <1>, 5 min, 50% loss of activity [8]) [1, 8]
86 <23> (<23>, melting temperature [33]) [33]
Additional information <17, 19, 24, 57, 61> (<17>, screening of thermostable alanine dehydrogenase in thermophilic Bacillus strains [19]; <24>, thermal stability is considerably lower in presence of KCl than in presence of NaCl [36]; <19>, stablilized by substrates. Enzyme undergoes a transition from heat-labile to thermostable within the growth temperature range between 44°C and 51°C [23]; <57,61>, thermal instability may result from relatively low numbers of salt bridges in the enzymes [58]) [19, 23, 36, 58]

Organic solvent stability

ethanol <2> (<2>, 10 min, 37°C, about 90% of activity remains after incubation with up to 35% ethanol, 50% activity with 48% ethanol [2]) [2]

General stability information

<2>, SDS, 10 min, 37°C, retains activity with up to 0.03% SDS, loss of activity with 0.05% [2]
<2>, Triton X-100, 10 min, 37°C, not inactivated by treatment with up to 4% [2]
<2>, stable to repeated freezing and thawing [2]
<6>, dialysis against 0.02 M Tris-HCl, pH 8.4, 75% loss of activity after 3 h, complete loss of activity after 24 h [6]
<6>, freezing and thawing completely destroys activity [6]

<11>, partial inactivation by dilution [28]

<11>, reversible deactivation by dilution [20]

<14>, urea, above 3 M, denaturation of soluble enzyme, resistance to denaturation increases by immobilization, 5 M, 30 min, 85% loss of activity of matrix-bound enzyme [30]

<24>, 6 M urea, complete inactivation and total loss of both α-helix and β-structure [34]

<24>, NaCl, required for stability, 2.0-4.3 M, 4°C, several weeks, stable [41]

<24>, inactivation after removal of salt [34]

<25>, 93% loss of activity after dialysis [35]

<26>, absolute requirement of high ionic strength for stability [38]

<52>, repeated freezing and thawing leads to significant loss of activity [46]

<55>, more than 75% loss of activity after two freeze-thaw cycles and storage at -20°C for 1 week, stabilized by addition of 50% glycerol [49]

<61>, enzyme is extremely unstable and easily lysed [58]

Storage stability

<1>, 4°C, 0.15 M KCl, stable for more than 2 months [1]

<2>, -20°C, pH 7.2, more than 5 months [2]

<2>, 4°C, 80% saturated ammonium sulfate suspension, pH 7-8, 2 years, stable [37]

<6>, 4°C, 0.02 M Tris-HCl buffer, pH 8.4, stable for 1 d, then the activity decreases gradually till the sixth day at which complete loss of activity occurs [6]

<8>, 4°C, stable [10]

<10>, -20°C, 0.2 M Tris/HCl buffer, pH 7.4, 2 months, stable [12]

<10>, -4°C, 0.2 M Tris/HCl buffer, pH 7.4, 24 h, stable [12]

<11>, 4°C, stable for many months [20]

<25>, 4°C, 75% saturated ammonium sulfate suspension, purified enzyme stable for at least 12 months [35]

<52>, -20°C, stable for at least 1 month [46]

<55>, -20°C, in presence of 50% glycerol, stable for up to 5 months [49]

<55>, 4°C or -20°C, less than 0.2 mg/ml, enzyme loses most of its activity in less than 2 weeks [49]

<55>, 4°C, pH 6-8, 50 mM phosphate buffer, stable for at least 3 days [49]

<62>, 4°C, 60% saturation with ammonium sulfate, pH 7.2, stable for over 6 months [60]

References

[1] Sakamoto, Y.; Nagata, S.; Esaki, N.; Tanaka, H.; Soda, K.: Gene cloning, purification and characterization of thermostable alanine dehydrogenase of Bacillus stearothermophilus. J. Ferment. Bioeng., **69**, 154-158 (1990)

[2] Ohshima, T.; Sakane, M.; Yamazaki, T.; Soda, K.: Thermostable alanine dehydrogenase from thermophilic Bacillus sphaericus DSM 462. Purification,

characterization and kinetic mechanism. Eur. J. Biochem., **191**, 715-720 (1990)

[3] Glasfeld, A.; Leanz, G.F.; Benner, S.A.: The stereospecificities of seven dehydrogenases from Acholeplasma laidlawii. The simplest historical model that explains dehydrogenase stereospecificity. J. Biol. Chem., **265**, 11692-11699 (1990)

[4] Honorat, A.; Monot, F.; Ballerini, D.: Synthesis of L-alanine and L-valine by enzyme systems from Bacillus megaterium. Enzyme Microb. Technol., **12**, 515-520 (1990)

[5] Schultz, N.A.; Benson, D.R.: Enzymes of ammonia assimilation in hyphae and vesicles of Frankia sp. strain CpI1. J. Bacteriol., **172**, 1380-1384 (1990)

[6] El-Awamry, Z.A.; El-Rahmany, T.A.: Partial purification and properties of Cunninghamella elegans L-alanine dehydrogenase. Zentralbl. Mikrobiol., **144**, 231-240 (1989)

[7] Flores, M.E.; Sanchez, S.: Ammonium-assimilating enzymes and erythromycin formation in Saccharolypospora erythrea. J. Gen. Appl. Microbiol., **35**, 203-211 (1989)

[8] Kuroda, S.; Tanizawa, K.; Sakamoto, Y.; Tanaka, H.; Soda, K.: Alanine dehydrogenases from two Bacillus species with distinct thermostabilities: molecular cloning, DNA and protein sequence determination, and structural comparison with other NAD(P)(+)-dependent dehydrogenases. Biochemistry, **29**, 1009-1015 (1990)

[9] Ohshima, T.; Wandrey, C.; Conrad, D.: Continous production of 3-fluoro-L-alanine dehydrogenase. Biotechnol. Bioeng., **34**, 394-397 (1989)

[10] Caballero, F.J.; Cardenas, J.; Castillo, F.: Purification and properties of L-alanine dehydrogenase of the phototrophic bacterium Rhodobacter capsulatus E1F1. J. Bacteriol., **171**, 3205-3210 (1989)

[11] Vancurova, I.; Vancura, A.; Volc, J.; Neuzil, J.; Behal, V.: A further characterization of alanine dehydrogenase from Streptomyces aureofaciens. J. Basic Microbiol., **29**, 185-189 (1989)

[12] Vancura, A.; Vancurova, I.; Volc, J.; Jones, S.K.T.; Flieger, M.; Basarova, G.; Behal, V.: Alanine dehydrogenase from Streptomyces fradiae. Purification and properties. Eur. J. Biochem., **179**, 221-227 (1989)

[13] Vancurova, I.; Vancura, A.; Volc, J.; Neuzil, J.; Flieger, M.; Basarova, G.; Behal, V.: Purification and partial characterization of alanine dehydrogenase from Streptomyces aureofaciens. Arch. Microbiol., **150**, 438-440 (1988)

[14] Weiss, P.M.; Chen, C.Y.; Cleland, W.W.; Cook, P.F.: Use of primary deuterium and 15N isotope effects to deduce the relative rates of steps in the mechanisms of alanine and glutamate dehydrogenases. Biochemistry, **27**, 4814-4822 (1988)

[15] Crow, V.L.: Properties of alanine dehydrogenase and aspartase from Propionibacterium freudenreichii subsp. shermanii. Appl. Environ. Microbiol., **53**, 1885-1892 (1987)

[16] Bellion, E.; Tan, F.: An NAD$^+$-dependent alanine dehydrogenase from a methylotrophic bacterium. Biochem. J., **244**, 565-570 (1987)

[17] Porumb, H.; Vancea, D.; Muresan, L.; Presecan, E.; Lascu, I.; Petrescu, I.; Porumb, T.; Pop, R.; Barzu, O.: Structural and catalytic properties of L-ala-

nine dehydrogenase from Bacillus cereus. J. Biol. Chem., **262**, 4610-4615 (1987)

[18] Stams, A.J.M.; Hansen, T.A.: Metabolism of L-alanine in Desulfotomaculum ruminis and two marine Desulfovibrio strains. Arch. Microbiol., **145**, 277-279 (1986)

[19] Ohshima, T.; Wandrey, C.; Sugiura, M.; Soda, K.: Screening of thermostable leucine and alanine dehydrogenases in thermophilic Bacillus strains. Biotechnol. Lett., **7**, 871-876 (1985)

[20] Yoshida, A.; Freese, E.: Enzymic properties of alanine dehydrogenase of Bacillus subtilis. Biochim. Biophys. Acta, **96**, 248-262 (1965)

[21] Itoh, N.; Morikawa, R.; Itoh, N.; Morikawa, R.: Crystallization and properties of L-alanine dehydrogenase from Streptomyces phaeochromogenes. Agric. Biol. Chem., **47**, 2511-2519 (1983)

[22] Bellion, E.; Bolbot, J.A.: Nitrogen assimilation in facultative methylotrophic bacteria. Curr. Microbiol., **9**, 37-44 (1983)

[23] Lauwers, A.M.; Heinen, W.: Thermal properties of enzymes from Bacillus flavothermus, grown between 34 and 70°C. Antonie Leeuwenhoek, **49**, 191-201 (1983)

[24] Grimshaw, C.E.; Cleland, W.W.: Kinetic mechanism of Bacillus subtilis L-alanine dehydrogenase. Biochemistry, **20**, 5650-5655 (1981)

[25] Grimshaw, C.E.; Cook, P.F.; Cleland, W.W.: Use of isotope effects and pH studies to determine the chemical mechanism of Bacillus subtilis L-alanine dehydrogenase. Biochemistry, **20**, 5655-5661 (1981)

[26] Germano, G.J.; Anderson, K.E.: Purification and properties of L-alanine dehydrogenase from Desulfovibrio desulfuricans. J. Bacteriol., **96**, 55-60 (1968)

[27] Alizade, M.A.; Bressler, R.; Brendel, K.: Stereochemistry of the hydrogen transfer to NAD catalyzed by (S)alanine dehydrogenase from Bacillus subtilis. Biochim. Biophys. Acta, **397**, 5-8 (1975)

[28] Sawa, Y.; Tani, M.; Murata, K.; Shibata, H.; Ochiai, H.: Purification and characterization of alanine dehydrogenase from a cyanobacterium, Phormidium lapideum. J. Biochem., **116**, 995-1000 (1994)

[29] Sulzman, F.M.; Edmunds, L.N.: Characterization of circadian oscillations in alanine dehydrogenase activity in non-dividing populations of Euglena gracilis (Z). Biochim. Biophys. Acta, **320**, 594-609 (1973)

[30] Muresan, L.; Vancea, D.; Presecan, E.; Porumb, H.; Lascu, I.; Oarga, M.; Matinca, D.; Abrudan, I.; Barzu, O.: Catalytic properties of Sepharose-bound L-alanine dehydrogenase from Bacillus cereus. Biochim. Biophys. Acta, **742**, 617-622 (1983)

[31] Mueller, P.; Werner, D.: Alanine dehydrogenase from bacteroids and free living cells of Rhizobium japonicum. Z. Naturforsch. C, **37**, 927-936 (1982)

[32] Tolxdorff-Neutzling, R.; Klemme, J.H.: Metabolic role and regulation of L-alanine dehydrogenase in Rhodopseudomonas capsulata. FEMS Microbiol. Lett., **13**, 155-159 (1982)

[33] Vali, Z.; Kilar, F.; Lakatos, S.; Venyaminov, S.A.; Zavodszky, P.: L-Alanine dehydrogenase from Thermus thermophilus. Biochim. Biophys. Acta, **615**, 34-47 (1980)

[34] Keradjopoulos, D.; Holldorf, A.W.: Salt-dependent conformational changes of alanine dehydrogenase from Halobacterium salinarium. FEBS Lett., 112, 183-185 (1980)

[35] Aharonowitz, Y.; Friedrich, C.G.: Alanine dehydrogenase of the β-lactam antibiotic producer Streptomyces clavuligerus. Arch. Microbiol., 125, 137-142 (1980)

[36] Keradjopoulos, D.; Holldorf, A.W.: Purification and properties of alanine dehydrogenase from Halobacterium salinarium. Biochim. Biophys. Acta, 570, 1-10 (1979)

[37] Ohashima, T.; Soda, K.: Purification and properties of alanine dehydrogenase from Bacillus sphaericus. Eur. J. Biochem., 100, 29-39 (1979)

[38] Kim, E.K.; Fitt, P.S.: Partial purification and properties of Halobacterium cutirubrum L-alanine dehydrogenase. Biochem. J., 161, 313-320 (1977)

[39] Rowell, P.; Stewart, W.D.P.: Alanine dehydrogenase of the N_2-fixing blue-green alga, Anbabaena cylindrica. Arch. Microbiol., 107, 115-124 (1976)

[40] Nitta, Y.; Yasuda, Y.; Tochikubo, K.; Hachisuka, Y.: L-Amino acid dehydrogenases in Bacillus subtilis spores. J. Bacteriol., 117, 588-592 (1974)

[41] Keradjopoulos, D.; Wulff, K.: Thermophilic alanine dehydrogenase from Halobacterium salinarium. Can. J. Biochem., 52, 1033-1037 (1974)

[42] McCowen, S.M.; Phibbs, P.V.: Regulation of alanine dehydrogenase in Bacillus (licheniformis). J. Bacteriol., 118, 590-597 (1974)

[43] Graefe, U.; Bocker, H.; Reinhardt, G.; Tkocz, H.; Thrum, H.: Activity of alanine dehydrogenase and production of antibiotic in cultures of Streptomyces hygroscopicus JA 6599. Z. Allg. Mikrobiol., 14, 181-192 (1974)

[44] Borris, D.P.; Aronson, J.N.: Relationship of L-alanine and L-glutamate dehydrogenases of Bacillus thuringienses. Biochim. Biophys. Acta, 191, 716-718 (1969)

[45] Keradjopoulos, D.; Holldorf, A.W.: Thermophilic character of enzymes from extreme halophilic bacteria. FEMS Microbiol. Lett., 1, 179-182 (1977)

[46] Laue, H.; Cook, A.M.: Purification, properties and primary structure of alanine dehydrogenase involved in taurine metabolism in the anaerobe Bilophila wadsworthia. Arch. Microbiol., 174, 162-167 (2000)

[47] Hutter, B.; Singh, M.: Host vector system for high-level expression and purification of recombinant, enzymatically active alanine dehydrogenase of Mycobacterium tuberculosis. Gene, 212, 21-29 (1998)

[48] Allaway, D.; Lodwig, E.M.; Crompton, L.A.; Wood, M.; Parsons, R.; Wheeler, T.R.; Poole, P.S.: Identification of alanine dehydrogenase and its role in mixed secretion of ammonium and alanine by pea bacteroids. Mol. Microbiol., 36, 508-515 (2000)

[49] Irwin, J.A.; Gudmundsson, H.M.; Marteinsson, V.T.; Hreggvidsson, G.O.; Lanzetti, A.J.; Alfredsson, G.A.; Engel, P.C.: Characterization of alanine and malate dehydrogenases from a marine psychrophile strain PA-43. Extremophiles, 5, 199-211 (2001)

[50] Williamson, D.H.: L-Alanin. Methods Enzym. Anal., 3rd Ed. (Bergmeyer, H.U., ed.), 2, 1724-1727 (1974)

[51] Smith, M.T.; Emerich, D.W.: Alanine dehydrogenase from soybean nodule bacteroids: purification and properties. Arch. Biochem. Biophys., 304, 379-385 (1993)

[52] Delforge, D.; Devreese, B.; Dieu, M.; Delaive, E.; Van Beeumen, J.; Remacle, J.: Identification of lysine 74 in the pyruvate binding site of alanine dehydrogenase from Bacillus subtilis. Chemical modification with 2,4,6-trinitrobenzenesulfonic acid, n-succinimidyl 3-(2-pyridyldithio)propionate, and 5'-(p-(fluorosulfonyl)benzoyl)adenosine. J. Biol. Chem., 272, 2276-2284 (1997)

[53] Hutter, B.; Singh, M.: Properties of the 40 kDa antigen of Mycobacterium tuberculosis, a functional L-alanine dehydrogenase. Biochem. J., 343, 669-672 (1999)

[54] Sedelnikova, S.; Rice, D.W.; Shibata, H.; Sawa, Y.; Baker, P.J.: Crystallization of the alanine dehydrogenase from Phormidium lapideum. Acta Crystallogr. Sect. D, 54, 407-408 (1998)

[55] Kim, S.J.; Lee, W.Y.; Kim, K.H.: Unusual allosteric property of L-alanine dehydrogenase from Bacillus subtilis. J. Biochem. Mol. Biol., 31, 25-30 (1998)

[56] Feng, Z.; Caceres, N.E.; Sarath, G.; Barletta, R.G.: Mycobacterium smegmatis L-alanine dehydrogenase (Ald) is required for proficient utilization of alanine as a sole nitrogen source and sustained anaerobic growth. J. Bacteriol., 184, 5001-5010 (2002)

[57] Vadas, A.J.; Schroder, I.; Monbouquette, H.G.: Room-temperature synthesis of L-alanine using the alanine dehydrogenase of the hyperthermophilic archaeon Archaeoglobus fulgidus. Biotechnol. Prog., 18, 909-911 (2002)

[58] Galkin, A.; Kulakova, L.; Ashida, H.; Sawa, Y.; Esaki, N.: Cold-adapted alanine dehydrogenases from two antarctic bacterial strains: gene cloning, protein characterization, and comparison with mesophilic and thermophilic counterparts. Appl. Environ. Microbiol., 65, 4014-4020 (1999)

[59] Usha, V.; Jayaraman, R.; Toro, J.C.; Hoffner, S.E.; Das, K.S.: Glycine and alanine dehydrogenase activities are catalyzed by the same protein in Mycobacterium smegmatis: upregulation of both activities under microaerophilic adaptation. Can. J. Microbiol., 48, 7-13 (2002)

[60] Chowdhury, E.K.; Saitoh, T.; Nagata, S.; Ashiuchi, M.; Misono, H.: Alanine dehydrogenase from Enterobacter aerogenes: purification, characterization, and primary structure. Biosci. Biotechnol. Biochem., 62, 2357-2363 (1998)

[61] Hutter B; Dick T: Increased alanine dehydrogenase activity during dormancy in Mycobacterium smegmatis. FEMS Microbiol. Lett., 167, 7-11 (1989)

[62] Siranosian, Kathryn Jaacks; Ireton, Keith; Grossman, Alan D.: Alanine dehydrogenase (ald) is required for normal sporulation in Bacillus subtilis. J. Bacteriol., 175, 6789-6796 (1993)

[63] Berberich, Robert; Kaback, Michael; Freese, Ernst.: D-Amino acids as inducers of L-alanine dehydrogenase in Bacillus subtilis. J. Biol. Chem., 243, 1008-1013 (1968)

[64] Brunhuber, N.M.W.; Blanchard, J.S.: The biochemistry and enzymology of amino acid dehydrogenases. Crit. Rev. Biochem. Mol. Biol., 29, 415-467 (1994)

Glutamate dehydrogenase

1 Nomenclature

EC number
1.4.1.2

Systematic name
L-glutamate:NAD$^+$ oxidoreductase (deaminating)

Recommended name
glutamate dehydrogenase

Synonyms
GDH
L-glutamate dehydrogenase
NAD-GDH
NAD-dependent glutamate dehydrogenase
NAD-dependent glutamic dehydrogenase
NAD-glutamate dehydrogenase
NAD-linked glutamate dehydrogenase
NAD-linked glutamic dehydrogenase
NAD-specific glutamate dehydrogenase
NAD-specific glutamic dehydrogenase
NAD:glutamate oxidoreductase
NADH-dependent glutamate dehydrogenase
NADH-linked glutamate dehydrogenase
surface-associated protein PGAG1
dehydrogenase, glutamate
glutamate dehydrogenase (NAD)
glutamate oxidoreductase
glutamic acid dehydrogenase
glutamic dehydrogenase

CAS registry number
9001-46-1

2 Source Organism

<1> *Neurospora crassa* [1, 10, 12, 16]
<2> *Saccharomyces cerevisiae* [1]
<3> *Apodachlya sp.* [1]
<4> *Pythium debaryanum* [1]

<5> *Achlya sp.* [1]
<6> *Blastocladiella emersonii* [1]
<7> *Peptococcus aerogenes* [1, 11]
<8> *Micrococcus aerogenes* [1]
<9> *Clostridium sp.* (SB4 [1]) [1]
<10> *Thiobacillus novellus* [1]
<11> *Pisum sativum* [1, 6, 28]
<12> *Synechocystis sp.* (PCC 6803 [2]) [2]
<13> *Lupinus luteus* (lupin [3,7]) [3, 7]
<14> *Crithidia fasciculata* [4]
<15> *Chlorella sorokiniana* [5]
<16> *Clostridium symbiosum* [8, 36]
<17> *Dictyostelium discoideum* [9, 45]
<18> *Vigna unguiculata* [13]
<19> *Mytilus edulis* [14]
<20> *Bacillus cereus* [15]
<21> *Arenicola marina* [17]
<22> *Loligo pealeii* [18]
<23> *Symbiodinium microadriaticum* [19]
<24> *Acanthamoeba culbertsoni* [20]
<25> *Vitis vinifera* [21]
<26> *Agave americana* [22]
<27> *Geobacillus stearothermophilus* [23]
<28> *Phycomyces blakesleeanus* [24]
<29> *Kluyveromyces fragilis* [25]
<30> *Clostridium botulinum* (113B [26]) [26]
<31> *Aspergillus nidulans* [27]
<32> *Lemna minor* [28]
<33> *Medicago sativa* [29]
<34> *Candida utilis* [30]
<35> *Peptostreptococcus asaccharolyticus* [31]
<36> *Zea mays* [32]
<37> *Cucurbita pepo* [33]
<38> *Pinus sylvestris* (scots pine [34]) [34]
<39> *Clostridium difficile* [35]
<40> *Bacteroides fragilis* (two enzymes GDHA and GDHB [37]) [37]
<41> *Bacillus acidocaldarius* [38]
<42> *Psychrobacter sp.* (TAD 1, two enzymes [39]) [39]
<43> *Sporosarcina urea* (DSM 320 [40]) [40]
<44> *Agaricus bisporus* [41]
<45> *Pyrococcus sp.* [42]
<46> *Laccaria bicolor* [43]
<47> *Thermus thermophilus* (HB8 [44]) [44]
<48> *Streptomyces clavuligerus* [46]
<49> *Streptomyces fradiae* [47]
<50> *Amphibacillus xylanus* [48]
<51> *Pyrobaculum islandicum* [49]

<52> *Thermotoga maritima* [50]
<53> *Bos taurus* [51]

3 Reaction and Specificity

Catalyzed reaction

L-glutamate + H_2O + NAD^+ = 2-oxoglutarate + NH_3 + NADH + H^+ (<13>
fully ordered reaction mechanism [3]; <13> NAD^+ binds first followed by L-
glutamate [3]; <13> formation of an enzyme-NAD-oxoglutarate dead end
complex [3]; <15> bi uni uni ping-pong addition sequence [5]; <11> partially random mechanism [6])

Reaction type

oxidation
redox reaction
reduction
reductive amination

Natural substrates and products

S L-glutamate + H_2O + NAD^+ <14, 22, 23> (<14> control point for amino
acid metabolism [4]; <22> primary role in squid mantle muscle is in reg-
ulating the catabolism of amino acids for energy production [18]) (Reversibility: ? <14, 22, 23> [4, 18, 19]) [4, 18, 19]

P 2-oxoglutarate + NH_3 + NADH

Substrates and products

S 2-oxoglutatate + NH_3 + NADH <16, 37, 39, 41, 44, 46-48, 50> (Reversibility: r <>16, 37, 39, 41, 44, 46-48, 50# [33, 35, 36, 38, 41, 43, 44, 46, 48]) [33,
35, 36, 38, 41, 43, 44, 46, 48]

P L-glutamate + H_2O + NAD^+ <16, 37, 39> [33, 35, 36]

S L-glutamate + H_2O + NAD^+ <1-51> (<12> strict substrate specificity [2];
<22> glutamate oxidation is favoured [18]) (Reversibility: ? <1-36, 38, 40,
42, 43, 45, 49, 51> [1-32, 34, 35, 37, 39, 40, 42, 47, 49]; r <16, 37, 39, 41, 44,
46-48, 50> [33, 35, 36, 38, 41, 43, 44, 46, 48]) [1-49]

P 2-oxoglutarate + NH_3 + NADH

S L-norvaline + H_2O + NAD^+ <13, 18> (<13> deamination at 5% the rate of
L-glutamate deamination [7]; <18> at 40% the rate [13]) (Reversibility: ?
<13, 18> [7, 13]) [7, 13]

P 2-oxopentanoate + NH_3 + NADH

S L-serine + H_2O + NAD^+ <18> (<18> deamination at 29% the rate of L-
glutamate deamination [13]) (Reversibility: ? <18> [13]) [13]

P 3-hydroxy-2-oxopropanoate + NH_3 + NADH

S Additional information <16> (<16> L-α-amino-γ-nitroaminobutyrate
deamination at 0.5% the rate of deamination of L-glutamate [8]) (Reversibility: ? <16> [8]) [8]

P ?

Inhibitors

5,5'-dithiobis(2-nitrobenzoate) <14, 30, 31> [4, 26, 27]

ADP <13, 15, 30, 35, 39, 46> [5, 7, 26, 31, 35, 43]

AMP <13, 21, 30, 35, 39, 46> [7, 17, 26, 31, 35, 43]

ATP <14, 21, 22, 26, 30, 35, 46, 48> (<26> weak [22]) [4, 17, 18, 22, 26, 31, 43, 46]

CMP <24> [20]

Ca^{2+} <11, 25, 40> (<11,25> activation of reductive amination [6,21]; <11> slight inhibition of oxidative deamination [6]; <25> no effect on oxidative deamination [21]) [6, 21, 37]

Cu^{2+} <46, 49> [43, 47]

DL-valine <18> [13]

EDTA <13, 18> [7, 13]

GDP <22> [18]

GTP <14, 21, 22, 35> [4, 17, 18, 31]

H_2O_2 <30> [26]

Hg^{2+} <30> [26]

KCN <13> [7]

L-aspartate <14, 18> (<14> weak [4]) [4, 13]

L-histidine <24> [20]

L-malic acid <14> [4]

Mg^{2+} <24, 40> [20, 37]

Mn^{2+} <28, 49> (<28,49> weak [24,47]) [24, 47]

N-acetylglutamate <1> [16]

N-α-p-tosyl-L-lysine chloromethyl ketone <37> (<37> TLCK [33]) [33]

N-carbamoylglutamate <1> [16]

N-ethylmaleimide <14, 31, 37> [4, 27, 33]

N-methylglutamate <1> [16]

NAD^+ <11, 24> [6, 20]

NADH <11, 48> [6, 46]

NH_4^+ <11, 48> [6, 46]

Ni^{2+} <49> [47]

Zn^{2+} <24, 26> [20, 22]

α,γ-diethyl glutamate <1> [16]

α-ketoglutarate <11> [6]

chlortetracyclin <24> [20]

citrate <13> (<13> inhibition of amination, no effect on deamination [7]) [7]

dithiothreitol <7> [11]

fumarate <30> (<30> amination [26]) [26]

glutamate <24, 30> (<30> amination [26]) [20, 26]

glutamine <30> (<30> amination [26]) [26]

iodoacetamide <49> [47]

iodoacetate <14> (<14> weak [4]) [4]

isocitrate <18, 24> [13, 20]

isophthalate <1, 31> [12, 27]

o-iodobenzoate <14> [4]

oxaloacetate <30> (<30> amination [26]) [26]

p-aminomercuribenzoate <31> [27]
p-chloromercuribenzoate <14> [4]
p-hydroxymercuribenzoate <18, 49> [13, 47]
phosphoenolpyruvate <14> (<14> weak [4]) [4]
pyridoxal 5'-phosphate <13, 24, 26> [7, 20, 22]
pyruvic acid <14, 18> [4, 13]
taurine <24> [20]
thiol reagents <31> [27]
Additional information <35> (<35> strong inhibition by increasing ionic strength [31]) [31]

Cofactors/prosthetic groups

NAD$^+$ <1-51> [1-49]
NADH <1-20, 22-36> (<21> no aminating activity either in presence of NADH or NADPH [17]) [1-16, 18-32]
NADP$^+$ <32, 35, 42> (<32> less than 1% of the activity with NAD$^+$ [28]; <35> about 4% of the activity with NAD$^+$ [31]; <12,13> no activity [2,7]) [28, 31, 39]
NADPH <11, 13, 16, 18, 19, 26, 32, 35> (<13> can replace NADH only at pH 6, not at pH 7-10 [7]; <12> no activity with NADPH [2]; <16> rate with NADPH is 300times lower than with NADH [8]; <18> activity with NADH, NADPH and NAD$^+$ in the ratio 126:2:1 [13]; <19> less than 1% of the activity with NADH [14]; <26> with NADPH less than 10% of the activity with NADH [22]; <35> NADPH can replace NADH [31]) [7, 8, 13, 14, 22, 28, 31]
acetyl-NAD$^+$ <21> [17]

Activating compounds

ADP <17, 19, 22, 28> (<17,22,28> stimulation [9,18,24]; <19,22> absolute requirement as cofactor [14,18]) [9, 14, 18, 24]
AMP <17, 22, 24, 28, 48> (<17,22,24,48> stimulation [9,18,20,45,46]; <28> required [24]) [9, 18, 20, 24, 45, 46]
L-cysteine <24> (<24> stimulates deamination [20]) [20]
acetonitrile <51> (<51> activates [49]) [49]
asparagine <48> (<48> activates [46]) [46]
aspartate <48> (<48> activates [46]) [46]
dAMP <17> (<17> stimulation [9]) [9]
guanidine hydrochloride <51> (<51> activates [49]) [49]
leucine <22> (<22> stimulation [18]) [18]
tetrahydrofuran <51> (<51> activates [49]) [49]

Metals, ions

Ca^{2+} <11, 25, 32, 33, 36-38, 46> (<11,32> required [28]; <11> activation of reductive amination [6]; <11> slight inhibition of oxidative deamination [6]; <25> activation of amination, no effect on deamination [21]; <33> activation of partially purified enzyme [29]; <36> amination is stimulated, deamination not influenced [32]; <37,38,46> activation [33,34,43]) [6, 21, 28, 29, 32-34, 43]
Mg^{2+} <46> (<46> stimulation [43]) [43]
Mn^{2+} <11> (<11> activation of reductive amination [6]) [6]

Zn^{2+} <11, 32> (<11> activation of reductive amination [6]; <11,32> can replace Ca^{2+}, requirement [28]) [6, 28]

Specific activity (U/mg)

1.8 <44> [41]

7.6 <45> [42]

21.2 <48> [46]

31.4 <37> [33]

38 <33> (<33> GDH-I [29]) [29]

54 <11, 33> (<33> GDH-II [29]) [6, 29]

64.5 <7> [11]

69.8 <31> [27]

70 <47> [44]

80 <41> [38]

104 <53> (<53> GDH II [51]) [51]

146 <39> [35]

167 <53> (<53> GDH I [51]) [51]

183 <40> [37]

200 <15> [5]

248 <42> (<42> NAD^+-dependent enzyme [39]) [39]

401 <11> [28]

441 <25> [21]

540 <32> [28]

822 <50> [48]

1092 <30> (<30> above [26]) [26]

Additional information <1-11, 18, 20, 34, 46, 49> [1, 10, 12, 13, 15, 30, 43, 47]

K_m-Value (mM)

0.005 <51> (NADH) [49]

0.025 <51> (NAD^+) [49]

0.033 <6> (NADH) [1]

0.037 <17> (NADH, <17> non-activated NAD-GDH [45]) [45]

0.046 <44> (NAD^+) [41]

0.05 <49> (NADH) [47]

0.055 <5> (NADH) [1]

0.06 <44> (NADH) [41]

0.065 <45> (NADPH) [42]

0.066 <51> (2-oxoglutarate) [49]

0.07 <41> (NADH) [38]

0.089 <46> (NADH) [43]

0.09 <17> (NADH, <17> activated NAD-GDH [45]) [45]

0.14 <45> (NADH) [42]

0.17 <51> (L-glutamate) [49]

0.2 <45> (2-oxoglutarate) [42]

0.24 <46> (L-glutamate) [43]

0.27 <47> (NAD^+) [44]

0.28 <46> (NAD^+) [43]

0.3 <49> (NAD^+) [47]

0.33 <1> (NAD$^+$) [1]
0.35 <41> (NAD$^+$) [38]
0.4 <6> (NAD$^+$) [1]
0.5 <42> (NAD$^+$, <42> NAD$^+$-dependent enzyme [39]) [39]
0.55 <1> (NADH) [1]
0.56 <6> (2-oxoglutarate) [1]
0.61 <5> (NAD$^+$) [1]
0.75 <17> (2-oxoglutarate, <17> activated NAD-GDH [45]) [45]
0.8 <39> (NAD$^+$) [35]
0.98 <45> (NH$_4^+$) [42]
1.35 <46> (2-oxoglutarate) [43]
1.4 <6> (L-glutamate) [1]
1.7 <38> (2-oxoglutarate) [34]
1.9 <17> (2-oxoglutarate, <17> non-activated NAD-GDH [45]) [45]
2.36 <42> (2-oxoglutarate, <42> NAD$^+$-dependent enzyme [39]) [39]
3.1 <5> (L-glutamate) [1]
3.3 <5, 37> (2-oxoglutarate) [1, 33]
3.4 <49> (2-oxoglutarate) [47]
4.3 <39> (L-glutamate) [35]
4.6 <1> (2-oxoglutarate) [1]
5.5 <1> (L-glutamate) [1]
6.5 <44> (NH$_4^+$) [41]
9.7 <51> (NH$_4^+$) [49]
14.2 <49> (NH$_4^+$) [47]
17 <1> (NH$_4^+$) [1]
19 <38> (NH$_4^+$) [34]
22.8 <17> (NH$_4^+$, <17> non-activated NAD-GDH [45]) [45]
24.6 <42> (NH$_4^+$, <42> NAD$^+$-dependent enzyme [39]) [39]
25 <41> (L-glutamate) [38]
25.6 <17> (NH$_4^+$, <17> activated NAD-GDH [45]) [45]
26 <5> (NH$_4^+$) [1]
28.6 <42> (L-glutamate, <42> NAD$^+$-dependent enzyme [39]) [39]
32.2 <49> (L-glutamte) [47]
33 <48> (NH$_4^+$) [46]
33.3 <37> (NH$_4^+$) [33]
37 <46> (NH$_4^+$) [43]
37.1 <44> (L-glutamate) [41]
40 <6> (NH$_4^+$) [1]
42.5 <38> (NADH) [34]
49 <47> (L-glutamate) [44]
60 <41> (NH$_4^+$) [38]
Additional information <1-15, 17-19, 22, 23, 25, 28, 30, 33-35, 53> (<53> different K$_m$ values for GDH I and GDH II [51]) [1, 2, 4, 5, 7, 9, 11, 13, 14, 16, 18, 19, 21, 24, 26, 29-31, 51]

pH-Optimum

7 <48> [46]

7.2 <37> [33]

7.3-7.5 <17> (<17> reductive amination [9]) [9]

7.4 <46> (<46> amination [43]) [43]

7.4-7.5 <23> (<23> reductive amination [19]) [19]

7.4-8.2 <19> (<19> reductive amination [14]) [14]

7.5 <17> [45]

7.8 <9> (<9> reductive amination [1]) [1]

7.9 <6> (<6> reductive amination [1]) [1]

8 <11, 15, 25, 40, 47> (<11,15,25,47> reductive amination [5,6,21,44]) [5, 6, 21, 37, 44]

8-8.3 <33> (<33> reductive amination [29]) [29]

8-8.5 <28> (<28> reductive amination [24]) [24]

8.2 <3, 11, 13> (<3,13> reductive amination [1,7]; <11> glutamate oxidation [1]) [1, 7]

8.4 <26, 41> (<26> reductive amination [22]; <41> forward reaction [38]) [22, 38]

8.5 <38, 39> [34, 35]

8.5-9 <12> (<12> reductive amination [2]) [2]

8.8 <6, 13, 46, 49> (<6> glutamate oxidation [1]; <13> oxidative deamination [7]; <46> deamination [43]; <49> reductive amination [47]) [1, 7, 43, 47]

9 <3, 15, 23, 28, 41> (<3> glutamate oxidation [1]; <15,23,28> oxidative deamination [5,19,24]; <41> reverse reaction [38]) [1, 5, 19, 24, 38]

9.2 <11, 33, 49> (<11,33,49> oxidative deamination [6,29,47]) [6, 29, 47]

9.3 <25> (<25> oxidative deamination [21]) [21]

9.4 <9> (<9> glutamate oxidation [1]) [1]

9.5 <19> (<19> oxidative deamination [14]) [14]

9.5-10 <12> (<12> oxidative deamination [2]) [2]

pH-Range

7-8.8 <11> (<11> about 50% of activity maximum at pH 7.0 and 8.8, reductive amination [6]) [6]

7.5-9 <26> (<26> about 65% of activity maximum at pH 7.5 and 9.0, reductive amination [22]) [22]

7.5-10.5 <37> [33]

7.9-8.5 <13> (<13> about 90% of activity maximum at pH 7.9 and 8.5, reductive amination [7]) [7]

8.4-9.2 <13> (<13> about 90% of activity maximum at pH 8.4 and 9.2, oxidative deamination [7]) [7]

8.4-10.1 <11> (<11> 50% of activity maximum at pH 8.4 and 10.1, oxidative deamination [6]) [6]

Temperature optimum (°C)

20 <42> (<42> NAD$^+$-dependent enzyme [39]) [39]

30 <48> [46]

30-35 <49> [47]

38 <40> [37]

40-45 <20> [15]
50-55 <7, 41> [11, 38]
85-90 <47> (<47> deamination [44]) [44]
90 <51> [49]

4 Enzyme Structure

Molecular weight
87000 <16> (<16> F187D mutant, gel filtration [36]) [36]
98000 <28> (<28> gel filtration [24]) [24]
110000-130000 <10> (<10> sucrose density gradient sedimentation [1]) [1]
165000-175000 <20> (<20> GDHA, gel filtration [15]) [15]
180000 <15> (<15> gel filtration [5]) [5]
210000 <11> (<11> gel filtration [1]) [1]
210000-250000 <6> (<6> sucrose density gradient sedimentation [1]) [1]
220000 <51> (<51> gel filtration [49]) [49]
225000 <4, 5> (<4,5> sucrose density gradient sedimentation [1]) [1]
230000 <11, 32> (<32> gel filtration [28]; <11> sedimentation equilibrium analysis [28]) [28]
240000-250000 <20> (<20> GDH B, gel filtration [15]) [15]
250000 <43> (<43> gel filtration [40]) [40]
250800 <30> (<30> gel filtration [26]) [26]
252000 <25> (<25> gel filtration [21]) [21]
260000 <50> (<50> gel filtration [48]) [48]
266000 <7, 35> (<7,35> gel filtration [11,31]) [11, 31]
268000-282000 <9> (<9> sedimentation equilibrium [1]) [1]
270000 <7, 13> (<7,13> gel filtration [1,7]) [1, 7]
284000 <45> (<45> gel filtration [42]) [42]
289000 <47> (<47> gel filtration [44]) [44]
290000 <41> (<41> gel filtration [38]) [38]
295000 <12> (<12> gel filtration [2]) [2]
300000 <39, 42> (<39> gel filtration [35]; <42> NAD$^+$-dependent enzyme, gel filtration [39]) [35, 39]
310000 <1> (<1> gel filtration [16]) [16]
330000 <1, 53> (<1> amino acid data [10]; <53> non-denaturing PAGE [51]) [10, 51]
350000-390000 <9> (<9> gel filtration [1]) [1]
356000 <17> (<17> gel filtration [9]) [9]
380000 <14> (<14> gel filtration [4]) [4]
460000 <34> (<34> sedimentation equilibrium [30]) [30]
470000 <46> (<46> gel filtration [43]) [43]
474000 <44> (<44> gel filtration [41]) [41]
480000 <1> (<1> sedimentation equilibrium analysis [12]) [12]
670000 <31> (<31> gel filtration [27]) [27]
1100000 <48> (<48> gel filtration [46]) [46]

Subunits

dimer <16, 28, 42> (<28> 2 * 54000, SDS-PAGE [24]; <16> F187D mutant, 2 * 47000 [36]; <42> NAD$^+$-dependent enzyme, 2 * 160000, SDS-PAGE [39]) [24, 36, 39]

hexamer <1, 12, 13, 17, 25, 30, 39, 41, 43, 45, 47, 48, 50, 51, 53> (<12> 6 * 48500, SDS-PAGE [2]; <13> 6 * 45000, SDS-PAGE [7]; <17> 6 * 54000, SDS-PAGE [9]; <1> 6 * 51500, SDS-PAGE [10]; <25> 6 * 42500, SDS-PAGE [21]; <30> 6 * 42500, SDS-PAGE [26]; <39> 6 * 45000, SDS-PAGE [35]; <41> 6 * 48000, SDS-PAGE [38]; <43> 6 * 42000, SDS-PAGE [40]; <45> 6 * 43000, SDS-PAGE [42]; <47> 6 * 48000, SDS-PAGE [44]; <48> 6 * 183000, SDS-PAGE [46]; <50> 6 * 45000, SDS-PAGE [48]; <51> 6 * 36000, SDS-PAGE [49]; <53> 6 * 57500, SDS-PAGE [51]) [2, 7, 9, 10, 21, 26, 35, 38, 40, 42, 44, 46, 48, 49, 51]

hexamer or pentamer <35> (<35> 5 * or 6 * 49000, SDS-PAGE [31]) [31]

tetramer <1, 11, 15, 32, 34, 44, 46> (<15> 4 * 45000, SDS-PAGE [5]; <1> 4 * 116000, sedimentation equilibrium analysis after treatment with 6 M guanidine HCl and 0.5% mercaptoethanol [12]; <11,32> 4 * 58500, SDS-PAGE [28]; <34> 4 * 116000, SDS-PAGE [30]; <44> 4 * 116000, SDS-PAGE [41]; <46> 4 * 116000, SDS-PAGE [43]) [5, 12, 28, 30, 41, 43]

Additional information <12> [2]

5 Isolation/Preparation/Mutation/Application

Source/tissue

alimentary canal <22> [18]
brain <53> [51]
cotyledon <37> [33]
heart <22> (<22> systemic [18]) [18]
hepatopancreas <19> [14]
leaf <25, 26> (<26> medulla, cortex [22]) [21, 22]
mantle muscle <22> [18]
muscle <21> [17]
mycelium <31> [27]
needle <38> [34]
nodule <13> [3, 7]
root <33> [29]
seed <11, 33> [28, 29]
seedling <18, 36> [13, 32]
shoot <36> [32]
stem <11> [6]

Localization

cytosol <31> [27]
membrane <40> [37]
mitochondrion <11, 22, 25, 36> [6, 18, 21, 32]

Purification

<1> [10, 12, 16]
<7> [11]
<11> [6, 28]
<13> (4 isoenzymes [7]) [7]
<14> [4]
<15> [5]
<16> (overexpression of F187D mutant in Escherichia coli [36]) [36]
<17> (activated and non-activated form [45]) [9, 45]
<18> [13]
<19> [14]
<20> (2 isoenzymes GDHA, GDHB [15]) [15]
<22> [18]
<23> [19]
<24> [20]
<25> (7 isoenzymes [21]) [21]
<28> [24]
<30> [26]
<31> [27]
<32> [28]
<33> (2 isoenzymes: GDH-I and GDH-II [29]) [29]
<34> [30]
<35> [31]
<37> [33]
<38> [34]
<39> (overexpression in Escherichia coli [35]) [35]
<40> (overexpression of GDHB in Escherichia coli [37]) [37]
<41> [38]
<42> (NAD$^+$-dependent and NADP$^+$-dependent GDH [39]) [39]
<43> [40]
<44> [41]
<45> (overexpression in Escherichia coli [42]) [42]
<46> [43]
<47> [44]
<48> [46]
<49> [47]
<50> [48]
<51> [49]
<52> (overexpression of mutants in Escherichia coli [50]) [50]
<53> (GDH I and II [51]) [51]

Crystallization

<52> [50]

Engineering

F187D <16> (<16> dimeric form of enzyme [36]) [36]
G376K <52> (<52> faster thermal inactivation, higher specific activity at 58°C [50]) [50]

N97D <52> (<52> faster thermal inactivation [50]) [50]
N97D/G376K <52> (<52> faster thermal inactivation, higher specific activity at 58°C [50]) [50]

6 Stability

pH-Stability
4 <19> (<19> 15 min, 30°C, inactivation [14]) [14]
6.5-7.2 <19> (<19> 15 min, 30°C, stable [14]) [14]
8 <19> (<19> 15 min, 30°C, inactivation [14]) [14]

Temperature stability
0-4 <43> (<43> complete inactivation [40]) [40]
50 <1> (<1> 6 min, complete loss of activity [12]) [12]
55 <23> (<23> half-life: 60 min [19]) [19]
65 <12> (<12> 20 min, 50% loss of activity [2]) [2]
70 <30> (<30> 10 min, 40% loss of activity [26]) [26]
75 <43> (<43> 15 min, complete inactivation [40]) [40]
80 <7> (<7> 10 min, complete inactivation [11]) [11]
85 <50> (<50> 120 min, 50% activity loss [48]) [48]
100 <51> (<51> no loss of activity after 2 h [49]) [49]

General stability information
<1>, DTT stabilizes [12]
<1>, NaCl, 0.1 M, stabilizes [10]
<1>, isophthalate stabilizes [12]
<1>, potassium phosphate buffer, high concentration, stabilizes [16]
<1>, presence of sulfhydryl groups in the environment stabilizes [16]
<18>, repeated freezing and thawing: loss of activity [13]
<19>, 2-oxoglutarate stabilizes [14]
<19>, ADP stabilizes [14]
<19>, NADH stabilizes [14]
<31>, guanidine hydrochloride, 1.0 M, complete denaturation [27]
<46>, ammonium sulfate improves stability [43]

Storage stability
<1>, -20°C, gradual loss of activity [12]
<1>, 4°C, suspension of 50% saturated ammonium sulfate, 0.1 mM DTT, 25 mM glutamate, stable for months [12]
<13>, -20°C, up to 1 year, stable [7]
<18>, -20°C, several months [13]
<18>, -5°C, 20% loss of activity after 3 months [13]
<20>, 4-5°C, 0.066 M phosphate buffer, pH 7.4 or 0.01 M arginine buffer, pH 9.5, 20 days, stable [15]
<23>, 4°C, more than 90% of original activity retained after 20 days [19]
<30>, -20°C, loss of activity overnight [26]
<30>, 4°C, 3 days, stable [26]

<39>, -15°C, 50 mM Tris-HCl, pH 7.5, 0.1 M KCl, stable for several months [35]

References

[1] Smith, E.L.; Austen, B.M.; Blumenthal, K.M.; Nyc, J.F.: Glutamate dehydrogenase. The Enzymes, 3rd Ed. (Boyer, P.D., ed.), 11, 293-367 (1975)

[2] Chavez, S.; Candau, P.: An NAD-specific glutamate dehydrogenase from cyanobacteria. Identification and properties. FEBS Lett., 285, 35-38 (1991)

[3] Stone, S.R.; Copeland, L.; Heyde, E.: Glutamate dehydrogenase of lupin nodules: kinetics of the deamination reaction. Arch. Biochem. Biophys., 199, 550-559 (1980)

[4] Higa, A.I.; de Cazzulo, B.M.F.; Cazzulo, J.J.: Some properties of the NAD-specific glutamate dehydrogenase from Crithidia fasciculata. J. Gen. Microbiol., 113, 429-432 (1979)

[5] Meredith, M.J.; Gronostajski, R.M.; Schmidt, R.R.: Physical and kinetic properties of the nicotinamide adenine dinucleotide-specific glutamate dehydrogenase purified from Chlorella sorokiniana. Plant Physiol., 61, 967-974 (1978)

[6] Garland, W.J.; Dennis, D.T.: Steady-state kinetics of glutamate dehydrogenase from Pisum sativum L. mitochondria. Arch. Biochem. Biophys., 182, 614-625 (1977)

[7] Stone, S.R.; Copeland, L.; Kennedy, I.R.: Glutamate dehydroenase of lupin nodules: Purification and properties. Phytochemistry, 18, 1273-1278 (1979)

[8] Syed, S.E.H.; Engel, P.C.; Parker, D.M.: Functional studies of a glutamate dehydrogenase with known three-dimensional structure: steady-state kinetics of the forward and reverse reactions catalysed by the NAD(+)-dependent glutamate dehydrogenase of Clostridium symbiosum. Biochim. Biophys. Acta, 1115, 123-130 (1991)

[9] Pamula, F.; Wheldrake, J.F.: The NAD-dependent glutamate dehydrogenase from Dictyostelium discoideum: purification and properties. Arch. Biochem. Biophys., 291, 225-230 (1991)

[10] Strickland, W.N.; Jacobson, J.W.; Strickland, M.: The amino acid composition and some properties of the NAD$^+$-specific glutamate dehydrogenase from Neurospora crassa. Biochim. Biophys. Acta, 251, 21-30 (1991)

[11] Johnson, W.M.; Westlake, D.W.S.: Purification and characterization of glutamic acid dehydrogenase and α-ketoglutaric acid reductase from Peptococcus aerogenes. Can. J. Microbiol., 18, 881-892 (1972)

[12] Veronese, F.M.; Nyc, J.F.; Degani, Y.; Brown, D.M.; Smith, E.L.: Nicotinamide adenine dinucleotide-specific glutamate dehydrogenase of Neurospora. I. Purification and molecular properties. J. Biol. Chem., 249, 7922-7928 (1974)

[13] Fawole, M.O.; Boulter, D.: Purification and properties of glutamate dehydrogenase from Vigna unguiculata (L.) Walp.. Planta, 134, 97-102 (1977)

[14] Ruano, A.R.; Riano, J.L.A.; Amil, M.R.; Santos, M.J. H.: Some enzymatic properties of NAD$^+$-dependent glutamate dehydrogenase of mussel hepato-

pancreas (Mytilus edulis L.) requirement of ADP. Comp. Biochem. Physiol. B, **82**, 197-202 (1985)

[15] Verma, N.S.; Sharma, D.; Gollakota, K.G.: Purification & properties of isozymes of glutamate dehydrogenase form Bacillus cereus T. Indian J. Biochem. Biophys., **13**, 344-346 (1976)

[16] Grover, A.K.; Kapoor, M.: Studies on the regulation, subunit structure, and some properties of NAD-specific glutamate dehydrogenase of Neurospora. J. Exp. Bot., **24**, 847-861 (1973)

[17] Batrel, Y.; Le Gal, Y.: Nitrogen metabolism in Arenicola marina. Characterization of a NAD dependent glutamate dehydrogenase. Comp. Biochem. Physiol. B, **78**, 119-124 (1984)

[18] Storey, K.B.; Fields, J.H.A.; Hochachka, P.W.: Purification and properties of glutamate dehydrogenase from the mantle muscle of the squid, Loligo pealeii. Role of the enzyme in energy production from amino acids. J. Exp. Zool., **205**, 111-118 (1978)

[19] Dudler, N.; Miller, D.J.: Characterization of two glutamate dehydrogenases from the symbiotic microalga Symbiodinium microadriaticum isolated from the coral Acropora formosa. Mar. Biol., **97**, 427-430 (1988)

[20] Singh, U.S.; Rao, V.K.M.: Characterization of L-glutamate dehydrogenase activity of axenically grown Acanthamoeba culbertsoni. Indian J. Biochem. Biophys., **20**, 146-148 (1983)

[21] Loulakakis, C.A.; Roubelakis-Angelakis, K.A.: Intracellular localization and properties of NADH-glutamate dehydrogenase from Vitis vinifera L.: Purification and characterization of the major leaf isoenzyme. J. Exp. Bot., **41**, 1223-1230 (1990)

[22] Ramirez, H.; Delgado, M.J.; Garcia-Peregrin, E.: Some properties of glutamate dehydrogenase from Agave americana L. leaves. Z. Pflanzenphysiol., **84**, 109-119 (1977)

[23] Schmidt, C.N.G.; Jervis, L.: Partial purification and characterization of glutamate synthase from a thermophilic bacillus. J. Gen. Microbiol., **128**, 1713-1718 (1982)

[24] Van Laere, A.J.: Purification and properties of NAD-dependent glutamate dehydrogenase from Phycomyces spores. J. Gen. Microbiol., **134**, 1597-1601 (1988)

[25] Nisbet, B.A.; Slaughter, J.C.: Glutamate dehydrogenase and glutamate synthase from the yeast Kluyveromyces fragilis: Variability in occurrence and properties. FEMS Microbiol. Lett., **7**, 319-321 (1980)

[26] Hammer, B.A.; Johnson, E.A.: Purification, properties, and metabolic roles of NAD$^+$-glutamate dehydrogenase in Clostridium botulinum 113B. Arch. Microbiol., **150**, 460-464 (1988)

[27] Stevens, L.; Duncan, D.; Robertson, P.: Purification and characterisation of NAD-glutamate dehydrogenase from Aspergillus nidulans. FEMS Microbiol. Lett., **57**, 173-178 (1989)

[28] Scheid, H.W.; Ehmke, A.; Hartmann, T.: Plant NAD-dependent glutamate dehydrogenase. Purification, molecular properties and metal ion activation of the enzymes from Lemna minor and Pisum sativum. Z. Naturforsch. C, **35**, 213-221 (1980)

[29] Nagel, M.; Hartmann, T.: Glutamate dehydrogenase from Medicago sativa L., purification and comparative kinetic studies of the organ-specific multiple forms. Z. Naturforsch. C, **35**, 406-415 (1980)

[30] Hemmings, B.A.: Purification and properties of the phospho and dephospho forms of yeast NAD-dependent glutamate dehydrogenase. J. Biol. Chem., **255**, 7925-7932 (1980)

[31] Hornby, D.P.; Engel, P.C.: Characterization of Peptostreptococcus asaccharolyticus glutamate dehydrogenase purified by dye-ligand chromatography. J. Gen. Microbiol., **130**, 2385-2394 (1984)

[32] Yamaya, T.; Oaks, A.; Matsumoto, H.: Characteristics of glutamate dehydrogenase in mitochondria prepared from corn shoots. Plant Physiol., **76**, 1009-1013 (1984)

[33] El-Shora, H.M.; Abo-Kassem, E.M.: Kinetic characterization of glutamate dehydrogenase of marrow cotyledons. Plant Sci., **161**, 1047-1053 (2001)

[34] Schlee, D.; Thoeringer, C.; Tintemann, H.: Purification and properties of glutamate dehydrogenase in Scots pine (Pinus sylvestris) needles. Physiol. Plant., **92**, 467-472 (1994)

[35] Anderson, B.M.; Anderson, C.D.; Van Tassell, R.L.; Lyerly, D.M.; Wilkins, T.D.: Purification and characterization of Clostridium difficile glutamate dehydrogenase. Arch. Biochem. Biophys., **300**, 483-488 (1993)

[36] Pasquo, A.; Britton, K.L.; Stillman, T.J.; Rice, D.W.; Coelfen, H.; Harding, S.E.; Scandurra, R.; Engel, P.C.: Construction of a dimeric form of glutamate dehydrogenase from Clostridium symbiosum by site-directed mutagenesis. Biochim. Biophys. Acta, **1297**, 149-158 (1996)

[37] Abrahams, G.L.; Iles, K.D.; Abratt, V.R.: The Bacteroides fragilis NAD-specific glutamate dehydrogenase enzyme is cell surface-associated and regulated by peptides at the protein level. Anaerobe, **7**, 135-142 (2001)

[38] Consalvi, V.; Chiaraluce, R.; Millevoi, S.; Pasquo, A.; Politi, L.; De Rosa, M.; Scandurra, R.: NAD-dependent glutamate dehydrogenase from the thermophilic eubacterium Bacillus acidocaldarius. Comp. Biochem. Physiol. B, **109**, 691-699 (1994)

[39] Camardella, L.; Di Fraia, R.; Antignani, A.; Ciardiello, M.A.; di Prisco, G.; Coleman, J.K.; Buchon, L.; Guespin, J.; Russell, N.J.: The Antarctic Psychrobacter sp. TAD1 has two cold-active glutamate dehydrogenases with different cofactor specificities. Characterisation of the NAD^+-dependent enzyme. Comp. Biochem. Physiol. A, **131**, 559-567. (2002)

[40] Jahns, T.; Kaltwasser, H.: Purification and properties of a heat-stable and cold-labile NAD-specific glutamate dehydrogenase from Sporosarcina ureae. Arch. Microbiol., **161**, 531-534 (1994)

[41] Kersten, M.A.S.H.; Muller, Y.; Baars, J.J.P.; Op den Camp, H.J.M.; Van der Drift, C.; Van Griensven, L.J.L.D.; Visser, J.; Schaap, P.J.: NAD^+-dependent glutamate dehydrogenase of the edible mushroom Agaricus bisporus: biochemical and molecular characterization. Mol. Gen. Genet., **261**, 452-462 (1999)

[42] Jongsareejit, B.; Fujiwara, S.; Takagi, M.; Imanaka, T.: Comparison of two glutamate producing enzymes from the hyperthermophilic archaeon Pyrococcus sp. KOD1. FEMS Microbiol. Lett., **158**, 243-248 (1998)

[43] Garnier, A.; Berredjem, A.; Botton, B.: Purification and characterization of the NAD-dependent glutamate dehydrogenase in the ectomycorrhizal fungus Laccaria bicolor (Maire) Orton. Fungal Genetics and Biology, **22**, 168-176 (1997)

[44] Ruiz, J.L.; Ferrer, J.; Camacho, M.; Bonete, M.J.: NAD-specific glutamate dehydrogenase from Thermus thermophilus HB8: purification and enzymic properties. FEMS Microbiol. Lett., **159**, 15-20 (1998)

[45] Pamula, F.; Wheldrake, J.F.: Kinetic properties and the mechanism of activation of NAD-dependent glutamate dehydrogenase from Dictyostelium discoideum. Biochem. Mol. Biol. Int., **38**, 729-738. (1996)

[46] Minambres, B.; Olivera, E.R.; Jensen, R.A.; Luengo, J.M.: A new class of glutamate dehydrogenases (GDH). J. Biol. Chem., **275**, 39529-39542 (2000)

[47] Nguyen, K.T.; Nguyen, L.T.; Kopecky, J.; Behal, V.: Properties of NAD-dependent glutamate dehydrogenase from the tylosin producer Streptomyces fradiae. Can. J. Microbiol., **43**, 1005-1010 (1997)

[48] Jahns, T.: Unusually stable NAD-specific glutamate dehydrogenase from the alkaliphile Amphibacillus xylanus. Antonie Leeuwenhoek, **70**, 89-95 (1996)

[49] Kujo, C.; Ohshima, T.: Enzymological characteristics of the hyperthermostable NAD-dependent glutamate dehydrogenase from the archaeon Pyrobaculum islandicum and effects of denaturants and organic solvents. Appl. Environ. Microbiol., **64**, 2152-2157 (1998)

[50] Lebbink, J.H.; Knapp, S.; van der Oost, J.; Rice, D.; Ladenstein, R.; de Vos, W.M.: Engineering activity and stability of Thermotoga maritima glutamate dehydrogenase. II: construction of a 16-residue ion-pair network at the subunit interface. J. Mol. Biol., **289**, 357-369 (1999)

[51] Cho, S.W.; Lee, J.; Choi, S.Y.: Two soluble forms of glutamate dehydrogenase isoproteins from bovine brain. Eur. J. Biochem., **233**, 340-346 (1995)

Glutamate dehydrogenase [NAD(P)+]   1.4.1.3

1 Nomenclature

EC number
1.4.1.3

Systematic name
L-glutamate:NAD(P)+ oxidoreductase (deaminating)

Recommended name
glutamate dehydrogenase [NAD(P)+]

Synonyms
GDH
L-glutamate dehydrogenase
L-glutamic acid dehydrogenase
Legdh1
MP50
membrane protein 50
NAD(P)-glutamate dehydrogenase
NAD(P)H-dependent glutamate dehydrogenase
NAD(P)H-utilizing glutamate dehydrogenase
dehydrogenase, glutamate (nicotinamide adenine dinucleotide (phosphate))
glutamic acid dehydrogenase
glutamic dehydrogenase

CAS registry number
9029-12-3

2 Source Organism

<1> *Pyrococcus furiosus* [32, 39]
<2> *Chlamydomonas reinhardtii* [31]
<3> *Sulfolobus solfataricus* [30, 44]
<4> *Bacteroides ovatus* [28]
<5> *Bacteroides distasonis* [28]
<6> *Bacteroides vulgatus* [28]
<7> *Azospirillum brasilense* [24]
<8> *Chlorella pyrenoidosa* [1]
<9> *Mycoplasma laidlawii* [1]
<10> *eel* [7]
<11> *Pisum sativum* (pea [3]) [3]
<12> *Homo sapiens* [1, 46]

43

<13> *Bos taurus* (brain isoenzymes 1 and 2, probably encoded by different genes [42]) [1, 2, 4-6, 8, 10-12, 21-23, 25, 26, 35, 37, 42, 45]

<14> *Rattus norvegicus* [1, 9, 22, 35, 43]

<15> *Rhodopseudomonas sphaeroides* (strain DSM 158 [18]) [18]

<16> *Bacillus subtilis* (strain PCI 219 [19]) [19]

<17> *dogfish* [20]

<18> *Bacteroides fragilis* (strictly anaerobic bacterium [27]; high activity-form and low-activity form [29]) [27-29]

<19> *Sus scrofa* (membrane-bound enzyme exhibits microtubule-binding activity [35]) [1, 35]

<20> *Gallus gallus* [1]

<21> *Rana catesbeiana* [1]

<22> *Thunnus thymnus* (tuna [1,15]) [1, 15]

<23> *Squalus acanthias* [1]

<24> *Lemna minor* [13]

<25> *Dictyostelium discoideum* [14]

<26> *Glycine max* (glutamate dehydrogenase 1 and 2 [16,17]) [16, 17]

<27> *Xenopus laevis* [33]

<28> *Tenebrio molitor* [34]

<29> *Pyrococcus sp.* (ES4, hyperthermophilic archeon [36]) [36]

<30> *Bacillus cereus* (strain DSM 31 [38]) [38]

<31> *Sphagnum fallax* (<31> enhanced ammonium concentrations and a reduced carbon supply induce the enzyme activity [40]) [40]

<32> *Pyrococcus* (sp. KOD1, hyperthermophilic archeon isolated from kodakara island, japan [41]) [41]

<33> *Spermophilus richardsonii* (Richardson's ground squirrel [45]) [45]

3 Reaction and Specificity

Catalyzed reaction

L-glutamate + H_2O + NAD(P)$^+$ = 2-oxoglutarate + NH_3 + NAD(P)H + H$^+$ (<13> mechanism, [1, 21, 33]; <13> compulsory order mechanism at pH 8.8 [4, 5]; <13> alternative order mechanism at pH 9.5 [4]; <17> kinetic data suggest either a rapid-equilibrium random mechanism or a compulsory mechanism with binding sequence NH$_4^+$, NAD(P)H, 2-oxoglutarate [20])

Reaction type

oxidation
redox reaction
reduction
reductive amination

Natural substrates and products

S L-glutamate + H_2O + NAD(P)$^+$ <18> (<18> enzyme for the main route for ammonia assimilation at low concentrations of ammonia [28]) (Reversibility: r <18> [28]) [28]

P 2-oxoglutarate + NH_3 + NAD(P)H <18> [28]

Substrates and products

S 2-aminobutyrate + H_2O + NADP$^+$ <18> (<18> 3% of activity with L-glutamate [27]) (Reversibility: ? <18> [27]) [27]

P 2-oxobutyrate + NH$_3$ + NADPH <18> [27]

S L-glutamate + H_2O + 2-azido-NAD$^+$ <13> (Reversibility: ? <13> [25])
[25]

P 2-oxoglutarate + NH$_3$ + 2-azido-NADH <13> [25]

S L-glutamate + H_2O + N^6-(2-aminoethyl)-NAD(P)$^+$ <13> (Reversibility: ?
<13> [26]) [26]

P 2-oxoglutarate + NH$_3$ + N^6-(2-aminoethyl)-NAD(P)H <13> [26]

S L-glutamate + H_2O + N^6-(2-hydroxy-3-trimethylammoniumpropyl)-NAD$^+$
<13> (Reversibility: ? <13> [26]) [26]

P 2-oxoglutarate + NH$_3$ + N^6-(2-hydroxy-3-trimethylammoniumpropyl)-
NADH <13> [26]

S L-glutamate + H_2O + N^6-(3-sulfonatopropyl)-NAD$^+$ <13> (Reversibility: ?
<13> [26]) [26]

P 2-oxoglutarate + NH$_3$ + N^6-(2 sulfonatopropyl)-NADH <13> [26]

S L-glutamate + H_2O + NAD(P)$^+$ <1-33> (<13> rate of glutamate synthesis
is several-fold higher than the rate for the reverse reaction [5]; <3,7>
highly specific for 2-oxoglutarate and glutamate [24,30]; <13> very low
activity with: leucine, α-aminobutyrate, valine, isoleucine and methionine
[1]) (Reversibility: r <1-33> [1-46]) [1-46]

P 2-oxoglutarate + NH$_3$ + NAD(P)H <1-33> [1-46]

S L-glutamate + H_2O + polyethylenglycol-N^6-(2-aminoethyl)-NAD(P)$^+$ <13>
(Reversibility: ? <13> [26]) [26]

P 2-oxoglutarate + NH$_3$ + polyethylenglycol-N^6-(2-aminoethyl)-NAD(P)H
<13> [26]

S alanine + H_2O + NAD(P)$^+$ <13> (<-1,8,9,12-14,19-23> very low activity
[1]) (Reversibility: ? <13> [1]) [1, 4]

P pyruvate + NH$_3$ + NAD(P)H <13> [1, 4]

S homocysteinesulfinate + H_2O + NAD(P)$^+$ <13> (Reversibility: ? <13> [1])
[1]

P ?

S norvaline + H_2O + NAD(P)$^+$ <13> (Reversibility: ? <13> [1]) [1]

P 2-oxopentanoate + NH$_3$ + NAD(P)H

S valine + H20 + NADP$^+$ <18> (<18> 3% of activity with L-glutamate [27])
(Reversibility: ? <18> [27]) [27]

P 2-oxovalerate + NH$_3$ + NADPH <18> [27]

Inhibitors

2-azido-NAD$^+$ <13> (<13> 0.1 mM, 60% inhibition after 3 min of photolabeling [25]) [25]

2-oxoglutarate <13, 18> [1, 21, 27]

2-oxovalerate <13> [21]

4-iodoacetamidosalicylic acid <13> [10]

5-bromofuroate <13> [1]

5-chlorofuroate <13> [1]

5-nitrofuroate <13> [1]

8-azidoguanosine 5'-triphosphate <13> (<13> used for affinity photolabeling, 0.1 mM, 95% inhibition [37]) [37]

ADP <13> (<13> above pH 7.0: allosteric activation, pH 6.0-7.0: strong inhibition [12]) [5, 12]

AMP <15> (<15> inhibits only NADPH-linked activity [18]) [18]

ATP <12, 13, 15, 19, 24, 25, 33> (<24> 1 mM, 65% inhibition of NADH reaction due to chelating properties of ATP [13]; <25> strong inhibition [14]; <19> membrane-bound enzyme form, inhibition of microtubule-binding activity [35]; <13,33> strong inhibition at 37°C, no inhibition at 5°C [45]; <12> wild-type: inhibition at 0.1 mM and between 0.5-1.0 mM and above, activation at 1 mM, H454Y and S448P mutant enzyme: activation between 0.1-1 mM, inhibition above, R463A mutant enzyme: progressive inhibition between 0.01 and 10 mM [46]) [13, 14, 18, 35, 45, 46]

D-glutamate <3, 18> (<18> 10 mM, 57% inhibition of NADP$^+$-linked activity, 30% inhibition of NADPH-linked activity [27]; <3> 2 mM, 11% inhibition, 6 mM, 34% inhibition [30]) [27, 30]

EDTA <2, 24> (<24> complete loss of NADH and NAD$^+$ activities, NADPH activity unaffected [13]; <2> 5 mM, 80-90% inhibition of isozymes 1-3 [31]) [13, 31]

GTP <12, 13, 14, 19, 22, 24, 33> (<13> inhibition at pH 9.0, activation in presence of electrolytes at pH 6.0 [12]; <24> 1 mM, 42% inhibition due to chelating properties of GTP [13]; <19> membrane-bound liver enzyme, complete inhibition [35]; <13> 0.06 mM, 95% inhibition [37]; <33> strong inhibition at 37°C, weak inhibition at 5°C [45]; <12> little inhibition of H454Y mutant enzyme [46]) [5, 12, 13, 15, 25, 35, 37, 45, 46]

KCN <2> (<2> 50 mM, strong inhibition of isozymes 1-3 [31]) [31]

L-aspartate <15> (<15> inhibition of NADPH linked reaction, activation of NAD(H) linked reaction [18]) [18]

N-(N'-acetyl-4-sulfamoylphenyl)maleimide <13> [10]

NADH <27> (<27> high concentration [33]) [33]

NADPH <13> [21]

NaCl <13, 18> (<18> 100 mM, 50% inhibition of NADH and NAD$^+$ dependent reactions [27]) [27, 45]

Ni^{2+} <2> (<2> 1 mM, moderate inhibition of isozymes 1-3 [31]) [31]

Zn^{2+} <2> (<2> 1 mM, strong inhibition of isozymes 1-3 [31]) [31]

alanine <7> (<7> weak inhibition at pH 8.5, strong inhibition at pH 10.0 [24]) [24]

α-ketoglutarate oxime <13> [1]

α-monofluoroglutarate <13> [1]

α-tetrazole <13> [1]

aminooxyacetate <2> (<2> 5 mM, weak inhibition of isozymes 1-3 [31]) [31]

cardiolipin <13> [11]

citrate <16> (<16> 10 mM, 60% inhibition of oxidative deamination [19]) [19]

diethylstilbestrol <12> [46]

ditetrazole <13> [1]

fumarate <13, 15> [1, 18]
γ-tetrazole <13> [1]
glutamate <7, 13, 18> (<7> weak inhibition at pH 8.5, strong inhibition at pH 10.0 [24]) [1, 24, 27]
glutarate <13> [1]
glyoxal <13> [1]
histidine <7> (<7> weak inhibition at pH 8.5, strong inhibition at pH 10.0 [24]) [24]
imidodiacetic acid <13> [1]
isophthalate <13, 22> (<22> competitive vs. glutamate [15]) [1, 15]
lysine <7> (<7> weak inhibition at pH 8.5, strong inhibition at pH 10.0 [24]) [24]
m-bromobenzoate <13> [1]
m-chlorobenzoate <13> [1]
m-iodobenzoate <13> [1]
m-nitrobenzoate <13> [1]
malate <15> (<15> 5 mM, complete inhibition of NADH-linked activity [18]) [18]
methylacetimidate <2> (<2> 100 mM, moderate inhibition of isozymes 1-3 [31]) [31]
o-phenanthroline <2> (<2> 5 mM, strong inhibition of isozymes 1-3 [31]) [31, 10]
o-phthalaldehyde <3> (<3> 0.1 mM, 98% inhibition after 5 min at 60°C, competitive vs. 2-oxoglutarate and NADH [44]) [44]
oxaloacetate <16, 18> (<16> 5 mM, 79% inhibition of oxidative deamination [19]; <18> 5 mM, 20-25% inhibition of NADH- and NAD$^+$-dependent activities [27]) [19, 27]
oxalylglycine <13> (<13> competitive vs. 2-oxoglutarate, uncompetitive vs. NADPH, noncompetitive vs. NH_4^+ [21]) [21]
oxydiglycolic acid <13> [1]
p-hydroxymercuribenzoate <2> (<2> 5 mM, moderate inhibition of isozymes 1-3 [31]) [31]
palmitoyl-CoA <12> [46]
phenylglyoxal <3> (<3> 4 mM, 75% inhibition, uncompetitive vs. 2-oxoglutarate, noncompetitive vs. NADH [44]) [44]
phosphate <13> (<13> pH 8.0-9.0: activation, pH 6.0-7.6: almost complete inhibition with 400 mM [12]) [12]
phosphatidylserine <13> (<13> assumed to be a simple non-competitive inhibition [11]) [11]
phosphoenolpyruvate <15> [18]
pyridoxal <13> (<13> NADH and NADPH protect from inactivation [10]) [10]
pyridoxal 5'-phosphate <2, 3, 17, 22> (<22> 0.11 mM, approx. 30% inactivation after 10 min, 0.78 mM, 80% inactivation, inactivation is completely reversed by dialysis [15]; <17> NAD$^+$ and NADP$^+$ protect from inactivation in the presence of sodium glutarate [20]; <3> 2 mM, complete loss of activity [30]; <2> 5 mM, 100% inhibition [31]) [15, 20, 30, 31]

sodium acetate <33> (<33> at 5°C only [45]) [45]

sodium dodecylsulfate <13> (<13> time-dependent irreversible inhibition, 0.2 mM, 37% inhibition, 0.15 mM, 50% inhibition after 30 min, in the presence of 2-oxoglutarate after 370 min [11]) [11]

succinate <15> (<15> 5 mM, complete inhibition of NADH-linked activity [18]) [18]

thiodiglycolic acid <13> [1]

Additional information <3, 13, 30, 33> (<30> no inhibition or activation in the presence of 500 mM AMP, ADP, ATP, cyclic-AMP, GMP, GDP or GTP [38]; <3> no inhibition or activation in the presence of ADP, GTP and leucine [44]; <13,33> complex inhibition pattern for ATP, GTP, NaCl, KCl, sodium acetate, NaI, and potassium nitrate of forward and reverse reaction at 5°C and 37°C [45]) [38, 44, 45]

Cofactors/prosthetic groups

2-azido-NAD$^+$ <13> [25]

NAD$^+$ <8, 9, 12-15, 17, 19-26, 30, 31> (<25> 2 times higher reaction rate than with NADP$^+$ [14]; <26> glutamate dehydrogenase 2 uses only NAD$^+$ in the deamination direction [16]; <30> no activity with NADP$^+$ [38]; <31> no activity with NADP$^+$ [40]) [1, 2, 13, 14, 16, 17, 18, 20, 26, 38, 40]

NADH <3, 8, 9, 11-15, 19-24, 26, 28, 31> (<11> reaction rate NADH/NADPH: 6 [3]; <22> reaction rate NADH/NADPH: 15 [15]; <26> reaction rate NADH/NADPH: 1.3 (GDH1) [17]; <26> reaction rate NADH/NADPH: 4.5 (GDH2) [17]; <28> preferred cofactor [34]; <31> ratio of NADH/NADPH-dependent activity: 1/1.6 [40]) [1-3, 13, 15, 17, 18, 30, 34, 40]

NADP$^+$ <1, 8, 9, 12-14, 17, 19-23, 26, 29> (<15> only 5% of the activity with NAD$^+$ [18]; <26> with NADP$^+$ initial rates are much slower than with NAD$^+$ [16]; <1> enzyme activity with NADP$^+$ is higher than the activity with NAD$^+$ [32]; <29> specific for [36]) [1, 17, 20, 26, 32, 36]

NADPH <3, 8, 9, 11-15, 18-24, 26, 28, 31, 32> (<11> reaction rate NADH/NADPH: 6 [3]; <22> reaction rate NADH/NADPH: 15 [15]; <26> glutamate dehydrogenase 1, reaction rate NADH/NADPH: 1.3 [17]; <26> glutamate dehydrogenase 2, reaction rate NADH/NADPH: 4.4 [17]; <16> NADPH/NADH ratio of reaction rate: 10/1, NAD$^+$/NADH ratio of reaction rate: 5/1 [19]; <18> amination activity is 5fold higher with NADPH than that with NADH [27]; <32> 3fold higher activity than with NADH [41]) [1, 3, 13, 15, 17, 18, 19, 27, 30, 34, 40, 41]

Additional information <26> (<26> 2 soybean isoenzymes with different coenzyme specificities, glutamate dehydrogenase 1: active with NADH, NAD$^+$, NADPH and NADP$^+$, glutamate dehydrogenase 2: active with NADH, NAD$^+$, NADPH but not NADP$^+$ [16]) [16]

Activating compounds

2-mercaptoethanol <26> (<26> 2fold activation [17]) [17]

8-azidoguanosine 5'-diphosphate <13> (<13> used for affinity photolabeling [37]) [37]

ADP <12, 13, 14, 19, 22, 25, 27, 28, 33> (<13> above pH 7.0: allosteric activation [12]; <13> pH 6.0-7.0: strong inhibition [12]; <22,25,27,28> stimulation

[14,15,33,34]; <13> 1.5fold activation, 0.08 mM NADH as cofactor [23]; <13> activation only if concentrations of both NAD(P)$^+$ and substrate are high [6]; <19> 1 mM, approx. 2fold activation of membrane-bound liver enzyme [35]; <12> no activation of R463A mutant enzyme [46]) [6, 12, 14, 15, 23, 25, 33, 34, 35, 45, 46]

AMP <13, 15, 25, 33> (<25> 1 mM, reaction velocity increases 15fold at saturating NAD$^+$ and glutamate levels [14]; <15> activates NADH-linked glutamate synthesis, inhibits NADPH-linked activity [18]; <15> activation only in biosynthetic direction [18]) [14, 18, 42, 45]

ATP <12, 13, 14, 19> (<13> 4.8fold activation, cofactor 0.08 mM NADH [23]; <19> approx. 1.5fold activation of membrane-bound liver enzyme [35]; <12> wild-type: activation at 1 mM, inhibition below and above, H454Y and S448P mutant enzymes: progressive increase in activity until 10 mM, inhibition above, [46]) [23, 35, 46]

GTP <13> (<13> inhibition at pH 9.0, activation in presence of electrolytes at pH 6.0 [12]) [12]

KCl <33> (<33> activates at 5°C and 37°C [45]) [45]

L-aspartate <15> (<15> 2fold activation of glutamate synthesis [18]) [18]

N^6-(2-aminoethyl)-NAD$^+$ <13> [26]

N^6-(2-aminoethyl)-NADP$^+$ <13> [26]

N^6-(2-hydroxy-3-trimethylammonium propyl)-NAD$^+$ <13> [26]

N^6-(3-sulfonatopropyl)-NAD$^+$ <13> [26]

NaCl <33> (<33> activates at 37°C only [45]) [45]

glutathione <26> (<26> 2fold activation [17]) [17]

leucine <12, 33> (<33> activates at 5°C only [45]; <12> wild-type, S448P, H454Y and R463A mutant enzymes [46]) [45, 46]

phosphate <13> (<13> activator at pH 8.0-9.0, inhibitor at pH 6.0-7.6 [12]) [12]

poly(ethylene glycol)-N^6-(2-aminoethyl)-NAD$^+$ <13> [26]

poly(ethylene glycol)-N^6-(2-aminoethyl)-NADP$^+$ <13> [26]

trypsin <10> (<10> limited trypsin proteolyis activates the purified enzyme 8fold if the peptide is absent from the assay mixture, the native enzyme is 3fold activated if the cleaved peptide is present, activation may therefore be induced by loss of the peptide from the subunit of the native enzyme [7]) [7]

Additional information <30> (<30> no inhibition or activation in the presence of 500 mM AMP, ADP, ATP, cyclic-AMP, GMP, GDP or GTP [38]) [38]

Metals, ions

Ca^{2+} <2> (<2> 1 mM, 1.5fold activation of isozymes 1 and 3, 2.5fold activation of isozyme 2 [31]) [31]

Turnover number (min^{-1})

200 <7> (L-glutamate, <7> cofactor NAD$^+$, value below 200.0 [24]) [24]

290 <13> (L-glutamate, <13> cofactor NAD$^+$ [2]) [2]

1500 <22> (glutamate, <22> oxidation [15]) [15]

2400 <32> (glutamate, <32> concentration range: 0.05-1.0 mM [41]) [41]

3000 <13> (2-oxoglutarate) [2]

3660 <7> (glutamate, <7> cofactor NADP$^+$ [24]) [24]

3800 <32> (NH$_4^+$, <32> concentration range: 0.2-5 mM [41]) [41]
5400 <32> (NADH, <32> concentration range: 0.04-0.1 mM [41]) [41]
6100 <32> (NADP$^+$, <32> concentration range: 0.02-0.2 mM [41]) [41]
6700 <32> (NADPH, <32> concentration range: 0.002-0.1 mM [41]) [41]
9700 <32> (glutamate, <32> concentration range: 1.0-10.0 mM [41]) [41]
10100 <32> (NH$_4^+$, <32> concentration range: 10-200 mM [41]) [41]
11000 <7> (2-oxoglutarate, <7> cofactor NADH [24]) [24]
11700 <32> (2-oxoglutarate, <32> concentration range: 0.3-4.0 mM [41]) [41]
12000 <7> (NADPH, <7> NADPH-dependent activity [24]) [24]
30600 <22> (2-oxoglutarate, <22> reduction [15]) [15]
42840 <28> (glutamate, <28> cofactor NADH, + 1 mM ADP [34]) [34]

Specific activity (U/mg)
0.00076 <13> (<13> activity in liver nuclei, cofactor NADH [5]) [5]
0.0028 <13> (<13> activity in liver mitochondria, cofactor NADH [5]) [5]
0.168 <4> (<4> NADH dependent activity in cells grown on medium containing 1 mM NH$_4$Cl [28]) [28]
0.257 <5> (<5> NADH dependent activity in cells grown on medium containing 1 mM NH$_4$Cl [28]) [28]
0.336 <4> (<4> NADH dependent activity in cells grown on medium containing 1 mM NH$_4$Cl [28]) [28]
0.36 <24> (<24> cofactor NADPH [13]) [13]
0.858 <4> (<4> NADPH dependent activity in cells grown on medium containing 1 mM NH$_4$Cl [28]) [28]
1.07 <18> (<18> NADH dependent activity in cells grown on medium containing 1 mM NH$_4$Cl [28]) [28]
1.2 <26> (<26> glutamate dehydrogenase 2 [17]) [17]
1.8 <26> (<26> glutamate dehydrogenase 1 [17]) [17]
1.93 <22> (<22> after 3rd crystallization [15]) [15]
1.96 <5> (<5> NADPH dependent activity in cells grown on medium containing 1 mM NH$_4$Cl [28]) [28]
2.22 <4> (<4> NADPH dependent activity in cells grown on medium containing 1 mM NH$_4$Cl [28]) [28]
3.5 <24> (<24> cofactor NADH [13]) [13]
4 <15> (<15> cofactor NADPH [18]) [18]
4.6 <19> (<19> membrane-bound liver enzyme [35]) [35]
4.8 <13> (<13> enzyme purified from liver nuclei [5]) [5]
5.18 <18> (<18> NADPH dependent activity in cells grown on medium containing 1 mM NH$_4$Cl [28]) [28]
5.9 <18> (<18> low-activity form of the enzyme, NADP$^+$ dependent deamination [29]) [29]
7.53 <10> [7]
13 <15> (<15> cofactor NADH [18]) [18]
17 <32> [41]
17.8 <14> (<14> liver enzyme [35]) [35]
18.4 <18> [27]
19 <12> (<12> recombinant S448P mutant enzyme, basal activity [46]) [46]

29.7 <14> [43]

32.3 <18> (<18> high-activity form of the enzyme, NADP$^+$-dependent deamination [29]) [29]

32.5 <1> (<1> at 85°C [32]) [32]

34 <13> (<13> liver enzyme [35]) [35]

37 <14> [22]

40 <13> [22]

40.86 <31> [40]

41.9 <7> [24]

47 <16> [19]

55 <12> (<12> recombinant S448P mutant enzyme, maximal activity in the presence of 6 mM leucine [46]) [46]

58 <2> (<2> isozyme 3 [31]) [31]

67 <2> (<2> isozyme 2 [31]) [31]

69 <12> (<12> recombinant S448P mutant enzyme, maximal activity in the presence of 0.2 mM ADP [46]) [46]

73 <12> (<12> recombinant R463A mutant enzyme, basal activity [46]) [46]

74 <12> (<12> recombinant H454Y mutant enzyme, basal activity [46]) [46]

83 <3> [44]

84 <12> (<12> recombinant wild-type enzyme, basal activity [46]) [46]

100 <12> (<12> recombinant R463A mutant enzyme, maximal activity in the presence of 6 mM leucine [46]) [46]

126.4 <28> (<28> cofactor NADH [34]) [34]

130 <12> (<12> recombinant S448P mutant enzyme, maximal activity in the presence of 6 mM leucine [46]) [46]

130 <12> (<12> recombinant wild-type enzyme, maximal activity in the presence of 0.2 mM ADP or 6 mM leucine [46]) [46]

140 <12> (<12> recombinant H454Y mutant enzyme, maximal activity in the presence of 0.2 mM ADP [46]) [46]

185 <14> [9]

190 <33> (<33> enzyme from euthermic animals [45]) [45]

201 <33> (<33> enzyme from hibernating animals [45]) [45]

230 <2> (<2> isozyme 1 [31]) [31]

286.4 <30> [38]

411 <1> (<1> native enzyme at 100°C [39]) [39]

419 <1> (<1> recombinant enzyme at 75°C [39]) [39]

423 <1> (<1> recombinant enzyme at 90°C [39]) [39]

423 <3> (<3> at 70°C [30]) [30]

Additional information <11> (<11> 0.263 Δ absorbance/min/mg, cofactor NADP$^+$ [3]) [3]

Additional information <11> (<11> 1.44 Δ absorbance/min/mg, cofactor NAD$^+$ [3]) [3]

K$_m$-Value (mM)

0.000175 <11> (NADH) [3]

0.00294 <11> (α-ketoglutarate) [3]

0.004 <15> (NADH, <15> 50 mM Tris-HCl, pH 7.2 [18]) [18]

0.004 <7> (NADP$^+$, <7> NADP$^+$-dependent activity [24]) [24]

0.006 <6> (NADPH, <6> NADPH-dependent amination [28]) [28]

0.007 <3> (NADH, <3> + 2-oxoglutarate [30]) [30]

0.0083 <31> (NADH) [40]

0.009 <26> (NADH, <26> glutamate dehydrogenase 1, 5 mM 2-oxoglutarate [16]) [16]

0.009 <26> (NADH, <26> glutamate dehydrogenase 2, 5 mM 2-oxoglutarate [16]) [16]

0.01 <15> (NADH, <15> 500 mM Tris-HCl, pH 8.2 [18]) [18]

0.01 <3> (NADPH, <3> + 2-oxoglutarate [30]) [30]

0.011 <31> (NADPH) [40]

0.011 <18> (NADPH, <18> reductive amination, low-activity form of the enzyme [29]) [29]

0.012 <15> (NADH, <15> 500 mM Tris-HCl, pH 7.2 [18]) [18]

0.012 <26> (NADH, <26> glutamate dehydrogenase 2, 80 mM NH$_4^+$ [16]) [16]

0.013 <18> (NADP$^+$, <18> oxidative deamination, low-activity form of the enzyme [29]) [29]

0.013 <18> (NADPH, <18> NADP-linked reductive amination [27]) [27]

0.013 <18> (NADPH, <18> reductive amination, high-activity form of the enzyme [29]) [29]

0.013 <7> (NADPH, <7> NADPH-dependent activity [24]) [24]

0.014 <31> (NAD$^+$) [40]

0.015 <27> (NADH, <27> liver enzyme [33]) [33]

0.016 <15> (NAD$^+$, <15> 50 mM Tris-HCl, pH 7.2 [18]) [18]

0.018 <5> (NADPH, <5> NADPH-dependent amination [28]) [28]

0.019 <18> (NADP$^+$, <18> NADP-linked oxidative deamination [27]) [27]

0.02 <13> (NADH) [23]

0.02 <26> (NADH, <26> glutamate dehydrogenase 1, 40 mM NH$_4^+$ [16]) [16]

0.02 <13> (NADPH) [23]

0.02 <4> (NADPH, <4> NADPH-dependent amination [28]) [28]

0.022 <27> (NADH, <27> kidney enzyme [33]) [33]

0.022 <13> (NADPH) [21]

0.025 <3> (NADP$^+$, <3> + L-glutamate [30]) [30]

0.028 <13> (NADP$^+$, <13> cosubstrate glutamate [21]) [21]

0.028 <26> (NADPH, <26> glutamate dehydrogenase 1, 10 mM 2-oxoglutarate [16]) [16]

0.028 <26> (NADPH, <26> glutamate dehydrogenase 2, 10 mM 2-oxoglutarate [16]) [16]

0.029 <18> (NADP$^+$, <18> oxidative deamination, high-activity form of the enzyme [29]) [29]

0.032 <27> (NAD$^+$, <27> kidney enzyme [33]) [33]

0.033 <30> (NH$_4^+$, <30> for 200 mM NH$_4^+$ [38]) [38]

0.04 <32> (NADPH, <32> concentration range: 0.002-0.1 mM [41]) [41]

0.0424 <11> (NH$_4^+$) [3]

0.048 <26> (NADH, <26> glutamate dehydrogenase 1, 400 mM NH$_4^+$ [16]) [16]

0.05 <26> (NADH, <26> glutamate dehydrogenase 1, 12.5 mM 2-oxoglutarate [16]) [16]

0.052 <13> (N^6-(3-sulfonatopropyl)-NAD$^+$) [26]

0.063 <3> (NAD$^+$, <3> + L-glutamate [30]) [30]

0.069 <26> (NADPH, <26> glutamate dehydrogenase 1, 40 mM NH$_4^+$ [16]) [16]

0.069 <26> (NADPH, <26> glutamate dehydrogenase 2, 80 mM NH$_4^+$ [16]) [16]

0.07 <2> (NADH, <2> isozyme 1, amination activity, cofactor NADH [31]) [31]

0.08 <32> (NADP$^+$, <32> concentration range: 0.02-0.2 mM [41]) [41]

0.09 <2> (NADH, <2> isozyme 3, amination activity, cofactor NADH [31]) [31]

0.09 <32> (NADH, <32> concentration range: 0.04-0.1 mM [41]) [41]

0.1 <33> (2-oxoglutarate, <33> euthermic animal, assay at 37°C [45]) [45]

0.11 <15> (NAD$^+$, <15> 500 mM Tris-HCl, pH 9.4 [18]) [18]

0.11 <24> (NADH) [13]

0.13 <18> (NADH, <18> reductive amination, low-activity form of the enzyme [29]) [29]

0.13 <24> (NADPH) [13]

0.14 <18> (2-oxoglutarate, <18> NADP-linked reductive amination [27]) [27]

0.14 <2> (NADH, <2> isozyme 2, amination activity, cofactor NADH [31]) [31]

0.148 <13> (N^6-(2-hydroxy-2-trimethylammoniumpropyl)-NAD$^+$) [26]

0.15 <15> (NAD$^+$, <15> 500 mM Tris-HCl, pH 7.2 [18]) [18]

0.16 <13> (NADP$^+$) [26]

0.17 <15> (NADPH, <15> 50 mM Tris-HCl, pH 7.2 [18]) [18]

0.175 <13> (NAD$^+$) [26]

0.18 <18> (2-oxoglutarate, <18> reductive amination, low-activity form of the enzyme [29]) [29]

0.18 <1> (NADP$^+$, <1> substrate L-glutamate [32]) [32]

0.2 <18> (2-oxoglutarate, <18> reductive amination, high-activity form of the enzyme [29]) [29]

0.2 <3> (2-oxoglutarate, <3> + NADPH [30]) [30]

0.2 <2> (NAD$^+$, <2> isozyme 1, deamination activity [31]) [31]

0.2 <18> (NADH, <18> NAD-linked reductive amination [27]) [27]

0.25 <32> (2-oxoglutarate, <32> concentration range: 0.3-4.0 mM [41]) [41]

0.25 <7> (2-oxoglutarate, <7> NADPH-dependent activity [24]) [24]

0.25 <33> (glutamate, <33> hibernating animal, assay at 5°C [45]) [45]

0.26 <2> (NAD$^+$, <2> isozyme 2, deamination activity [31]) [31]

0.27 <2> (NADPH, <2> isozyme 1, amination activity, cofactor NADPH [31]) [31]

0.273 <13> (N^6-(2-aminoethyl)-NADP$^+$) [26]

0.291 <13> (N^6-(2-aminoethyl)-NAD$^+$) [26]

0.3 <3> (L-glutamate, <3> + NAD$^+$ [30]) [30]

0.31 <13> (NADP^1, <13> cosubstrate norvaline [21]) [21]

0.32 <2> (NAD$^+$, <2> isozyme 3, deamination activity [31]) [31]

0.32 <32> (glutamate, <32> concentration range: 0.05-1.0 mM [41]) [41]

0.36 <13> (2-oxoglutarate, <13> NADH + NH$_4$Cl [23]) [23]

0.36 <2> (2-oxoglutarate, <2> isozyme 1, amination activity, cofactor NADH [31]) [31]

0.364 <14> (NAD$^+$, <14> mitochondrial enzyme [43]) [43]

0.38 <7> (NH$_4$Cl, <7> biphasic Lineweaver-Burk plot suggests 2 K$_m$ values [24]) [24]

0.4 <15> (2-oxoglutarate, <15> 50 mM Tris-HCl, pH 7.2 [18]) [18]

0.4 <27> (L-glutamate, <27> kidney enzyme [33]) [33]

0.425 <13> (poly(ethyleneglycol)-N^6-(2-aminoethyl)-NADP$^+$) [26]

0.43 <33> (2-oxoglutarate, <33> hibernating animal, assay at 5°C [45]) [45]

0.43 <18> (NAD$^+$, <18> oxidative deamination, low-activity form of the enzyme [29]) [29]

0.443 <14> (NADP$^+$, <14> enzyme from rough ER [43]) [43]

0.444 <13> (poly(ethyleneglycol)-N^6-(2-aminoethyl)-NAD$^+$) [26]

0.46 <24> (NAD$^+$) [13]

0.46 <2> (NADPH, <2> isozyme 3, amination activity, cofactor NADPH [31]) [31]

0.47 <13> (2-oxoglutarate, <13> NADPH + NH$_4$Cl [23]) [23]

0.47 <33> (glutamate, <33> assay at 5°C [45]) [45]

0.5 <7> (NADH, <7> NADH-dependent activity [24]) [24]

0.5 <33> (glutamate, <33> euthermic animal, assay at 5°C [45]) [45]

0.52 <18> (NADH, <18> reductive amination, high-activity form of the enzyme [29]) [29]

0.53 <1> (NAD$^+$, <1> substrate L-glutamate [32]) [32]

0.54 <15> (NADPH, <15> 500 mM Tris-HCl, pH 7.2 [18]) [18]

0.56 <30> (NAD$^+$) [38]

0.56 <1> (NADPH, <1> substrate 2-oxoglutarate [32]) [32]

0.56 <1> (NADPH, <1> substrate NH$_3$ [32]) [32]

0.59 <2> (2-oxoglutarate, <2> isozyme 3, amination activity, cofactor NADH [31]) [31]

0.6 <3> (2-oxoglutarate, <3> + NADH [30]) [30]

0.62 <26> (2-oxoglutarate, <26> glutamate dehydrogenase 2, 0.1 mM NADH [16]) [16]

0.63 <6> (2-oxoglutarate, <6> NADPH-dependent amination [28]) [28]

0.637 <14> (NADP$^+$, <14> mitochondrial enzyme [43]) [43]

0.64 <4> (2-oxoglutarate, <4> NADPH-dependent amination [28]) [28]

0.64 <2> (L-glutamate, <2> isozyme 1, deamination activity [31]) [31]

0.7 <31> (2-oxoglutarate, <31> cofactor NADH [40]) [40]

0.7 <5> (2-oxoglutarate, <5> NADPH-dependent amination [28]) [28]

0.74 <13> (L-glutamate) [21]

0.75 <15> (NADPH, <15> 500 mM Tris-HCl, pH 8.2 [18]) [18]

0.78 <2> (NADPH, <2> isozyme 2, amination activity, cofactor NADPH [31]) [31]

0.924 <14> (NAD$^+$, <14> enzyme from rough ER [43]) [43]

0.98 <33> (2-oxoglutarate, <33> hibernating animal, assay at 37°C [45]) [45]

0.98 <1> (NADH, <1> substrate 2-oxoglutarate [32]) [32]

1 <27> (2-oxoglutarate, <27> liver enzyme [33]) [33]

1 <13> (L-glutamate, <13> 1 mM, NAD$^+$ [23]) [23]

1.1 <26> (2-oxoglutarate, <26> glutamate dehydrogenase 2, 80 mM NH$_4^+$ [16]) [16]

1.1 <3> (L-glutamate, <3> + NADP$^+$ [30]) [30]

1.2 <15> (2-oxoglutarate, <15> 500 mM Tris-HCl, pH 7.2 [18]) [18]

1.23 <2> (2-oxoglutarate, <2> isozyme 2, amination activity, cofactor NADH [31]) [31]

1.25 <18> (NH$_4$Cl, <18> reductive amination, high-activity form of the enzyme [29]) [29]

1.3 <32> (NH$_4^+$, <32> concentration range: 0.2-5 mM [41]) [41]

1.31 <18> (NH$_4$Cl, <18> reductive amination, low-activity form of the enzyme [29]) [29]

1.4 <13> (2-oxoglutarate, <33> assay at 37°C [45]) [45]

1.5 <6> (NH$_4$Cl, <6> NADPH-dependent amination [28]) [28]

1.51 <26> (2-oxoglutarate, <26> glutamate dehydrogenase 1, 0.1 mM NADH [16]) [16]

1.51 <26> (2-oxoglutarate, <26> glutamate dehydrogenase 1, 400 mM NH$_4^+$ [16]) [16]

1.54 <2> (2-oxoglutarate, <2> isozyme 1, amination activity, cofactor NADPH [31]) [31]

1.6 <1> (L-glutamate, <1> cofactor NADP$^+$ [32]) [32]

1.66 <1> (L-glutamate, <1> cofactor NAD$^+$ [32]) [32]

1.7 <15> (L-glutamate, <15> 50 mM Tris-HCl, pH 7.2 [18]) [18]

1.7 <18> (NH$_4$Cl, <18> NADP-linked reductive amination [27]) [27]

1.73 <2> (2-oxoglutarate, <2> isozyme 3, amination activity, cofactor NADH [31]) [31]

1.85 <2> (2-oxoglutarate, <2> isozyme 2, amination activity, cofactor NADPH [31]) [31]

2.03 <33> (glutamate, <33> euthermic animal, assay at 37°C [45]) [45]

2.1 <24> (2-oxoglutarate, <24> + NADPH [13]) [13]

2.1 <26> (2-oxoglutarate, <26> glutamate dehydrogenase 1, 40 mM NH$_4^+$ [16]) [16]

2.1 <31> (glutamate) [40]

2.3 <27> (2-oxoglutarate, <27> kidney enzyme [33]) [33]

2.4 <13> (2-oxoglutarate) [21]

2.4 <18> (L-glutamate, <18> NADP-linked oxidative deamination [27]) [27]

2.4 <5> (NH$_4$Cl, <5> NADPH-dependent amination [28]) [28]

2.44 <18> (NAD$^+$, <18> oxidative deamination, high-activity form of the enzyme [29]) [29]

2.6 <26> (2-oxoglutarate, <26> glutamate dehydrogenase 1, 0.2 mM NADPH [16]) [16]

2.6 <33> (glutamate, <33> assay at 37°C [45]) [45]

2.9 <4> (NH$_4$Cl, <4> NADPH-dependent amination [28]) [28]

3 <13> (L-glutamate, <13> 1 mM, NADP$^+$ [23]) [23]

3 <2> (L-glutamate, <2> isozyme 3, deamination activity [31]) [31]

3 <18> (NAD$^+$, <18> NAD-linked oxidative deamination [27]) [27]

3.2 <13> (NH$_4^+$) [11]

3.3 <24> (2-oxoglutarate, <24> + NADH [13]) [13]

3.52 <2> (L-glutamate, <2> isozyme 2, deamination activity [31]) [31]

3.65 <18> (L-glutamate, <18> oxidative deamination, low-activity form of the enzyme [29]) [29]

3.66 <33> (2-oxoglutarate, <33> euthermic animal, assay at 5°C [45]) [45]

3.7 <26> (2-oxoglutarate, <26> glutamate dehydrogenase 1, 0.2 mM NADH [16]) [16]

3.7 <26> (2-oxoglutarate, <26> glutamate dehydrogenase 2, 0.2 mM NADPH [16]) [16]

3.83 <18> (L-glutamate, <18> oxidative deamination, high-activity form of the enzyme [29]) [29]

4 <1> (2-oxoglutarate, <1> cofactor NADPH [32]) [32]

4.1 <13> (NH$_4^+$, <33> assay at 5°C [45]) [45]

4.5 <1> (2-oxoglutarate, <1> cofactor NADH [32]) [32]

4.61 <14> (glutamate, <14> mitochondrial enzyme, cofactor NAD$^+$ [43]) [43]

4.9 <18> (NH$_4$Cl, <18> NAD-linked reductive amination [27]) [27]

5 <7> (2-oxoglutarate, <7> NADH-dependent activity [24]) [24]

5.1 <18> (NH$_4$Cl, <18> NADP-linked reductive amination [27]) [27]

5.2 <33> (glutamate, <33> hibernating animal, assay at 37°C [45]) [45]

5.5 <32> (glutamate, <32> concentration range: 1.0-10.0 mM [41]) [41]

5.8 <26> (NH$_4^+$, <26> glutamate dehydrogenase 1, 0.2 mM NADPH [16]) [16]

5.93 <14> (glutamate, <14> enzyme from rough ER, cofactor NAD$^+$ [43]) [43]

6.3 <31> (2-oxoglutarate, <31> cofactor NADPH [40]) [40]

6.5 <13> (NH$_4^+$) [21]

7 <26> (NH$_4^+$, <26> glutamate dehydrogenase 1, 0.2 mM NADPH [16]) [16]

7 <33> (NH$_4^+$, <33> euthermic animal, assay at 37°C [45]) [45]

7.1 <18> (2-oxoglutarate, <18> NAD-linked reductive amination [27]) [27]

7.3 <18> (L-glutamate, <18> NAD-linked oxidative deamination [27]) [27]

7.4 <30> (glutamate) [38]

9.5 <1> (NH$_3$, <1> cofactor NADPH [32]) [32]

10 <7> (L-glutamate, <7> NADP$^+$-dependent activity [24]) [24]

10.4 <26> (NH$_4^+$, <26> glutamate dehydrogenase 1, 10 mM 2-oxoglutarate [16]) [16]

10.5 <13> (2-oxoglutarate, <33> assay at 5°C [45]) [45]

12 <24> (glutamate, <24> + NAD$^+$ [13]) [13]

12.1 <33> (NH$_4^+$, <33> hibernating animal, assay at 5°C [45]) [45]

12.9 <26> (NH$_4^+$, <26> glutamate dehydrogenase 2, 5 mM 2-oxoglutarate [16]) [16]

13 <2> (NH$_4^+$, <2> isozyme 2, amination activity, cofactor NADH [31]) [31]

13 <32> (NH$_4^+$, <32> concentration range: 10-200 mM [41]) [41]

14 <15> (2-oxoglutarate, <15> 500 mM Tris-HCl, pH 8.2 [18]) [18]

15 <24> (NH$_4^+$, <24> + NADPH + 2-oxoglutarate [13]) [13]

15.8 <26> (NH$_4^+$, <26> glutamate dehydrogenase 1, 5 mM 2-oxoglutarate [16]) [16]

15.8 <33> (NH$_4^+$, <33> hibernating animal, assay at 37°C [45]) [45]

17.8 <13> (NH$_4^+$, <33> assay at 37°C [45]) [45]

19 <26> (NH$_4^+$, <26> glutamate dehydrogenase 1, 10 mM 2-oxoglutarate [16]) [16]

19.5 <15> (L-glutamate, <15> 500 mM Tris-HCl, pH 9.4 [18]) [18]

20 <27> (NH$_4$Cl, <27> kidney enzyme [33]) [33]

20.7 <14> (glutamate, <14> mitochondrial enzyme, cofactor NADP$^+$ [43]) [43]

21 <27> (NH$_4$Cl, <27> liver enzyme [33]) [33]

21.7 <26> (NH$_4^+$, <26> glutamate dehydrogenase 1, 0.1 mM NADH [16]) [16]

22.8 <26> (NH$_4^+$, <26> glutamate dehydrogenase 1, 0.1 mM NADH [16]) [16]

23.8 <14> (glutamate, <14> enzyme from rough ER, cofactor NADP$^+$ [43]) [43]

24.7 <33> (NH$_4^+$, <33> euthermic animal, assay at 5°C [45]) [45]

25 <13> (L-glutamate, <13> 0.004 mM, NAD$^+$ [23]) [23]

27 <24> (NH$_4^+$, <24> + NADH + 2-oxoglutarate [13]) [13]

28 <31> (NH$_4$Cl, <31> cofactor NADH [40]) [40]

30 <2> (NH$_4^+$, <2> isozyme 1, amination activity, cofactor NADH [31]) [31]

31 <15> (L-glutamate, <15> 500 mM Tris-HCl, pH 7.2 [18]) [18]

36 <2> (NH$_4^+$, <2> isozyme 1, amination activity, cofactor NADPH [31]) [31]

41 <2> (NH$_4^+$, <2> isozyme 3, amination activity, cofactor NADPH [31]) [31]

41 <31> (NH$_4$Cl, <31> cofactor NADPH [40]) [40]

44 <2> (NH$_4^+$, <2> isozyme 3, amination activity, cofactor NADH [31]) [31]

44 <13> (NH$_4$Cl, <13> NADPH + 2-oxoglutarate [23]) [23]

49 <13> (norvaline) [21]

50 <13> (NH$_4$Cl, <13> NADH + 2-oxoglutarate [23]) [23]

53 <2> (NH$_4^+$, <2> isozyme 2, amination activity, cofactor NADPH [31]) [31]

58 <15> (NH$_4^+$, <15> 500 mM Tris-HCl, pH 8.2 [18]) [18]

66 <7> (NH$_4$Cl, <7> NADH-dependent activity [24]) [24]

77 <15> (NH$_4^+$, <15> 500 mM Tris-HCl, pH 7.2 [18]) [18]

96 <30> (NH$_4^+$, <30> from double-reciprocal plots for concentrations between 5 mM and 200 mM [38]) [38]

100 <7> (NH$_4$Cl, <7> biphasic Lineweaver-Burk plot suggests 2 K$_m$ values [24]) [24]

106 <26> (NH$_4^+$, <26> glutamate dehydrogenase 1, 12.5 mM 2-oxoglutarate [16]) [16]

115.1 <26> (NH$_4^+$, <26> glutamate dehydrogenase 1, 0.2 mM NADH [16]) [16]

160 <15> (NH$_4^+$, <15> 50 mM Tris-HCl, pH 7.2 [18]) [18]

K$_i$-Value (mM)

0.000013 <12> (palmitoyl-coA, <12> inhibition of recombinant S448P mutant enzyme [46]) [46]

0.000015 <12> (palmitoyl-CoA, <12> inhibition of recombinant H454Y mutant enzyme [46]) [46]

0.00003 <12> (palmitoyl-CoA, <12> inhibition of recombinant wild-type enzyme [46]) [46]

0.000042 <12> (GTP, <12> inhibition of recombinant wild-type enzyme [46]) [46]

0.0002 <12> (palmitoyl-CoA, <12> inhibition of recombinant R463A mutant enzyme [46]) [46]

0.00022 <12> (GTP, <12> inhibition of recombinant R463A mutant enzyme [46]) [46]

0.00036 <12> (diethylstilbestrol, <12> inhibition of recombinant H454Y mutant enzyme [46]) [46]

0.0006 <13> (GTP) [37]

0.00074 <12> (diethylstilbestrol, <12> inhibition of recombinant S448P mutant enzyme [46]) [46]

0.0009-0.061 <13> (phosphatidylserine) [11]

0.0017 <12> (diethylstilbestrol, <12> inhibition of recombinant wild-type enzyme [46]) [46]

0.0031 <12> (GTP, <12> inhibition of recombinant S448P mutant enzyme [46]) [46]

0.005 <13> (8-azidoguanosine 5'-triphosphate) [37]

0.0051 <12> (diethylstilbestrol, <12> inhibition of recombinant R463A mutant enzyme [46]) [46]

0.0065 <13> (GTP, <13> 1 mM NAD$^+$ or 80 mM NADH [23]) [23]

0.009 <13> (GTP, <13> 0.08 mM NADH [23]) [23]

0.011 <27> (NADH, <27> liver enzyme [33]) [33]

0.012 <13> (GTP, <13> 0.016 mM NADH [23]) [23]

0.016 <27> (NADH, <27> kidney enzyme [33]) [33]

0.018 <13> (GTP, <13> 1 mM NADP$^+$ [23]) [23]

0.028 <13> (NADPH) [21]

0.03 <3> (o-phthalaldehyde, <3> vs. 2-oxoglutarate [44]) [44]

0.1 <3> (o-phthalaldehyde, <3> vs. NADH [44]) [44]

0.21 <12> (GTP, <12> inhibition of recombinant H454Y mutant enzyme [46]) [46]

0.24 <13> (NADP$^+$) [21]

0.28 <22> (isophthalate) [15]

0.3 <27> (NAD$^+$, <27> kidney enzyme [33]) [33]

0.315 <13> (2-oxoglutarate) [21]

0.36 <13> (oxalylglycine, <13> vs. 2-oxoglutarate [21]) [21]

0.54 <13> (oxalylglycine, <13> vs. NADPH [21]) [21]

0.76 <19> (ATP, <19> inhibition of microtubules to membrane-bound liver enzyme [35]) [35]

0.9 <13> (oxalylglycine, <13> vs. NH$_4^+$ [21]) [21]

1.1 <7> (glutamate, <7> at pH 10.0 [24]) [24]

2.6 <27> (2-oxoglutarate, <27> liver enzyme [33]) [33]

2.9 <13> (NH$_4^+$) [21]

3.7 <7> (lysine, <7> at pH 10.0 [24]) [24]

4 <13> (ADP) [6]

4.1 <7> (glutamate, <7> at pH 8.5 [24]) [24]

5 <27> (2-oxoglutarate, <27> kidney enzyme [33]) [33]

5 <3> (phenylglyoxal, <3> vs. 2-oxoglutarate [44]) [44]

6 <3> (phenylglyoxal, <3> vs. NADH [44]) [44]
6.1 <7> (histidine, <7> at pH 10.0 [24]) [24]
6.2 <7> (alanine, <7> at pH 10.0 [24]) [24]
6.3 <13> (L-glutamate) [21]
9.3 <7> (lysine, <7> at pH 8.5 [24]) [24]
23.2 <7> (alanine, <7> at pH 8.5 [24]) [24]
24.4 <13> (2-oxovalerate) [21]
31 <13> (2-oxobutyrate) [21]
46 <27> (NH₄Cl, <27> liver enzyme [33]) [33]
60 <13> (2-oxoglutarate, <13> substrate inhibition vs. NADPH [21]) [21]
69 <27> (NH₄Cl, <27> kidney enzyme [33]) [33]
70 <13> (2-oxoglutarate, <13> substrate inhibition vs. NH_4^+ [21]) [21]
70 <18> (NaCl) [27]
78 <13> (norvaline) [21]
80 <18> (L-glutamate, <18> reductive amination with NADPH, not with NADH is inhibited [27]) [27]

pH-Optimum

6.4 <16> (<16> NADPH + NH₃ + 2-oxoglutarate [19]) [19]
6.75 <28> (<28> NADPH + NH₃ + 2-oxoglutarate [34]) [34]
7.4 <18> (<18> NADH + 2-oxoglutarate + NH₃ [27]) [27]
7.5 <28> (<28> NADP⁺ + L-glutamate, NADH + NH₃ + 2-oxoglutarate [34]) [34]
7.6 <3, 13, 16, 26> (<3> NAD(P)H + 2-oxoglutarate + NH₃ [30]; <26> NADPH [17]; <16> NADP⁺ + glutamate [19]; <13> in the absence of phosphate [12]) [12, 17, 19, 30]
7.7 <2, 24, 30, 31> (<2,24> NADPH + 2-oxoglutarate + NH₃ [13,31]; <30> reductive amination of 2-oxoglutarate [38]; <31> with NAD(P)H [40]) [13, 31, 38, 40]
7.8 <12, 13, 16, 20> (<12,13,20> 2-oxoglutarate + NH₃ + NAD(P)H, liver [1]; <16> NADH + NH₃ + 2-oxoglutarate [19]) [1, 19]
7.8-8.4 <11> [3]
7.95 <4, 5> (<4,5> NADPH-dependent amination [28]) [28]
8 <14, 26, 28> (<18> NADPH + 2-oxoglutarate + NH₃ [27]; <28> NAD⁺ + L-glutamate [34]; <14> reductive amination [43]) [17, 34, 43]
8.05 <6> (<6> NADPH-dependent amination [28]) [28]
8.2 <15, 16, 24> (<24> NADH + NH₃ + 2-oxoglutarate [13]; <15> NADH + NH₃ + 2-oxoglutarate [18]; <16> NAD⁺ + H₂O + glutamate [19]) [13, 18, 19]
8.3 <26> (<26> 2-oxoglutarate + NH₃ + NAD(P)H [17]) [17]
8.3-8.6 <13> (<13> glutamate + NAD(P)⁺ + H₂O [1]) [1]
8.4-10 <7> (<7> NADPH + 2-oxoglutarate + NH₃, potassium phosphate buffer [24]) [24]
8.5 <2> (<2> NADH + 2-oxoglutarate + NH₃ [31]) [31]
8.5-8.6 <7> (<7> NADPH + NH₃ + 2-oxoglutarate, Tris-HCl buffer [24]) [24]
8.6 <31> (<31> with NAD⁺ [40]) [40]
8.7 <26> (<26> NADP⁺ + glutamate [17]) [17]

8.8 <24> (<24> NAD$^+$ + glutamate [13]) [13]

8.9 <26> (<26> NAD$^+$ + glutamate [17]) [17]

9 <2, 13, 18> (<13> homocysteinesulfinate + NAD(P)$^+$ [1]; <18> NADH + NH$_3$ + 2-oxoglutarate [27]; <2> NAD$^+$ + glutamate [31]; <13> in the presence of 330 mM phosphate [12]) [1, 12, 27, 31]

9-9.5 <14> (<14> glutamte deamination [43]) [43]

9.3 <15> (<15> NAD$^+$ + glutamate [18]) [18]

9.4 <30> (<30> oxidative deamination of glutamate [38]) [38]

9.5 <13, 18> (<13> norvaline + NAD(P)$^+$ [1]; <18> glutamate + NADP$^+$ [27]) [1, 27]

9.5-10 <13> (<13> α-aminobutyrate, valine, norleucine, isoleucine, methionine or alanine + NAD(P)$^+$ [1]) [1]

9.7 <13> (<13> leucine + NAD(P)$^+$ [1]) [1]

pH-Range

7-8 <3> (<3> 90-100% activity [30]) [30]

7.5-8 <14> (<14> half activity of glutamate deamination compared to pH 9.5 [43]) [43]

7.5-8.8 <26> (<26> NAD(P)H + 2-oxoglutarate + NH$_3$, more than 50% of activity maximum at pH 7.5 and 8.8 [17]) [17]

8-9.2 <26> (<26> deamination, more than 50% of activity maximum at pH 8.0 and 9.2 [17]) [17]

Temperature optimum (°C)

25 <28> (<28> NAD$^+$ + L-glutamate [34]) [34]

35 <28> (<28> NADH + NH$_3$ + 2-oxoglutarate [34]) [34]

60-62 <2> [31]

85 <1, 3> (<1> native and recombinant enzyme [39]) [30, 39]

95 <1> [32]

Temperature range (°C)

30-90 <1, 3> (<3> 30°C: about 90% of activity maximum, 90°C: about 35% of activity maximum [30]) [30, 39]

4 Enzyme Structure

Molecular weight

210000 <31> (<31> gel filtration [40]) [40]

240000-260000 <9> (<9> sucrose density gradient sedimentation [1]) [1]

250000 <19> (<19> liver enzyme, sedimentation velocity, light scattering [1]) [1]

266000-269000 <2> (<2> gel filtration [31]) [31]

270000 <1, 3, 16, 29, 30> (<16> gel filtration, sucrose density gradient centrifugation [19]; <3> gel filtration [30]; <1> gel filtration [32]; <29> gel filtration [36]; <30> gel filtration [38]; <1> recombinant enzyme, only hexameric form is enzymatically active, gel filtration [39]) [19, 30, 32, 36, 38, 39]

284000 <32> (<32> recombinant and native enzyme, gel filtration [41]) [41]

285000 <7> (<7> gel filtration [24]) [24]

290000 <18> (<18> hig- and low-activity form, gel filtration [29]) [29]

300000-350000 <13, 14, 17, 18, 19, 20, 22, 26> (<18> polyacrylamide disc gel electrophoresis [27]; <13> liver, light scattering, sedimentation equilibrium [1]; <26> isoenzymes GDH1, GDH2, disc gel electrophoresis [17]; <20> liver, sedimentation velocity [1]; <19> liver, light scattering [1]; <17> liver, sedimentation equilibrium [20]; <22> liver, sedimentation equilibrium [1]; <14> liver, sedimentation equilibrium [1]) [1, 17, 20, 27]

320000 <33> (enzyme from hibernating animals, gel filtration [45]) [45]

330000 <19> (<19> membrane-bound liver enzyme, velocity sedimentation on sucrose gradient [35]) [35]

331000 <13> (<13> gel filtration [45]) [45]

333000 <22> (<22> sedimentation equilibrium [1]) [1, 15]

335000 <33> (<33> enzyme from euthermic animals, gel filtration [45]) [45]

340000 <10, 28> (<10> gel filtration [7]; <28> gel filtration [34]) [7, 34]

Subunits

? <14> (<14> x * 56000, enzymes from mitochondria and endoplasmic reticulum [43]) [43]

? <16> (<16> x * 57000, SDS-PAGE [19]) [19]

hexamer <1-3, 7, 10, 18, 19, 22, 28, 29, 32, 33> (<10> 6 * 54000, SDS-PAGE [7]; <22> 6 * 53900, sedimentation equilibrium of enzyme treated with 6 M guanidinium and 0.5% 2-mercaptoethanol [15]; <7> 6 * 48000, SDS-PAGE [24]; <18> SDS-PAGE [27]; <18> 6 * 49000, SDS-PAGE [29]; <3> 6 * 44000, SDS-PAGE [30]; <3> 6 * 45000, SDS-PAGE [44]; <2> isoenzyme GDH 1, SDS-PAGE [31]; <2> 4 * 44000 + 2 * 46000, isoenzyme GDH 2, SDS-PAGE [31]; <1> 6 * 46000, SDS-PAGE [32]; <28> 6 * 57000, SDS-PAGE [34]; <19> 6 * 50000, membrane-bound liver enzyme, SDS-PAGE [35]; <29> 6 * 46000, SDS-PAGE [36]; <1> 6 * 46000, recombinant enzyme, SDS-PAGE [39]; <32> 6 * 47300, recombinant enzyme, second peak in gel filtration corresponding to catalytically inactive monomer, SDS-PAGE [41]; <33> 6 * 59500, enzyme from both euthermic and hibernating animals [45]) [7, 15, 24, 27, 29-32, 34, 35, 36, 39, 41, 45]

tetramer <31> (<31> 4 * 52000, SDS-PAGE [40]) [40]

Posttranslational modification

proteolytic modification <10> (<10> limited trypsin proteolysis removes a 39 amino acid peptide from the enzyme [7]) [7]

5 Isolation/Preparation/Mutation/Application

Source/tissue

brain <12-14, 19-23> [1, 23, 35, 42]

fat body <28> [34]

heart <12-14, 19-23> [1]

intestinal mucosa <12-14, 19-23> [1]

kidney <12-14, 19-23, 27> [1, 33]

larva <28> [34]
leaf <11, 26> [3, 17]
liver <10, 12-14, 17, 19-23, 27, 33> (<33> from euthermic, 37°C body temperature, and hibernating, 5°C body temperature, animals [45]) [1, 2, 4-7, 9-11, 15, 20, 22, 23, 25, 33, 35, 43, 45]
seed <26> (<26> glutamate dehydrogenase 1 is found only in the developing seed [17]) [16, 17]

Localization

chloroplast <11, 26> (<26> glutamate dehydrogenase 2 [16]) [3, 16, 17]
membrane <19> (<19> peripheral membrane protein [35]) [35]
mitochondrion <12-14, 19-25, 31> [1, 9, 13, 14, 40]
nucleus <13> [5]
rough endoplasmic reticulum <14> (<14> peripheral membrane protein [43]) [43]

Purification

<1> (native and recombinant enzyme [39]) [32, 39]
<2> (isoenzymes, glutamate dehydrogenases 1, 2 and 3 [31]) [31]
<3> [30, 44]
<7> [24]
<10> (hydroxylapatite, DEAE-Toyopearl, GTP-Sepharose [7]) [7]
<11> (partial [3]) [3]
<12> (wild-type and mutant enzyme, Q-Sepharose, ω-aminopentyl column, GTP-agarose [46]) [46]
<13> (ammonium sulfate, DEAE-cellulose, affinity precipitation with adipo-N^2,N^2-dihydrazido-bis(N^6-carboxymethyl-NAD⁺) [22]) [8, 22]
<13> (brain isoenzymes 1 and 2 [42]) [5, 22, 23, 42]
<14> (DEAE-Sepharose, ATP-agarose, Resource Q [43]) [9, 22]
<16> [19]
<18> [27, 29]
<19> (affinity-purified [35]) [35]
<22> [15]
<24> (partial [3]) [13]
<25> [14]
<27> (partial [33]) [33]
<28> [34]
<29> [36]
<30> (18-22°C, 20% glycerol, heat treatment, ammonium sulfate, gel filtration, ion-exchange chromatography [38]) [38]
<31> (ammonium sulfate, phenyl Sepharose, Superdex 200, Q Sepharose [40]) [40]
<32> (HitrapQ, Hitrap blue, recombinant and native enzyme [41]) [41]
<33> (enzyme from euthermic and hibernating animals [45]) [45]

Renaturation

<13, 14> (complete loss of activity after incubation with 6 M guanidine hydrochloride or 7 M urea, no renaturation after dilution into 200 mM phosphate buffer [22]) [22]

Crystallization

<8, 9, 12-14, 19, 20, 22, 23> [1, 2, 9, 15]

Cloning

<1> (expression in Escherichia coli [39]) [39]

<12> (coexpression of wild-type and mutant enzymes with GroES and GroEl in Escherichia coli [46]) [46]

<29> (expression by in vitro transcription/translation in rabbit reticulocytes [36]) [36]

<32> (expression in Escherichia coli [41]) [41]

Engineering

H454Y <12> (<12> lower basal activity but comparable maximal activity as wild-type [46]) [46]

R463A <12> (<12> stimulatory effect of ADP is eliminated [46]) [46]

S448P <12> (<12> unstable in Tris-buffer especially in the absence of allosteric activators, basal and maximal specific activity is lower than that from wild-type [46]) [46]

6 Stability

pH-Stability

3-9 <3> (<3> at least 65% residual activity after 14 h [30]) [30]

7-9 <7> (<7> room temperature, 20% v/v glycerol, stable, room temperature, 1% v/v glycerol, 25% loss of activity after 48 h [24]) [24]

Temperature stability

0 <7, 30> (<7> complete loss of activity after 5 h, 20% glycerol protects from inactivation [24]) [24, 38]

5 <7> (<7> moderately stable above [24]) [24]

20-25 <30> (<30> no activity in cell extracts prepared at 0 to 4°C, high activities between 20°C and 25°C, 70% activity is retained after 2 h at 0°C in the presence of 20% glycerol [38]) [38]

37 <13> (<13> complete loss of activity [11]) [11]

42 <14> (<14> mitochondrial enzyme loses 80% activity after 20 min, enzyme from endoplasmic reticulum loses 20% activity [43]) [43]

55 <24> (<24> short periods [13]) [13]

66 <16> (<16> 10 min, 50% loss of activity [19]) [19]

70-100 <3> (<3> no loss in activity after 1 h at 90°C, 20% residual activity after 1 h at 100°C [30]) [30]

74 <16> (<16> 10 min, complete loss of activity [19]) [19]

100 <1> (<1> half-life: 2.3 h at 0.053 mg/ml protein concentration, 10 h at 1.06 mg/ml protein concentration [32]) [32]

100 <1> (<1> native enzyme, half-life at 100°C: 10.5 h, recombinant enzyme, half-life at 75°C: 7 h, at 90°C: 8.1 h [39]) [39]

100-105 <29> (<29> 1 mg/ml enzyme, half-life of 10.5 h at 100°C, 3.5 h at 105°C, and 20 h at 90°C, thermal denaturation at 113°C [36]) [36]

Organic solvent stability

ethanol <3> (50% v/v, complete loss of activity) [30]

isopropanol <3> (50% v/v, complete loss of activity) [30]

methanol <3> (50% v/v, complete loss of activity) [30]

General stability information

<3>, SDS: 0.1% v/v, no loss of activity after 12 h, 0.5%, half-life 5 h [30]

<3>, urea: 4 M, 10% loss of activity after 12 h, 7.5 M, half-life 9 h [30]

<7>, glycerol, 20% v/v, protects against cold inactivation [24]

<15>, AMP stabilizes [18]

<15>, L-aspartate stabilizes [18]

<16>, freeze-thawing decreases activity [19]

<16>, glycerol, 5-20% v/v, stabilizes [19]

<26>, 2-mercaptoethanol stabilizes [17]

<26>, glutathione stabilizes [17]

<29>, absolutely stable at 4°C or at room temperature for at least 6 months [36]

<30>, extremely unstable at 0°C to 4°C due to the dissociation of the holoenzyme into catalytically inactive subunits [38]

Storage stability

<1>, 4°C, 6 months, stable [32]

<3>, -20°C, 4°C, 25°C, 6 months, 5% loss of activity [44]

<3>, -70°C, stable for 6 months [30]

<3>, 4°C, 20% loss of activity after 40 days [30]

<3>, 4°C, more than 20 days, no loss of activity [30]

<13>, 0°C, 200 mM phosphate buffer, pH 7.4, several weeks, very little loss of activity [2]

<13>, 5°, 28% Na_2SO_4 solution, several weeks, very little loss of activity [2]

<15>, -18°C, several months, no loss of activity [18]

<16>, 0°C, 50 mM Na,K phosphate buffer, pH 5.5, stable [19]

<17>, -15°, stable for several months [20]

<22>, suspended in ammonium sulfate, stable for months [15]

References

[1] Smith, E.L.; Austen, B.M.; Blumenthal, K.M.; Nyc, J.F.: Glutamate dehydrogenase. The Enzymes, 3rd Ed. (Boyer, P.D., ed.), **11**, 293-367 (1975)

[2] Olson, J.A.; Anfinsen, C.B.: The crystallization and characterization of L-glutamate acid dehydrogenase. J. Biol. Chem., **197**, 67-79 (1952)

[3] Tsenova, E.N.: Isolation and properties of glutamate dehydrogenase from pea chloroplasts. Enzymologia, **43**, 397-408 (1972)

[4] Silverstein, E.: Equilibrium kinetic study of bovine liver glutamate dehydrogenase at high pH. Biochemistry, **13**, 3750-3754 (1974)

[5] Di Prisco, G.; Garofano, F.: Purification and some properties of glutamate dehydrogenase from ox liver nuclei. Biochem. Biophys. Res. Commun., **58**, 683-689 (1974)

[6] Hornby, D.P.; Aitchison, M.J.; Engel, P.C.: The kinetic mechanism of ox liver glutamate dehydrogenase in the presence of the allosteric effector ADP. The oxidative deamination of L-glutamate. Biochem. J., **223**, 161-168 (1984)

[7] Tang, M.Q.; Ando, S.; Yamada, S.; Hayashi, S.: The trypsin-catalyzed activation of glutamate dehydrogenase purified from eel liver. J. Biochem., **111**, 655-661 (1992)

[8] Beattie, R.E.; Graham, L.D.; Griffin, T.O.; Tipton, K.F.: Purification of NAD+-dependent dehydrogenases by affinity precipitation with adipo-N^2,N^2-dihydrazzidobis-(N^6-carboxymethyl-NAD+) (bis-NAD+). Biochem. Soc. Trans., **12**, 433 (1984)

[9] Arnold, H.; Maier, K.P.: Glutamate dehydrogenase from rat liver. Biochim. Biophys. Acta, **251**, 133-140 (1971)

[10] Piszkiewicz, D.; Smith, E.L.: Bovine liver glutamate dehydrogenase. Equilibria and kinetics of inactivation by pyridoxal. Biochemistry, **10**, 4538-4544 (1971)

[11] Nemat-Gorgani, M.; Dodd, G.: The interaction of phospholipid membranes and detergents with glutamate dehydrogenase. Eur. J. Biochem., **74**, 129-137 (1977)

[12] Di Prisco, G.: Effect of pH and ionic strength on the catalytic and allosteric properties of native and chemically modified ox liver mitochondrial glutamate dehydrogenase. Arch. Biochem. Biophys., **171**, 604-612 (1975)

[13] Ehmke, A.; Hartmann, T.: Properties of glutamate dehydrogenase from Lemna minor. Phytochemistry, **15**, 1611-1617 (1976)

[14] Langridge, W.H.R.; Komuniecki, P.; DeToma, F.J.: Isolation and regulatory properties of two glutamate dehydrogenases from the cellular slime mold Dictyostelium discoideum. Arch. Biochem. Biophys., **178**, 581-587 (1977)

[15] Veronese, F.M.; Bevilacqua, R.; Boccu, E.; Brown, D. M.: Purification, characteristics and sequence of a peptide containing an essential lysine residue. Biochim. Biophys. Acta, **445**, 1-13 (1976)

[16] Mc Kenzie, E.A.; Copeland, L.; Lees, E.M.: Glutamate dehydrogenase activity in developing soybean seed: kinetic properties of three forms of the enzyme. Arch. Biochem. Biophys., **212**, 298-305 (1981)

[17] Mc Kenzie, E.A.; Lees, E.M.: Glutamate dehydrogenase activity in developing soybean seed: isolation and characterization of three forms of the enzyme. Arch. Biochem. Biophys., **212**, 290-297 (1981)

[18] Engelhardt, H.; Klemme, J.H.: Characterization of an allosteric, nucleotide-unspecific glutamate dehydrogenase from Rhodopseudomonas sphaeroides. FEMS Microbiol. Lett., **3**, 287-290 (1978)

[19] Kimura, K.; Miyakawa, A.; Imai, T.; Sasakawa, T.: Glutamate dehydrogenase from Bacillus subtilis PCI 219. I. Purification and properties. J. Biochem., **81**, 467-476 (1977)

[20] Electricwala, A.H.; Dickinson, F.M.: Kinetic studies of dogfish liver glutamate dehydrogenase. Biochem. J., **177**, 449-459 (1979)

[21] Rife, J.E.; Cleland, W.W.: Kinetic mechanism of glutamate dehydrogenase. Biochemistry, **19**, 2321-2328 (1980)

[22] Graham, L.D.; Griffin, T.O.; Beatty, R.E.; McCarthy, A.D.; Tipton, K.F.: Purification of liver glutamate dehydrogenase by affinity precipitation and studies on its denaturation. Biochim. Biophys. Acta, **828**, 266-269 (1985)

[23] McCarthy, A.D.; Tipton, K.F.: Ox glutamate dehydrogenase. Comparison of the kinetic properties of native and proteolysed preparations. Biochem. J., **230**, 95-99 (1985)

[24] Maulik, P.; Ghosh, S.: NADPH/NADH-dependent cold-labile glutamate dehydrogenase in Azospirillum brasilense. Purification and properties. Eur. J. Biochem., **155**, 595-602 (1986)

[25] Kim, H.; Haley, B.E.: Synthesis and properties of 2-azido-NAD$^+$. A study of interaction with glutamate dehydrogenase. J. Biol. Chem., **265**, 3636-3641 (1990)

[26] Ottolina, G.; Carrea, G.; Riva, S.: Coenzymatic properties of low molecular-weight and macromolecular N^6-derivatives of NAD$^+$ and NADP$^+$ with dehydrogenases of interest for organic synthesis. Enzyme Microb. Technol., **12**, 596-602 (1990)

[27] Yamamoto, I.; Abe, A.; Ishimoto, M.: Properties of glutamate dehydrogenase purified from Bacteroides fragilis. J. Biochem., **101**, 1391-1397 (1987)

[28] Yamamoto, I.; Saito, H.; Ishimoto, M.: Comparison of properties of glutamate dehydrogenases in members of the Bacteroides fragilis group. J. Gen. Appl. Microbiol., **33**, 429-436 (1987)

[29] Saito, H.; Yamamoto, I.; Ishimoto, M.: Reversible inactivation of glutamate dehydrogenase in Bacteroides fragilis: purification and characterization of high activity- and low activity-enzymes. J. Gen. Appl. Microbiol., **34**, 377-385 (1988)

[30] Schinkinger, M.F.; Redl, B.; Stöffler, G.: Purification and properties of an extreme thermostable glutamate dehydrogenase from the archaebacterium Sulfolobus solfataricus. Biochim. Biophys. Acta, **1073**, 142-148 (1991)

[31] Moyano, E.; Cardenas, J.; Munoz-Blanco, J.: Purification and properties of three NAD(P)$^+$ isozymes of L-glutamate dehydrogenase of Chlamydomonas reinhardtii. Biochim. Biophys. Acta, **1119**, 63-68 (1992)

[32] Robb, F.T.; Park, J.B.; Adams, M.W.W.: Characterization of an extremely thermostable glutamate dehydrogenase: a key enzyme in the primary metabolism of the hyperthermophilic archaebacterium, Pyrococcus furiosus. Biochim. Biophys. Acta, **1120**, 267-272 (1992)

[33] Lee, A.R.; Balinsky, J.B.: A kinetic study of glutamate dehydrogenase from Xenopus laevis. Int. J. Biochem., **5**, 795-805 (1974)

[34] Teller, J.K.: Purification and some properties of glutamate dehydrogenase from the mealworm fat body. Insect Biochem., **18**, 101-106 (1988)

[35] Rajas, F.; Rousset, B.: A membrane-bound form of glutamate dehydrogenase possesses an ATP-dependent high-affinity microtubule-binding activity. Biochem. J., **295**, 447-455 (1993)

[36] DiRuggiero, J.; Robb, F.T.; Jagus, R.; Klump, H.H.; Borges, K.M.; Kessel, M.; Mai, X.; Adams, M.W.W.: Characterization, cloning, and in vitro expression of the extremely thermostable glutamate dehydrogenase from the hyperthermophilic archaeon, ES4. J. Biol. Chem., 268, 17767-17774 (1993)

[37] Shoemaker, M.T.; Haley, B.E.: Identification of a guanine binding domain peptide of the GTP binding site of glutamate dehydrogenase: Isolation with metal-chelate affinity chromatography. Biochemistry, 32, 1883-1890 (1993)

[38] Jahns, T.; Kaltwasser, H.: Properties of the cold-labile NAD$^+$-specific glutamate dehydrogenase from Bacillus cereus DSM 31. J. Gen. Microbiol., 139, 775-780 (1993)

[39] Diruggiero, J.; Robb, F.T.: Expression and in vitro assembly of recombinant glutamate dehydrogenase from the hyperthermophilic archaeon Pyrococcus furiosus. Appl. Environ. Microbiol., 61, 159-164 (1995)

[40] Heeschen, V.; Gerendas, J.; Richter, C.P.; Rudolph, H.: Glutamate dehydrogenase of Sphagnum. Phytochemistry, 45, 881-887 (1997)

[41] Rahman, R.N.Z.A.; Fujiwara, S.; Takagi, M.; Imanaka, T.: Sequence analysis of glutamate dehydrogenase (GDH) from the hyperthermophilic archaeon Pyrococcus sp. KOD1 and comparison of the enzymic characteristics of native and recombinant GDHs. Mol. Gen. Genet., 257, 338-347 (1998)

[42] Choi, S.Y.; Hong, J.W.; Song, M.S.; Jeon, S.G.; Bahn, J.H.; Lee, B.R.; Ahn, J.Y.; Cho, S.W.: Different antigenic reactivities of bovine brain glutamate dehydrogenase isoproteins. J. Neurochem., 72, 2162-2169 (1999)

[43] Lee, W.K.; Shin, S.; Cho, S.S.; Park, J.S.: Purification and characterization of glutamate dehydrogenase as another isoprotein binding to the membrane of rough endoplasmic reticulum. J. Cell. Biochem., 76, 244-253 (1999)

[44] Ahn, J.Y.; Lee, K.S.; Choi, S.Y.; Cho, S.W.: Regulatory properties of glutamate dehydrogenase from Sulfolobus solfataricus. Mol. Cells, 10, 25-31 (2000)

[45] Thatcher, B.J.; Storey, K.B.: Glutamate dehydrogenase from liver of euthermic and hibernating Richardson's ground squirrels: Evidence for two distinct enzyme forms. Biochem. Cell Biol., 79, 11-19 (2001)

[46] Fang, J.; Hsu, B.Y.L.; MacMullen, C.M.; Poncz, M.; Smith, T.J.; Stanley, C.A.: Expression, purification and characterization of human glutamate dehydrogenase (GDH) allosteric regulatory mutations. Biochem. J., 363, 81-87 (2002)

Glutamate dehydrogenase (NADP⁺) 1.4.1.4

1 Nomenclature

EC number
1.4.1.4

Systematic name
L-glutamate:NADP⁺ oxidoreductase (deaminating)

Recommended name
glutamate dehydrogenase (NADP⁺)

Synonyms
L-glutamate dehydrogenase
NAD(P)-glutamate dehydrogenase
NAD(P)H-dependent glutamate dehydrogenase
NADP-GDH
dehydrogenase, glutamate (nicotinamide adenine dinucleotide phosphate)
glutamic acid dehydrogenase
glutamic dehydrogenase

CAS registry number
9029-11-2

2 Source Organism

<1> *Aeropyrum pernix K1* (strain JCM 9820 [1]) [1, 5, 7]
<2> *Thermococcus litoralis* [2, 21]
<3> *Archaeoglobus fulgidus* (strain DSM 8774 [3]; strain 7324 [11]) [3, 11]
<4> *Saccharomyces cerevisiae* (contains two NADPH-dependent glutamate dehydrogenases [4]) [4, 47]
<5> *Psychrobacter sp.* (strain TAD1 [6]) [6]
<6> *Escherichia coli* (strain DH5α [6]; strain K2 [30]; strain K12 [38]; strain MRE 600 [48]) [6, 22, 25, 30, 38, 48, 49]
<7> *Thermotoga maritima* [8]
<8> *Pyrococcus kodakarensis* (strain KOD1 [9]) [9]
<9> *Giardia intestinalis* [10]
<10> *Bryopsis maxima* [12]
<11> *Haloferax mediterranei* [13]
<12> *Bacillus sp.* (strain KSM-635 [14]) [14]
<13> *Penicillium chrysogenum* (strain NCAIM 00237 [15]) [15]
<14> *Plasmodium falciparum* [16]
<15> *Thermococcus profundus* [17]

<16> *Agaricus bisporus* [18]
<17> *Pyrococcus furiosus* (strain DSM 3638 [23]) [19, 23]
<18> *Clostridium difficile* [19]
<19> *Schizosaccharomyces pombe* (strains 975h- and 972h+ [20]) [20]
<20> *Pyrococcus woesei* (strain DSM 3773 [23]) [23]
<21> *Salmonella typhimurium* [51, 52]
<22> *Thermococcus sp.* (isolate AN1 [24]) [24]
<23> *Dictyostelium discoideum* [26]
<24> *Halobacterium halobium* (strain NRC 36014 [27]) [27]
<25> *Bacillus fastidiosus* [28]
<26> *Streptomyces fradiae* [29]
<27> *Phormidium laminosum* [31]
<28> *Trichomonas vaginalis* [32]
<29> *Bifidobacterium bifidum* (hosts human adult and human infant [33]) [33]
<30> *Bifidobacterium infantis* (strain infantis a and liberorum [33]) [33]
<31> *Bifidobacterium breve* (strain breve a and parvolorum a [33]) [33]
<32> *Bifidobacterium adolescentis* [33]
<33> *Bifidobacterium thermophilum* (host pig [33]) [33]
<34> *Bifidobacterium longum* (hosts rat and calf [33]) [33]
<35> *Bifidobacterium pseudolongum* (hosts pig and chicken [33]) [33]
<36> *Chlorella sorokiniana* (two isoenzymes α-subunits at 2 mM ammonia and β-subunits at 29 mM ammonia [34]) [34]
<37> *Nitrobacter hamburgensis* (strain X14 [35]) [35]
<38> *Synechocystis sp.* (strain PCC 6803 [36]) [36]
<39> *Acropora formosa* [37]
<40> *Lactobacillus fermentum* (strain IFO 3071 [39]) [39]
<41> *Streptococcus equinus* (strain IFO 12553 [39]) [39]
<42> *Streptococcus faecalis* (strain IFO 12964 [39]) [39]
<43> *Streptococcus lactis* (strain IFO 12546 [39]) [39]
<44> *Aspergillus ochraceus* (strain ATCC [40]) [40]
<45> *Sphaerostilbe repens* [41]
<46> *Mycobacterium smegmatis* (strain CDC46 [42]) [42]
<47> *Proteus inconstans* [43]
<48> *Bacillus megaterium* [44]
<49> *Halobacterium sp.* [45]
<50> *Trypanosoma cruzi* [46]
<51> *Thermophilic bacillus* [50]

3 Reaction and Specificity

Catalyzed reaction
  L-glutamate + H_2O + NADP$^+$ = 2-oxoglutarate + NH_3 + NADPH + H^+

Reaction type
  oxidation
  oxidative deamination

redox reaction
reduction
reductive amination

Natural substrates and products

S L-glutamate + NADP$^+$ + H$_2$O <1-51> (Reversibility: r <1-51> [1-52]) [1-52]

P 2-oxoglutarate + NADPH + NH$_3$

S Additional information <2> (<2> glutamate dehydrogenase represents an enzymatic link between major catabolic and biosynthetic pathways via the tricarboxylic acid cycle intermediate 2-oxoglutarate [2]) [2]

P ?

S Additional information <3> (<3> glutamate dehydrogenase functions physiologically for the synthesis of L-glutamate from 2-oxoglutarate and ammonia [3]) [3]

P ?

Substrates and products

S L-glutamate + NADP$^+$ + H$_2$O <1-51> (Reversibility: r <1-51> [1-52]) [1-52]

P 2-oxoglutarate + NADPH + NH$_3$ <1-51> (<1,9,25,39,45,50> biphasic kinetic behavior for ammonia [1,10,28,37,41,46]; <39> biphasic kinetic behavior for L-glutamate [37]) [1-52]

Inhibitors

2-oxoglutarate <28> (<28> competitive inhibitor of the deamination [32]) [32]

4-chloromercuribenzoate <12, 26, 46> (<12> at 1 mM, 86% inhibition [14]; <26> at 0.01 mM, 35% inhibition [29]; <46> at 1 mM and 10 mM, 30% inhibition and 100% inhibition [42]) [14, 29, 42]

ADP <16, 26, 47> (<16> at 4 mM slight inhibitory [18]; <26> at 1 mM, 57% inhibition for oxidative deamination and 23% inhibition for reductive amination respectively [29]; <47> at 0.3 mM, 22% inhibition of oxidative deamination [43]) [18, 29, 43]

AMP <6, 23, 26, 47> (<23,26> slight inhibitory [26,29]; <47> at 0.3 mM 33% inhibition of oxidative deamination [43]; <6> inhibitory at higher concentrations than 1 mM [48]) [26, 29, 43, 48]

ATP <16> (<16> at 4 mM slight inhibitory [18]) [18]

Ag^{2+} <12> (<12> at 1 mM, 60% inhibition [14]) [14]

AlCl$_3$ <17, 20> (<17> at 1 mM, 30% inhibition [23]; <20> at 1 mM, 21% inhibition [23]) [23]

Ca^{2+} <6> (<6> at 1 mM 27% inhibition [25]) [25]

Hg^{2+} <10, 12, 15, 46, 47> (<10,47> at 1 mM, 100% activity loss [12,43]; <12> at 1 mM, 70% inhibition [14]; <15> at 0.1 mM, complete activity loss [17]; <46> at 1 mM, 10% inhibition [42]) [12, 14, 17, 42, 43]

HgCl$_2$ <17, 20, 22, 26> (<17> at 1 mM, 64% inhibition [23]; <20> at 1 mM, 45% inhibition [23]; <22> at 1 mM, no activity, oxidative deamination [24]; <26> at 0.01 mM, 27% inhibition [29]) [23, 24, 29]

KCl <50> (<50> more than 50 mM [46]) [46]

L-glutamic acid <27> (<27> at 20 mM 25% inhibition [31]) [31]

L-homoserine <6> (<6> competitive inhibitor with respect to both ammonia and glutamine [49]) [49]

L-tryptophan <27> (<27> at 20 mM, 15% inhibition [31]) [31]

Mg^{2+} <10> (<10> at 1 mM, 16% activity loss [12]) [12]

Mn^{2+} <10> (<10> at 1 mM, 19% activity loss [12]) [12]

$MnCl_2$ <17, 20> (<17> at 1 mM, 63% inhibition [23]; <20> at 1 mM, 20% inhibition [23]) [23]

N-ethylmaleinimide <47> (<47> at 0.8 mM, 44% inhibition [43]) [43]

NaCl <50> (<50> more than 50 mM [46]) [46]

O_2 <3> (<3> inactivation [11]) [11]

$Pb(CH_3COO)_2$ <17, 20> (<17> at 1 mM, 59% inhibition [23]; <20> at 1 mM, 64% inhibition [23]) [23]

Zn^{2+} <6, 10, 46> (<10> at 1 mM, 64% activity loss [12]; <6> at 1 mM, 40% inhibition [25]; <46> over 0.1 mM [42]) [12, 25, 42]

fumarate <25, 37> (<25> at 5 mM 20% inhibition [28]; <37> at 20 mM 30% inhibition [35]) [28, 35]

glutamate <28> (<28> competitive inhibitor of the amination reaction [32]) [32]

glutaric acid <27> (<27> at 20 mM, 25% inhibition [31]) [31]

glyoxylate <37> (<37> at 20 mM, 30% inhibition [35]) [35]

guanidine hydrochloride <1, 6> (<1,6> complete loss of activity [5,49]) [5, 49]

hydroxylamine <24> (<24> competitive inhibitor with ammonia and uncompetitive inhibitor with both 2-oxoglutarate and NADPH [27]) [27]

iodoacetamide <47> (<47> at 4 mM, complete inactivation [43]) [43]

malate <25, 37> (<25> at 5 mM, 20% inhibition [28]; <37> at 20 mM, 30% inhibition [35]) [28, 35]

oxaloacetate <16, 37> (<16> at 20 mM slight inhibitory [18]; <37> at 20 mM, 60% inhibition [35]) [18, 35]

p-chloromercuriphenyl sulfonate <47> (<47> inactivetes, can be reversed by addition of cysteine [43]) [43]

p-hydroxymercuribenzoic acid <23, 21> (<23> at 0.08 mM, 50% inhibition [26]; <21> at 10 mM, 90% inhibition after 40 min [51]) [26, 51]

potassium phosphate <4> (<4> over 0.1 M at oxidative deamination [47]) [47]

pyridoxal 5'-phosphate <6, 17, 20> (<17> at 1 mM, 35% inhibition [23]; <20> at 1 mM, 45% inhibition [23]; <6> at 10 mM, 90% inhibition, complete protection when 16.8 mM 2-oxoglutarate and 1.68 mM NADP⁺ are added [30]) [23, 30]

pyruvate <16> (<16> at 10 mM slight inhibitory [18]) [18]

sodium dodecylsulfate <21> (<21> at 0.7% w/v after 1 h 5% activity [51]) [51]

succinate <24> (<24> competitive inhibitor with 2 oxoglutarate, uncompetitive with NADPH and non-competitive with ammonia [27]) [27] urea <16, 17, 20, 27, 6> (<16> at 0°C 1 h at 2 M 65% inhibition [18]; <17> inactivation with 2 mM urea [23]; <20> fully active at 20°C in 8 mM urea [23]; <27> no activity at 6 mM [31]; <6> at 8 M stable for 10 min [48]) [18, 23, 31, 48]

Cofactors/prosthetic groups
$NADP^+$ <1-51> [1-52]
NADPH <1-51> [1-52]

Activating compounds
ADP <23, 47> (<23> at 1 mM, 11% activation [26]; <47> at 0.3 mM 40% activation [43]) [26, 43]
AMP <6, 21> (<6,21> slight activation up to 1 mM [48,52]) [48, 52]
ATP <6, 47, 21> (<47> at 0.3 mM, 60% activation of reductive amination [43]; <6> at 2 mM, 50% activation [48]; <21> at 0.8 M, 68% activation [52]) [43, 48, 52]
$CaCl_2$ <22, 38> (<22> at 5 mM, 135% activation, oxidative deamination [24]; <38> at 20 mM, 25% and 108% activation of reductive amination and oxidative deamination respectively [36]) [24, 36]
GDP <28> (<28> at 1 mM five-fold Michaelis Menten constant of 2-oxoglutarate [32]) [32]
GTP <23> (<23> at 0.083 mM 22% activation [26]) [26]
IDP <28> (<28> at 1 mM five-fold Michaelis Menten constant of 2-oxoglutarate [32]) [32]
K_2HPO_4 <3> (<3> at 0.15 M three to fourfold stimulation of activity [3]) [3]
K_2SO_4 <1> (<1> 280-300% activity at 150-200 mM [7]) [7]
K_3PO_4<1> (<1> less effective than K_2SO_4 and Na_3PO_4 [7]) [7]
KCl <1, 3, 11, 17, 20, 50> (<1> 170-200% activity at 50-100 mM [7]; <3> at 0.2 M three to fourfold stimulation of activity [3,11]; <11> optimal activity at 1-1.5 M [13]; <17> at concentrations below 2 mM [23]; <20> at concentrations below 1 mM [23]; <50> up to 50 mM [46]) [3, 7, 11, 13, 23, 46]
$MgCl_2$ <22> (<22> at 5 mM, 104% activity, oxidative deamination [24]) [24]
$MnCl_2$ <22> (<22> at 5 mM, 250% activity, oxidative deamination [24]) [24]
Na_2SO_4 <1> (<1> less effective than K_2SO_4 and Na_3PO_4 [7]) [7]
Na3PO4 <1> (<1> 280-300% activity at 150-200 mM [7]) [7]
NaCl <1, 3> (<1> 170-200% activity at 50-100 mM [7]; <3> at 0.2 M three to fourfold stimulation of activity [3,11]; <11> optimal activity at 1-1.5 M [13]; <50> up to 50 mM [46]) [3, 7, 11, 13, 46]
UTP <6> (<6> activates up to 70% [38]) [38]
Zn^{2+} <46> (<46> activates up to 0.1 mM [42]) [42]
acetonitrile <1> (<1> activates [5]) [5]
ethanol <1> (<1> activates up to 40% v/v [5]) [5]
methanol <1> (<1> activates [5]) [5]
potassium phosphate <4> (<4> up to 0.1 M oxidative deamination [47]) [47]

Turnover number (min⁻¹)
66000 <45> (NADPH) [41]

Specific activity (U/mg)

0.29 <29> (<29> host human adult, reductive amination at 30°C and pH 6.5 [33]) [33]

0.31 <29> (<29> host human infant, reductive amination at 30°C and pH 6.5 [33]) [33]

0.34 <35> (<35> host chicken, reductive amination at 30°C and pH 6.5 [33]) [33]

0.4 <33> (<33> reductive amination at 30°C and pH 6.5 [33]) [33]

0.45 <34> (<34> host calf, reductive amination at 30°C and pH 6.5 [33]) [33]

0.59 <34> (<34> host rat, reductive amination at 30°C and pH 6.5 [33]) [33]

0.62 <31> (<31> strain parvolorum, reductive amination at 30°C and pH 6.5 [33]) [33]

0.7 <32, 35> (<32,35> reductive amination at 30°C and pH 6.5 [33]; <35> host pig [33]) [33]

0.8 <30> (<30> strain liberorum, reductive amination at 30°C and pH 6.5 [33]) [33]

1.1 <31> (<31> strain breve a, reductive amination at 30°C and pH 6.5 [33]) [33]

1.16 <30> (<30> strain infantis a, reductive amination at 30°C and pH 6.5 [33]) [33]

1.68 <26> (<26> oxidative deamination [29]) [29]

2.5 <7> [8]

2.54 <4> (<4> reductive amination [47]) [47]

8.2 <16> (<16> reductive amination at 33°C [18]) [18]

9.2 <37> (<37> reductive amination [35]) [35]

10.3 <17> (<17> oxidative deamination at 50°C and pH 8.2 [23]) [23]

10.92 <26> (<26> reductive amination [29]) [29]

18.4 <10> (<10> oxidative deamination at 25°C [12]) [12]

24.3 <20> (<20> oxidative deamination at 50°C and pH 8.2 [23]) [23]

32.9 <27> (<27> reductive amination at 30°C [31]) [31]

40 <8> (<8> T138E-mutant [9]) [9]

46.2 <51> (<51> reductive amination at 55°C [50]) [50]

51.2 <50> (<50> reductive amination [46]) [46]

57.6 <40> (<40> reductive amination [39]) [39]

60 <11> (<11> reductive amination at 40°C [13]) [13]

74.5 <14> (<14> oxidative deamination at pH 7.0 and 70°C [16]) [16]

78 <5> (<5> reductive amination at 20°C [6]) [6]

80 <8> (<8> E158Q-mutant [9]) [9]

80.2 <46> (<46> reductive amination [42]) [42]

85 <2> (<2> oxidative deamination at 80°C and pH 8.0 [21]) [21]

121 <23> (<23> reductive amination at 30°C [26]) [26]

133 <28> (<28> reductive amination at 37°C and pH 6.7 [32]) [32]

180 <6> (<6> oxidative deamination at 25°C [49]) [49]

194.3 <2> (<2> oxidative deamination at 75°C [2]) [2]

204.1 <5> (<5> oxidative deamination at 20°C [6]) [6]

214 <21> (<21> reductive amination at 25°C [52]) [52]

231 <47> (<47> oxidative deamination at 25°C [43]) [43]

250 <6> (<6> oxidative deamination [48]) [48]
388 <25> (<25> reductive amination [28]) [28]
416 <12> (<12> reductive amination at 30°C [14]) [14]
464 <22> (<22> oxidative deamination at 80°C and pH 7.54 [24]) [24]
930 <9> (<9> reductive amination at 37°C [10]) [10]
1140 <12> (<12> reductive amination at pH 7.5 and 60°C [14]) [14]
2500 <8> (<8> wild-type [9]) [9]
2940 <22> (<22> reductive amination at 80°C and pH 7.54 [24]) [24]

K_m-Value (mM)

0.00056 <2> (NH$_3$, <2> recombinant enzyme, reductive amination [2]) [2]
0.00065 <2> (NH$_3$, <2> wild-type enzyme, reductive amination [2]) [2]
0.003 <47> (NADPH, <47> reductive amination [43]) [43]
0.004 <50> (NADPH, <50> reductive amination [46]) [46]
0.0064 <50> (NADP$^+$, <50> oxidative deamination [46]) [46]
0.0087 <48> (NADPH, <48> reductive amination [44]) [44]
0.0097 <23> (NADPH, <23> reductive amination [26]) [26]
0.0098 <2> (NADP$^+$, <2> recombinant enzyme, oxidative deamination [2]) [2]
0.01 <10> (NADP$^+$, <10> oxidative deamination at 25°C [12]) [12]
0.01 <28> (NADPH, <28> reductive amination [32]) [32]
0.0102 <2> (NADP$^+$, <2> wild-type enzyme, oxidative deamination [2]) [2]
0.0105 <4> (NADP$^+$, <4> oxidative deamination, Gdh3p gene [4]) [4]
0.0113 <4> (NADPH, <4> reductive amination, Gdh1p gene [4]) [4]
0.013 <25, 21> (NADP$^+$, <25,21> oxidative deamination [28,52]) [28, 52]
0.0141 <4> (NADP$^+$, <4> oxidative deamination, Gdh1p gene [4]) [4]
0.017 <9> (NADPH, <9> reductive amination [10]) [10]
0.018 <11> (NADPH, <11> reductive amination at 25°C [13]) [13]
0.019 <21> (NADPH, <21> reductive amination [52]) [52]
0.02 <3, 38> (NADPH, <3> reductive amination at 60°C [3,11]; <38> reductive amination [36]) [3, 11, 36]
0.022 <1> (NADPH, <1> reductive amination at 50°C [7]) [7]
0.023 <47> (NADP$^+$, <47> oxidative deamination [43]) [43]
0.024 <49> (NADPH, <49> reductive amination [45]) [45]
0.027 <25, 44> (NADPH, <25,44> reductive amination [28,40]) [28, 40]
0.028 <4> (NADPH, <4> reductive amination [47]) [47]
0.029 <12, 2> (NADP$^+$, <12> oxidative deamination at 30°C [14]; <2> oxidative deamination at 80°C [21]) [14, 21]
0.03 <39> (NADP$^+$, <39> oxidative deamination [37]) [37]
0.03 <10> (NADPH, <10> reductive amination at 25°C [12]) [12]
0.031 <36> (NADP$^+$, <36> oxidative deamination [34]) [34]
0.0331 <4> (NADPH, <4> reductive amination, Gdh3p gene [4]) [4]
0.035 <6> (NADPH, <6> reductive amination [6]) [6]
0.038 <22> (NADP$^+$, <22> oxidative deamination at 80°C [24]) [24]
0.039 <1> (NADP$^+$, <1> oxidative deamination [1]) [1]
0.039 <1> (NADP$^+$, <1> oxidative deamination at 50°C [7]) [7]
0.04 <28, 36> (NADP$^+$, <28,36> oxidative deamination [32,34]) [32, 34]

0.04 <6> (NADPH, <6> reductive amination [49]) [49]
0.042 <6> (NADP$^+$, <6> oxidative deamination [49]) [49]
0.043 <5, 45> (NADP$^+$, <5,45> oxidative deamination [6,41]) [6, 41]
0.044 <40> (NADP$^+$, <40> oxidative deamination [39]) [39]
0.044 <12> (NADPH, <12> reductive amination at 30°C [14]) [14]
0.045 <6> (NADP$^+$, <6> oxidative deamination [6]) [6]
0.049 <1, 45> (NADPH, <1,45> reductive amination [5,41]) [5, 41]
0.05 <48> (NADP$^+$, <48> oxidative deamination [44]) [44]
0.053 <1> (NADP$^+$, <1> oxidative deamination [5]) [5]
0.053 <51> (NADPH, <51> reductive amination [50]) [50]
0.06 <3> (NADP$^+$, <3> oxidative deamination at 60°C [3,11]) [3, 11]
0.061 <49> (NADP$^+$, <49> oxidative deamination [45]) [45]
0.064 <27> (NADPH, <27> reductive amination [31]) [31]
0.066 <22> (NADPH, <22> reductive amination at 80°C [24]) [24]
0.07 <26> (NADPH, <26> reductive amination [29]) [29]
0.074 <16> (NADPH, <16> reductive amination at 33°C [18]) [18]
0.075 <1> (NADPH, <1> reductive amination at 70°C [1]) [1]
0.078 <40> (NADPH, <40> reductive amination [39]) [39]
0.083 <6> (NADPH, <6> reductive amination [25]) [25]
0.095 <29> (NADPH, <29> reductive amination [33]) [33]
0.11 <6> (NADP$^+$, <6> oxidative deamination [25]) [25]
0.11 <39> (NADPH, <39> reductive amination [37]) [37]
0.117 <16> (NADP$^+$, <16> oxidative deamination at 33°C [18]) [18]
0.12 <26> (NADP$^+$, <26> oxidative deamination [29]) [29]
0.12 <4> (NADP$^+$, <4> oxidative deamination [47]) [47]
0.14 <2> (NADPH, <2> reductive amination at 80°C [21]) [21]
0.16 <2> (2-oxoglutarate, <2> reductive amination at 80°C [21]) [21]
0.2 <6> (2-oxoglutarate, <6> reductive amination [6]) [6]
0.22 <2> (L-glutamate, <2> oxidative deamination at 80°C [21]) [21]
0.28 <9> (2-oxoglutarate, <9> reductive amination [10]) [10]
0.29 <21> (NH$_4^+$, <21> reductive amination [52]) [52]
0.3 <51> (NADP$^+$, <51> oxidative deamination [50]) [50]
0.32 <1> (2-oxoglutarate, <1> reductive amination [5]) [5]
0.34 <11> (2-oxoglutarate, <11> reductive amination at 25°C [13]) [13]
0.36 <48> (2-oxoglutarate, <48> reductive amination [44]) [44]
0.37 <47> (NH$_4^+$, <47> reductive amination [43]) [43]
0.4 <4> (2-oxoglutarate, <4> reductive amination [47]) [47]
0.5 <25> (2-oxoglutarate, <25> reductive amination [28]) [28]
0.5 <3> (2-oxoglutarate, <3> reductive amination at 60°C [11]) [11]
0.63 <2> (NH$_4^+$, <2> reductive amination at 80°C [21]) [21]
0.64 <6, 45> (2-oxoglutarate, <6,45> reductive amination [41,49]) [41, 49]
0.77 <29> (2-oxoglutarate, <29> reductive amination [33]) [33]
0.93 <39> (2-oxoglutarate, <39> reductive amination [37]) [37]
1 <50> (2-oxoglutarate, <50> reductive amination [46]) [46]
1.1 <51> (L-glutamate, <51> oxidative deamination [50]) [50]
1.1 <6> (NH$_4^+$, <6> reductive amination [49]) [49]
1.2 <23> (2-oxoglutarate, <23> reductive amination [26]) [26]

1.25 <27> (2-oxoglutarate, <27> reductive amination [31]) [31]
1.3 <51> (2-oxoglutarate, <51> reductive amination [50]) [50]
1.3 <6> (L-glutamate, <6> oxidative deamination [49]) [49]
1.5 <38> (2-oxoglutarate, <38> reductive amiantion [36]) [36]
1.54 <26> (2-oxoglutarate, <26> reductive amination [29]) [29]
1.7 <1> (2-oxoglutarate, <1> reductive amination at 50°C [7]) [7]
1.7 <22> (2-oxoglutarate, <22> reductive amination at 80°C [24]) [24]
1.7 <50> (NH$_4^+$, <50> reductive amination [46]) [46]
2 <1> (2-oxoglutarate, <1> reductive amination at 0.6 M NH$_4$Cl [1]) [1]
2 <19> (2-oxoglutarate, <19> reductive amination at 56°C [20]) [20]
2 <50> (L-glutamate, <50> reductive amination [46]) [46]
2 <29> (NH$_4^+$, <29> reductive amination [33]) [33]
2.1 <16> (NH$_4^+$, <16> reductive amination at 33°C [18]) [18]
2.2 <6> (L-glutamate, <6> oxidative deamination [25]) [25]
2.2 <10> (NH$_4^+$, <10> reductive amination at 25°C [12]) [12]
2.2 <23> (NH$_4^+$, <23> reductive amination [26]) [26]
2.3 <6, 47> (L-glutamate, <6,47> oxidative deamination [6,43]) [6, 43]
2.41 <44> (2-oxoglutarate, <44> reductive amination [40]) [40]
2.5 <6> (NH$_4$Cl, <6> reductive amination [6]) [6]
2.6 <45> (NH$_4^+$, <45> reductive amination, biphasic kinetics [41]) [41]
3 <10, 49> (2-oxoglutarate, <10,49> reductive amination at 25°C [12,45]) [12, 45]
3.13 <12> (2-oxoglutarate, <12> reductive amination at 30°C [14]) [14]
3.2 <16> (2-oxoglutarate, <16> reductive amination at 33°C [18]) [18]
3.2 <10> (L-glutamate, <10> oxidative deamination at 25°C [12]) [12]
3.25 <6> (2-oxoglutarate, <6> reductive amination [25]) [25]
3.3 <1> (L-glutamate, <1> oxidative deamination at 50°C [7]) [7]
3.3 <36> (NH4, <36> reductive amination α-isoenzyme, depending on NADPH-concentration [34]) [34]
3.7 <47> (L-glutamate, <47> reductive amination [43]) [43]
3.7 <38> (NH$_4^+$, <38> reductive amination [36]) [36]
3.9 <3> (L-glutamate, <3> oxidative deamination at 60°C [3,11]) [3, 11]
4 <21> (2-oxoglutarate, <21> reductive amination [52]) [52]
4 <3> (NH$_4$Cl, <3> reductive amination at 60°C [3,11]) [3, 11]
4 <5> (NH$_4$Cl, <5> reductive amination [6]) [6]
4.2 <11> (NH$_4^+$, <11> reductive amination at 25°C [13]) [13]
4.56 <1> (NH$_4$Cl, <1> reductive amination [5]) [5]
5 <37> (2-oxoglutarate, <37> reductive amination [35]) [35]
5 <4> (NH$_3$, <4> reductive amination, Gdh3p gene [4]) [4]
5.18 <1> (L-glutamate, <1> oxidative deamination [5]) [5]
5.5 <45> (L-glutamate, <45> oxidative deamination [41]) [41]
5.6 <40> (2-oxoglutarate, <40> reductive amination [39]) [39]
5.96 <4> (NH$_3$, <4> reductive amination, Gdh1p gene [4]) [4]
5.96 <12> (NH$_4$Cl, <12> reductive amination at 30°C [14]) [14]
6.06 <12> (L-glutamate, <12> oxidative deamination at 30°C [14]) [14]
6.36 <4> (L-glutamate, <4> oxidative deamination, Gdh3p gene [4]) [4]
6.5 <25> (NH$_4^+$, <25> reductive amination [28]) [28]

6.76 <40> (NH$_4^+$, <40> reductive amination [39]) [39]
7.5 <19> (NH$_4^+$, <19> reductive amination at 56°C [20]) [20]
7.7 <44> (NH$_4^+$, <44> reductive amination [40]) [40]
9 <1> (2-oxoglutarate, <1> reductive amination at 1 M NH$_4$Cl [1]) [1]
9 <37> (NH$_4^+$, <37> reductive amination [35]) [35]
9.12 <22> (L-glutamate, <22> oxidative deamination at 80°C [24]) [24]
9.2 <39> (NH$_4^+$, <39> reductive amination [37]) [37]
9.79 <4> (L-glutamate, <4> oxidative deamination, Gdh1p gene [4]) [4]
10 <4> (L-glutamate, <4> oxidative deamination [47]) [47]
10 <4> (NH$_4^+$, <4> reductive amination [47]) [47]
14.2 <49> (L-glutamate, <49> oxidative deamination [45]) [45]
15.5 <22> (NH$_4^+$, <22> reductive amination at 80°C [24]) [24]
16.6 <49> (NH$_4^+$, <49> reductive amination [45]) [45]
18 <39> (L-glutamate, <39> oxidative deamination, biphasic kinetics [37]) [37]
20 <25> (NH$_4^+$, <25> reductive amination [28]) [28]
21 <51> (NH$_4^+$, <51> reductive amination [50]) [50]
21.2 <45> (NH$_4^+$, <45> reductive amination; biphasic kinetics [41]) [41]
22 <48> (NH$_4^+$, <48> reductive amination [44]) [44]
27 <16> (L-glutamate, <16> oxidative deamination at 33°C [18]) [18]
28.6 <26> (L-glutamate, <26> oxidative deamination [29]) [29]
29 <48> (L-glutamate, <48> oxidative deamination [44]) [44]
30.8 <26> (NH$_4^+$, <26> reductive amination [29]) [29]
32.3 <36> (L-glutamate, <36> oxidative deamination [34]) [34]
36 <6> (NH$_4^+$, <6> reductive amination [25]) [25]
38.2 <36> (L-glutamate, <36> oxidative deamination [34]) [34]
44 <25> (L-glutamate, <25> oxidative deamination [28]) [28]
66 <50> (NH$_4^+$, <50> reductive amination [46]) [46]
67.4 <5> (L-glutamate, <5> oxidative deamination [6]) [6]
75 <36> (NH$_4^+$, <36> reductive amination β-isoenzyme [34]) [34]
79 <40> (L-glutamate, <40> oxidative deamination [39]) [39]
81 <39> (L-glutamate, <39> oxidative deamination, biphasic kinetics [37]) [37]
83 <1> (NH$_4^+$, <1> reductive amination at 50°C [7]) [7]
119 <1> (NH$_4$Cl, <1> reductive amination at 70°C [1]) [1]
225 <1> (L-glutamate, <1> oxidative deamination [1]) [1]
416 <39> (NH$_4^+$, <39> reductive amination [37]) [37]

K$_i$-Value (mM)
0.45 <28> (2-oxoglutarate, <28> competitive [32]) [32]
5.6 <28> (glutamate, <28> competitive inhibitor [32]) [32]

pH-Optimum
6.7 <4, 28> (<4,28> reductive amination [32,47]) [32, 47]
6.9 <45> (<45> reductive amination [41]) [41]
7 <1, 15, 50> (<1> reductive amination and oxidative deamination [5]; <15,50> oxidative deamination [17,46]) [5, 17, 46]
7.2 <15, 51> (<15,51> reductive amination [17,50]) [17, 50]

7.4 <17, 20> (<17,20> reductive amination at 50°C [23]) [23]
7.4-7.6 <39> (<39> oxidative deamination [37]) [37]
7.5 <9, 10, 12, 19, 27, 29, 37> (<9,10,12,19,29,37> reductive amination [10,12,14,20,33,35]; <27> reductive amination in potassium phosphate buffer [31]) [10, 12, 14, 20, 31, 33, 35]
7.5-8 <38> (<38> reductive amination [36]) [36]
7.6 <16> (<16> reductive amination [18]) [18]
7.7 <25> (<25> reductive amination [28]) [28]
7.8 <23> (<23> reductive amination [26]) [26]
7.9 <48> (<48> reductive amination [44]) [44]
8 <3, 5, 6, 2, 27, 28, 40, 44, 46, 50> (<3> oxidative deamination at 60°C [3,11]; <5> oxidative deamination at 20°C [6]; <2> oxidative deamination at 80°C [21]; <6,40,44,46,50> reductive amination [30,39,40,42,46,48]; <27> reductive amination in Tris hydrochloride buffer [31]; <28> oxidative deamination [32]) [3, 6, 11, 21, 30-32, 39, 40, 42, 46, 48]
8.1-8.4 <1> (<1> reductive amination at 50°C [7]) [7]
8.2 <17, 20> (<17,20> oxidative deamination at 50°C [23]) [23]
8.3 <22> (<22> reductive amination and oxidative deamination [24]) [24]
8.3-8.7 <1> (<1> oxidative deamination at 50°C [7]) [7]
8.4 <3, 26, 45, 51> (<3> reductive amination at 60°C [3,11]; <26> reductive amination [29]; <45,51> oxidative deamination [41,50]) [3, 11, 29, 41, 50]
8.5 <11, 12, 47> (<11,47> reductive amination [13,43]; <12> oxidative deamination [14]) [13, 14, 43]
8.6 <10, 25, 21> (<10,25> oxidative deamination [12,28]; <21> reductive amination [51]) [12, 28, 51]
8.8-9.8 <46> (<46> oxidative deamination [42]) [42]
9 <4, 5, 6, 16, 40, 48> (<5> reductive amination at 20°C [6]; <4,6,16,40,48> oxidative deamination [18,30,39,44,47,48]) [6, 18, 30, 39, 44, 47, 48]
9.3 <9, 26> (<9,26> oxidative deamination [10,29]) [10, 29]
9.6-10 <39> (<39> reductive amination [37]) [37]
10 <47> (<47> oxidative deamination [43]) [43]

pH-Range
4-9 <1> (<1> above pH 9 or below 4.0 no activity [5]) [5]
4-10 <15, 17, 20> (<15> fully active after 60 min [17]) [17, 23]
5-10 <1> (<1> fully active [7]) [7]

Temperature optimum (°C)
25 <5> [6]
34 <16> (<16> reductive amination [18]) [18]
40 <47> [43]
55 <6, 7> (<7> T158E-mutant [8]) [6, 8]
56 <19> (<19> reductive amination [20]) [20]
58 <7> (<7> S128R-mutant [8]) [8]
60 <7, 8, 11, 12, 25, 27, 46> (<7> S128R/T158E/S160E-mutant [8]; <8> E158Q-mutant [9]; <11,27> reductive amination [13,31]; <12,46> reductive amination and oxidative deamination [14,42]) [8, 9, 13, 14, 28, 31, 42]
63 <7> (<7> N117R-mutant and S128R/T158E/N117R/S160E-mutant [8]) [8]

65 <7> (<7> wild-type and S128R/T158E-mutant [8]) [8]
68 <7> (<7> S128R/T158E/N117R-mutant [8]) [8]
80 <1> (<1> reductive amination [5]) [5]
85 <8, 15> (<8> T138E-mutant [9]) [9, 17]
95 <1, 3, 2> (<3> reductive amination [3,11]; <1,2> oxidative deamination [5,21]) [3, 5, 11, 21]
100 <1, 17, 20> (<1,17,20> oxidative deamination [7,23]) [7, 23]

Temperature range (°C)
Additional information <1> (<1> no temperature dependence [1]) [1]

4 Enzyme Structure

Molecular weight
200000 <26> (<26> gel filtration [29]) [29]
204000 <22> (<22> gel filtration [24]) [24]
208000 <38> (<38> gel filtration [36]) [36]
212000 <49> (<49> sedimentation equilibrium and sedimentation velocity and diffusion coefficient [45]) [45]
230000 <28> (<28> gel filtration [32]) [32]
245000 <6> (<6> gel filtration [48]) [48]
245500 <46> (<46> gel filtration [42]) [42]
263000 <3, 15> (<3> gel filtration [3,11,17]) [3, 11, 17]
270000 <1, 2, 9, 48> (<1,2> gel filtration, expressed in Escherichia coli [2,7]; <9> gel filtration [10]; <48> sucrose-density-gradient centrifugation [44]) [2, 7, 10, 38, 44]
275000 <6> (<6> gel filtration [25]) [25]
280000 <10, 27, 45, 50, 21> (<10,27,50,21> gel filtration [12,31,46,51]; <45> non-denaturating PAGE [41]) [12, 31, 41, 46, 51]
284000 <8> (<8> gel filtration [9]) [9]
285000 <1> (<1> gel filtration [5]) [5]
290000 <5> (<5> gel filtration [6]) [6]
291000 <21> (<21> sucrose-density-gradient centrifugation [51]) [51]
294000 <23> (<23> gel filtration [26]) [26]
300000 <6, 17, 20, 40> (<17,20,40> gel filtration [23,39]; <6> sedimentation equilibrium [49]) [23, 39, 49]
310000 <37> (<37> gel filtration [35]) [35]
315000 <12> (<12> gel filtration [14]) [14]
320000 <11> (<11> gel filtration [13]) [13]
330000 <16> (<16> gel filtration [18]) [18]
339000 <13> (<13> gel filtration [15]) [15]
350000 <2> (<2> gel filtration [21]) [21]
360000 <39> (<39> gel filtration [37]) [37]
400000 <47> (<47> gel filtration [43]) [43]

Subunits

? <1> (<1> x * 46170, amino acid analysis [7]) [7]

? <2> (<2> x * 47000, gel filtration [2]) [2]

? <2> (<2> x * 47169, amino acid analysis [2]) [2]

? <36> (<36> x * 53000, SDS-PAGE, β-holoenzyme [34]) [34]

? <36> (<36> x * 55500, SDS-PAGE, α-holoenzyme [34]) [34]

? <47> (<47> x * 40000, SDS-PAGE [43]) [43]

? <6> (<6> x * 46000, SDS-PAGE [48]) [48]

hexamer <14> [16]

homohexamer <1, 10> (<1,10> α_6, 6 * 46000, SDS-PAGE [7,12]) [7, 12]

homohexamer <1, 3, 5, 17, 20> (<1,3,5,17,20> α_6, 6 * 47000, SDS-PAGE [3,5,6,23]) [3, 5, 6, 23]

homohexamer <11> (<11> α_6, 6 * 48000, CTAB-PAGE [13]) [13]

homohexamer <11> (<11> α_6, 6 * 55000, SDS-PAGE [13]) [13]

homohexamer <12, 23> (<12,23> α_6, 6 * 52000, SDS-PAGE [14,26]) [14, 26]

homohexamer <13, 39> (<13,39> α_6, 6 * 56000, SDS-PAGE [15,37]) [15, 37]

homohexamer <15> (<15> α_6, 6 * 43000, SDS-PAGE [17]) [17]

homohexamer <16, 25, 37, 45> (<16,25,37,45> α_6, 6 * 48000, SDS-PAGE [18,28,35,41]) [18, 28, 35, 41]

homohexamer <2> (<2> α_6, 6 * 45000, SDS-PAGE [21]) [21]

homohexamer <2> (<2> α_6, 6 * 45000, SDS-PAGE, expressed in Escherichia coli [2]) [2]

homohexamer <46> (<46> α_6, 6 * 40000, SDS-PAGE [42]) [42]

homohexamer <48> (<48> α_6, 6 * 47000, SDS-PAGE [44]) [44]

homohexamer <5> (<5> α_6, 6 * 49285, amino acid analysis [6]) [6]

homohexamer <6> (<6> α_6, 6 * 44500, SDS-PAGE [25]) [25]

homohexamer <6, 40> (<6,40> α_6, 6 * 50000, SDS-PAGE [39,49]) [39, 49]

homohexamer <7> (<7> α_6, 6 * ?, crystallization [8]) [8]

homohexamer <8> (<8> α_6, 6 * 47300, SDS-PAGE [9]) [9]

homohexamer <9> (<9> α_6, 6 * 46500, SDS-PAGE [10]) [10]

homohexamer <9> (<9> α_6, 6 * 50120, MALDI⁻TOF [10]) [10]

homotetramer <22> (<22> α_4, 4 * 47000, SDS-PAGE [24]) [24]

homotetramer <26> (<26> α_4, 4 * 49000, SDS-PAGE [29]) [29]

homotetramer <38> (<38> α_4, 4 * 46000, SDS-PAGE [36]) [36]

homotetramer <49> (<49> α_4, 4 * 53500, SDS-PAGE [45]) [45]

homotetramer <50> (<50> 4 * 64000, SDS-PAGE [46]) [46]

5 Isolation/Preparation/Mutation/Application

Localization

chloroplast <36> [34]

nucleus <4> [47]

soluble <28> [32]

Purification

<1> [7]

<2> [2, 21]

<3> (under semianoxically conditions [3]) [3, 11]
<4> [4]
<5> [6]
<6> [25, 48]
<9> [10]
<10> [12]
<11> [13]
<12> [14]
<13> [15]
<14> (co-purification with glutathione disulfide reductase [16]) [16]
<15> [17]
<21> [51]
<22> [24]
<23> [26]
<26> [29]
<27> [31]
<28> (partially [32]) [32]
<36> [34]
<37> [35]
<38> [36]
<40> [39]
<46> [42]
<47> [43]
<50> [46]
<51> [50]

Crystallization

<6> (three non-isomorphous crystal forms, all belong to orthorhombic system, homohexamers, one grown from ammonium sulfate, two from L-glutamate, 3.0 A resolution [22]) [22]
<7> (quadruple mutant S128R/T158E/N117R/S160E, homohexamer, 2.9 A resolution [8]) [8]
<40> [39]
<47> (dialysis against ammonium sulfate solution [43]) [43]

Cloning

<1> (expression in Escherichia coli JM109 [7]) [7]
<17, 18> (expression in Escherichia coli, hybrid proteins containing the Pyrococcus furiosus glutamate and the Clostridium difficile cofactor binding domain with reduced substrate binding affinity [19]) [19]
<2, 8> (expression in Escherichia coli BL21 [2,9]) [2, 9]

Engineering

E158Q <8> (<8> 3.3% as active as wild-type [9]) [9]
N117R <7> (<7> 80% activity of wild-type at optimum temperature for catalysis [8]) [8]
S128R <7> (<7> same activity as wild-type at optimum temperature for catalysis [8]) [8]

S128R/T158E <7> (<7> 120% activity of wild-type at optimum temperature for catalysis [8]) [8]
S128R/T158E/N117R <7> (<7> same activity as wild-type at optimum temperature for catalysis [8]) [8]
S128R/T158E/N117R/S160E <7> (<7> same activity as wild-type at optimum temperature for catalysis [8]) [8]
S128R/T158E/S160E <7> (<7> same activity as wild-type at optimum temperature for catalysis [8]) [8]
T138E <8> (<8> 1.6% as active as wild-type [9]) [9]
T158E <7> (<7> 60% activity of wild-type at optimum temperature for catalysis [8]) [8]

6 Stability

pH-Stability
4.7 <51> (<51> inactivation [50]) [50]
5.2-8 <10> [12]
5.8-9 <51> (<51> highly thermostable [50]) [50]
6-9 <47> (<47> unstable below pH 5.0 or above pH 10.0 [43]) [43]

Temperature stability
0 <6> (<6> inactivated by freezing [49]) [49]
20 <45> (<45> stable for 15 h [41]) [41]
25 <16> (<16> 12% activity loss after 1 h [18]) [18]
37 <16> (<16> 72% activity loss after 1 h [18]) [18]
45 <10> (<10> maximal temperature for activity, 60 min stable [12]) [12]
50 <5, 10, 47, 21> (<5,47> irreversible heat inactivation [6,43]; <10> 10 min stable [12]; <21> 5 min stable [51]) [6, 12, 43, 51]
55 <16, 25> (<16> complete activity loss after 1 h [18]; <25> 9% activity loss after 5 min [28]) [18, 28]
57 <28> (<28> after 1 h stable [32]) [32]
60 <12> (<12> irreversible inactivation [14]) [14]
61 <6> (<6> 50% activity after 5 min [6]) [6]
65 <49, 21> (<49> at 4.3 M NaCl stable for several h [45]; <21> complete inactivation [51]) [45, 51]
66 <5> (<5> 50% activity after 5 min [6]) [6]
70 <25, 39, 40, 51> (<25> complete inactivation [28]; <39> after 25 min 10% activity [37]; <40> after 10 min in phosphate buffer pH 7.4 100% activity [39]; <51> after 300 min 60% activity [50]) [28, 37, 39, 50]
72 <27> (<27> complete inactivation [31]) [31]
75 <39> (<39> after 2 min 50% activity [37]) [37]
80 <15> (<15> 50% activity after 4 h [17]) [17]
85 <1, 2, 7> (<2> wild-type enzyme retains 100% activity after 3 h of incubation [2]; <2> recombinant enzyme retains 90% activity after 3 h of incubation [2]; <1> recombinant enzyme fully active after 30 min [7]; <7> quadru-

ple mutant 50% activity after 5 min [8]; <7> wild-type 50% activity after 209 min [8]; <2> no activity loss after 5 h [21]) [2, 7, 8]

90 <3, 15, 22> (<3> fully active after 5 h [3]; <15> 50% activity after 1 h [17]; <22> 50% activity after 12.5 h in 0.1 M NaH_2PO_4/Na_2HPO_4 [24]) [3, 17, 24]

95 <1> (<1> fully active after 30 min [7]) [7]

98 <2> (<2> recombinant enzyme retains 50% activity after 1.5 h of incubation [2]) [2]

98 <2> (<2> wild-type enzyme retains 50% activity after 6.4 h of incubation [2]; <2> 50% activity after 2 h [21]) [2, 21]

100 <1, 3, 8> (<1> 74% activity after 5 h [5]; <3> 50% activity after 140 min [3,11]; <1> fully active after 30 min [7]; <8> wild-type and T138E 50% activity after 2 h and 3 h respectively [9]) [3, 5, 7, 9, 11]

102 <3> (<3> 50% activity after 55 min [11]) [11]

105 <3, 17, 20> (<3> 50% activity after 20 min [3,11]; <17,20> full activity after 10 min, 50% activity after 90 min [23]) [3, 11, 23]

107 <3> (<3> 50% activity after 10 min [11]) [11]

115 <1, 17, 20> (<1,17,20> inactivation after 10 min [7,23]) [7, 23]

General stability information

<1>, 1 mg/ml bovine serum albumin stabilizes [5]

<3>, K_2HPO_4 enhances the thermostability optimally at 1 mM concentration [11]

<3>, potassium phosphate enhances the thermostability [3]

<6>, inactivated by freezing/thawing [49]

<11>, NaCl and KCl markedly increase the thermostability [13]

<22>, 10 mM MOPS-NaOH pH 7.1 and 0.3-0.4 M NaCl is the best buffer [24]

<27>, unstable when frozen at -20°C after a few h even with addition of 50% v/v glycerol [31]

<36>, purification under anoxically conditions, 2 mM dithiothreitol stabilizes [34]

<39>, inactivated by freezing, less than 5% activity after one freeze/thaw cycle [37]

<44>, 50% v/v glycerol and or bovine serum albumin in buffer enhances the stability, inactivated by freezing/thawing [40]

<45>, 2-mercaptoethanol and dithiothreitol stabilizes [41]

<49>, in absence of salt it is irreversibly inactivated, optimal NaCl concentration is 1.1 M [45]

Storage stability

<1>, 4°C, 2 months, no activity loss [7]

<5>, -20°C, 40% v/v glycerol, no activity loss [6]

<6>, 4°C, 0.1 M phosphate buffer, several days, no loss of activity [48]

<6>, 4°C, water, 20 mM potassium phosphate or sodium phosphate, pH 7.2, at least 3 months [49]

<6>, room temperature, several months stable [49]

<11>, 25°C, 20% v/v glycerol, 476 h, 50% activity [13]

<11>, fully active at 4 M concentration NaCl at room temperature for months [13]

<12>, -20°C, phosphate buffer at pH 7.0, 0.2 mM 2-mercaptoethanol, 20 h, complete activity loss [14]

<12>, 5°C, phosphate buffer at pH 7.0, 2 mM 2-mercaptoethanol, 6 months, no activity loss [14]

<26>, 4°C and -25°C, 0.2 M Tris/HCl buffer, pH 7.4, 24 h or 2 months respectively, 100% activity [29]

<38>, 0-4°C, Tris-HCl buffer, pH 9.0, 2 months, 100% activity [36]

<39>, 4°C after 7 days, 55% activity [37]

<40>, -20°C, potassium phosphate buffer, pH 7.4, containing 0.01% 2-mercaptoethanol and 10% glycerol, 100% activity [39]

<44>, 0-4°C, after 15 or 30 days, 18% or 2% activity [40]

<45>, 4°C, ten days and 25°C, 15 h, both 100% activity [41]

<17, 20>, 4°C, 10 mM potassium phosphate buffer, pH 7.2, 0.1 mM EDTA, 0.1 mM DTT, 5 months, no activity loss [23]

References

[1] Bhuiya, M.W.; Sakuraba, H.; Ohshima, T.: Temperature dependence of kinetic parameters for hyperthermophilic glutamate dehydrogenase from Aeropyrum pernix K1. Biosci. Biotechnol. Biochem., **66**, 873-876 (2002)

[2] Robb, F.T.; Maeder, D.L.; DiRuggiero, J.; Borges, K.M.; Tolliday, N.: Glutamate dehydrogenases from hyperthermophiles. Methods Enzymol., **331**, 26-41 (2001)

[3] Steen, I.H.; Hvoslef, H.; Lien, T.; Birkeland, N.K.: Isocitrate dehydrogenase, malate dehydrogenase, and glutamate dehydrogenase from Archaeoglobus fulgidus. Methods Enzymol., **331**, 13-26 (2001)

[4] DeLuna, A.; Avendano, A.; Riego, L.; Gonzalez, A.: NADP-glutamate dehydrogenase isoenzymes of Saccharomyces cerevisiae. Purification, kinetic properties, and physiological roles. J. Biol. Chem., **276**, 43775-43783 (2001)

[5] Helianti, I.; Morita, Y.; Yamamura, A.; Murakami, Y.; Yokoyama, K.; Tamiya, E.: Characterization of native glutamate dehydrogenase from an aerobic hyperthermophilic archaeon Aeropyrum pernix K1. Appl. Microbiol. Biotechnol., **56**, 388-394 (2001)

[6] Di Fraia, R.; Wilquet, V.; Ciardiello, M.A.; Carratore, V.; Antignani, A.; Camardella, L.; Glansdorff, N.; Di Prisco, G.: NADP⁺-dependent glutamate dehydrogenase in the antarctic psychrotolerant bacterium Psychrobacter sp. TAD1. Characterization, protein and DNA sequence, and relationship to other glutamate dehydrogenases. Eur. J. Biochem., **267**, 121-131 (2000)

[7] Bhuiya, M.W.; Sakuraba, H.; Kujo, C.; Nunoura-Kominato, N.; Kawarabayasi, Y.; Kikuchi, H.; Ohshima, T.: Glutamate dehydrogenase from the aerobic hyperthermophilic archaeon Aeropyrum pernix K1: enzymatic characterization, identification of the encoding gene, and phylogenetic implications. Extremophiles, **4**, 333-341 (2000)

[8] Lebbink, J.H.; Knapp, S.; van der Oost, J.; Rice, D.; Ladenstein, R.; de Vos, W.M.: Engineering activity and stability of Thermotoga maritima glutamate

dehydrogenase. II: construction of a 16-residue ion-pair network at the subunit interface. J. Mol. Biol., **289**, 357-369 (1999)

[9] Rahman, R.N.Z.A.; Fujiwara, S.; Nakamura, H.; Takagi, M.; Imanaka, T.: Ion pairs involved in maintaining a thermostable structure of glutamate dehydrogenase from a hyperthermophilic archaeon. Biochem. Biophys. Res. Commun., **248**, 920-926 (1998)

[10] Park, J.H.; Schofield, P.J.; Edwards, M.R.: Giardia intestinalis: characterization of a NADP-dependent glutamate dehydrogenase. Exp. Parasitol., **88**, 131-138 (1998)

[11] Aalen, N.; Steen, I.H.; Birkeland, N.K.; Lien, T.: Purification and properties of an extremely thermostable NADP⁺-specific glutamate dehydrogenase from Archaeoglobus fulgidus. Arch. Microbiol., **168**, 536-539 (1997)

[12] Inokuchi, R.; Itagaki, T.; Wiskich, J.T.; Nakayama, K.; Okada, M.: An NADP-glutamate dehydrogenase from the green alga Bryopsis maxima. Purification and properties. Plant Cell Physiol., **38**, 327-335 (1997)

[13] Ferrer, J.; Perez-Pomares, F.; Bonete, M.J.: NADP-glutamate dehydrogenase from the halophilic archaeon Haloferax mediterranei: enzyme purification, N-terminal sequence and stability studies. FEMS Microbiol. Lett., **141**, 59-63 (1996)

[14] Koike, K.; Hakamada, Y.; Yoshimatsu, T.; Kobayashi, T.; Ito, S.: NADP-specific glutamate dehydrogenase from alkalophilic Bacillus sp. KSM-635: purification and enzymic properties. Biosci. Biotechnol. Biochem., **60**, 1764-1767 (1996)

[15] Bogati, M.S.; Pocsi, I.; Maticsek, J.; Boross, P.; Tozser, J.; Szentirmai, A.: NADP-specific glutamate dehydrogenase of Penicillium chrysogenum has a homohexamer structure. J. Basic Microbiol., **36**, 371-375. (1996)

[16] Krauth-Siegel, R.L.; Mueller, J.G.; Lottspeich, F.; Schirmer, R.H.: Glutathione reductase and glutamate dehydrogenase of Plasmodium falciparum, the causative agent of tropical malaria. Eur. J. Biochem., **235**, 345-350 (1996)

[17] Kobayashi, T.; Higuchi, S.; Kimura, K.; Kudo, T.; Horikoshi, K.: Properties of glutamate dehydrogenase and its involvement in alanine production in a hyperthermophilic archaeon, Thermococcus profundus. J. Biochem., **118**, 587-592 (1995)

[18] Baars, J.J.P.; Op den Camp, H.J.M.; van Hoek, A.H.A.M.; van der Drift, C.; Van Griensven, L.J.L.D.; Visser, J.; Vogels, G.D.: Purification and characterization of NADP-dependent glutamate dehydrogenase from the commercial mushroom Agaricus bisporus. Curr. Microbiol., **30**, 211-217 (1995)

[19] Lebbink, J.H.G.; Eggen, R.I.L.; Geerling, A.C.M.; Consalvi, V.; Chiaraluce, R.; Scandurra, R.; de Vos, W.M.: Exchange of domains of glutamate dehydrogenase from the hyperthermophilic archaeon Pyrococcus furiosus and the mesophilic bacterium Clostridium difficile: Effects on catalysis, thermoactivity and stability. Protein Eng., **8**, 1287-1294 (1995)

[20] Perysinakis, A.; Kinghorn, J.R.; Drainas, C.: Biochemical and genetical studies of NADP-specific glutamate dehydrogenase in the fission yeast Schizosaccharomyces pombe. Curr. Genet., **26**, 315-320 (1994)

[21] Ma, K.; Robb, F.T.; Adams, M.W.: Purification and characterization of NADP-specific alcohol dehydrogenase and glutamate dehydrogenase from

the hyperthermophilic archaeon Thermococcus litoralis. Appl. Environ. Microbiol., **60**, 562-568 (1994)

[22] Korber, F.C.; Rizkallah, P.J.; Attwood, T.K.; Wootton, J.C.; McPherson, M.J.; North, A.C.; Geddes, A.J.; Abeysinghe, I.S.; Baker, P.J.; Dean, J.L.; et al.: Crystallization of the NADP(+)-dependent glutamate dehydrogenase from Escherichia coli. J. Mol. Biol., **234**, 1270-1273. (1993)

[23] Ohshima, T.; Nishida, N.: Purification and properties of extremely thermostable glutamate dehydrogenase from two hyperthermophilic archaebacteria, Pyrococcus woesei and Pyrococcus furiosus. Biosci. Biotechnol. Biochem., **57**, 945-951 (1993)

[24] Hudson, R.C.; Ruttersmith, L.D.; Daniel, R.M.: Glutamate dehydrogenase from the extremely thermophilic archaebacterial isolate AN1. Biochim. Biophys. Acta, **1202**, 244-250 (1993)

[25] Lin, H.P.P.; Reeves, H.C.: Purification and characterization of NADP⁺-specific glutamate dehydrogenase from Escherichia coli. Curr. Microbiol., **22**, 371-376 (1991)

[26] Pamula, F.; Wheldrake, J.F.: Purification and properties of the NADP-dependent glutamate dehydrogenase from Dictyostelium discoideum. Mol. Cell. Biochem., **105**, 85-92 (1991)

[27] Bonete, M.J.; Camacho, M.L.; Candenas, E.: Analysis of the kinetic mechanism of halophilic NADP-dependent glutamate dehydrogenase. Biochim. Biophys. Acta, **1041**, 305-310 (1990)

[28] Op den Camp, H.J.M.; Liem, K.D.; Meesters, P.; Hermans, J.M.H.; van der Drift, C.: Purification and characterization of the NADP-dependent glutamate dehydrogenase from Bacillus fastidiosus. Antonie Leeuwenhoek, **55**, 303-311 (1989)

[29] Vancurova, I.; Vancura, A.; Volc, J.; Kopecky, J.; Neuzil, J.; Basarova, G.; Behal, V.: Purification and properties of NADP-dependent glutamate dehydrogenase from Streptomyces fradiae. J. Gen. Microbiol., **135**, 3311-3318 (1989)

[30] Lilley, K.S.; Hornby, D.P.; Engel, P.C.: Purification, chemical modification and analysis of pH-dependence of the NADP⁺-dependent glutamate dehydrogenase of Escherichia coli. Biochem. Soc. Trans., **16**, 876-877 (1988)

[31] Martinez-Bilbao, M.; Martinez, A.; Urkijo, I.; Llama, M.J.; Serra, J.L.: Induction, isolation, and some properties of the NADPH-dependent glutamate dehydrogenase from the nonheterocystous cyanobacterium Phormidium laminosum. J. Bacteriol., **170**, 4897-4902 (1988)

[32] Turner, A.C.; Lushbaugh, W.B.: Trichomonas vaginalis: characterization of its glutamate dehydrogenase. Exp. Parasitol., **67**, 47-53 (1988)

[33] Hatanaka, M.; Tachiki, T.; Kumagai, H.; Tochikura, T.: Distribution and some properties of glutamine synthetase and glutamate dehydrogenase in bifidobacteria. Agric. Biol. Chem., **51**, 251-252 (1987)

[34] Bascomb, N.F.; Schmidt, R.R.: Purification and partial kinetic and physical characterization of two chloroplast-localized NADP-specific glutamate dehydrogenase isoenzymes and their preferential accumulation in Chlorella sorokiniana cells cultured at low or high ammonium levels. Plant Physiol., **83**, 75-84 (1987)

[35] Weining, S.; Nicholas, D.J.D.: Purification and properties of NADP$^+$-dependent glutamate dehydrogenase from Nitrobacter hamburgensis strain X14. Phytochemistry, **26**, 2151-2153 (1987)

[36] Florencio, F.J.; Marques, S.; Candau, P.: Identification and characterization of a glutamate dehydrogenase in the unicellular cyanobacterium Synechocystis PCC 6803. FEBS Lett., **223**, 37-41 (1987)

[37] Catmull, J.; Yellowlees, D.; Miller, D.J.: NADP$^+$-dependent glutamate dehydrogenase from Acropora formosa. Mar. Biol., **95**, 559-563 (1987)

[38] Cioni, P.; Cavaliere, M.G.; Fissi, A.; Balestreri, E.; Felicioli, R.: Glutamate dehydrogenase from E. coli kinetic structural and regulation studies. Ital. J. Biochem., **35**, 175A-177A (1986)

[39] Misono, H.; Goto, N.; Nagasaki, S.: Purification, crystallization and properties of NADP$^+$-specific glutamate dehydrogenase from Lactobacillus fermentum. Agric. Biol. Chem., **49**, 117-123 (1985)

[40] Agrawal, A.K.; Rao, V.K.M.: Kinetic studies on glutamate dehydrogenase of Aspergillus ochraceus. Indian J. Biochem. Biophys., **21**, 386-388 (1984)

[41] Botton, B.; Msatef, Y.: Purification and properties of NADP-dependent glutamate dehydrogenase from Sphaerostilbe repens. Physiol. Plant., **59**, 438-444 (1983)

[42] Sarada, K.V.; Rao, N.A.; Venkitasubramanian, T.A.: Isolation and characterisation of glutamate dehydrogenase from Mycobacterium smegmatis CDC 46. Biochim. Biophys. Acta, **615**, 299-308 (1980)

[43] Shimizu, H.; Kuratsu, T.; Hirata, F.: Purification and some properties of glutamate dehydrogenase from Proteus inconstans. J. Ferment. Technol., **57**, 428-433 (1979)

[44] Hemmilä, I.A.; Mäntsälä, P.I.: Purification and properties of glutamate synthase and glutamate dehydrogenase from Bacillus megaterium. Biochem. J., **173**, 45-52 (1978)

[45] Leicht, W.; Werber, M.M.; Eisenberg, H.: Purification and characterization of glutamate dehydrogenase from Halobacterium of the Dead Sea. Biochemistry, **17**, 4004-4010 (1978)

[46] Juan, S.M.; Segura, E.L.; Cazzulo, J.J.: Purification and some properties of the NADP-linked glutamate dehydrogenase from Trypanosoma cruzi. Int. J. Biochem., **9**, 395-400 (1978)

[47] Camardella, L.; di Prisco, G.; Garofano, F.; Guerrini, A.M.: Purification and properties of NADP-dependent glutamate dehydrogenase from yeast nuclear fractions. Biochim. Biophys. Acta, **429**, 324-330 (1976)

[48] Veronese, F.M.; Boccu, E.; Conventi, L.: Glutamate dehydrogenase from Escherichia coli: induction, purification and properties of the enzyme. Biochim. Biophys. Acta, **377**, 217-228 (1975)

[49] Sakamoto, N.; Kotre, A.M.; Savageau, M.A.: Glutamate dehydrogenase from Escherichia coli: purification and properties. J. Bacteriol., **124**, 775-783 (1975)

[50] Epstein, I.; Grossowicz, N.: Purification and properties of glutamate dehydrogenase from a thermophilic bacillus. J. Bacteriol., **122**, 1257-1264 (1975)

[51] Coulton, J.W.; Kapoor, M.: Purification and some properties of the glutamate dehydrogenase of Salmonella typhimurium. Can. J. Microbiol., **19**,
427-438 (1973)

[52] Coulton, J.W.; Kapoor, M.: Studies on the kinetics and regulation of glutamate dehydrogenase of Salmonella typhimurium. Can. J. Microbiol., **19**,
439-450 (1973)

L-Amino-acid dehydrogenase

1 Nomenclature

EC number
1.4.1.5

Systematic name
L-amino-acid:NAD$^+$ oxidoreductase (deaminating)

Recommended name
L-amino-acid dehydrogenase

Synonyms
dehydrogenase, L-amino acid

CAS registry number
9029-13-4

2 Source Organism

<1> *Bacillus subtilis* [1]

3 Reaction and Specificity

Catalyzed reaction
an L-amino acid + H$_2$O + NAD$^+$ = a 2-oxo acid + NH$_3$ + NADH + H$^+$

Reaction type
oxidative deamination
redox reaction

Natural substrates and products
S   L-amino acid + H$_2$O + NAD$^+$ <1> [1]

Substrates and products
S   L-amino acid + H$_2$O + NAD$^+$ <1> (aliphatic amino acids, e.g. L-valine, L-leucine, L-isoleucine) [1]
P   2-oxo acid + NH$_3$ + NADH

Inhibitors
Zn^{2+} <1> [1]

Cofactors/prosthetic groups
NAD$^+$ <1> (cannot be replaced by NADP$^+$) [1]

K$_m$-Value (mM)
0.16 <1> (NAD$^+$, (+ L-isoleucine)) [1]
2.2 <1> (L-isoleucine) [1]
25 <1> (L-valine) [1]
33 <1> (NAD$^+$, (+ L-valine)) [1]

pH-Optimum
10.7 <1> (L-valine, L-isoleucine) [1]

Temperature optimum (°C)
37 <1> (assay at) [1]

4 Enzyme Structure

Molecular weight
290000 <1> (gel filtration) [1]

5 Isolation/Preparation/Mutation/Application

Source/tissue
spore <1> (<1>, resting [1]) [1]

Purification
<1> (partial) [1]

6 Stability

Temperature stability
20 <1> (10 min, 25-35% loss of activity) [1]

References

[1] Nitta, Y.; Yasuda, Y.; Tochikubo, K.; Hachisuka, Y.: L-Amino acid dehydrogenases in Bacillus subtilis spores. J. Bacteriol., **117**, 588-592 (1974)

D-Proline reductase 1.4.1.6

1 Nomenclature

EC number
1.4.1.6 (deleted, included in EC 1.4.4.1)

Recommended name
D-proline reductase

Serine 2-dehydrogenase 1.4.1.7

1 Nomenclature

EC number
1.4.1.7

Systematic name
L-serine:NAD$^+$ 2-oxidoreductase (deaminating)

Recommended name
serine 2-dehydrogenase

Synonyms
dehydrogenase, serine
serine dehydrogenase

CAS registry number
9038-55-5

2 Source Organism

<-13> no activity in *Arthrobacter globiformis* (IFO 12956 [2]) [2]
<-12> no activity in *Brevibacterium linens* (IFO 12142 [2]) [2]
<-11> no activity in *Brevibacterium ammoniagenes* (IFO 12072 [2]) [2]
<-10> no activity in *Corynebacterium glutamicum* (ATCC 13032 [2]) [2]
<-9> no activity in *Micrococcus luteus* (IFO 12708 [2]) [2]
<-8> no activity in *Micrococcus roseus* (IFO 3764 [2]) [2]
<-7> no activity in *Bacillus licheniformis* (IFO 12200 [2]) [2]
<-6> no activity in *Bacillus subtilis* (IFO 3022 [2]) [2]
<-5> no activity in *Bacillus meganterium* (ICR 1340 [2]) [2]
<-4> no activity in *Bacillus cereus* (IFO 3001 [2]) [2]
<-3> no activity in *Achromobacter polymorph* (ICR 0880 [2]) [2]
<-2> no activity in *Flavobacterium lutescens* (IFO 3084 [2]) [2]
<-1> no activity in *Serratia plumuthicum* (ICR 0520 [2]) [2]
<1> *Agrobacterium tumefaciens* (ICR 1600 [1,2]; IFO 3058 [2]) [1, 2]
<2> *Escherichia coli* (JM109 [1]; Crookes ICR 0010 [2]) [1, 2]
<3> *Klebsiella pneumoniae* (IFO 12059 [2]) [2]
<4> *Erwinia carotovora* (IFO 3830 [2]) [2]
<5> *Proteus vulgaris* (IFO 3167 [2]) [2]
<6> *Proteus rettgeri* (ICR 0650 [2]) [2]
<7> *Alcaligenes faecalis* (ICR 0810 [2]) [2]
<8> *Bacillus sphaericus* (IFO 3525 [2]) [2]

<9> *Pseudomonas aereofaciens* (IFO 3521 [2]) [2]
<10> *Agrobacterium radiobacter* (IFO 12664 [2]) [2]
<11> *Corynebacterium pseudodiphtheriticum* (ICR 2210 [2]) [2]
<12> *Bacterium mycoides* (ICR 2410 [2]) [2]
<13> *Hafnia alvei* (IFO 3731 [2]) [2]
<14> *Pseudomonas fragi* (IFO 3458 [2]) [2]
<15> *Pseudomonas aeruginosa* (IFO 3080 [2]) [2]
<16> *Pseudomonas fluorescens* (IFO 3081 [2]) [2]
<17> *Pseudomonas putida* (IFO 3738 [2]) [2]

3 Reaction and Specificity

Catalyzed reaction

L-serine + H_2O + NAD^+ = 3-hydroxypyruvate + NH_3 + NADH + H^+ (<1> belongs to short-chain alcohol dehydrogenase family [1])

Reaction type

oxidation
oxidative deamination
redox reaction
reduction
reductive amination

Natural substrates and products

S L-serine + $NADP^+$ <1> (<1> oxidation of 3-hydroxyl group of serine [1,2]) (Reversibility: ? <1> [1]; r <1> [2]) [1, 2]
P 2-aminomalonate-semialdehyde + NADPH <1> (<1> ammonia is not formed, product is spontaneously converted into 2-aminoacetaldehyde [2]) [1, 2]

Substrates and products

S L-serine + $NADP^+$ <1> (<1> oxidation of 3-hydroxyl group of serine [1,2]) (Reversibility: ? <1> [1]; r <1> [2]) [1, 2]
P 2-aminomalonate-semialdehyde + NADPH <1> (<1> ammonia is not formed, product is spontaneously converted into 2-aminoacetaldehyde [2]) [1, 2]
S Additional information <1> (<1> L-serine, D-serine, L-glycerate, D-glycerate, 2-methyl-DL-serine, L-3-hydroxyisobutyrate, D-3-hydroxyisobutyrate and 3-hydroxypropionate are poor substrates, o-methyl-DL-serine and L-threonine no substrates [2]) [2]
P ?

Inhibitors

2-methylmalonate <1> (<1> 50% inhibition at 50 mM [2]) [2]
D-cysteine <1> (<1> 71% inhibition at 10 mM [2]) [2]
$HgCl_2$ <1> [2]
L-cysteine <1> (<1> complete inhibition at 10 mM [2]) [2]
malonate <1> (<1> 50% inhibition at 50 mM [2]) [2]

p-chloromercuribenzoate <1> [2]

tartronate <1> (<1> 65% inhibition at 50 mM [2]) [2]

Cofactors/prosthetic groups

$NADP^+$ <1> [1, 2]

NADPH <1> [2]

Additional information <1> (<1> NAD^+ is inert [2]) [2]

Metals, ions

Additional information <1> (<1> enzyme is not inhibited by metal-chelating reagents and does not require metal ions for its activity [2]) [2]

Specific activity (U/mg)

0.00009 <10> [2]

0.0001 <6> [2]

0.00013 <11> [2]

0.0002 <7, 8> [2]

0.00023 <13> [2]

0.00023 <17> [2]

0.00028 <16> [2]

0.00073 <12> [2]

0.00076 <14> [2]

0.00115 <2> [2]

0.0013 <2, 9> [1, 2]

0.00134 <1> (<1> IFO 3058 [2]) [2]

0.00147 <15> [2]

0.00163 <5> [2]

0.00213 <4> [2]

0.0024 <3> [2]

0.00357 <1> (<1> ICR 1600 [2]) [2]

0.0433 <1> (<1> cell extract, recombinant protein [1]) [1]

45.3 <1> [2]

K_m-Value (mM)

0.029 <1> ($NADP^+$) [2]

42 <1> (L-serine) [2]

44 <1> (D-serine) [2]

54 <1> (L-glycerate) [2]

56 <1> (D-glycerate) [2]

pH-Optimum

9.1 <1> (<1> oxidation of L-serine [2]) [2]

Temperature optimum (°C)

30 <1> (<1> assay at [2]) [2]

4 Enzyme Structure

Molecular weight
100000 <1> (<1> gel filtration [2]) [1, 2]

Subunits
tetramer <1> (<1> 4 * 25000, SDS-PAGE [2]) [1, 2]

5 Isolation/Preparation/Mutation/Application

Purification
<1> (ammonium sulfate precipitation, DEAE-cellulose column, hydroxyapatite column, 2',5'-ADP-Sepharose 4B column, Mono Q HR 5/5 anion-exchange column [2]) [2]

Cloning
<1> (expressed in Escherichia coli JM109 [1]) [1]

6 Stability

pH-Stability
7-8 <1> [2]
7-10 <1> (<1> 35°C, stable for 10 min [2]) [2]

Temperature stability
35 <1> (<1> up to, stable for 10 min in 10 mM potassium phosphate buffer, pH 7.4, 0.01% 2-mercaptoethanol [2]) [2]

Storage stability
<1>, -20°C, 10 mM Tris-HCl buffer, pH 8.0, 160 mM KCl, 40% glycerol [2]

References

[1] Fujisawa, H.; Nagata, S.; Chowdhury, E.K.; Matsumoto, M.; Misono, H.: Cloning and sequencing of the serine dehydrogenase gene from Agrobacterium tumefaciens. Biosci. Biotechnol. Biochem., **66**, 1137-1139. (2002)
[2] Chowdhury, E.K.; Higuchi, K.; Nagata, S.; Misono, H.: A novel NADP+ dependent serine dehydrogenase from Agrobacterium tumefaciens. Biosci. Biotechnol. Biochem., **61**, 152-157 (1997)

Valine dehydrogenase (NADP⁺)   1.4.1.8

1 Nomenclature

EC number
1.4.1.8

Systematic name
L-valine:NADP⁺ oxidoreductase (deaminating)

Recommended name
valine dehydrogenase (NADP⁺)

Synonyms
VDH
ValDH
dehydrogenase, valine (nicotinamide adenine dinucleotide phosphate)
valine dehydrogenase (nicotinamide adenine dinucleotide phosphate)

CAS registry number
37255-39-3

2 Source Organism

<1> *Streptomyces fradiae* (strain 30/3 [1,10,14]; crude extract [10]; tylosin producer [12,14]; two enzymes: VDH1 and VDH₂ [14]) [1, 10, 12, 14]
<2> *Triticum aestivum* [8, 9]
<3> *Planococcus citreus* (strains HK 709, HK706, HK 707, HK 708; VDH activity detectable [2]) [2]
<4> *Planococcus sp.* (strains HK 701, HK 702, HK 703, HK 704, HK 705; VDH activity detectable [2]) [2]
<5> *Bacillus megaterium* (strain ATCC 39118 [3]) [3]
<6> *Streptomyces coelicolor* (strain A3 [4]; actinorhodin producer [12]) [4, 12]
<7> *Streptomyces aureofaciens* (strain 50/137 [6] ; strain BS-5 [10,13]; crude extract [10]; spiramycin producer [12]; two isoenzymes: VDH1 and VDH2 [13]) [6, 10, 12, 13]
<8> *Pisum sativum* [7-9]
<9> *Streptomyces cinnamonensis* (strains A 3823.5 and ATCC 15413 [5]; monensin producer [12]) [5, 12]
<10> *Glycine max* [8, 9]
<11> *Phaseolus vulgaris* [8, 9]
<12> *Zea mays* [8, 9]
<13> *Plantago sp.* [8, 9]
<14> *Streptomyces albus* (strain ATCC 21838 [12] ; salinomycin producer [11, 12]) [11, 12]

3 Reaction and Specificity

Catalyzed reaction

L-valine + H$_2$O + NADP$^+$ = 3-methyl-2-oxobutanoate + NH$_3$ + NADPH + H$^+$
(<9> kinetic mechanism [5])

Reaction type

amination
deamination
oxidation
redox reaction
reduction

Natural substrates and products

S L-valine + NAD(P)$^+$ + H$_2$O <1, 6-9, 14> (<1> regulatory enzyme involved
in biosynthesis of n-butyrate, a building unit of the oligoketide antibiotic
tylosin, NAD$^+$ as natural cofactor for oxidative deamination [1]; <1, 6, 7,
9> first step in valine catabolism [4, 5, 10, 13]; <14> first catabolic step of
L-valine and some other branched chain L-amino acids, important role in
providing precursors for biosynthesis of polyether antibiotic, salinomycin,
that is produced at a high level [11, 12]; <1, 7> VDH is required for utilization of branched chain amino acids, the catabolism appears to be an
alternative source of n-butyrate, 2-methylmalonate, and propionate
needed for biosynthesis of macrolide and polyether antibiotics [13, 14];
<1, 6, 7, 14> metabolic connection between valine catabolism and biosyn-
thesis of macrolide or polyether antibiotics [4, 10-12]; <1> VDH2 rather
than VDH1 plays a role in metabolism of branched chain amino acids and
thus in tylosin biosynthesis, expression of VDH1 is unstable [14]; <8> two
forms of enzyme activity: enzyme associated with subcellular structures
in shoots or leaves probably plays a biosynthetic role for valine production and the soluble enzyme in roots plays a degradative role in valine
catabolism [8, 9]) (Reversibility: ?<1, 6-9, 14> [1, 4, 5, 8-14]) [1, 4, 5, 8-14]

P 2-oxoisovalerate + NH$_3$ + NAD(P)H <6, 9> [4, 5]

Substrates and products

S L-2-aminobutyrate + H$_2$O + NAD$^+$ <1, 5-7, 9, 14> (<1> 68.9% of the activity with L-valine [1]; <1> VDH2: preferred substrate, 158% of the activity with L-valine [14]; <7> 16.5% of the activity with L-valine [6]; <7>
VDH2: 125% of the activity with L-valine [13]; <6> 130% of the activity
with L-valine [4]; <5> 9% of the activity with L-leucine [3]; <14> preferred substrate, 92.34% of the activity with L-valine [12]; L-2-aminobuty-
rate is identical with L-α-aminobutyrate) (Reversibility: r <1, 5, 7, 9, 14>
[1, 3, 5, 6, 12-14]; ? <5, 6> [3, 4]) [1, 3, 4-6, 12-14]

P 2-oxobutyrate + NADH + NH$_3$ <1, 5, 7, 9, 14> (<1> reductive amination:
78% of the activity with 2-oxoisovalerate [1]; <1> VDH2: reductive amination, 100% of the activity with 2-oxoisovalerate [14]; <5> reductive
amination: 28.1% of the activity with 2-oxoisovalerate [3]; <7> reductive
amination: 48.2% of the activity with 2-oxoisovalerate [6]; <7> VDH2:

reductive amination, 116% of the activity with 2-oxoisovalerate [13]; <14> reductive amination: 91.8% of the activity with 2-oxoisovalerate [12]; 2-oxobutyrate is identical with α-ketobutyrate) [1, 3, 5, 12-14]

S L-alanine + H_2O + NAD⁺ <1, 5-7, 9, 14> (<1> 14.7% of the activity with L-valine [1]; <1> VDH2: 8.3% of the activity with L-valine [14]; <5> 3% of the activity with L-valine [3]; <6> 5% of the activity with L-valine [4]; <7> very poor substrate: 1.2% of the activity with L-valine [6]; <7> VDH2: 4.5% of the activity with L-valine [13]; <14> 13.63% of the activity with L-valine [12]) (Reversibility: r <1, 5-7, 14> [1, 3, 4, 6, 12-14]) [1, 3-6, 12-14]

P pyruvate + NH_3 + NADH <1, 5, 7, 9> (<1> reductive amination: 31.6% of the activity with 2-oxoisovalerate [1]; <1> VDH2: reductive amination, 23.3% of the activity with 2-oxoisovalerate [14]; <5> reductive amination: 2% of the activity with 2-oxoisovalerate [3]; <7> reductive amination: very poor substrate, 1.3% of the activity with 2-oxoisovalerate [6]; <7> VDH2: reductive amination, 20.8% of the activity with 2-oxoisovalerate [13]; pyruvate is identical with α-ketopropanoate and 2-oxopropanoate) [1, 3, 5, 6, 13, 14]

S L-cysteine + H_2O + NAD⁺ <6, 14> (<6> 3% of the activity with L-valine [4]; <14> 3.75% of the activity with L-valine, A124G mutant: 0.063% [11]; <14> 1.87% of the activity with L-valine [12]) (Reversibility: r <14> [11, 12]; ? <6> [4]) [4, 11, 12]

P 3-mercapto-2-oxopropanoate + NH_3 + NADH

S L-isoleucine + H_2O + NAD⁺ <1, 5-9, 14> (<1> 29.4% of the activity with L-valine [1]; <1> VDH2: 22% of the activity with L-valine [14]; <6> 28% of the activity with L-valine [4]; <5> 61.3% of the activity with L-leucine [3]; <7> 46.5% of the activity with L-valine [6]; <7> VDH2: 34% of the activity with L-valine [13]; <14> 27.5% of the activity with L-valine, A124G mutant: 9.98% [11]; <14> 12.46% of the activity with L-valine [12]) (Reversibility: r <1, 5-7, 9, 14> [1, 3, 5, 6, 11-14]; ? <6> [4]) [1, 3-7, 9, 11-14]

P 2-oxo-3-methylvalerate + NADH + NH_3 <1, 5, 7, 8, 14> (<1> reductive amination: 23.2% of the activity with 2-oxoisovalerate [1]; <1> VDH2: reductive amination, 16.6% of the activity with 2-oxoisovalerate [14]; <5> reductive amination: 21.9% of the activity with 2-oxoisovalerate [3]; <7> reductive amination: 22.3% of the activity with 2-oxoisovalerate [6]; <7> VDH2: reductive amination, 24.5% of the activity with 2-oxoisovalerate [13]; <8> reductive amination: 59% of the activity of 2-oxoisovalerate, but no oxidative deamination of L-isoleucine [7,9]; <14> reductive amination: 35.6% of the activity with 2-oxoisovalerate [12]; 2-oxo-3-methylvalerate is identical with α-keto-β-methylpentanoate) [1, 3, 6, 7, 9, 12-14]

S L-leucine + H_2O + NAD⁺ <1, 5-7, 9, 14> (<1> 24.5% of the activity with L-valine [1]; <1> VDH2: 24% of the activity with L-valine [14]; <5> preferred substrate [3]; <7> 36% of the activity with L-valine [6]; <7> VDH2: 52.8% of the activity with L-valine [13]; <6> 8% of the activity with L-valine [4]; <14> 15% of the activity with L-valine, A124G mutant: 5.03%

[11]; <14> 7.46% of the activity with L-valine [12]) (Reversibility: r <1, 5, 7, 9, 14> [1, 3, 5, 6, 11-14]; ? <6> [4]) [1, 3-6, 11-14]

P   2-oxoisocaproate + NADH + NH$_3$ <1, 5, 7, 9, 14> (<1> reductive amination: 20% of the activity with 2-oxoisovalerate [1]; <1> VDH2: reductive amination, 20% of the activity with 2-oxoisovalerate [14]; <5> reductive amination: 21.1% of the activity with 2-oxoisovalerate [3]; <7> reductive amination: 38.1% of the activity with 2-oxoisovalerate [6]; <7> VDH2: reductive amination, 26.5% of the activity with 2-oxoisovalerate [13]; <14> reductive amination: 26.85% of the activity with 2-oxoisovalerate [12]; 2-oxoisocaproate is identical with 2-oxo-4-methylpentanoate, 2-oxoisohexanoate and α-ketoisocaproate) [1, 3, 5, 6, 12-14]

S   L-methionine + H$_2$O + NAD$^+$ <7> (<7> VDH2: 2% of the activity with L-valine [13]) (Reversibility: r <7> [13]) [13]

P   4-(methylthio)-2-oxobutanoic acid + NH$_3$ + NADH

S   L-norleucine + H$_2$O + NAD$^+$ <1, 5, 7, 9, 14> (<1> 51.5% of the activity with L-valine [1]; <1> VDH2: 34.5% of the activity with L-valine [14]; <5> 4.7% of the activity with L-valine [3]; <7> 10.5% of the activity with L-valine [6]; <7> VDH2: 18.5% of the activity with L-valine [13]; <14> 15% of the activity with L-valine, A124G mutant: 15% [11]) (Reversibility: r <1, 5, 7, 9, 14> [1, 3, 5, 6, 11, 13, 14]) [1, 3, 5, 6, 11, 13, 14]

P   2-oxocaproate + NADH + NH$_3$ <1, 5, 9> (<1> reductive amination: 43.8% of the activity with 2-oxoisovalerate [1]; <5> reductive amination: 9.7% of the activity with 2-oxoisovalerate [3]; 2-oxocaproate is identical with 2-oxohexanoate and α-keto-n-hexanoate) [1, 3, 5]

S   L-norvaline + H$_2$O + NAD$^+$ <1, 5, 7, 14> (<1> 98% of the activity with L-valine [1]; <1> VDH2: 96% of the activity with L-valine [14]; <5> 47.2% of the activity with L-leucine [3]; <7> 43% of the activity with L-valine [6]; <7> VDH2: 100% of the activity with L-valine [13]; <14> 100.5% of the activity of L-valine, A124G mutant: 28.77% [11]) (Reversibility: r <1, 5, 7, 9, 14> [1, 3, 5, 6, 11, 13, 14]) [1, 3, 5, 6, 11, 13, 14]

P   2-oxovalerate + NH$_3$ + NADH <1, 5, 9> (<1> reductive amination: 79.5% of the activity with 2-oxoisovalerate [1]; <5> reductive amination: 37% of the activity with 2-oxoisovalerate [3]; <9> optimal chain length for reductive amination [5]; 2-oxovalerate is identical with 2-oxopentanoate and α-ketovalerate) [1, 3, 5]

S   L-phenylalanine + H$_2$O + NAD$^+$ <14> (<14> 1.25% of the activity with L-valine, A124G mutant: 2.5% [11]; <14> only 0.95% of the activity with L-valine [12]) (Reversibility: r <14> [11, 12]) [11, 12]

P   2-oxo-3-phenylpropanoic acid + NH$_3$ + NADH

S   L-valine + NAD(P)$^+$ + H$_2$O <1, 2, 5-14> (<1,2,8,10-13> NADP$^+$ [1,7-9]; <1,5-7,9,14> NAD$^+$ [1,3-6,11-14]; <6> L-valine oxidized efficiently [4]; <1> reductive amination rate is twice the oxidative deamination rate [14]; <5> 88.9% of the activity with L-leucine [3]; <8> rate of reductive amination of ketoanalogue of valine 5times greater than rate of oxidative deamination of L-valine [7]; <8> soluble enzyme in roots: ratio of rate of reductive amination of valine ketoanalogue to the rate of oxidative deamination of L-valine is 1:1; in the enzyme associated with subcellular structures in shoots it is 2.5:1 [8]; <8> soluble enzyme in roots: ratio of rate of reductive amination of valine ketoanalogue to the rate of oxidative deamination of L-valine is 1:1; in the enzyme associated with subcellular structures in shoots it is 5:1 [9]; <14> A124G mutant: 3.75% of the activity with L-valine in wild type [11]; <1,14> preferred substrate [12,14]) (Reversibility: r <1, 2, 5-14> [1, 3-9, 11-14]; ? <1, 7> [10]) [1, 3-14]

P 2-oxoisovalerate + NH_3 + NAD(P)H <1, 2, 5-14> (<2,8,10-13> NADPH [7,8]; <1,5-7,9,14> NADH [1,3-6,11-14]; <1,5,7,9,14> reductive amination: preferred substrate [1,3,5,12-14]; <7> highest activity with 2-oxoisovalerate [6]; 2-oxoisovalerate is identical with 3-methyl-2-oxobutanoate and α-ketoisovalerate) [1, 3-8, 11-14]

S Additional information <1, 2, 5, 6, 8-14> (<9> pro-S hydrogen at C-4' of NADH is transferred to substrate, B-stereospecific [5]; <5> no activity with the D-isomers of the substrates [3]; <6,14> not: D-valine [4,12]; <1,9> NH_3 is the sole substrate as amino donor [5,14]; <8> no oxidative deamination of L-isoleucine and L-alanine [7,9]; <14> Lys-79, Lys-91 are involved in substrate binding and catalysis [11,12]; <14> Ala-124 is involved in substrate binding, but less of nicotinamide coenzyme [11]) [3-5, 7, 9, 11, 12, 14]

P Additional information <8, 9> (<9> reductive amination: sequential ordered ternary-binary mechanism [5]; <8> no reductive amination of 2-oxobutyrate [7,9]) [5, 7, 9]

Inhibitors

$(NH_4)_2SO_4$ <6, 9> (<9> enzyme is sensitive to high concentrations [5]) [4, 5]

2-oxoisovalerate <5> (<5> 65 mM, 50% loss of specific activity [3]) [3]

Ag^+ <9> (<9> 0.01 mM in assay buffer [5]) [5]

Ca^{2+} <9> (<9> 1 mM, partial inhibition [5]) [5]

$CoCl_2$ <5> [3]

Cu^{2+} <1, 5, 9> (<9> 0.01 mM in assay buffer [5]; <1> VDH2 : 1 mM, 35.5% inhibition [14]; <5> $CuSO_4$: less inhibition compared with $HgCl_2$ and $CoCl_2$ [3]) [3, 5, 14]

DEPC <14> (<14> 20 mM, 60% loss of activity [11]) [11]

Hg+ <9> [5]

Hg^{2+} <1, 5, 7, 9> (<1> $HgCl_2$: strong inhibition, partially abolished in presence of 2-mercaptoethanol [1]; <7> $HgCl_2$: 0.01 mM, 91% loss of activity [6]) [1, 3, 5, 6]

L-valine <1, 5, 6> (<5> 10 mM, about 50% loss of specific activity [3]; <6> substrate inhibition at higher concentrations than 20 mM [4]; <1> VDH2: 20 mM instead of 10 mM in reaction mixture caused 22% loss of activity, 50 mM 40% loss of activity [14]) [3, 4, 14]

Mg^{2+} <9> (<9> 1 mM, partial inhibition [5]) [5]

Mn^{2+} <5, 7> (<5> $MnCl_2$: less inhibition compared with $HgCl_2$ and $CoCl_2$ [3]; <7> VDH2: 1 mM, 60% loss of activity [13]) [3, 13]

N-bromosuccinimide <14> (<14> 0.5 mM, complete inhibition [11]) [11]

Zn^{2+} <9> (<9> 1 mM, partial inhibition [5]) [5]

iodoacetamide <1, 7> (<7> VDH2: 1 mM, 50% loss of activity [13]; <1> VDH2: 1 mM, complete inhibition [14]) [13, 14]

p-chloromercuribenzoate <1, 7, 9, 14> (<1> strong inhibition, partially abolished in presence of 2-mercaptoethanol [1]; <9> 0.01 mM in assay buffer [5]; <7> 0.01 mM, 83% loss of activity [6]; <14> 0.5 mM, complete inhibition [11]) [1, 5, 6, 11]

p-hydroxymercuribenzoate <1, 7> (<7> VDH2: 0.01 mM, complete inhibition [13]; <1> VDH2: 0.01 mM, 70% loss of activity [14]) [13, 14]

phenyl glyoxal <14> (<14> 5 mM, 60% loss of activity [11]) [11]

pyridoxal 5'-phosphate <14> (<14> 40% loss of activity, forms Schiff base with a lysine residue [12]) [12]

urea <9> (<9> 3.5 M, complete and irreversible inactivation [5]) [5]

Additional information <7, 9> (<7,9> no inhibition by nucleotide mono-, di-, triphosphates or the free bases adenine, cytosine, thymine, guanine and uracil [5,6]) [5, 6]

Cofactors/prosthetic groups

1,N^6-etheno-NAD$^+$ <1, 7> (<1> 64.4% of the activity with NAD$^+$ [1]; <1> VDH2: 70.5% of the activity with NAD$^+$ [14]; <7> 58.4% of the activity with NAD$^+$ [6]; <7> VDH2: 62.4% of the activity with NAD$^+$ [13]) [1, 6, 13, 14]

3-acetylpyridine-NAD$^+$ <1, 7> (<1> 64.5% of the activity with NAD$^+$ [1]; <7> 100% of the activity with NAD$^+$ [6]) [1, 6]

3-pyridinealdehyde-NAD$^+$ <1, 7> (<1> VDH2: 47.2% of the activity with NAD$^+$, VDH1: no activity [14]; <7> 37.4% of the activity with NAD$^+$ [6]; <7> VDH2: 37.5% of the activity with NAD$^+$ [13]) [6, 13, 14]

NAD$^+$ <1, 5-7, 9, 14> (<1,7> natural cofactor for oxidative deamination [1,13,14]; <1,6,7,9> NAD$^+$ required, no use of NADP$^+$ [4-6,13,14]) [1, 3-6, 11-14]

NADH <1, 5-7, 9, 14> (<1,5-7,9> NADH can not be replaced by NADPH for the reductive amination [1,3-5,13]) [1, 3-6, 11-14]

NADP$^+$ <1, 2, 7, 8, 10-13> (<2,8,10-13> in plants enzyme is coupled exclusively to NADP$^+$ [7-9]; <1> 10.8% of the activity with NAD$^+$ [1]; <7> only 8.4% of the activity with NAD$^+$ [6]; <8> shoots: enzyme affinity to NADP$^+$ less than to NADPH [9]) [1, 6-9]

NADPH <2, 8, 10-13> (<2,8,10-13> in plants valine dehydrogenase coupled exclusively to NADPH [7-9]) [7-9]

α-NAD$^+$ <1, 7> (<1> 3.6% of the activity with NAD$^+$ [1]; <1> VDH2: 5.8% of the activity with NAD$^+$ [14]; <7> VDH2: 5% of the activity with NAD$^+$, VDH1: no activity [13]) [1, 13, 14]

deamino-NAD$^+$ <1, 7> (<1> 93.9% of the activity with NAD$^+$ [1]; <1> VDH2: 124% of the activity with NAD$^+$; amino group in the adenine moiety of NAD$^+$ is not essential [14]; <7> 83.7% of the activity with NAD$^+$ [6]; <7> VDH2: 375.5% of the activity with NAD$^+$ [13]) [1, 6, 13, 14]

nicotinamide guanine dinucleotide <1, 7> (<1> 92.9% of the activity with NAD$^+$ [1]; <7> 72.3% of the activity with NAD$^+$ [6]; <7> VDH2: 80% of the activity with NAD$^+$ [13]) [1, 6, 13]

Additional information <1, 2, 6-13> (<2,8,10-13> plant enzyme not coupled to NAD⁺ and NADH [8]; <6,7,9> not with NADP⁺ and NADPH [4,5,13]; <7> VDH2: not with 3-acetylpyridine-NAD⁺, compared to VDH1 with 100% of the activity with NAD⁺ [13]; <1> VDH2: not with NADP⁺, compared to VDH1 with 10.8% of the activity with NAD⁺ [14]) [4, 5, 8, 13, 14]

Activating compounds

2-mercaptoethanol <1> (<1> presence of 2-mercaptoethanol or dithiothreitol at 10 mM required for maximal activity [1]) [1]

L-valine <1, 7, 9> (<1,7> L-valine in growth medium induces VDH activity, not interfered by ammonia [10]; <1,7> VDH synthesis induced by L-valine, but severely repressed by ammonia [13,14]; <9> enzyme induced 100fold by L-valine [5]) [5, 10, 13, 14]

dithiothreitol <1> (<1> presence of 2-mercaptoethanol or dithiothreitol at 10 mM required for maximal activity [1]) [1]

Turnover number (min⁻¹)

0.029 <14> (L-valine, <14> K79A mutant [12]) [12]
1.51 <14> (L-valine, <14> K91A mutant [12]) [12]
342 <14> (L-valine, <14> A124G mutant [11]) [11]
345 <9> (L-isoleucine) [5]
433 <14> (2-oxoisovalerate, <14> K79A mutant [12]) [12]
714 <9> (L-leucine) [5]
782 <14> (2-oxoisovalerate, <14> K91A mutant [12]) [12]
1074 <9> (L-norleucine) [5]
1080 <9> (L-norvaline) [5]
1698 <9> (L-valine) [5]
2848 <14> (L-valine) [11, 12]
4314 <9> (2-oxoisocaproate) [5]
5940 <9> (2-oxobutyrate) [5]
6240 <9> (L-2-aminobutyrate) [5]
6420 <9> (2-oxocaproate) [5]
6840 <9> (pyruvate) [5]
9840 <9> (2-oxovalerate) [5]
13560 <9> (2-oxoisovalerate) [5]
18870 <14> (2-oxoisovalerate) [12]

Specific activity (U/mg)

0.00534 <14> (<14> L-phenylalanine [12]) [12]
0.7 <6> (<6> L-alanine [4]) [4]
1.056 <14> (<14> L-cysteine [12]) [12]
1.2 <6> (<6> L-leucine [4]) [4]
3.8 <6> (<6> L-isoleucine [4]) [4]
4.2 <14> (<14> L-leucine [12]) [12]
7.02 <14> (<14> L-isoleucine [12]) [12]
7.68 <14> (<14> L-alanine [12]) [12]
7.8 <5> [3]
14 <6> (<6> L-valine [4]) [4]

19 <6> (<6> L-2-aminobutyrate [4]) [4]
19.56 <1> (<1> oxidative deamination [14]) [14]
19.6 <9> [5]
21.16 <14> (<14> L-valine [11]) [11]
40.92 <1> (<1> reductive amination [14]) [14]
45.01 <7> (<7> NAD⁺ [13]) [13]
52.01 <14> (<14> 2-aminobutyrate [12]) [12]
56.33 <14> (<14> L-valine [12]) [12]
64.31 <7> (<7> oxidative deamination [13]) [13]
81.8 <14> (<14> 2-oxoisocaproate [12]) [12]
107.6 <14> (<14> 2-oxo-3-methylvalerate [12]) [12]
278 <14> (<14> 2-oxobutyrate [12]) [12]
302.4 <14> (<14> 2-oxoisovalerate [12]) [12]
366.5 <7> (<7> reductive amination [13]) [13]
Additional information <1, 2, 7, 8, 10-14> (<2,8,10-13> values for reductive amination at pH 8.5 [8,9]; <1,7> effect of L-valine and ammonium in different growth media on the VDH specific activity in crude extract [10]; <14> activities of wild type and A124G mutant enzymes [11]) [1, 6-11]

Kₘ-Value (mM)

0.022 <14> (NADH, <14> K91A mutant [12]) [12]
0.023 <7> (NADH) [6]
0.028 <1> (NADH, <1> VDH2 [14]) [14]
0.029 <1> (NAD⁺) [1]
0.029 <14> (NADH, <14> K79A mutant [12]) [12]
0.04 <1> (NAD⁺, <1> VDH2 [14]) [14]
0.05 <1, 5> (NADH) [1, 3]
0.0576 <14> (NADH) [12]
0.074 <9> (NADH) [5]
0.093 <6> (NADH) [4]
0.1 <7> (NAD⁺) [6]
0.13 <9> (2-oxobutyrate) [5]
0.13 <14> (NAD⁺) [11, 12]
0.14 <8> (NADPH, <8> shoots [7,9]) [7, 9]
0.17 <6> (NAD⁺) [4]
0.18 <9> (NAD⁺) [5]
0.19 <14> (NAD⁺, <14> K91A mutant [12]) [12]
0.2 <14> (NAD⁺, <14> K79A mutant [12]) [12]
0.235 <8> (NADP⁺, <8> shoots [7,9]) [7, 9]
0.31 <1> (2-oxoisovalerate, <1> VDH2 [14]) [14]
0.43 <1> (L-valine, <1> VDH2 [14]) [14]
0.43 <14> (NAD⁺, <14> A124G mutant [11]) [11]
0.438 <14> (2-oxoisovalerate) [12]
0.46 <7> (2-oxoisovalerate, <7> VDH2 [13]) [13]
0.48 <9> (2-oxovalerate) [5]
0.5 <9> (2 oxoisocaproate) [5]
0.538 <8> (2-oxoisovalerate, <8> shoots [7,9]) [7, 9]

0.6 <6> (2-oxoisovalerate) [4]
0.67 <1> (L-leucine) [1]
0.75 <7> (L-valine, <7> VDH2 [13]) [13]
0.8 <7> (2-oxoisocaproate) [6]
0.8 <1> (2-oxoisovalerate) [1]
0.81 <9> (2-oxocaproate) [5]
0.81 <9> (2-oxoisovalerate) [5]
0.82 <1> (2-oxoisocaproate) [1]
1 <1> (2-oxobutyrate) [1]
1 <1> (L-valine) [1]
1.06 <7> (2-oxo-3-methylvalerate, <7> VDH2 [13]) [13]
1.1 <8> (L-valine, <8> shoots [7,9]) [7, 9]
1.18 <1> (L-isoleucine) [1]
1.2 <7> (2-oxoisocaproate, <7> VDH2 [13]) [13]
1.25 <7> (2-oxoisovalerate) [6]
1.3 <9> (L-valine) [5]
1.33 <1> (L-norvaline) [1]
1.5 <7> (2-oxobutyrate) [6]
1.5 <5> (2-oxoisovalerate) [3]
1.5 <14> (L-valine) [11, 12]
1.66 <7> (L-norvaline, <7> VDH2 [13]) [13]
1.72 <7> (L-isoleucine, <7> VDH2 [13]) [13]
1.85 <7> (L-leucine, <7> VDH2 [13]) [13]
1.94 <1> (2-oxovalerate) [1]
2.08 <14> (L-valine, <14> K91A mutant [12]) [12]
2.23 <1> (2-oxocaproate) [1]
2.5 <7> (L-valine) [6]
3.07 <1> (L-2-aminobutyrate) [1]
3.2 <9> (L-norvaline) [5]
3.33 <1> (2-oxo-3-methylvalerate) [1]
4 <9> (L-leucine) [5]
4.35 <7> (L-2-aminobutyrate, <7> VDH2 [13]) [13]
4.5 <14> (2-oxoisovalerate, <14> K91A mutant [12]) [12]
5 <7> (L-isoleucine) [6]
5.7 <7> (L-norvaline) [6]
6.3 <7> (L-leucine) [6]
6.68 <1> (L-norleucine) [1]
6.7 <7> (2-oxo-3-methylvalerate) [6]
10 <6> (L-valine) [4]
10.8 <9> (L-isoleucine) [5]
12.3 <9> (pyruvate) [5]
14.8 <7> (L-2-aminobutyrate) [6]
15.3 <7> (pyruvate) [6]
15.5 <8> (NH₄⁺, <8> shoots [7,9]) [7, 9]
15.6 <7> (L-norleucine) [6]
17.2 <7> (pyruvate, <7> VDH2 [13]) [13]
18.2 <7> (NH₄⁺) [6]

19.6 <9> (L-2-aminobutyrate) [5]
22 <1> (NH$_4^+$) [1]
22.5 <7> (L-norleucine, <7> VDH2 [13]) [13]
25 <1> (pyruvate) [1]
25.6 <1> (NH$_4^+$, <1> VDH2 [14]) [14]
28.2 <1> (L-alanine) [1]
30 <9> (L-norleucine) [5]
50 <14> (2-oxoisovalerate, <14> K79A mutant [12]) [12]
55 <9> (NH$_3$) [5]
55.06 <14> (L-valine, <14> A124G mutant [11]) [11]
117.6 <7> (L-methionine, <7> VDH2 [13]) [13]
153 <5> (NH$_4^+$) [3]
333.3 <7> (L-alanine, <7> VDH2 [13]) [13]
333.6 <7> (L-alanine) [6]
714 <14> (L-valine, <14> K79A mutant [12]) [12]

K$_i$-Value (mM)

36.8 <6> (L-valine) [4]

pH-Optimum

8 <14> (<14> reductive amination of 2-oxoisovalerate [11]) [11]
8-8.5 <14> (<14> reductive amination of 2-oxoisovalerate, 2-oxoisocaproate, 2-oxobutyrate, and DL-2-oxo-3-methylvalerate [11]) [11]
8.4-8.6 <8> (<8> shoots, reductive amination of valine ketoanalogue [8,9]) [8, 9]
8.5 <5> (<5> reductive amination [3]) [3]
8.7-8.9 <1> (<1> reductive amination of 2-oxoisovalerate, in presence of 0.3 M Tris-HCl buffer [1]) [1]
8.8 <1> (<1> VDH2: reductive amination of 2-oxoisovalerate, 0.3 M Tris-HCl buffer [14]) [14]
9 <6, 7> (<6> reductive amination of 2-oxoisovalerate [4]; <7> reductive amination of 2-oxoisovalerate, Tris-HCl buffer [6]; <7> VDH2: reductive amination of 2-oxoisovalerate, 0.3 M Tris-HCl buffer [13]) [4, 6, 13]
9.3 <8> (<8> roots, reductive amination of valine ketoanalogue [8,9]) [8, 9]
9.4-9.6 <8> (<8> shoots, oxidative deamination of L-valine [8,9]) [8, 9]
9.5 <8, 9> (<9> reductive amination [5]; <8> oxidative deamination of L-valine [7,9]) [5, 7, 9]
10 <9> (<9> oxidative deamination [5]) [5]
10-10.5 <14> (<14> oxidative deamination of L-valine, L-norvaline, L-leucine, and L-norleucine [11]) [11]
10.4 <1, 7> (<1,7> VDH2: oxidative deamination of L-valine, 0.3 M glycine-KCl-KOH buffer [13,14]) [13, 14]
10.5 <5, 6, 14> (<5,6,14> oxidative deamination of L-valine [3,4,11]) [3, 4, 11]
10.5-10.7 <1> (<1> oxidative deamination of L-valine, in presence of 0.3 M glycine-KCl-KOH buffer [1]) [1]
10.7 <7> (<7> oxidative deamination of L-valine, glycine-KCl-KOH buffer [6]) [6]
11 <8> (<8> roots, oxidative deamination of L-valine [8,9]) [8, 9]

pH-Range

8.5-9.5 <6> (<6> pH 8.5: 40% loss of activity, pH 9.5: 10% loss of activity [4]) [4]

10.5 <9> (<9> above rates of reductive amination and oxidative deamination decrease dramatically [5]) [5]

Temperature optimum (°C)

21 <9> (<9> assay at [5]) [5]

25 <14> (<14> assay at [11]) [11]

30 <1, 7> (<1,7> assay at [1,6]) [1, 6]

40 <8> (<8> enzyme in shoots [9]) [9]

47 <5> [3]

65 <1, 7> [1, 6, 13, 14]

4 Enzyme Structure

Molecular weight

67000 <14> (<14> gel filtration [11]) [11]

70000 <6> (<6> gel filtration [4]) [4]

80000 <1> (<1> VDH2, gel filtration [14]) [14]

88000 <9> (<9> gel filtration [5]) [5]

116000 <7> (<7> equilibrium ultracentrifugation [6]) [6]

118000 <7> (<7> gel filtration [6]) [6]

215000 <1> (<1> equilibrium ultracentrifugation [1]) [1]

218000 <1> (<1> gel filtration [1]) [1]

240000 <7> (<7> VDH2, gel filtration [13]) [13]

Subunits

? <14> (<14> x * 38000, SDS-PAGE [12]) [12]

dimer <1, 6, 9, 14> (<9> 2 * 41200, SDS-PAGE [5]; <6> 2 * 41000, SDS-PAGE [4]; <1> 2 * 41000, VDH2, SDS-PAGE [14]; <14> 2 * 38000, SDS-PAGE [11]) [4, 5, 11, 14]

dodecamer <1> (<1> 12 * 18000, SDS-PAGE [1]) [1]

hexamer <7> (<7> 6 * 41000, VDH2, SDS-PAGE [13]) [13]

tetramer <7> (<7> 4 * 29000, SDS-PAGE [6]) [6]

5 Isolation/Preparation/Mutation/Application

Source/tissue

leaf <2, 12, 13> (<2,12,13> subcellular particles [8,9]) [8, 9]

root <8, 10, 11> (<8,10,11> soluble enzyme [8,9]) [8, 9]

seedling <2, 8, 10-12> (<2,8,10-12> enzyme associated with subcellular particles [8]; <8,10,11> subcellular structures [9]) [8, 9]

sprout <8> [7]

Additional information <2, 12> (<2,12> enzyme activity observed in seedling shoots, but not in roots [8]) [8]

Localization

particle-bound <2, 8, 10-13> (<2,8,10-13> leaves, seedling shoots [8,9]; <8> sprout [7]) [7-9]
soluble <8, 10, 11, 14> (<8,10,11> roots [7-9]; <14> soluble enzyme expressed in Escherichia coli [11,12]) [7-9, 11, 12]

Purification

<1> (1508fold purification [1]; VDH2, 346fold purification [14]) [1, 14]
<5> [3]
<6> (143fold purification [4]) [4]
<7> (785fold purification [6]; VDH2, 52.6fold purification [13]) [6, 13]
<8> (21fold purification [7]; soluble enzyme from roots, 3.9fold purification [8]; shoots: 21fold purification, roots: 3.9fold purification [9]) [7-9]
<9> (180fold purification [5]) [5]
<14> (VDH and A124G mutant, His-bind resin column [11] ; His-bind resin column, 95% purity [12]) [11, 12]

Cloning

<1> (vdh gene [12]) [12]
<1> (vdh gene encoding VDH2, only one gene responsible for VDH activity, VDH₂ is the only active VDH [14]) [14]
<6> (vdh gene [12]) [12]
<7> (vdh gene [12]) [12]
<9> (vdh gene [12]) [12]
<14> (valDH gene cloned to identify the active site, coding region expressed in Escherichia coli using a pET expression system [11,12]) [11, 12]

Engineering

A124G <14> (<14> lower activity toward aliphatic amino acid substrates, e.g. catalytic rate constant for oxidative deamination of L-valine is only 12% of the wild type, but higher activities toward L-phenylalanine, L-tyrosine, and L-methionine [11]) [11]
K79A <14> (<14> lower affinity for L-valine and 2-oxoisovalerate and lower catalytic rate constants compared to the wild type enzyme [12]) [12]
K91A <14> (<14> lower affinity for L-valine and 2-oxoisovalerate and lower catalytic rate constants compared to the wild type enzyme [12]) [12]
Additional information <14> (<14> active site lysine residue modified with pyridoxal 5'-phosphate [12]) [12]

6 Stability

pH-Stability

6.3 <1> (<1> VDH2: stable at pH 6.3 during purification step Reactive-Blue 2 Sepharose chromatography [14]) [14]
7 <9> (<9> below markedly unstable [5]) [5]
7.4 <9> (<9> most stable [5]) [5]

Temperature stability

30 <5> (<5> 0.75 M ammonia buffer pH 8.5, half-life: 12 days [3]) [3]

General stability information

<6>, glycerol stabilizes [4]

Storage stability

<1>, -20°C, 0.2 M Tris-HCl buffer, pH 7.4, 3 months, no loss of activity [1]

<1>, -20°C, 50 mM Tris-HCl buffer, pH 7.4, 5 mM 2-mercaptoethanol, 1 mM EDTA, 30% v/v glycerol, 24 h, 90% loss of activity [14]

<1>, -20°C, 70% $(NH_4)_2SO_4$ suspension in 0.2 M Tris-HCl buffer, 5 months, stable [1]

<1>, 4°C, 0.2 M Tris-HCl buffer, pH 7.4, 24 h, no loss of activity [1]

<1>, 4°C, 50 mM Tris-HCl buffer, pH 7.4, 5 mM 2-mercaptoethanol, 1 mM EDTA, 10% v/v glycerol, 48 h, stable [14]

<6>, -80°C, 50 mM potassium phosphate pH 7.5, 10% v/v glycerol, 1 mM EDTA, 5 mM 2-mercaptoethanol, 1 month, loss of specific activity [4]

<7>, -25°C, 0.2 M Tris-HCl buffer, pH 7.4, 2 months, no loss of activity [6]

<7>, -80°C, 100 mM Tris-HCl buffer, pH 7.4, 20% v/v glycerol, 2 months, stable [13]

<7>, 4°C, 70% $(NH_4)_2SO_4$ suspension, 3 weeks, no loss of activity [6]

<9>, -70°C, 25 mM sodium phosphate buffer pH 7, equal volume of glycerol, 3 months, stable [5]

References

[1] Vancura, A.; Vancurova, I.; Volc, J.; Fussey, S.P.M.; Flieger, M.; Neuzil, J.; Marsalek, J.; Behal, V.: Valine dehydrogenase from Streptomyces fradiae: purification and properties. J. Gen. Microbiol., 134, 3213-3219 (1988)

[2] Van Hao, M.; Komagata, K.: A new species of Planococcus, P. kocurii isolated from fish, frozen foods, and fish curing brine. J. Gen. Appl. Microbiol., 31, 441-455 (1985)

[3] Honorat, A.; Monot, F.; Ballerini, D.: Synthesis of L-alanine and L-valine by enzyme systems from Bacillus megaterium. Enzyme Microb. Technol., 12, 515-520 (1990)

[4] Navarrete, R.M.; Vara, J.A.; Hutchinson, C.R.: Purification of an inducible L-valine dehydrogenase of Streptomyces coelicolor A3(2). J. Gen. Microbiol., 136, 273-281 (1990)

[5] Priestley, N.D.; Robinson, J.A.: Purification and catalytic properties of L-valine dehydrogenase from Streptomyces cinnamonensis. Biochem. J., 261, 853-861 (1989)

[6] Vancurova, I.; Vancura, A.; Volc, J.; Neuzil, J.; Flieger, M.; Basarova, G.; Behal, V.: Isolation and characterization of valine dehydrogenase from Streptomyces aureofaciens. J. Bacteriol., 170, 5192-5196 (1988)

[7] Kagan, Z.S.; Polyakov, V.A.; Kretovich, V.L.: Purification and properties of valine dehydrogenase. Biokhimiya, 34, 59-65 (1969)

[8] Kagan, Z.S.; Polyakov, V.A.; Kretovich, V.L.: Soluble valine dehydrogenase from roots of plant seedlings. Biokhimiya, **33**, 89-96 (1968)

[9] Kagan, Z.S.; Polyakov, V.A.; Kretovich, V.L.: Further studies on plant valine dehydrogenase. Enzymologia, **38**, 201-224 (1970)

[10] Nguyen, K.T.; Nguyen, L.T.; Behal, V.: The introduction of valine dehydrogenase activity from Streptomyces by L-valine is not repressed by ammonium. Biotechnol. Lett., **17**, 31-34 (1995)

[11] Hyun, C.G.; Kim, S.S.; Lee, I.H.; Suh, J.W.: Alteration of substrate specificity of valine dehydrogenase from Streptomyces albus. Antonie Leeuwenhoek, **78**, 237-242 (2000)

[12] Hyun, C.G.; Kim, S.S.; Park, K.H.; Suh, J.W.: Valine dehydrogenase from Streptomyces albus: gene cloning, heterologous expression and identification of active site by site-directed mutagenesis. FEMS Microbiol. Lett., **182**, 29-34 (2000)

[13] Nguyen, L.T.; Nguyen, K.T.; Kopecky, J.; Nova, P.; Novotna, J.; Behal, V.: Purification and characterization of a novel valine dehydrogenase from Streptomyces aureofaciens. Biochim. Biophys. Acta, **1251**, 186-190 (1995)

[14] Nguyen, L.T.; Nguyen, K.T.; Spizek, J.; Behal, V.: The tylosin producer, Streptomyces fradiae, contains a second valine dehydrogenase. Microbiology, **141**, 1139-1145 (1995)

1 Nomenclature

EC number
1.4.1.9

Systematic name
L-leucine:NAD$^+$ oxidoreductase (deaminating)

Recommended name
leucine dehydrogenase

Synonyms
L-leucine dehydrogenase
L-leucine:NAD$^+$ oxidoreductase, deaminating
LeuDH
dehydrogenase, leucine

CAS registry number
9082-71-7

2 Source Organism

<1> Bacillus cereus [1, 2, 3, 4, 8, 13, 21, 22, 27]
<2> Bacillus subtilis (168M [14]; mutant which is constitutive for L-Leu dehydrogenase [9]) [3, 6, 8, 9, 14, 40, 41, 42]
<3> Bacillus niger [3]
<4> Bacillus mycoides [3]
<5> Bacillus megaterium [3, 8]
<6> Bacillus pumilus [3]
<7> Bacillus sphaericus (DSM 462 [20]) [3, 8, 10, 11, 15, 16, 18, 20, 31, 35, 36, 40]
<8> Geobacillus stearothermophilus (recombinant enzyme expressed in Escherichia coli JM109 cells carrying the vector plasmid pICD212 [19]) [5, 12, 17, 19, 26, 30, 32, 33, 34, 37]
<9> Bacillus natto [6]
<10> Bacillus caldolyticus [7]
<11> Thermoactinomyces intermedius [28, 29, 35]
<12> Bacillus brevis [8]
<13> Bacillus mesentericus (var. flavus [8]) [8]
<14> Bacillus sp. (DSM 405, DSM 730 and DSM 1521 [20]; DSM 730 [24]) [20, 21, 24, 38, 39]
<15> Clostridium thermoaceticum [23]
<16> Bacillus licheniformis [25]

3 Reaction and Specificity

Catalyzed reaction

L-leucine + H_2O + NAD^+ = 4-methyl-2-oxopentanoate + NH_3 + NADH + H^+
(<8>, ordered bi-ter mechanism, in which NAD^+ and L-Leu are bound and
NH_4^+, 2-oxoisohexanoate, and NADH are released in that order [30])

Reaction type

oxidation
redox reaction
reduction

Natural substrates and products

S L-Leu + H_2O + NAD^+ <2> (<2>, the enzyme may have a function in
   spore germination [14]) (Reversibility: ? <2> [14]) [14]
P 4-methyl-2-oxopentanoate + NH_3 + NADH <2> [14]
S Additional information <2> (<2>, enzyme liberates ammonium ions
   from branched chain amino acids when supplied as the sole nitrogen
   source. It synthesizes from L-Ile, L-Leu, L-Val the branched chain α-oxo
   acids which are precursors of branched chain fatty acid biosynthesis [9])
   [9]
P ?

Substrates and products

S 2-oxo-3,3-dimethylbutanoate + NH_3 + NADH <10> (Reversibility: ? <10>
   [7]) [7]
P L-2-amino-3,3-dimethylbutanoate + NAD^+ + H_2O
S 2-oxo-4-methylselenobutyrate + NH_3 + NADH <8> (Reversibility: ? <8>
   [19]) [19]
P L-selenomethionine + H_2O + NAD^+ <8> [19]
S 4-methyl-2-oxo-5,5,5-trifluoropentanoate + NH_3 + NADH <11> (Reversi-
   bility: ? <11> [28]) [28]
P 2-amino-4-methyl-5,5,5-trifluoropentanoate + NAD^+ + H_2O
S L-2-aminobutanoate + H_2O + NAD^+ <1, 7, 10, 11, 15, 16> (<7>, 14% of
   the activity with L-Leu [8]; <15>, 19% of the activity with L-Leu [23];
   <16>, 32% of the activity with L-Leu [25]) (Reversibility: r <1, 7, 10, 11,
   15, 16> [4, 7, 8, 23, 25, 28]) [4, 7, 8, 23, 25, 28]
P 2-oxobutanoate + NH_3 + NADH <1, 7, 10, 11, 15, 16> [4, 7, 8, 23, 25, 28]
S L-2-aminobutanoate + H_2O + NAD^+ <1, 2, 7, 10, 11, 15> (<7>, 41% of the
   activity with L-Leu [8]; <15>, 11% of the activity with L-Leu [23]) (Reversibility: r <1, 2, 7, 10, 11, 15> [3, 4, 7, 8, 18, 23]; ? <1> [1]) [1, 3, 4, 6, 7, 8,
   18, 23, 28]
P 2-oxopentanoate + NH_3 + NADH <1, 10, 15> [4, 7, 23]
S L-Ala + H_2O + NAD^+ <16> (<16>, 10% of the activity with L-Leu [25])
   (Reversibility: ? <16> [25]) [25]
P 2-oxopropanoate + NH_3 + NADH
S L-Ile + H_2O + NAD^+ <1, 2, 7, 8, 10, 11, 14, 15, 16> (<8>, 73% of the
   activity with L-Leu [5]; <7>, 58% of the activity with L-Leu [8]; <15>,

88% of the activity with L-Leu [23]; <16>, 72% of the activity with L-Leu [25]) (Reversibility: r <1, 2, 7, 8, 10, 11, 14, 15, 16> [1, 3, 4, 5, 6, 7, 8, 18, 23, 24, 25, 28, 40]) [1, 3, 4, 5, 6, 7, 8, 18, 23, 24, 25, 28, 40]

P 3-methyl-2-oxopentanoate + NH_3 + NADH <1, 2, 10, 11, 15, 16> [1, 4, 5, 6, 7, 23, 25, 28]

S L-Leu + H_2O + 3-acetylpyridine-NAD^+ <7> (<7>, 166% of the activity with NAD^+ [8]) (Reversibility: ? <7> [8]) [8]

P 4-methyl-2-oxopentanoate + NH_3 + 3-acetylpyridine-NADH

S L-Leu + H_2O + 3-acetylpyridine-deamino-NAD^+ <7> (<7>, as effective as NAD^+ [8]) (Reversibility: ? <7> [8]) [8]

P 4-methyl-2-oxopentanoate + NH_3 + 3-acetylpyridine-deamino-NADH

S L-Leu + H_2O + 3-pyridinealdehyde-NAD^+ <7> (<7>, 19% of the activity with NAD^+ [8]) (Reversibility: ? <7> [8]) [8]

P 4-methyl-2-oxopentanoate + NH_3 + 3-pyridinealdehyde-NADH

S L-Leu + H_2O + NAD^+ <1-16> (<1>, the equilibrium favors synthesis of L-Leu [2]) (Reversibility: r <1-16> [1-42]) [1-42]

P 4-methyl-2-oxopentanoate + NH_3 + NADH <1-16> [1-40]

S L-Leu + H_2O + deamino-NAD^+ <7> (<7>, 81% of the activity with NAD^+ [8]) (Reversibility: ? <7> [8]) [8]

P 4-methyl-2-oxopentanoate + NH_3 + deamino-NADH

S L-Leu + H_2O + thionicotinamide-NAD^+ <7> (<7>, 21% of the activity with NAD^+ [8]) (Reversibility: ? <7> [8]) [8]

P 4-methyl-2-oxopentanoate + NH_3 + thionicotinamide-NADH

S L-Met + H_2O + NAD^+ <1, 2, 10> (<8>, no activity [5]) (Reversibility: r <1, 2> [4, 6]; ? <10> [7]) [4, 6, 7]

P 3-methylthio-2-oxobutanoate + NH_3 + NADH <1, 2> [4, 6]

S L-S-methylcysteine + H_2O + NAD^+ <16> (<16>, 19% of the activity with L-Leu [25]) (Reversibility: ? <16> [25]) [25]

P 3-methylthio-2-oxopropionate + NH_3 + NADH

S L-Val + H_2O + NAD^+ <1, 2, 7, 8, 10, 11, 14, 15, 16> (<8>, 98% of the activity with L-Leu [5]; <7>, 74% of the activity with L-Leu [8]; <15>, 93% of the activity with L-Leu [23]; <16>, 59% of the activity with L-Leu [25]) (Reversibility: r <1, 2, 7, 8, 10, 11, 15, 16> [1, 3, 4, 5, 6, 7, 16, 23, 25, 28]) [1, 3, 4, 5, 6, 7, 8, 16, 18, 23, 24, 25, 28, 40]

P 3-methyl-2-oxobutanoate + NH_3 + NADH <1, 8, 10, 11, 15, 16> [1, 4, 5, 6, 7, 16, 23, 25, 28]

S L-norleucine + H_2O + NAD^+ <1, 2, 7, 10, 15, 16> (<7>, 74% of the activity with L-Leu [8]; <15>, 15% of the activity with L-Leu [23]; <16>, 7% of the activity with L-Leu [25]) (Reversibility: r <1, 7, 10, 15, 16> [4, 6, 7, 8, 18, 23, 25]; ? <2> [3]) [3, 4, 6, 7, 8, 18, 23, 25]

P 2-oxohexanoate + NH_3 + NADH <1, 7, 10, 15, 16> [4, 7, 8, 23, 25]

S L-tert-Leu + H_2O + NAD^+ <7, 10> (Reversibility: ? <7, 10> [7, 8]) [7, 8]

P 3,3-dimethyl-2-oxobutanoate + NH_3 + NADH

S Additional information <7, 8, 15> (<7,15>, the enzyme is B-stereospecific [8,23]; <7>, reductive amination proceeds through a sequential ordered ternary-binary mechanism [8]; <7>, antitumor activity, the enzyme is highly inhibitory to Ehrlich ascites carcinoma cells [15]; <8>, substrate specificity of the chimeric enzyme consisting of an amino-terminal domain of phenylalanine dehydrogenase and a carboxy-terminal domain of leucine dehydrogenase [26]) [8, 15, 23, 26]

P ?

Inhibitors

4-methyl-2-pentanone <8> (<8>, competitive inhibition of wild-type enzyme, noncompetitive inhibition of mutant enzyme K80A [34]) [34]

4-methylpentanoate <8> (<8>, competitive inhibition of wild-type enzyme and mutant enzyme K80A [30]) [30, 34]

Co^{2+} <7, 14> (<7>, $CoCl_2$ [8]) [8, 21]

$Cu(CH_3COO)_2$ <8> (<8>, 1 mM, 20% inhibition [5]) [5]

Cu^{2+} <7, 8, 14> (<7>, $CuSO_4$ [8]; <8>, 1 mM $CuCl_2$, 22% inhibition [5]) [5, 8, 21]

$HgCl_2$ <1, 7, 8, 10> (<1>, 1 mM, 92% inhibition [4]; <8>, 0.01 M, 65% inhibition [5]) [4, 5, 7, 8, 10]

KCN <1> (<1>, 1 mM, 43% inhibition [4]) [4]

L-Leu <2> (<2>, competitive inhibition of the reductive amination of 4-methylthio-2-oxobutanoate [6]) [6]

$MgCl_2$ <7> [8]

Mn^{2+} <2> (<2>, inhibits oxidative deamination [6]) [6]

NAD^+ <7> (<7>, product inhibition [8]) [8]

NADH <16> (<16>, noncompetitive inhibition with respect to L-Leu [25]) [25]

NH_4^+ <2> (<2>, inhibits oxidative deamination [6]) [6]

Na_2S <7> [8]

PCMB <1, 7, 8> (<1>, reversed by L-Cys [2]; <8>, 0.01 M, 50% inhibition [5]) [1, 2, 5, 8, 10]

$Pb(CH_3COO)_2$ <8> (<8>, 1 mM, 20% inhibition [5]) [5]

p-mercuribenzoate <1, 10> (<1>, 1 mM, 75% inhibition [4]) [4, 7]

pyridoxal 5'-phosphate <7, 8, 10> (<7>, fully protected from inactivation only by NADH [16]) [7, 8, 16, 30]

sulfide <1> (<1>, Na_2S [2]) [1, 2]

Cofactors/prosthetic groups

NAD^+ <1-16> (<7>, the nicotinamide ring of the NAD^+ cofactor binds deeply in the cleft that separates the two domains of each subunit [31]) [1-42]

NADH <1-16> [1-42]

$NADP^+$ <11> (<11>, wild-type enzyme is inactive, mutant enzyme D203A exhibits dual specificity for NAD^+ and $NADP^+$, mutant enzymes D203A/I204R and D203A/I204R/D210R show high affinity for $NADP^+$ [29]) [29]

Additional information <1, 8, 14, 15, 16> (<1,8,14,15,16>, no activity with NADPH [2,4,5,23,24,25]; <1,7,8,14,15,16>, no activity with $NADP^+$ [2,4,5,18,23,24,25]) [2, 4, 5, 18, 23, 24, 25]

Turnover number (min^{-1})

780 <11> ($NADP^+$, <11>, mutant enzyme D203A/I204R and D203A/I204R/D210R [29]) [29]

5400 <11> (NAD⁺, <11>, wild-type enzyme [29]) [29]

Additional information <8> (<8>, turnover numbers of wild-type enzyme and mutant enzymes [30,33,34]) [30, 33, 34]

Specific activity (U/mg)

10 <7> [11]

15.23 <2> [6]

42.3 <7> [8]

57 <1> [4]

102 <11> [28]

110 <14> [24]

112 <8> [12]

115 <15> [23]

120 <8> [5, 17]

122 <11> (<11>, recombinant enzyme [28]) [28]

Additional information <1, 7, 10> [1, 2, 7, 18]

K_m-Value (mM)

0.021 <8> (NADH, <8>, mutant enzyme K68R [33]) [33]

0.022 <8> (NADH, <8>, mutant enzyme K80A [34]) [34]

0.034 <8> (NAD⁺, <8>, mutant enzyme K80Q [34]) [34]

0.034 <1> (NADH) [4]

0.035 <8> (NADH, <8>, wild-type enzyme [30,33,34]) [30, 33, 34]

0.036 <10> (NADH) [7]

0.037 <8> (NADH, <8>, mutant enzyme K80R [34]) [34]

0.038 <8> (NADH, <8>, mutant enzyme G77A [30]) [30]

0.039 <8> (NAD⁺, <8>, mutant enzyme G77A [30]) [30]

0.042 <11> (NADH) [28]

0.045 <8> (NADH, <8>, mutant enzyme K80Q [34]) [34]

0.053 <8> (NADH, <8>, mutant enzyme K68A [33]) [33]

0.054 <8> (NADH, <8>, mutant enzyme G78A [30]) [30]

0.055 <8> (NAD⁺, <8>, mutant enzyme K68R [33]) [33]

0.055 <8> (NADH, <8>, mutant enzyme G79A [30]) [30]

0.059 <8> (NAD⁺, <8>, mutant enzyme G78A [30]) [30]

0.063 <8> (NAD⁺, <8>, wild-type enzyme [30,33,34]) [30, 33, 34]

0.071 <8> (NAD⁺, <8>, mutant enzyme G79A [30]) [30]

0.074 <8> (NAD⁺, <8>, mutant enzyme K80R [34]) [34]

0.1 <1> (NADH) [2]

0.11 <8> (NAD⁺, <8>, mutant enzyme K68A [33]) [33]

0.12 <1, 2> (NADH) [1, 3]

0.14 <8> (NAD⁺, <8>, mutant enzyme K80A [34]) [34]

0.16 <1, 2> (NAD⁺) [1, 2, 3]

0.18 <10> (NAD⁺) [7]

0.31 <7> (4-methyl-2-oxopentanoate) [8]

0.34 <1> (NAD⁺) [4]

0.36 <11> (NAD⁺) [28]

0.39 <7> (NAD⁺) [8]

0.4 <1> (2-oxopentanoate) [4]

0.4 <11> (L-Ile) [28, 29]
0.4 <7> (NAD$^+$) [18]
0.45 <1> (4-methyl-2-oxopentanoate) [4]
0.6 <2> (4-methyl-2-oxopentanoate) [6]
0.63 <11> (4-methyl-2-oxopentanoate) [28]
0.7 <10> (4-methyl-2-oxopentanoate) [7]
0.7 <10> (L-Ile) [7]
0.77 <7> (3-acetylpyridine-NAD$^+$) [8]
0.8 <15> (4-methyl-2-oxopentanoate) [23]
0.81 <10> (2-oxopentanoate) [7]
0.88 <8> (2-oxoisohexanoate, <8>, wild-type enzyme [30,33,34]) [30, 33, 34]
0.9 <1, 2> (2-oxo-3-methylpentanoate) [4, 6]
0.91 <11> (2-oxopentanoate) [28]
0.99 <8> (2-oxoisohexanoate, <8>, mutant enzyme K80R [34]) [34]
1 <15> (2-oxopentanoate) [23]
1 <16> (4-methyl-2-oxopentanoate) [25]
1 <1> (L-Ile) [4]
1 <7> (L-Leu) [8]
1.1 <10> (L-Leu) [7]
1.2 <1> (2-oxohexanoate) [4]
1.2 <8> (4-methyl-2-oxopentanoate, <8>, mutant enzyme K68R [33]) [33]
1.2 <7> (L-Leu) [18]
1.2 <14> (NAD$^+$) [24]
1.33 <2> (L-Leu) [6]
1.4 <11> (2-oxohexanoate) [28]
1.4 <7> (3-methyl-2-oxobutanoate) [8]
1.5 <1> (2-oxobutanoate) [4]
1.5 <15> (L-2-aminobutanoate) [23]
1.5 <1> (L-Leu) [4]
1.5 <1> (L-norleucine) [4]
1.5 <2> (NAD$^+$) [6]
1.7 <7> (2-oxopentanoate) [8]
1.7 <7> (L-Val) [8]
1.8 <7> (L-Ile) [8]
1.8 <10> (L-tert-Leu) [7]
1.9 <10> (2-oxo-3-methylpentanoate) [7]
2 <11> (L-Leu) [28]
2 <1, 10> (L-Val) [1, 7]
2.1 <1> (3-methyl-2-oxobutanoate) [4]
2.1 <1> (3-methylthio-2-oxobutanoate) [4]
2.1 <16> (L-Leu) [25]
2.1 <10> (L-norleucine) [7]
2.2 <11> (2-oxo-3-methylpentanoate) [28]
2.2 <1, 2> (3-methyl-2-oxobutanoate) [1, 2, 3]
2.3 <15> (L-norleucine) [23]
2.3 <10> (L-norvaline) [7]
2.4 <16> (2-oxopentanoate) [25]

2.4 <10> (3-methyl-2-oxobutanoate) [7]
2.4 <2> (L-Ile) [6]
2.4 <11> (L-Val) [28]
2.41 <7> (3-acetylpyridine-deamino-NAD$^+$) [8]
2.5 <1> (L-Val) [4]
2.5 <11> (NADP$^+$, <11>, mutant D203A/I204R/D210R [29]) [29]
2.7 <10> (2-oxohexanoate) [7]
2.7 <11> (4-methyl-2-oxo-5,5,5-trifluoropentanoate) [28]
2.8 <14> (L-Leu) [24]
2.8 <8> (NAD$^+$, <8>, mutant enzyme K80R [34]) [34]
2.8 <11> (NADP$^+$, <11>, mutant enzyme D203A/I204R [28]) [28]
2.9 <1> (L-norvaline) [4]
3.2 <8> (L-Leu, <8>, mutant enzyme K68R [33]) [33]
3.3 <2> (3-methyl-2-oxobutanoate) [1, 3, 6]
3.3 <16> (L-Ile) [25]
3.4 <2> (L-Val) [6]
3.5 <8> (L-Leu, <8>, mutant enzyme G77A [30]) [30]
3.5 <7> (L-norvaline) [8]
3.6 <8> (L-Leu, <8>, mutant enzyme G79A [30]) [30]
3.7 <8> (L-Leu, <8>, mutant enzyme K80A [34]) [34]
3.8 <16> (3-methyl-2-oxobutanoate) [25]
3.9 <15> (L-Ile) [23]
4 <15> (3-methyl-2-oxobutanoate) [23]
4.3 <8> (L-Leu, <8>, mutant enzyme G78A [30]) [30]
4.4 <11> (3-methyl-2-oxobutanoate) [28]
5.1 <8> (L-Leu, <8>, wild-type enzyme [30,33,34]) [30, 33, 34]
5.2 <1, 2> (L-Ile) [1, 3]
5.5 <8> (2-oxoisohexanoate, <8>, mutant enzyme G77A [30]) [30]
5.6 <8> (2-oxoisohexanoate, <8>, mutant enzyme G79A [30]) [30]
5.6 <15> (L-norvaline) [23]
6.2 <15> (2-oxohexanoate) [23]
6.2 <1, 2> (L-Leu) [1, 2, 3]
6.3 <7> (L-norleucine) [8]
6.7 <2> (4-methylthio-2-oxobutanoate) [6]
6.8 <15> (L-Val) [23]
6.9 <10> (2-oxobutanoate) [7]
7 <7> (2-oxohexanoate) [8]
7.2 <11> (2-oxobutanoate) [28]
7.3 <15> (2-oxobutanoate) [23]
7.7 <7> (2-oxobutanoate) [8]
7.7 <8> (2-oxoisohexanoate, <8>, mutant enzyme G78A [30]) [30]
8.1 <10> (2-oxo-4-methylthiobutanoate) [7]
8.5 <15> (L-Leu) [23]
8.9 <10> (L-2-aminobutanoate) [7]
9.8 <8> (2-oxoisohexanoate, <8>, mutant enzyme K80A [34]) [34]
10 <7> (L-2-aminobutanoate) [8]
12.5 <16> (L-Val) [25]

13 <8> (2-oxo-4-methylselenobutanoate) [19]
13 <1, 2> (NH_4^+) [1, 2, 3]
17 <8> (NAD^+, <8>, mutant enzyme K80Q [34]) [34]
19 <10> (2-oxo-3,3-dimethylbutanoate) [7]
20 <2> (L-Val) [3]
22 <1> (L-2-aminobutanoate) [4]
23 <1, 10> (L-Met) [4, 7]
25 <8> (2-oxoisohexanoate, <8>, mutant enzyme K80Q [34]) [34]
25 <2> (L-Met) [6]
75 <8> (NH_4^+, <8>, wild-type enzyme [30]) [30]
130 <8> (L-Leu, <8>, mutant enzyme K68A [33]) [33]
140 <8> (4-methyl-2-oxopentanoate, <8>, mutant enzyme K68A [33]) [33]
210 <8> (NH_4^+, <8>, mutant enzyme G77A [30]) [30]
290 <8> (NH_4^+, <8>, mutant enzyme G79A [30]) [30]
330 <16> (NH_3) [25]
750 <8> (NH_4^+, <8>, mutant enzyme G78A [30]) [30]

K_i-Value (mM)

1 <2> (L-Leu, <2>, competitive inhibition of the reductive amination of 4-methylthio-2-oxobutanoate [6]) [6]
2.2 <7> (D-Leu, <7>, inhibition of deamination of D-Leu [8]) [8]
3 <7> (D-norvaline, <7>, inhibition of deamination of D-Leu [8]) [8]
7.9 <7> (D-alloisoleucine, <7>, inhibition of deamination of D-Leu [8]) [8]
9.4 <7> (D-2-aminobutanoate, <7>, inhibition of deamination of D-Leu [8]) [8]
10.8 <7> (D-Val, <7>, inhibition of deamination of D-Leu [8]) [8]
20 <8> (4-methylpentanoate, <8>, mutant enzyme G77A [30]) [30]
25 <8> (4-methylpentanoate, <8>, wild-type enzyme [30,34]; <8>, mutant enzyme G78A [30]) [30,34]
30 <8> (4-methylpentanoate, <8>, mutant enzyme G79A [30]) [30]
33 <8> (4-methylpentanoate, <8>, mutant enzyme K80A [34]) [34]
52.6 <7> (D-norleucine, <7>, inhibition of deamination of D-Leu [8]) [8]
140 <8> (4-methyl-2-pentanone, <8>, wild-type enzyme [34]) [34]

pH-Optimum

8.5 <1> (<1>, reductive amination of 2-oxopentanoate and 3-methyl-2-oxo-butanoate [4]) [4]
8.6 <16> (<16>, reductive amination of 3-methyl-2-oxobutanoate and 2-oxopentanoate [25]) [25]
8.8 <8> (<8>, reductive amination of 3-methyl-2-oxobutanoate [5]) [5]
9 <11> (<11>, reductive amination of 3-methyl-2-oxobutanoate [28]) [28]
9-9.5 <1, 7> (<1>, reductive amination of 4-methyl-2-oxopentanoate, 2-oxopentanoate and 2-oxomethionine [4]; <7>, reductive amination of 4-methyl-2-oxopentanoate and 3-methyl-2-oxobutanoate, phosphopyridoxylated enzyme [16]) [4, 16]
9.3 <11, 16> (<16>, reductive amination of 4-methyl-2-oxopentanoate [25]; <11>, reductive amination of 2-oxo-3-methylpentanoate [28]) [25, 28]

9.5 <2, 15> (<2>, reductive amination of 4-methyl-2-oxobutanoate [6]; <15>, reductive amination of 3-methyl-2-oxobutanoate [23]) [6, 23]

9.6 <11, 14> (<11,14>, reductive amination of 4-methyl-2-oxopentanoate [24,28]) [24, 28]

9.7 <8> (<8>, reductive amination of 4-methyl-2-oxopentanoate [5]; <8>, reductive amination of 2-oxo-4-methylselenobutanoate [19]) [5, 19]

10 <11> (<11>, deamination of L-Val [28]) [28]

10.3 <16> (<16>, oxidative deamination of L-Leu in presence of 1.0 M NaCl [25]) [25]

10.5 <11> (<11>, deamination of L-Leu [28]) [28]

10.5-10.8 <7> (<7>, oxidative deamination of L-Leu and L-Val, phosphopyridoxylated enzyme [16]) [16]

11 <2, 8, 14> (<2,8,14>, oxidative deamination of L-Leu [5,6,21]) [5, 6, 21]

11.3 <1, 14> (<1>, oxidative deamination [1,2]; <14>, oxidative deamination of L-Leu [24]) [1, 2, 24]

Additional information <8> (<8>, pH-optima for the chimeric enzyme consisting of an amino-terminal domain of phenylalanine dehydrogenase and a carboxy-terminal domain of leucine dehydrogenase [26]) [26]

pH-Range

8.7-10.7 <8> (<8>, pH 8.7: about 65% of maximal activity, pH 10.7: about 45% of maximal activity, reductive amination of 2-oxo-4-methylselenobutanoate [19]) [19]

10-11.5 <14> (<14>, pH 10.0: about 40% of maximal activity, pH 11.5: about 80% of maximal activity, oxidative deamination of L-Leu [21]) [21]

Temperature optimum (°C)

50 <14> [21]

68-70 <8> (<8>, oxidative amination of L-Leu [5]) [5]

Temperature range (°C)

30-52 <14> (<14>, 30°C: about 45% of maximal activity, 48-52°C: maximal activity [21]) [21]

50-80 <8> (<8>, 50°C: 37% of maximal activity, 80°C: 55% of maximal activity, oxidative amination of L-Leu [5]) [5]

4 Enzyme Structure

Molecular weight

245000 <7> (<7>, equilibrium sedimentation [8]) [8]

280000 <7, 14> (<7>, equilibrium sedimentation [18]; <14>, gel filtration [24]) [18, 24]

300000 <8> (<8>, gel filtration [5]) [5]

310000 <1> (<1>, gel filtration, equilibrium sedimenation [4]) [4]

313000 <10> (<10>, non-denaturing PAGE [7]) [7]

325000 <10> (<10>, gel filtration [7]) [7]

340000 <11> (<11>, gel filtration [28]) [28]

350000 <15> (<15>, gel filtration [23]) [23]

360000 <15, 16> (<15>, equilibrium sedimentation [23]; <16>, gel filtration [25]) [23, 25]

Additional information <8> (<8>, MW of the chimeric enzyme consisting of an amino-terminal domain of phenylalanine dehydrogenase and a carboxy-terminal domain of leucine dehydrogenase is 72000 Da, determined by gel filtration [26]) [26]

Subunits

? <8> (<8>, x * 46903, calculation from nucleotide sequence [12]) [12]

dimer <8> (<8>, 2 * 40000, SDS-PAGE [26]) [26]

hexamer <7, 8, 14, 15> (<7>, 6 * 41000, SDS-PAGE [8]; <14>, 6 * 47000, SDS-PAGE [24]; <8>, 6 * 49000, SDS-PAGE [5]; <15>, 6 * 56000, SDS-PAGE [23]) [5, 8, 23, 24]

octamer <1, 10, 16> (<1,10>, 8 * 39000, SDS-PAGE [4,7]; <11>, 4 * 42000, SDS-PAGE [28]; <16>, 8 * 44000, SDS-PAGE [25]) [4, 7, 25, 28]

5 Isolation/Preparation/Mutation/Application

Source/tissue

spore <1, 2> [3, 14, 22]

vegetative <1, 2> [3, 14, 22]

Purification

<1> (affinity extraction [13]) [1, 2, 4, 13, 27]

<2> [3, 6]

<7> (large-scale [11]) [8, 11]

<8> (one-step purification of recombinant enzyme [17]; chimeric enzyme consisting of an amino-terminal domain of phenylalanine dehydrogenase and a carboxy-terminal domain of leucine dehydrogenase [26]; mutant enzymes K68A and K68R [33]) [5, 12, 17, 26, 30, 33, 34]

<10> [7]

<11> [28]

<15> [23]

<16> [25]

Crystallization

<1> [4]

<7> (crystallized by addition of ammonium sulfate [8]; hanging drop method of vapour diffusion, using ammonium sulfate as the precipitant [36]) [8, 18, 36]

<11> (crystals of the binary complex with 4-methyl-2-oxopentanoate, hanging-drop vapour-diffusion method using PEG 4000 as precipitant [35]) [35]

Cloning

<1> (production of recombinant L-leucine dehydrogenase from Bacillus cereus in pilot scale using the runaway replication system Escherichia coli[-pIET98] [27]) [27]

<8> (expression in Escherichia coli [12,17]; chimeric enzyme consisting of an amino-terminal domain of phenylalanine dehydrogenase and a carboxy-terminal domain of leucine dehydrogenase [26]) [12, 17, 26]
<11> [28]
<14> (expression in Escherichia coli [24]) [24]
<15> (expression in Escherichia coli C600 [23]) [23]
<16> (cloned into Escherichia coli JM 109 with a vector plasmid pUC18 [25]) [25]

Engineering

D203A <11> (<11>, dual specificity for NAD$^+$ and NADP$^+$ [29]) [29]
D203A/I204R <11> (<11>, high affinity for NADP$^+$ [29]) [29]
D203A/I204R/D210R <11> (<11>, high affinity for NADP$^+$ [29]) [29]
G77A <8> (<8>, turnover numver in oxidative deamination of L-Leu is 36% of that of the wild-type enzyme. In reductive amination the turnover number is comparable to that of the wild-type enzyme. The K_m-value for 2-oxoisohexanoate is 6.3fold higher and the K_m-value for NH_4^+ is 2.8fold higher than that of the wild-type enzyme. Mutant enzyme shows lowered unfolding temperature compared with the wild-type enzyme. Faster degradation than wild-type enzyme after incubation at 37°C for 15 h with trypsin or subtilisin at a protease-to-substrate ratio of 1:1 [30]) [30]
G78A <8> (<8>, turnover number in oxidative deamination of L-Leu is 5.4% of that of the wild-type enzyme. In reductive amination the turnover number is comparable to that of the wild-type enzyme. The K_m-value for 2-oxoisohexanoate is 8.8fold higher and the K_m-value for NH_4^+ is 10fold higher than that of the wild-type enzyme. Mutant enzyme shows lowered unfolding temperature compared with the wild-type enzyme. Faster degradation than wild-type enzyme after incubation at 37°C for 15 h with trypsin or subtilisin at a protease-to-substrate ratio of 1:1 [30]) [30]
G79A <8> (<8>, turnover number in oxidative deamination of L-Leu is 40% of that of the wild-type enzyme. In reductive amination the turnover number is comparable to that of the wild-type enzyme. The K_m-value for 2-oxoisohexanoate is 6.4fold higher and the K_m-value for NH_4^+ is 3.9fold higher than that of the wild-type enzyme. Mutant enzyme shows lowered unfolding temperature compared with the wild-type enzyme [30]) [30]
K68A <8> (<8>, nearly complete loss of activity in the oxidative deamination, marked increase in K_m-values for both amino acid substrates and oxo acid substrates. An ionizable group in the wild-type enzyme with a pKa value of 10.1-10.7, which must be protonated for binding of substrate and competitive inhibitor with an α-carbopxyl group, is unobservable in mutant enzyme [33]) [33]
K68R <8> (<8>, nearly complete loss of activity in the oxidative deamination. An ionizable group in the wild-type enzyme with a pKa value of 10.1-10.7, which must be protonated for binding of substrate and competitive inhibitor with an α-carboxyl group, is unobservable in mutant enzyme [33]) [33]

K80A <8> (<8>, markedly reduced activity in oxidative deamination, nearly 90% of the wild-type activity in reductive amination. K_m-value for 2-oxoiso-hexanoate is 11fold higher than that of the wild-type enzyme, K_m-value for L-Leu is lower than that of the wild-type enzyme [34]) [34]

K80Q <8> (<8>, markedly reduced activity in oxidative deamination. K_m-value for 2-oxoisohexanoate is 28fold higher than that of the wild-type enzyme, K_m-value for L-Leu is about 3times larger than that of the wild-type enzyme [34]) [34]

K80R <8> (<8>, markedly reduced activity in oxidative deamination, 0.6% of the wild-type activity in reductive amination, K_m-value for L-Leu is lower than that of the wild-type enzyme [34]) [34]

Additional information <8> (<8>, chimeric enzyme consisting of an amino-terminal domain of phenylalanine dehydrogenase and a carboxy-terminal domain of leucine dehydrogenase containing the NAD^+-binding region, the substrate specificity of the chimeric enzyme in the reductive amination is a mixture of those of the two parent enzymes [26]; <8>, construction of a fragmentary enzyme form consisting of an N-terminal polypeptide fragment corresponding to the substrate-binding domain including an N-terminus, and a C-terminal fragment corresponding to the NAD^+-binding domain [37]) [26, 37]

Application

analysis <2, 7, 8, 14> (<14>, flow-injection determination of branched-chain L-amino acids with immobilized leucine dehydrogenase [21]; <8>, postcolumn co-immobilized leucine dehydrogenase-NADH oxidase reactor for the determination of branched-chain amino acids by high-performance liquid chromatography with chemiluminescence detection [32]; <2,7>, cheap and rapid determination of branched-chain amino and oxo acids [40]; <2>, assay of serum and urine for urea with use of urease and leucine dehydrogenase [41]; <2>, high-performance liquid chromatographic determination of branched-chain α-keto acids in serum using immobilized leucine dehydrogenase as post-column reactor [42]) [21, 32, 40, 41, 42]

synthesis <8, 14> (<8>, synthesis of L-selenomethionine from 2-oxo-4-methylselenobutanoate [19]; <14>, production of L-Leu, L-Val and L-Ile by artificial cells containing a glucose dehydrogenase and leucine dehydrogenase [38]; <14>, conversion of ammonia or urea into essential amino acids, L-Leu, L-Val, and L-Ile, using artificial cells containing an immobilized multienzyme system that consists of EC 1.1.1.1, EC 1.4.1.9, EC 3.5.1.5 and dextran-NAD^+ [39]) [19, 38, 39]

6 Stability

pH-Stability

5.4-10.3 <14> (<14>, 55°C, 10 min, stable [24]) [24]

5.5-10 <8> (<8>, 55°C, 5 min, stable [5]) [5]

5.6-9.8 <1> (<1>, 25°C, 24 h, maximal loss of 12% of the activity [4]) [4]

6-10 <15> (<15>, 55°C, 10 min, stable [23]) [23]
6-10.9 <11> (<11>, 50°C, 5 min, stable [28]) [28]
6-11 <8> (<8>, 55°C, 10 min, stable [17]) [17]
6.5-10.5 <10> (<10>, 30 min, 50% loss of activity [7]) [7]
7-11.2 <8> (<8>, 25°C, 30 min, wild-type enzyme and mutant enzymes K80A, K80R and K80Q [34]) [34]
7.5 <11> (<11>, 70°C, 5 min, most stable at pH 7.5 [28]) [28]

Temperature stability
30 <16> (<16>, 10 mM potassium phosphate buffer, pH 7.4, 0.01% 2-mercaptoethanol, 1 mM EDTA, 1.0 M NaCl, stable for more than 6 months [25]) [25]
50 <8> (<8>, 60 min, enzyme retains more than 75% of its activity [37]) [37]
52 <14> (<14>, denaturation above [21]) [21]
53 <8> (<8>, unfolding temperature of mutant enzyme G78A [30]) [30]
54 <8> (<8>, pH 7.0-9.5, 60 min, chimeric enzyme consisting of an amino-terminal domain of phenylalanine dehydrogenase and a carboxy-terminal domain of leucine dehydrogenase, stable [26]) [26]
55 <14, 15> (<8>, pH 6.0-11.0, 10 min, stable [17]; <15>, pH 6.0-10.0, 10 min, stable [23]; <14>, pH 5.4-10.3, 10 min, stable [24]) [17, 23, 24]
58 <8> (<8>, pH 7.0-9.5, 60 min, chimeric enzyme consisting of an amino-terminal domain of phenylalanine dehydrogenase and a carboxy-terminal domain of leucine dehydrogenase, loss of activity [26]) [26]
60 <8> (<8>, unfolding temperature of mutant enzyme G77A [30]) [30]
65 <11, 16> (<16>, 1 h, in presence of 2.5 M NaCl, stable [25]; <11>, 10 min, retains its full activity [28]) [25, 28]
70 <1, 8, 11, 16> (<8>, 20 min, enzyme retains full activity [5]; <8>, 30 min, retains full activity [17]; <1>, 10 min, 50% loss of activity, enzyme from cell extract [22]; <16>, 1 h, complete loss of activity [25]; <11>, 40 min, 3 M NaCl, less than 10% loss of activity [28]) [5, 17, 22, 25, 28]
75 <7, 8, 11, 14, 15> (<7,14>, 5 min, stable [20]; <15>, 30 min, stable [23]; <11>, 10 min, about 75% loss of activity [28]; <8>, unfolding temperature of mutant enzyme K80A [34]) [20, 23, 28, 34]
76 <8> (<8>, unfolding temperature of mutant enzyme G79A [30]) [30]
79 <8> (<8>, unfolding temperature of mutant enzyme K80R and K80Q [34]) [34]
80 <8, 10, 14> (<8>, 5 min, substantial loss of activity [5]; <10>, pH 7.5, 30 min, 50% loss of activity [7]; <14>, 10 min, stable [24]; <8>, unfolding temperature of the wild-type enzyme [30,34]) [5, 7, 24, 30, 34]
83 <7, 14> (<7,14>, 5 min, 50% loss of activity [20]) [20]
85 <15> (<15>, 5 min, substantial loss of activity [23]) [23]
94 <1> (<1>, 10 min, 50% loss of activity, enzyme from spores [22]) [22]
Additional information <2> [3]

Organic solvent stability
2-propanol <10> (<10>, 2 M, 4°C, 5% loss of activity after 2 months [7]) [7]
acetonitrile <10> (<10>, 2 M, 4°C, 22% loss of activity after 3 d [7]) [7]

General stability information

<8>, mutant enzymes G77A and G78A show faster degradation than wild-type enzyme after incubation at 37°C for 15 h with trypsin or subtilisin at a protease-to-substrate ratio of 1:1. Wild-type enzyme and mutant enzyme G79A are degraded at almost the same rate [30]

Storage stability

<1>, -20°C, 50% glycerol, stable for 1 year [1]

<1>, -20°C, stable for over 1 month [1]

<8>, 4°C, buffer containing 0.02% sodium azide, stable for more than 1 year [17]

<8>, 4°C, stable for at least 1 month [5]

<10>, 4°C, 50 mM potassium phosphate, 1 mM dithioerythritol, pH 7.5, 15% loss of activity after 2 months [7]

<11>, 4°C, stable for over two years [28]

<14>, 4°C, 5 mM tetrasodium EDTA solution in phosphate buffer, 0.1 M, pH 7.0, stable for at least 2 months [21]

<15>, 4°C, buffer containing 0.02% azide, stable for more than 1 year [23]

References

[1] Zink, M.W.; Sanwal, B.D.: L-Leucine dehydrogenase (Bacillus cereus). Methods Enzymol., 17A, 799-802 (1970)

[2] Sanwal, B.D.; Zink, M.W.: L-Leucine dehydrogenase of Bacillus cereus. Arch. Biochem. Biophys., 94, 430-435 (1961)

[3] Zink, M.W.; Sanwal, B.D.: The distribution and substrate specificity of L-Leucine dehydrogenase. Arch. Biochem. Biophys., 99, 72-77 (1962)

[4] Schuette, H.; Hummel, W.; Tsai, H.; Kula, M.R.: L-Leucine dehydrogenase from Bacillus cereus. Appl. Microbiol. Biotechnol., 22, 306-317 (1985)

[5] Ohshima, T.; Nagata, S.; Soda, K.: Purification and characterization of thermostable leucine dehydrogenase from Bacillus stearothermophilus. Arch. Microbiol., 141, 407-411 (1985)

[6] Livesey, G.; Lund, P.: Isolation and characterization of leucine dehydrogenase from Bacillus subtilis. Methods Enzymol., 166, 282-288 (1988)

[7] Kaerst, U.; Schuette, H.; Baydoun, H.; Tsai, H.: Purification and characterization of leucine dehydrogenase from the thermophile Bacillus caldolyticus. J. Gen. Microbiol., 135, 1305-1313 (1989)

[8] Ohshima, T.; Misono, H.; Soda, K.: Properties of crystalline leucine dehydrogenase from Bacillus sphaericus. J. Biol. Chem., 253, 5719-5725 (1978)

[9] Obermeier, N.; Poralla, K.: Some physiological functions of the L-leucine dehydrogenase in Bacillus subtilis. Arch. Microbiol., 109, 59-63 (1976)

[10] Ohshima, T.; Yamamoto, T.; Misono, H.; Soda, K.: Leucine dehydrogenase of bacillus sphaericus: sulfhydryl groups and catalytic sites. Agric. Biol. Chem., 42, 1739-1743 (1978)

[11] Hummel, W.; Schuette, H.; Kula, M.R.: Leucine dehydrogenase from Bacillus sphaericus. Optimized production conditions and efficient method for

its large-scale purification. Eur. J. Appl. Microbiol. Biotechnol., **12**, 22-27 (1981)

[12] Nagata, S.; Tanizawa, K.; Esaki, N.; Sakamoto, Y.; Ohshima, T.; Tanaka, H.; Soda, K.: Gene cloning and sequence determination of leucine dehydrogenase from Bacillus stearothermophilus and structural comparison with other NAD(P)$^+$-dependent dehydrogenases. Biochemistry, **27**, 9056-9062 (1988)

[13] Schustolla, D.; Hustedt, H.: Gene cloning and sequence determination of leucine dehydrogenase from Bacillus stearothermophilus and structural comparison with other NAD(P)$^+$-dependent dehydrogenases. DECHEMA Biotechnol. Conf., **3**, 1097-1101 (1989)

[14] Obermeier, N.; Poralla, K.: Experiments on the role of leucine dehydrogenase in initiation of Bacillus subtilis spore germination. FEMS Microbiol. Lett., **5**, 81-83 (1979)

[15] Oki, T.; Shirai, M.: Antitumor activities of bacterial leucine dehydrogenase and glutaminase A. FEBS Lett., **33**, 286-288 (1973)

[16] Ohshima, T.; Soda, K.: Modification of leucine dehydrogenase by pyridoxal 5'-phosphate. Agric. Biol. Chem., **48**, 349-354 (1984)

[17] Oka, M.; Yang, Y.S.; Nagata, S.; Esaki, N.; Tanaka, H.; Soda, K.: Overproduction of thermostable leucine dehydrogenase of Bacillus stearothermophilus and its one-step purification from recombinant cells of Escherichia coli. Biotechnol. Appl. Biochem., **11**, 307-311 (1989)

[18] Soda, K.; Misono, H.; Mori, K.; Sakato, H.: Crystalline L-leucine dehydrogenase. Biochem. Biophys. Res. Commun., **44**, 931-935 (1971)

[19] Esaki, N.; Shimoi, H.; Yang, Y.S.; Tanaka, H.; Soda, K.: Enantioselective synthesis of L-selenomethionine with leucine dehydrogenase. Biotechnol. Appl. Biochem., **11**, 312-317 (1989)

[20] Ohshima, T.; Wandrey, C.; Sugiura, M.; Soda, K.: Screening of thermostable leucine and alanine dehydrogenases in thermophilic Bacillus strains. Biotechnol. Lett., **7**, 871-876 (1985)

[21] Kiba, N.; Hori, S.; Furusawa, M.: Flow-injection determination of branched-chain L-amino acids with immobilized leucine dehydrogenase. Anal. Chim. Acta, **218**, 161-166 (1989)

[22] Warth, A.D.: Heat stability of Bacillus cereus enzymes within spores and in extracts. J. Bacteriol., **143**, 27-34 (1980)

[23] Shimoi, H.; Nagata, S.; Esaki, N.; Tanaka, H.; Soda, K.: Leucine dehydrogenase of a thermophilic anaerobe, Clostridium thermoaceticum: gene cloning, ourification and characterization. Agric. Biol. Chem., **51**, 3375-3381 (1987)

[24] Nagata, S.; Misono, H.; Nagasaki, S.; Esaki, N.; Tanaka, H.; Soda, K.: Gene cloning, purification, and characterization of the highly thermostable leucine dehydrogenase of Bacillus sp.. J. Ferment. Bioeng., **69**, 199-203 (1990)

[25] Nagata, S.; Bakthavatsalam, S.; Galkin, A.G.; Asada, H.; Sakai, S.; Esaki, N.; Soda, K.; Ohshima, T.; Nagasaki, S.; Misono, H.: Gene cloning, purification, and characterization of thermostable and halophilic leucine dehydrogenase from a halophilic thermophile, Bacillus licheniformis TSN9. Appl. Microbiol. Biotechnol., **44**, 432-438 (1995)

[26] Kataoka, K.; Takada, H.; Tanizawa, K.; Yoshimura, T.; Esaki, N.; Ohshima, T.; Soda, K.: Construction and characterization of chimeric enzyme con-

sisting of an amino-terminal domain of phenylalanine dehydrogenase and a carboxy-terminal domain of leucine dehydrogenase. J. Biochem., **116**, 931-936 (1994)

[27] Ansorge, M.B.; Kula, M.R.: Production of recombinant L-leucine dehydrogenase from Bacillus cereus in pilot scale using the runaway replication system E. coli[pIET98]. Biotechnol. Bioeng., **68**, 557-562 (2000)

[28] Ohshima, T.; Nishida, N.; Bakthavatsalam, S.; Kataoka, K.; Takada, H.; Yoshimura, T.; Esaki, N.; Soda, K.: The purification, characterization, cloning and sequencing of the gene for a halostable and thermostable leucine dehydrogenase from Thermoactinomyces intermedius. Eur. J. Biochem., **222**, 305-312 (1994)

[29] Galkin, A.; Kulakova, L.; Ohshima, T.; Esaki, N.; Soda, K.: Construction of a new leucine dehydrogenase with preferred specificity for NADP$^+$ by site-directed mutagenesis of the strictly NAD$^+$-specific enzyme. Protein Eng., **10**, 687-690 (1997)

[30] Sekimoto, T.; Fukui, T.; Tanizawa, K.: Role of the conserved glycyl residues located at the active site of leucine dehydrogenase from Bacillus stearothermophilus. J. Biochem., **116**, 176-182 (1994)

[31] Baker, P.J.; Turnbull, A.P.; Sedelnikova, S.E.; Stillman, T.J.; Rice, D.W.: A role for quaternary structure in the substrate specificity of leucine dehydrogenase. Structure, **3**, 693-705 (1995)

[32] Kiba, N.; Oyama, Y.; Kato, A.; Furusawa, M.: Postcolumn co-immobilized leucine dehydrogenase-NADH oxidase reactor for the determination of branched-chain amino acids by high-performance liquid chromatography with chemiluminescence detection. J. Chromatogr. A, **724**, 354-357 (1996)

[33] Sekimoto, T.; Fukui, T.; Tanizawa, K.: Involvement of conserved lysine 68 of Bacillus stearothermophilus leucine dehydrogenase in substrate binding. J. Biol. Chem., **269**, 7262-7266 (1994)

[34] Sekimoto, T.; Matsuyama, T.; Fukui, T.; Tanizawa, K.: Evidence for lysine 80 as general base catalyst of leucine dehydrogenase. J. Biol. Chem., **268**, 27039-27045 (1993)

[35] Muranova, T.A.; Ruzheinikov, S.N.; Sedelnikova, S.E.; Baker, P.J.; Pasquo, A.; Galkin, A.; Esaki, N.; Ohshima, T.; Soda, K.; Rice, D.W.: Crystallization and preliminary X-ray analysis of substrate complexes of leucine dehydrogenase from Thermoactinomyces intermedius. Acta Crystallogr. Sect. D, **58**, 1059-1062. (2002)

[36] Turnbull, A.P.; Ashford, S.R.; Baker, P.J.; Rice, D.W.; Rodgers, F.H.; Stillman, T.J.; Hanson, R.L.: Crystallization and quaternary structure analysis of the NAD$^+$-dependent leucine dehydrogenase from Bacillus sphaericus. J. Mol. Biol., **236**, 663-665 (1994)

[37] Oikawa, T.; Kataoka, K.; Jin, Y.; Suzuki, S.; Soda, K.: Fragmentary form of thermostable leucine dehydrogenase of Bacillus stearothermophilus: its construction and reconstitution of active fragmentary enzyme. Biochem. Biophys. Res. Commun., **280**, 1177-1182. (2001)

[38] Gu, K.F.; Chang, M.S.: Production of essential L branched chain amino acids in bioreactors containing artificial cells immobilized multienzyme system and dextran-NAD$^+$. Biotechnol. Bioeng., **36**, 263-269 (1990)

[39] Gu, K.F.; Chang, M.S.: Conversion of ammonia or urea into essential amino acids, L-leucine, L-valine, and L-isoleucine, using artificial cells containing as immobilized multienzyme system and dextran-NAD$^+$. Biotechnol. Appl. Biochem., 12, 227-236 (1990)

[40] Livesey, G.; Lund, P.: Determination of branched-chain amino and keto acids with leucine dehydrogenase. Methods Enzymol., 166, 3-10 (1988)

[41] Morishita, Y.; Nakane, K.; Fukatsu, T.; Nakashima, N.; Tsuji, K.; Soya, Y.; Yoneda, K.; Asano, S.; Kawamura, Y.: Kinetic assay of serum and urine for urea with use of urease and leucine dehydrogenase. Clin. Chem., 43, 1932-11936 (1997)

[42] Kiba, N.; Muto, M.; Furusawa, M.: High-performance liquid chromato-graphic determination of branched-chain α-keto acids in serum using im-mobilized leucine dehydrogenase as post-column reactor. J. Chromatogr., 497, 236-242 (1989)

Glycine dehydrogenase

1.4.1.10

1 Nomenclature

EC number
1.4.1.10

Systematic name
glycine:NAD$^+$ oxidoreductase (deaminating)

Recommended name
glycine dehydrogenase

Synonyms
dehydrogenase, glycine

CAS registry number
37255-40-6

2 Source Organism

<-3> no activity in *Arthrobacter globiformis* [2]
<-2> no activity in *Methylophius methylovora* (M8-5, M12-4 [2]) [2]
<-1> no activity in *Methylophius methanolovous* [2]
<1> *Mycobacterium tuberculosis* (H37Ra strain [1]) [1, 3]
<2> *Pseudomonas methylica* [2]
<3> *Pseudomonas sp.* (AM1 [2]) [2]
<4> *Hyphomicrobium vulgare* [2]
<5> *Blastobacter viscosus* [2]
<6> *Blastobacter aminovorus* [2]
<7> *Mycobacterium smegmatis* (ATCC 607 [4]) [4]

3 Reaction and Specificity

Catalyzed reaction
glycine + H$_2$O + NAD$^+$ = glyoxylate + NH$_3$ + NADH$^+$ H$^+$

Reaction type
redox reaction

Natural substrates and products
S glyoxylate + NH$_3$ + NADH <1> (<1> alternate pathway for formation of glycine in absence of transamination [1]) (Reversibility: r <1> [1]) [1]
P glycine + H$_2$O + NAD$^+$

Substrates and products

S glyoxylate + NH_3 + NADH <1, 7> (Reversibility: r <1, 7> [1, 3, 4]) [1, 3, 4]

P glycine + H_2O + NAD^+

S Additional information <1> (<1> not: acetic acid, oxalic acid [1]) [1]

P ?

Inhibitors

glycine <1> [1]

oxygen <1, 7> [3, 4]

serine <1> (<1> non-competitive [1]) [1]

Cofactors/prosthetic groups

NADH <2-6> [2]

NADH <1> [3]

Specific activity (U/mg)

7.25 <1> [1]

Additional information <1> (<1> in vitro under anaerobic conditions approximately 10fold greater than under aerobic conditions [3]) [3]

Additional information <7> (<7> 4-5fold increased in cells under microaerophilic conditions as compared with an aerobically grown culture [4]) [4]

K_m-Value (mM)

0.0087 <1> (NADH, <1> anaerobic resting bacilli - RB [3]) [3]

0.22 <1> (glyoxylate) [1]

4.3 <1> (glyoxylate, <1> aerobic aerated bacilli - AB [3]) [3]

5.4 <1> (glyoxylate, <1> anaerobic resting bacilli - RB [3]) [3]

120 <1> (NH_3, <1> anaerobic resting bacilli - RB [3]) [3]

Additional information <1> [1]

pH-Optimum

6.4 <1, 7> (<1,7> assay at [1,4]) [1, 4]

Temperature optimum (°C)

22 <1, 7> (<1,7> assay at [1,4]) [1, 4]

5 Isolation/Preparation/Mutation/Application

Purification

<1> (H37Ra strain, partial [1]) [1]

<7> (ATCC 607, partial from microaerophilic adapted cultures [4]) [4]

6 Stability

Storage stability

<1>, -17°C, 72 h, 30-50% loss of activity [1]

<7>, 37°C, 12 d under microaerophilic conditions [4]

References

[1] Goldman, D.S.; Wagner, M.J.: Enzyme systems in the Mycobacteria. Biochim. Biophys. Acta, **65**, 297-306 (1962)

[2] Loginova, N.V.; Govorukhina, N.I.; Trotsenko, Y.A.: Enzymes of the assimilation of ammonium in bacteria with various pathways of C1-metabolism. Mikrobiologiya, **51**, 38-42 (1981)

[3] Wayne, L.G.; Lin, K.Y.: Glyoxylate metabolism and adaptation of Mycobacterium tuberculosis to survival under anaerobic conditions. Infect. Immun., **37**, 1042-1049 (1982)

[4] Usha, V.; Jayaraman, R.; Toro, J.C.; Hoffner, S.E.; Das, K.S.: Glycine and alanine dehydrogenase activities are catalyzed by the same protein in Mycobacterium smegmatis: upregulation of both activities under microaerophilic adaptation. Can. J. Microbiol., **48**, 7-13 (2002)

L-Erythro-3,5-diaminohexanoate dehydrogenase 1.4.1.11

1 Nomenclature

EC number
1.4.1.11

Systematic name
L-erythro-3,5-diaminohexanoate:NAD^+ oxidoreductase (deaminating)

Recommended name
L-erythro-3,5-diaminohexanoate dehydrogenase

Synonyms
L-3,5-diaminohexanoate dehydrogenase
dehydrogenase, L-3,5-diaminohexanoate

CAS registry number
37377-90-5

2 Source Organism

<1> *Brevibacterium sp.* (strain L5 [1]) [1]
<2> *Clostridium sp.* (strain SB4 and strain M-E [2]; strain SB4 with an active tetramer form and a less active dimer form of enzyme [3]) [2, 3]
<3> *Clostridium sticklandii* [2]

3 Reaction and Specificity

Catalyzed reaction
L-erythro-3,5-diaminohexanoate + H_2O + NAD^+ = (S)-5-amino-3-oxo-hexanoate + NH_3 + NADH + H^+

Reaction type
amination
deamination
oxidation
redox reaction
reduction

Natural substrates and products
S L-erythro-3,5-diaminohexanoate + H_2O + $NAD(P)^+$ <1-3> (<1,2> NAD^+ [1-3]; <3> $NADP^+$ [2]; <1-3> metabolism of L-erythro-3,5-diaminohex-

anoate, that is an intermediate in the lysine fermentation [1,2]; <3> important regulatory enzyme of the L-lysine fermentation pathway [2]; <2> enzyme is involved in lysin degradation [3]) (Reversibility: r <1-3> [1-3]) [1-3]

P (S)-5-amino-3-oxohexanoate + NH_3 + NAD(P)H <1-3> [1-3]

Substrates and products

S L-erythro-3,5-diaminohexanoate + H_2O + NAD(P)$^+$ <1-3> (<1,2> NAD$^+$ [1,3]; <2> strain SB4: relatively specific for NAD$^+$, strain M-E: NAD$^+$ and NADP$^+$ with lower activity than with NAD$^+$ [2]; <3> NADP$^+$ and NAD$^+$ with lower activity than with NADP$^+$ [2]; <1,2> enzyme is highly specific for its substrates [1,3]; <2> equilibrium is more favorable for amino acid oxidation [3]) (Reversibility: r <1-3> [1-3]) [1-3]

P (S)-5-amino-3-oxohexanoate + NH_3 + NAD(P)H <1-3> [1-3]

Inhibitors

ATP <3> (<3> when NAD$^+$ is substrate, enzyme is inhibited by ATP [2]) [2]
D-erythro-3,5-diaminohexanoate <1, 2> (<1> substrate inhibition observed with DL-erythro-3,5-diaminohexanoate caused by D-isomer [1]; <2> noncompetitive inhibitor of L-erythro-3,5-diaminohexanoate [3]) [1, 3]
DL-threo-3,5-diaminohexanoate <1, 2> (<1> inhibitory effect [1]; <2> partial inhibition by concentrations above 0.3 mM, pH 6.8 [3]) [1, 3]
NADH <2> (<2> dimeric and tetrameric form of enzyme: product inhibition, competitive inhibitor to NAD$^+$ and uncompetitive inhibitor to 5-amino-3-oxohexanoate [3]) [3]

Cofactors/prosthetic groups

3-acetylpyridine-NAD$^+$ <1, 2> (<1> 34% as effective as NAD$^+$ [1]; <2> 70% of activity with NAD$^+$ [3]) [1, 3]
3-pyridinealdehyde-NAD$^+$ <1> (<1> 4.5% as effective as NAD$^+$ [1]) [1]
NAD$^+$ <1-3> (<1,2> NAD$^+$ is most active cofactor [1,3]; <2> strain SB4: relatively specific for NAD$^+$, strain M-E: NAD$^+$ and NADP$^+$ with lower activity than with NAD$^+$ [2]; <3> NAD$^+$: lower activity than with NADP$^+$ [2]) [1-3]
NADP$^+$ <2, 3> (<2> strain M-E: NADP$^+$ with lower activity than with NAD$^+$ [2]; <3> NADP$^+$ and NAD$^+$ with lower activity than with NADP$^+$, enzyme with two different NADP$^+$ binding sites [2]) [2]

Specific activity (U/mg)

7.47 <1> [1]
34 <2> [3]
Additional information <2> (<2> more [3]) [3]

K_m-Value (mM)

0.03 <3> (NADP$^+$, <3> pH 7, two NADP$^+$ binding sites with different K_m-values [2]) [2]
0.05 <3> (NADP$^+$, <3> pH 8.8-9.2, two NADP$^+$ binding sites with different K_m-values [2]) [2]
0.074 <2> (NADH) [3]
0.13 <2> (NAD$^+$, <2> dimeric form of enzyme [3]) [3]

0.16 <3> (NADP⁺, <3> pH 7, two NADP⁺ binding sites with different K_m-values [2]) [2]

0.18 <2> (L-erythro-3,5-diaminohexanoate, <2> tetrameric form of enzyme, pH 6.8 [3]) [3]

0.22 <2> (L-erythro-3,5-diaminohexanoate, <2> tetrameric form of enzyme, higher pH levels [3]) [3]

0.25 <2> (L-erythro-3,5-diaminohexanoate, <2> dimeric form of enzyme, pH 6.6 [3]) [3]

0.26 <2> (5-amino-3-oxohexanoate) [3]

0.28 <2> (NAD⁺, <2> strain SB4, pH 7 [2]; <2> tetrameric form of enzyme [3]) [2, 3]

0.32 <3> (NAD⁺, <3> pH 7 [2]) [2]

0.75 <1> (L-erythro-3,5-diaminohexanoate) [1]

0.77 <2> (L-erythro-3,5-diaminohexanoate, <2> tetrameric form of enzyme, pH 7.6 and 8.9 [3]) [3]

1.25 <3> (NADP⁺, <3> pH 8.8-9.2, two NADP⁺ binding sites with different K_m-values [2]) [2]

2 <2, 3> (NAD⁺, <2> strain SB4, pH 8.8-9.2 [2]; <3> pH 8.8-9.2 [2]) [2]

140 <2> (NH₄Cl) [3]

K_i-Value (mM)

0.004 <2> (NADH, <2> tetrameric form of enzyme [3]) [3]

0.5 <2> (D-erythro-3,5-diaminohexanoate) [3]

1.2 <2> (DL-threo-3,5-diaminohexanoate) [3]

16 <2> (NADH, <2> dimeric form of enzyme [3]) [3]

pH-Optimum

7 <2> (<2> reductive amination [3]) [3]

8.7-9 <3> (<3> oxidative deamination [2]) [2]

8.9 <2> (<2> strain SB4, oxidative deamination [2,3]) [2, 3]

9.3 <1> (<1> oxidative deamination [1]) [1]

pH-Range

5.8-9.8 <2> (<2> oxidative deamination [3]) [3]

5.8-10.5 <1> (<1> oxidative deamination [1]) [1]

4 Enzyme Structure

Molecular weight

71000 <1> (<1> gel filtration [1]) [1]

78000-80000 <2, 3> (<2,3> strain M-E, gel filtration [2]) [2]

135000-144000 <2> (<2> gel filtration [3]) [3]

140000 <2> (<2> strain SB4, gel filtration [2]) [2]

Subunits

dimer <2, 3> (<2> strain M-E [2]; <3> 2 * 39000, SDS-PAGE [2]; <2> 2 * 68000, strain SB4, dimeric form one-third as active as tetrameric form, dimer is built by dissociation at high pH or in solution of low ionic strength [3]) [2, 3]

tetramer <2> (<2> 4 * 37000, strain SB4, SDS-PAGE [2,3]) [2, 3]

5 Isolation/Preparation/Mutation/Application

Purification

<1> (70fold purification [1]) [1]
<2> (strain SB4, 30fold purification [2,3]) [2, 3]
<3> (100fold purification [2]) [2]

6 Stability

pH-Stability

6-7 <2> (<2> most stable [3]) [3]
7-8.5 <1> (<1> most stable [1]) [1]
8 <2> (<2> unstable above [3]) [3]

General stability information

<1>, DL-erythro-3,5-diaminohexanoate, NAD$^+$ and glycerol stabilizes the enzyme, higher concentrations increase the effect [1]

<1>, enzyme is rapidly inactivated during dialysis or upon dilution into buffer [1]

<2>, DL-erythro-3,5-diaminohexanoate, Tris-chloride, NaCl, 100 mM EDTA, 2.7 M glycerol stabilizes [3]

<2>, enzyme is very unstable in crude extracts and to dilution in buffer, purified enzyme is only partially inactivated by dilution in a solution of low ionic strength [3]

<2, 3>, strain SB4: enzyme is very unstable in crude extracts, Clostridia sticklandii and strain M-E with a stable enzyme at all stages of purity [2]

Storage stability

<1>, -20°C, 20 mM potassium phosphate buffer, pH 7, 5% glycerol, 20 mM DL-erythro-3,5-diaminohexanoate, 0.1 mM NAD$^+$, per month, 3-4% loss of activity [1]

<2>, -12°C, 100 mM EDTA, pH 6.8, 5 months, 50% loss of activity [3]

<2>, 4°C, suspension in ammonium sulfate, 6 months, no detectable loss of activity [3]

References

[1] Hong, S.C.L.; Barker, H.A.: Aerobic metabolism of 3,5-diaminohexanoate in a Brevibacterium. Purification of 3,5-diaminohexanoate dehydrogenase and degradation of 3-keto-5-aminohexanoate. J. Biol. Chem., **248**, 41-49 (1973)

[2] Stadtman, T.C.: Lysine metabolism by Clostridia. Adv. Enzymol. Relat. Areas Mol. Biol., **38**, 413-448 (1973)

[3] Baker, J.J.; Jeng, I.; Barker, H.A.: Purification and properties of L-erythro-3,5-diaminohexanoate dehydrogenase from a lysine-fermenting Clostridium. J. Biol. Chem., **247**, 7724-7734 (1972)

2,4-Diaminopentanoate dehydrogenase 1.4.1.12

1 Nomenclature

EC number
1.4.1.12

Systematic name
2,4-diaminiopentanoate:NAD(P)$^+$oxidoreductase (deaminating)

Recommended name
2,4-diaminopentanoate dehydrogenase

Synonyms
2,4-diaminopentanoic acid C_4 dehydrogenase

CAS registry number
39346-26-4

2 Source Organism

<1> *Clostridium sticklandii* [1-3]
<2> *Clostridium sp.* (M-E, SB4 [2]) [2]

3 Reaction and Specificity

Catalyzed reaction
2,4-diaminopentanoate + H_2O + NAD(P)$^+$ = 2-amino-4-oxopentanoate + NH_3 + NAD(P)H + H^+

Reaction type
oxidation
redox reaction
reduction

Natural substrates and products
S 2,4-diaminopentanoate + H_2O + NAD(P)$^+$ <1, 2> (Reversibility: r <1, 2> [2, 3]) [1-3]
P 2-amino-4-oxopentanoate + NH_3 + NAD(P)H <1, 2> [1-3]
S 2,5-diaminohexanoate + H_2O + NAD(P)$^+$ <1, 2> (Reversibility: r <1, 2> [2]) [2]
P 2-amino-5-oxohexanoate + NH_3 + NAD(P)H <1, 2> [2]

Substrates and products

S 2,4-diaminopentanoate + H_2O + $NAD(P)^+$ <1, 2> (Reversibility: r <1, 2> [2, 3]) [1-3]

P 2-amino-4-oxopentanoate + NH_3 + $NAD(P)H$ <1, 2> [1-3]

S 2,5-diaminohexanoate + H_2O + $NAD(P)^+$ <1, 2> (Reversibility: r <1, 2> [2]) [2]

P 2-amino-5-oxohexanoate + NH_3 + $NAD(P)H$ <1, 2> (<1,2> 2-amino-5-oxohexanoate cyclizes non-enzymically to 1-pyrroline-2-methyl-5-carboxylate [2]) [2]

Inhibitors

5,5'-dithiobis(2-nitrobenzoate) <1> [1]
iodoacetamide <1> [2]
iodoacetate <1> [1]
n-ethylmaleimide <1> [1]
organic mercurials <1> [2]
p-chloromercuribenzoate <1> [1, 2]

Cofactors/prosthetic groups

NAD^+ <1> [1-3]
$NADP^+$ <1> [1-3]

Metals, ions

Additional information <1> (<1> no stimulation of activity by Mg^{2+}, Zn^{2+}, Mn^{2+}, Fe^{2+}, Cd^{2+}, Ni^{2+}, Cu^{2+} [3]) [3]

Specific activity (U/mg)

68.5 <1> [3]
272.5 <1> [1]

K_m-Value (mM)

0.11-0.28 <1> ($NADP^+$) [2]
0.6-3.3 <1> (NAD^+) [2, 3]
1.2-1.8 <1> (2,4-diaminopentanoate) [2, 3]
2.5 <1> (2,5-diaminohexanoate) [2]

pH-Optimum

8.8 <1> (<1> 2,4-diaminopentanoate + NAD^+ + H_2O [3]) [3]

pH-Range

8-10 <1> (<1> half maximal activity at pH 8.0 and 10.0, 2,4-diaminopentanoate + NAD^+ + H_2O [3]) [3]

4 Enzyme Structure

Molecular weight

72000-80000 <1> (<1> sedimentation equilibrium centrifugation [1,2]) [1, 2]
80000 <1> (<1> gel electrophoresis [1]) [1]

Subunits
    dimer <1> (<1> 2 * 35400 SDS-PAGE, sedimentation centrifugation in 6 M
    guanidine HCl [1]; <1> 2 * 40000 SDS-PAGE, gel electrophoresis [1]) [1]

5 Isolation/Preparation/Mutation/Application

Purification
    <1> [1, 3]

6 Stability

pH-Stability
    9 <1> (<1> unstable above [2]) [2]

Storage stability
    <1>, 4°C, phosphate buffer, pH 7,5, 10-20% glycerol, one month [3]

References

[1] Somack, R.; Costilow, R.N.: 2,4-Diaminopentanoic acid C_4 dehydrogenase,
    purification and properties of the protein. J. Biol. Chem., **247**, 385-388 (1973)
[2] Stadtman, T.C.: Lysine metyabolim by clostridia, XIIB 2,5-diaminohexanoate
    dehydrogenase (2,4-diaminopentanoate dehydrogenase). Adv. Enzymol. Re-
    lat. Areas Mol. Biol., **38**, 441-445 (1973)
[3] Tsuda, Y.; Friedmann, H.C.: Ornithine metabolism by Clostridium sticklan-
    dii. Oxidation of ornithine to 2-amino-4-ketopentanoic acid via 2,4-diamino-
    pentanoic acid; participation of B_{12} coenzyme, pyridoxal phosphate, and
    pyridine nucleotide. J. Biol. Chem., **245**, 5914-5926 (1970)

Glutamate synthase (NADPH) 1.4.1.13

1 Nomenclature

EC number
1.4.1.13

Systematic name
L-glutamate:NADP$^+$ oxidoreductase (transaminating)

Recommended name
glutamate synthase (NADPH)

Synonyms
EC 2.6.1.53 (formerly)
L-glutamate synthase
L-glutamate synthetase
L-glutamine:2-oxoglutarate aminotransferase, NADPH oxidizing
NADPH-GOGAT
NADPH-dependent glutamate synthase
NADPH-glutamate synthase
NADPH-linked glutamate synthase
glutamate synthetase (NADP)
glutamine amide-2-oxoglutarate aminotransferase (oxidoreductase, NADP)
glutamine-ketoglutaric aminotransferase
synthase, glutamate (reduced nicotinamide adenine dinucleotide phosphate)

CAS registry number
37213-53-9

2 Source Organism

<1> *Brevibacterium flavum* [10]
<2> *Klebsiella aerogenes* [23]
<3> *Escherichia coli* [1, 4, 5, 19, 20, 23]
<4> *Aerobacter aerogenes* [2]
<5> *Nocardia mediterranei* [3]
<6> *Pseudomonas aeruginosa* [6]
<7> *Azospirillum brasilense* [7-9, 26, 28-32]
<8> *Gluconobacter suboxydans* [11]
<9> *Bacillus licheniformis* [12]
<10> *Oryza sativa* [13]
<11> *Geobacillus stearothermophilus* [14]

<12> *Rhodospirillum rubrum* [15, 25]
<13> *Kluyveromyces fragilis* [16]
<14> *Bacillus megaterium* [17]
<15> *Thiobacillus thioparus* [18]
<16> *Bradyrhizobium sp.* [21]
<17> *Bacillus subtilis* [22]
<18> *Sclerotinia sclerotiorum* [24]
<19> *Pyrococcus sp.* [27]

3 Reaction and Specificity

Catalyzed reaction
2 L-glutamate + NADP$^+$ = L-glutamine + 2-oxoglutarate + NADPH + H$^+$
(<3>, kinetic mechanism [5]; <7,17>, mechanism [8,22]; <1>, hexa-uni ping
pong mechanism [10])

Reaction type
oxidation
redox reaction
reduction

Natural substrates and products
S L-glutamine + 2-oxoglutarate + NADPH <12, 18> (<12>, glutamate biosynthesis [15]; <12,18>, main route for assimilation of ammonium compounds [24,25]) (Reversibility: ? <12, 18> [15, 24, 25]) [15, 24, 25]
P L-glutamate + NADP$^+$ <12, 18> [15, 24, 25]

Substrates and products
S L-glutamine + 2-oxoglutarate + NADPH + H$^+$ <1-19> (<2, 3, 5, 9, 11, 14,
15>, in the reverse reaction ammonia can act instead of glutamine, but
more slowly [3, 12, 14, 17, 18, 20, 23]; <1, 6, 16>, ammonia does not replace L-glutamine as amino donor [6, 10, 21]; <15>, NH$_3$-dependent activity is increased approximately 5-fold in apoglutamate synthase lacking
flavin and non-heme iron [18]; <12>, no activity with NH$_4^+$ [25]; <1, 3, 8,
9>, highly specific [1, 10, 11, 12]; <3>, L-glutamine, 2-oxoglutarate and
NADPH are all required for catalytic activity [1]; <7>, when L-glutamine
is replaced by ammonia as the amino-group donor, the catalytic activity is
less than 1% [9]; <1>, glyoxylate shows 3% reactivity compared with α-
ketoglutarate [10]; <3>, both native and apoglutamate synthase catalyze
NADP$^+$ reduction at approximately 12% the rate of NADPH oxidation
[19]; <7>, 2-oxoglutarate promotes electron transfer from FAD to 3Fe-4S
cluster of the holoenzyme [31]) (Reversibility: ir <3, 7, 11> [1, 8, 14]; r
<3> [19]; ? <1, 2, 4-6, 8-10, 12-19> [2-7, 9-13, 15-18, 20-32]) [1-32]
P L-glutamate + NADP$^+$ <1-19> [1-32]
S NH$_3$ + 2-oxoglutarate + NADPH + H$^+$ <2, 3, 5, 11, 12, 14, 15, 17, 19>
(<5>, 14% relative activity to L-glutamine [3]; <11>, the rate is only
10% to 15% that of the L-glutamine-dependent reaction [14]; <14>, 2%

to 4% relative activity to L-glutamine [17]; <15>, ammonia activity with 100 mM NH_4Cl is about 6% of the glutamine activity with 5 mM L-glutamine [18]; <3>, the specific activity of native enzyme using NH_3 varies between 5% and 7% of the glutamine-dependent activity [20]; <17>, 24% relative activity to L-glutamine with NADPH as electron donor and 6.3% relative activity to L-glutamine with NADH as electron donor [22]; <12>, the activity with 10 mM NH_4^+ ions is less than 2% that with L-glutamine [25]) (Reversibility: ? <2, 3, 5, 11, 12, 14, 15, 17, 19> [3, 14, 17-20, 22, 23, 25, 27]) [3, 14, 17-20, 22, 23, 25, 27]

P L-glutamate + $NADP^+$ <2, 3, 5, 11, 12, 14, 15, 17, 19> [3, 14, 17-20, 22, 23, 25, 27]

S Additional information <2, 3, 7, 8, 15> (<3>, NH_4Cl, L-asparagine, D-glutamine, or alkylated glutamine analogues do not substitute for L-glutamine, pyruvate or oxalacetate do not substitute for α-ketoglutarate [1]; <7>, under conditions of physiological pH the enzyme exhibits a reversible half-reaction, but overall catalysis is essentially irreversible [8]; <8>, amino acids, amines and ammonium chloride do not substitute for L-glutamine, other α-keto acids including oxalacetate, pyruvate, glyoxylate and α-ketobutyrate do not support the activity [11]; <15>, α-ketoglutarate can not be replaced with pyruvate or oxalacetate [18]; <3>, NH_3-dependent activity is increased approximately 5fold in apoglutamate synthase lacking flavin and non-heme iron [19]; <2>, glutamine binding site of the enzyme is located on the heavy subunit of the enzyme, preparations of the enzyme that lack flavins or the flavins and iron sulfide catalyze NH_3-dependent reaction but not glutamine-dependent reaction [23]; <7>, the enzyme β subunit is devoid of glutamate synthase activity in either direction at both pH 7.5 and 9.5, but it can oxidize NADPH and transfer electrons to synthetic electron acceptors like iodonitrotetrazolium, ferricyanide, menadione, dichloroindophenol, the β subunit is highly specific toward NADPH, the rate of oxidation of NADH in the presence of electron acceptors is less than 5% of that measured with NADPH [26]; <7>, the α subunit catalyzes the synthesis of glutamate from L-glutamine and 2-oxoglutarate, provided that a reducing system is present, reducing system: dithionite and methyl viologen [28]; <7>, the recombinant enzyme has diaphorase activity, it can oxidize NADPH and transfer electrons to synthetic electron acceptors like iodonitrotetrazolium and ferricyanide [30]) [1, 8, 11, 18-19, 23, 26, 28, 30]

P ?

Inhibitors

2',5'-ADP <3> (<3>, acts as a competitive ihibitor [4]) [4]

2',5'-diphosphoadenylic acid <7> (<7>, competitive inhibitor with respect to NADPH [30]) [30]

2'-adenylic acid <3> (<3>, competitive inhibitor of the forward reaction with NADPH as the varied substrate, uncompetitive inhibitor with α-ketoglutarate or L-glutamine [5]) [5]

2'-phosphoadenosine-5'-diphospho-5'-β-D-ribose <7> (<7>, for the NADPH: acceptor oxidoreductase activity of the β subunit of the enzyme, uncompetitive inhibition with ferricyanide or iodonitrotetrazolium as substrate, competitive inhibition with NADPH as substrate [26]; <7>, inhibitor of the NADPH: iodonitrotetrazolium oxidoreductase reaction of the G298A-β subunit, competitive with respect to NADPH [29]) [26, 29, 30]

2'-phosphoadenylic acid <7> (<7>, competitive inhibitor with respect to NADPH [30]) [30]

2,2'-bipyridyl <15, 18> (<15>, 100% inhibition at 3 mM [18]; <18>, 88% inhibition at 10 mM [24]) [18, 24]

3-aminopyridine adenine dinucleotide phosphate <7> (<7>, good inhibitor of the holoenzyme, competitive with respect to NADPH [28]; <7>, inhibitor of the NADPH: iodonitrotetrazolium oxidoreductase reaction of the G298A-β subunit, competitive with respect to NADPH [29]) [28-30]

3-bromo-2-oxoglutarate <3> [19, 20]

4-iodoacetamidosalicylate <3> (<3>, inactivation is faster at pH 4.6 to 5.5 compared to pH 6.5 to 7.2 [19]) [19, 20]

6-diazo-5-oxo-L-norleucine <7, 12> (<7>, potent inhibitor, complete loss of activity at 0.1 mM after 10 min of preincubation [9]; <12>, 35% inhibition at 0.0025 mM, when reaction is initiated with 0.5 mM L-glutamine, 73% inhibition at 0.0025 mM, when reaction is initiated with 2.5 mM 2-oxoglutarate [25]) [9, 25]

ADP <3, 18> (<3>, 15% inhibition at 50 mM [1]) [1, 24]

AMP <18> [24]

ATP <1, 3, 15, 18> (<3>, 19% inhibition at 25 mM and 43% inhibition at 50 mM [1]; <1>, at 3.5 mM 12% inhibition [10]; <15>, 33% inhibition at 20 mM [18]; <18>, restricts activity by 47% at 5 mM [24]) [1, 10, 18, 24]

AgNO$_3$ <1> (<1>, at 0.001 mM completely inhibits [10]) [10]

BaCl$_2$ <3> (<3>, 61% inhibition at 50 mM [1]) [1]

CTP <3> (<3>, 9% inhibition at 25 mM and 21% inhibition at 50 mM [1]) [1]

CaCl$_2$ <3, 7, 9> (<3>, 74% inhibition at 50 mM [1]; <9>, 27% inhibition at 50 mM [12]) [1, 9, 12]

CdCl$_2$ <3> (<3>, produces more than 60% inhibition at 5 mM [1]) [1]

CoCl$_2$ <3> (<3>, produces more than 60% inhibition at 5 mM [1]) [1]

D-glutamate <1, 3, 5, 7, 8, 15, 18> (<3>, 50% inhibition at concentrations below 7 mM [1]; <16>, no inhibition: 50-200 mM [21]; <7>, competitive inhibitor with respect to 2-oxoglutarate, noncompetitive with L-glutamine and uncompetitive with NADPH [8]; <1>, at 3.5 mM 10% inhibition [10]) [1, 3, 8, 10, 11, 18, 24]

D-glutamine <8> [11]

D-lysine <1> (<1>, significantly inhibits [10]) [10]

D-methionine <3> (<3>, 19% inhibition at 50 mM [1]) [1]

DL-methionine DL-sulfoximine <8, 17, 18> (<8>, potent inhibitor [11]) [11, 22, 24]

DL-methionine sulfoxide <9> (<9>, potent inhibitor [12]) [12]

DL-theanine <8> [11]

GTP <3> (<3>, 16% inhibition at 25 mM and 43% inhibition at 50 mM [1])
[1]

HgCl$_2$ <1> (<1>, at 0.001 mM remains 8% of activity [10]) [10]

ITP <3> (<3>, 17% inhibition at 25 mM and 29% inhibition at 50 mM [1])
[1]

KBr <3, 7> (<3>, 43% inhibition at 100 mM [1]) [1, 9]

KCN <15> (<15>, 89% inhibition at 20 mM [18]) [18]

KCl <3> (<3>, 32% inhibition at 100 mM [1]) [1]

KI <7> [9]

KNO$_3$ <3> (<3>, 47% inhibition at 100 mM [1]) [1]

L-2-amino-4-oxo-5-chloropentanoic acid <2, 3> (<2,3>, selective inhibitor of
glutamine-dependent activity [20,23]) [20, 23]

L-alanine <3, 5, 8, 9> (<3>, significant inhibitor [1]; <5>, less than 50% inhibition at 5 mM [3]) [1, 3, 11, 12]

L-alanine <9> (<9>, 5% inhibition at 5 mM [12]) [12]

L-asparagine <3, 5, 7> (<3>, significant inhibitor [1]; <5>, less than 50%
inhibition at 5 mM [3]; <7>, activity decreases to about 0.5% of the original
at 25 mM [9]) [1, 3, 9]

L-aspartate <3, 8, 12> (<3>, 50% inhibition at concentrations below 7 mM
[1]) [1, 11, 25]

L-cysteine <3> (<3>, significant inhibitor [1]) [1]

L-fumarate <3, 17> (<3>, 21% inhibition at 50 mM [1]; <17>, potent inhibitor of the glutamine-dependent activity [22]) [1, 22]

L-glutamate <1, 3, 5, 15-17> (<3>, 50% inhibition at concentrations between
17 and 35 mM [1]; <5>, potent inhibitor [3]; <3>, competitive inhibitor of the
forward reaction with L-glutamine as the varied substrate, noncompetitive
inhibitor with α-ketoglutarate as the varied substrate [5]; <1>, at 3.5 mM
13% inhibition [10]; <15>, more than 40% inhibition at 50 mM [18]; <16>,
30% inhibition at 500 mM [21]; <17>, noncompetitive inhibitor with respect
to both 2-oxoglutarate and L-glutamine [22]) [1, 3, 5, 10, 18, 21, 22]

L-histidine <1, 3, 5, 8, 9, 15> (<1,3>, significant inhibitor [1,10]; <5>, less
than 50% inhibition at 5 mM [3]; <9>, 5% inhibition at 5 mM [12]; <15>,
more than 40% inhibition at 50 mM [18]) [1, 3, 10-12, 18]

L-homoserine <1, 3, 8> (<1,3>, significant inhibitor [1,10]) [1, 10, 11]

L-leucine <5, 8, 9, 18> (<5>, less than 50% inhibition at 5 mM [3]; <9>, 16%
inhibition at 5 mM [12]; <18>, less than 45% inhibition [24]) [3, 11, 12, 24]

L-malate <1, 3, 5, 17, 18> (<3>, 19% inhibition at 50 mM [1]; <5>, more than
50% inhibition at 5 mM [3]; <1>, at 3.5 mM 67% inhibition [10]; <17>, potent inhibitor of the glutamine-dependent activity [22]; <18>, 40% inhibition
at 5 mM [24]) [1, 3, 10, 22, 24]

L-methionine <1, 3, 7-9, 12, 15> (<3>, 50% inhibition at concentrations below 7 mM [1]; <5>, shows no effect [3]; <7>, competitive inhibitor with respect to L-glutamine [9]; <9>, 16% inhibition at 5 mM [12]; <15>, more than
40% inhibition at 50 mM [18]) [1, 9-12, 18, 25]

L-methionine DL-sulfoximine <9, 17> (<9>, potent inhibitor [12]; <17>, inhibits the glutamine-dependent reaction [22]) [12, 22]

L-methionine sulfone <3, 9, 11> (<3>, competitive inhibitor vs. L-glutamine, noncompetitive inhibitor vs. α-ketoglutarate and uncompetitive vs. NADPH [5]; <9>, potent inhibitor [12]) [5, 12, 14]

L-serine <3, 5, 9, 15> (<3>, significant inhibitor [1]; <5>, less than 50% inhibition at 5 mM [3]; <9>, 10% inhibition at 5 mM [12]; <15>, more than 40% inhibition at 10 mM and 50 mM [18]) [1, 3, 12, 18]

L-tryptophan <5> (<5>, strong inhibitor [3]) [3]

MgCl$_2$ <3, 9> (<3>, 70% inhibition at 50 mM [1]; <9>, 31% inhibition at 50 mM [12]) [1, 12]

MnCl$_2$ <3, 9> (<3>, 82% inhibition at 50 mM [1]; <9>, 64% inhibition at 50 mM [12]) [1, 12]

N-ethylmaleimide <3, 15> (<15>, 83% inhibition at 10 mM [18]) [18, 20]

NAD$^+$ <17> (<17>, competitive inhibition with NADH, noncompetitive inhibition with 2-oxoglutarate or L-glutamine [22]) [22]

NADH <1, 8> (<1>, at 3.5 mM 31% inhibition [10]; <8>, partially inhibits [11]) [10, 11]

NADP$^+$ <1, 3, 5, 7, 8, 9, 12, 15, 17> (<3>, 50% inhibition at concentrations below 7 mM [1]; <5>, potent inhibitor [3]; <3>, competitive vs. NADPH and noncompetitive vs. L-glutamine [5]; <1>, 76% inhibition [10]; <8,9>, potent inhibitor [11,12]; <9>, competitive inhibition with respect to NADPH [12]; <15>, 59% inhibition at 1 mM [18]; <17>, competitive inhibition with NADPH, noncompetitive inhibition with 2-oxoglutarate or L-glutamine [22]; <12>, competitive inhibitor with NADPH, 54% inhibition at 1 mM [25]; <7>, inhibitor of the NADPH: iodonitrotetrazolium oxidoreductase reaction of the G298A-β subunit, competitive with respect to NADPH [29]) [1, 3, 5, 10, 11, 12, 18, 22, 25, 29, 30]

NADPH <3, 9, 11> (<11>, rapid inactivation in absence of L-glutamine and 2-oxoglutarate, capable of causing 50% inactivation in 5 min at a concentration of 1 mM [14]; <3>, inactivation of glutamine-dependent activity, 50% inactivation at 0.36 mM in 10 min [20]) [12, 14, 20]

NaClO$_4$ <7> [9]

NaN$_3$ <15> (<15>, 100% inhibition at 50 mM, 11% inhibition at 20 mM [18]) [18]

NiCl$_2$ <3> (<3>, produces more than 60% inhibition at 5 mM [1]) [1]

O-carbamoylserine <3> (<3>, competitive inhibitor vs. L-glutamine [5]) [5]

UTP <3> (<3>, 17% inhibition at 25 mM and 33% inhibition at 50 mM [1]) [1]

acetyl-NADP$^+$ <7> [30]

α-aminobutyrate <8> [11]

arginine <18> (<18>, less than 45% inhibition [24]) [24]

atebrin <15> (<15>, 100% inhibition at 1 mM [18]) [18]

azaserine <11, 12, 18> (<11>, in both the L-glutamine- and NH$_4$Cl-dependent reactions [14]; <18>, 68% inhibition at 1 mM [24]; <12>, 38% inhibition at 0.0125 mM when reaction is initiated with 0.5 mM L-glutamine and 69% inhibition at 0.0125 mM when reaction is initiated with 2.5 mM 2-oxoglutarate [25]) [14, 24, 25]

bromopyruvate <3> [19, 20]

Cibacron blue 3GA <11> [14]

cis-aconitate <3> (<3>, 22% inhibition at 50 mM [1]) [1]

citrate <18> (<18>, slight [24]) [24]

dimethyl suberimidate <14> (<14>, inactivates glutamine-dependent activity [17]) [17]

glycine <3, 5, 9> (<3>, significant inhibitor [1]; <5>, less than 50% inhibition at 5 mM [3]; <9>, 5% inhibition at 5 mM [12]) [1, 3, 12]

hydroxylamine <8> [11]

iodoacetamide <3> (<3>, more than 50% loss of activity after 60 min at 0.5 mM [1]; <3>, selective inactivation of glutamine-dependent activity [20]) [1, 19, 20]

iodoacetate <1, 15> (<1>, completely inhibits at 1 mM, 2-mercaptoethanol protects against inhibition [10]; <15>, 27% inhibition at 25 mM [18]) [10, 18, 19]

iodonitrotetrazolium <7> (<7>, inhibitor of the L-glutamate: iodonitrotetrazolim oxidoreductase activity of the α subunit at concentrations above 0.1 mM [28]) [28]

isocitrate <3, 5, 15> (<3>, 35% inhibition at 50 mM [1]; <5>, 35% inhibition at 5 mM [3]; <15>, 32% inhibition at 20 mM [18]) [1, 3, 18]

lysine <1, 8> (<1>, significantly inhibits [10]) [10, 11]

o-phenanthroline <15, 18> (<15>, 37% inhibition at 10 mM [18]; <18>, complete inhibition at 10 mM [24]) [18, 24]

oxalate <18> (<18>, 48% inhibition at 5 mM [24]) [24]

oxalglycine <1> [10]

oxaloacetate <3, 8, 15, 18> (<15>, more than 50% inhibition at 20 mM [18]) [1, 11, 18, 24]

oxalylglycine <3> (<3>, competitive inhibitor vs. α-ketoglutarate, noncompetitive inhibitor vs. L-glutamine and uncompetitive vs. NADPH [5]) [5]

p-chloromercuribenzoate <1, 3, 17, 18> (<1>, completely inhibits at 1 mM, 2-mercaptoethanol protects against inhibition [10]; <17>, inhibits the glutamine-dependent reaction [22]; <18>, complete inhibition at 0.1 mM [24]) [10, 20, 22, 24]

p-chloromercuriphenylsulfonate <3> (<3>, more than 50% loss in activity in 60 min at 0.05 mM [1]) [1]

p-hydroxymercuribenzoate <15> (<15>, complete inhibition at 1 mM, 5 min [18]) [18]

phenylalanine <1, 18> (<1>, significantly inhibits [10]; <18>, 78% inhibition at 10 mM [24]) [10, 24]

putrescine <8> [11]

pyridoxal-5'-phosphate <17> (<17>, 52% inhibition of the glutamine-dependent reaction at 5 mM [22]) [22]

pyruvate <3, 8, 18> (<3>, 20% inhibition at 50 mM [1]) [1, 11, 24]

sodium arsenite <15, 18> (<15>, 100% inhibition at 50 mM , 10% inhibition at 20 mM [18]; <18>, 57% inhibition at 10 mM [24]) [18, 24]

succinate <3, 5, 18> (<3>, 16% inhibition at 50 mM [1]; <5>, 25% inhibition at 5 mM [3]) [1, 3, 24]

thio-NADP$^+$ <7> [30]

valine <8, 18> (<18>, less than 45% inhibition [24]) [11, 24]
Additional information <9, 15> (<9>, relatively insensitive to inhibition by amino acids, keto acids or various nucleotides, 38% inhibition with a combination of each 5 mM L-methionine, L-leucine, L-serine, L-histidine, L-glycine and L-alanine [12]; <15>, a combination of L-serine, L-methionine, L-alanine, L-glycine, L-histidine and L-asparagine inhibits 73% of the acitivity at 7.5 mM each [18]) [12, 18]

Cofactors/prosthetic groups

NADH <17> (<1-3, 5, 6, 7, 11, 12, 15, 18>, no activity with NADH [1, 3, 6, 9, 10, 14, 15, 18, 23, 24, 25]; <17>, 21% relative activity to NADPH [22]) [22]
NADPH <1-19> (<7>, oxidizes NADPH stereospecifically at the 4S position [8]; <7>, the β-subunit contains the NADPH binding site of the enzyme [26, 28]; <7>, the C-terminal potential ADP-binding fold of the β-subunit is the NADPH-binding site of the enzyme [29]) [1-32]
flavin <1, 3, 5, 7, 9, 12, 15, 17> (<3, 7, 12, 15>, flavoenzyme [1, 7, 9, 15, 18, 25, 26, 30-32]; <3, 5, 7>, contains FAD and FMN [1, 3, 7, 9, 30-32]; <7>, ratio FAD:FMN is 1 [9, 31]; <3, 9>, ratio FAD:FMN is 1.2 [1, 12]; <3>, enzyme contains 7.8 mol of flavin [1]; <1>, enzyme contains 2 mol of FMN and 2 mol of FAD [10]; <9>, 2 flavin moieties per 220000 MW [12]; <17>, 0.83 mol of FMN and 0.81 mol of FAD per mol of enzyme [22]; <5>, ratio FAD:FMN is 1.1 [3]; <7>, the β-subunit contains the binding site for FAD, 0.83 mol FAD per mol β-subunit [26, 28]; <7>, the α-subunit contains FMN as flavin cofactor, 0.94 FMN bound per α-subunit [28]; <7>, 0.86 mol FAD per mol of mutant β-subunit, 0.83 mol FAD per mol of the wild type species [29]; <7>, 0.83 mol FAD and 0.86 mol FMN per mol $\alpha\beta$-protomer [30]) [1, 3, 7, 9, 10, 12, 15, 18, 22, 25, 26, 28-32]
Additional information <3, 5, 7, 9, 17> (<3>, purified enzyme contains 30.4 mol of labile sulfide per 800,000 g of protein [1]; <5>, the enzyme contains 7.2 mol of acid-labile sulfur per 200,000 g of protein [3]; <7>, the FMN and FAD prosthetic groups are demonstrated to be nonequivalent with respect to their reactivities with sulfite, sulfite reacts with only one of the two flavins forming an N(5)-sulfite adduct [7]; <7>, the enzyme contains 7.9 sulfur atoms per protomer with a molecular weight of 185000 Da [9]; <9>, the enzyme contains 8.1 acid-labile sulfur atoms per 220000-dalton dimer [12]; <7>, thio-NADPH and acetylpyridine-NADPH can be used as electron donors but are less efficient than NADPH [8]; <17>, the enzyme contains 8.7 g atoms per mol [22]; <7>, the enzyme contains three distinct ion-sulfur centers per $\alpha\beta$-protomer [26]; <7>, the α-subunit contains the [3Fe-4S] cluster of the enzyme [28]; <7>, the enzyme contains three different iron-sulfur clusters, one 3Fe-4S center on the α-subunit and two 4Fe-4S clusters of unknown location, 11.7 mol sulfur per mol $\alpha\beta$-protomer [30]; <7>, midpoint potential value of the FMN cofactor: approximately -240 mV, midpoint potential value of the 3Fe-4S cluster: approximately -270 mV, midpoint potential value of the FAD cofactor: approximately -340 mV for the β-subunit and -300 mV for the holoenzyme [31]) [1, 3, 7-9, 12, 22, 26, 28, 31]

Activating compounds

dimethyl suberimidate <14> (<14>, increases NH_3-dependent activity [17])
[17]

Metals, ions

iron <1, 3, 5, 7, 9, 12, 17> (<3, 5, 7, 9, 12, 17>, iron-sulfur protein [1, 3, 7, 9,
12, 15, 22, 25, 26, 30-32]; <3>, contains 38.4 mol of iron [1]; <7>, contains
(Fe-S)II and (Fe-S)III as [4Fe-4S] clusters [7]; <7>, 12.1 mol of non-heme iron
per enzyme protomer [7]; <7>, 8.1 iron atoms per MW 185000 [9]; <9>, 8.1
iron atoms per MW 220000 Da [12]; <1>, enzyme contains 8 mol of non-
heme iron [10]; <17>, 7.4 g atoms of non-heme iron per mol of enzyme
[22]; <5>, contains 7.5 mol of nonheme iron [3]; <7>, contains 12.05 mol
Fe^{2+} per mol $\alpha\beta$-protomer [30]) [1, 3, 7, 9, 10, 12, 15, 22, 25, 26, 30-32]

Turnover number (min^{-1})

1098 <7> (NADPH, <7>, of NADPH: acceptor oxidoreductase activity of the
β-subunit of the enzyme, co-substrate: iodonitrotetrazolium, inhibitor: 2'-
phosphoadenosine-5'-diphospho-5'-β-D-ribose [26]) [26]
1200 <7> (iodonitrotetrazolium, <7>, of NADPH: acceptor oxidoreductase
activity of the β-subunit of the enzyme, co-substrate: NADPH, inhibitor: 2'-
phosphoadenosine-5'-diphospho-5'-β-D-ribose [26]) [26]
1518 <7> (menadione, <7>, of NADPH: acceptor oxidoreductase acitivity of
the β-subunit of the enzyme [26]) [26]
1848 <7> (NADPH, <7>, of NADPH: acceptor oxidoreductase activity of the
β-subunit of the enzyme, co-substrate: ferricyanide, inhibitor: 2'-phosphoa-
denosine-5'-diphospho-5'-β-D-ribose [26]) [26]
1920 <7> (iodonitrotetrazolium, <7>, of NADPH: acceptor oxidoreductase
acitivity of the β-subunit of the enzyme [26]) [26]
2370 <7> (ferricyanide, <7>, of NADPH: acceptor oxidoreductase activity of
the β-subunit of the enzyme, co-substrate: NADPH, inhibitor: 2'-phosphoa-
denosine-5'-diphospho-5'β-D-ribose [26]) [26]
2514 <7> (ferricyanide, <7>, of NADPH: acceptor oxidoreductase acitivity of
the β-subunit of the enzyme [26]) [26]

Specific activity (U/mg)

0.24 <3> (<3>, apoenzyme lacking flavin and non-heme iron, glutamine-dependent activity [19]) [19]
0.92 <3> (<3>, native enzyme, NH_3-dependent activity [19]) [19]
1.48 <5> [3]
4.45 <3> (<3>, apoenzyme lacking flavin and non-heme iron, NH_3-depen-
dent activity [19]) [19]
5.72 <8> [11]
7.6 <18> (<18>, partially purified enzyme [24]) [24]
9.3 <15> [18]
13.33 <3> [23]
14.3 <3> (<3>, native enzyme [20]) [20]
17.9 <7> [30]
18.64 <3> (<3>, native enzyme, glutamine-dependent activity [19]) [19]

19.5 <7> [9]
20.83 <2> [23]
21 <12> [25]
22.9 <9> [12]
23.8 <14> [17]
26.2 <3> [1]
30.5 <1> [10]
1910 <19> [27]
Additional information <2, 3, 7, 11, 12, 15, 16> [4, 14, 15, 18, 19, 21, 23, 30]

K_m-Value (mM)

0.0022 <3> (NADPH, <3>, in the presence of 1 mM α-ketoglutarate and 5 mM L-glutamine [5]) [5]

0.003 <15> (NADPH) [18]

0.0035 <7> (NADPH, <7>, of NADPH: acceptor oxidoreductase activity of the β-subunit of the enzyme, acceptor: iodonitrotetrazolium [26]) [26]

0.0047 <3> (α-ketoglutarate, <3>, in the presence of 0.04 mM NADPH and 5 mM L-glutamine [5]) [5]

0.006 <17> (NADPH, <17>, in the presence of a fixed concentration of L-glutamine, and varied concentrations of α-ketoglutarate [22]) [22]

0.00625 <7> (NADPH) [9]

0.007 <17> (NADPH, <17>, in the presence of a fixed concentration of α-ketoglutarate, and varied concentrations of L-glutamine [22]) [22]

0.007 <17> (α-ketoglutarate, <17>, in the presence of a fixed concentration of NADPH, and varied concentrations of L-glutamine [22]) [22]

0.0071 <14> (NADPH) [17]

0.0073 <3> (α-ketoglutarate) [1]

0.0077 <3> (NADPH) [1]

0.0086 <8> (α-ketoglutarate) [11]

0.009 <14> (2-oxoglutarate) [17]

0.0098 <7> (NADPH, <7>, of NADPH: acceptor oxidoreductase activity of the β-subunit of the enzyme, acceptor: menadione [26]) [26]

0.01 <12> (2-oxoglutarate) [25]

0.01 <17> (α-ketoglutarate, <17>, in the presence of a fixed concentration of NADH, and varied concentrations of L-glutamine [22]) [22]

0.01 <7> (thio-NADPH, <7>, in the presence of 2.5 mM 2-oxo-glutarate and 5 mM L-glutamine, the apparent maximal velocity is 54% that obtained in the presence of NADPH [8]) [8]

0.0118 <7> (NADPH, <7>, of NADPH: acceptor oxidoreductase activity of the β-subunit of the enzyme, acceptor: ferricyanide [26]) [26]

0.012 <2> (NADPH) [23]

0.012 <17> (α-ketoglutarate, <17>, in the presence of a fixed concentration of L-glutamine, and varied concentrations of NADH [22]) [22]

0.013 <9> (NADPH) [12]

0.013 <17> (α-ketoglutarate, <17>, in the presence of a fixed concentration of L-glutamine, and varied concentrations of NADPH [22]) [22]

0.014 <8> (NADPH) [11]

0.014 <3> (NADPH, <3>, NH$_3$-dependent glutamate synthase and apoglutamate synthase [19]) [19]
0.015 <11> (2-oxoglutarate) [14]
0.015 <12> (NADPH) [15]
0.016 <12> (NADPH) [25]
0.017 <7> (acetylpyridine-NADPH, <7>, in the presence of 2.5 mM 2-oxo-
glutarate and 5 mM L-glutamine, the apparent maximal velocity is 3.7% that
obtained in the presence of NADPH [8]) [8]
0.022 <11> (NADPH) [14]
0.029 <7> (2-oxoglutarate) [9]
0.029 <11> (L-glutamine) [14]
0.035 <12> (2-oxoglutarate) [15]
0.035 <18> (NADPH) [24]
0.035 <7> (menadione, <7>, of NADPH: acceptor oxidoreductase activity of
the β-subunit of the enzyme [26]) [26]
0.05 <15> (α-ketoglutarate) [18]
0.05 <7> (iodonitrotetrazolium, <7>, of NADPH: acceptor oxidoreductase
activity of the β-subunit of the enzyme [26]) [26]
0.053 <5> (α-ketoglutarate) [3]
0.06 <1> (NADPH) [10]
0.06 <1> (α-ketoglutarate) [10]
0.065 <12> (L-glutamine) [25]
0.077 <5> (L-glutamine) [3]
0.084 <7> (NADPH, <7>, of NADPH: iodonitrotetrazolium oxidoreductase
activity of the G298A-β-subunit [29]) [29]
0.091 <17> (NADH, <17>, in the presence of a fixed concentration of L-glutamine, and varied concentrations of α-ketoglutarate [22]) [22]
0.102 <17> (L-glutamine, <17>, in the presence of a fixed concentration of
NADPH, and varied concentrations of α-ketoglutarate [22]) [22]
0.108 <17> (L-glutamine, <17>, in the presence of a fixed concentration of
NADH, and varied concentrations of α-ketoglutarate [22]) [22]
0.11 <5> (NADPH) [3]
0.113 <17> (NADH, <17>, in the presence of a fixed concentration of α-ke-
toglutarate, and varied concentrations of L-glutamine [22]) [22]
0.115 <17> (L-glutamine, <17>, in the presence of a fixed concentration of α-
ketoglutarate, and varied concentrations of NADPH [22]) [22]
0.12 <17> (L-glutamine, <17>, in the presence of a fixed concentration of α-
ketoglutarate, and varied concentrations of NADH [22]) [22]
0.13 <12> (L-glutamine) [15]
0.14 <7> (ferricyanide, <7>, of NADPH: acceptor oxidoreductase activity of
the β-subunit of the enzyme [26]) [26]
0.2 <14> (L-glutamine) [17]
0.23 <7> (L-glutamine, <7>, holoenzyme, buffer: Hepes, pH 8.5 or 7.5 [32])
[32]
0.24 <3> (2-oxoglutarate, <3>, NH$_3$-dependent glutamate synthase and apoglutamate synthase [19]) [19]
0.24 <1> (L-glutamine) [10]

0.25 <3> (L-glutamine) [1]
0.3 <2> (L-glutamine) [23]
0.3 <2> (α-ketoglutarate) [23]
0.35 <18> (α-ketoglutarate) [24]
0.45 <7> (L-glutamine) [9]
0.5-2 <2> (NH_4Cl) [23]
0.59 <8> (L-glutamine) [11]
0.63 <7> (L-glutamine, <7>, α-subunit, buffer: HEPES, pH 7.5 [32]) [32]
0.73 <7> (L-glutamine, <7>, holoenzyme, buffer: CAPS, pH 9.5 [32]) [32]
0.92 <7> (L-glutamine, <7>, α-subunit, buffer: HEPES, pH 8.5 [32]) [32]
1.1 <15> (L-glutamine) [18]
1.5 <7> (L-glutamine, <7>, α-subunit, buffer: CAPS, pH 9.5 [32]) [32]
1.7 <6> (L-glutamine) [6]
2 <6> (2-oxoglutarate) [6]
2.6 <18> (L-glutamine) [24]
22 <14> (NH_4Cl) [17]
44 <11> (NH_4Cl) [14]
Additional information <9, 7> [12, 28, 30]

K_i-Value (mM)
0.00023 <7> (thio-$NADP^+$) [30]
0.0024 <7> (3-aminopyridine adenine dinucleotide phosphate) [30]
0.0024 <7> (3-aminopyridine adenine dinucleotide phosphate, <7>, inhibitor of the holoenzyme, competitive with NADPH [28]) [28]
0.0028 <7> (2'-phosphoadenosine 5'-diphospho-5-β-D-ribose) [30]
0.0058 <7> (acetyl-$NADP^+$) [30]
0.0059 <7> ($NADP^+$) [30]
0.01 <1> (methionine sulfone, <1>, competitive inhibitor with respect to glutamine [10]) [10]
0.011 <7> (D-glutamate, <7>, good competitive inhibitor with respect to 2-oxoglutarate [8]) [8]
0.0112 <7> (3-aminopyridine adenine dinucleotide phosphate, <7>, inhibitor of the NADPH:iodonitrotetrazolium oxidoreductase reaction of wild type-β-subunit [29]) [29]
0.012 <11> (Cibacron Blue 3GA) [14]
0.02 <8> (DL-methionine DL-sulfoximine, <8>, competitive inhibitor with respect to α-keto-glutarate [11]) [11]
0.02 <9> ($NADP^+$, <9>, competitive inhibition with NADPH [12]) [12]
0.021 <7> (2'-phosphoadenosine-5'-diphospho-5'-β-D-ribose, <7>, for the NADPH: acceptor oxidoreductase activity of the β-subunit of the enzyme, substrate: NADPH, cosubstrate: ferricyanide [26]) [26]
0.022 <11> (L-methionine sulphone) [14]
0.0227 <7> (2'-phosphoadenosine-5'-diphosphoribose, <7>, inhibitor of the NADPH:iodonitrotetrazolium oxidoreductase reaction of wild type-β-subunit [29]) [29]
0.025 <3> (L-2-amino-4-oxo-5-chloropentanoic acid) [20]

0.037 <7> (NADP$^+$, <7>, inhibitor of the NADPH:iodonitrotetrazolium oxidoreductase reaction of G298A-β-subunit [29]) [29]

0.041 <1> (methionine sulfone, <1>, uncompetitive inhibitor with respect to α-ketoglutarate [10]) [10]

0.044 <7> (D-glutamate, <7>, uncompetitive inhibitor with NADPH [8]) [8]

0.045 <1> (methionine sulfone, <1>, uncompetitive inhibitor with respect to NADPH [10]) [10]

0.048 <7> (2',5'-diphosphoadenylic acid) [30]

0.056 <7> (D-glutamate, <7>, noncompetitive inhibitor with L-glutamine [8]) [8]

0.07 <8> (NADP$^+$, <8>, competitive inhibition with NADPH [11]) [11]

0.08 <7> (2'-phosphoadenosine-5'-diphospho-5'-β-D-ribose, <7>, for the NADPH:acceptor oxidoreductase activity of the β-subunit of the enzyme, substrate: NADPH, cosubstrate: iodonitrotetrazolium [26]) [26]

0.106 <3> (2',5'-ADP) [4]

0.12 <8> (NADP$^+$, <8>, mixed-type inhibition with α-ketoglutarate [11]) [11]

0.128 <7> (3-aminopyridine adenine dinucleotide phosphate, <7>, inhibitor of the NADPH:iodonitrotetrazolium oxidoreductase reaction of G298A-β-subunit [29]) [29]

0.14 <8> (DL-methionine DL-sulfoximine, <8>, competitive inhibitor with respect to glutamine [11]) [11]

0.18 <7> (2'-phosphoadenosine-5'-diphospho-5'-β-D-ribose, <7>, for the NADPH:acceptor oxidoreductase activity of the β-subunit of the enzyme, substrate: ferricyanide, cosubstrate: NADPH [26]) [26]

0.186 <7> (2'-phosphoadenylic acid) [30]

0.233 <7> (2'-phosphoadenosine-5'-diphosphoribose, <7>, inhibitor of the NADPH:iodonitrotetrazolium oxidoreductase reaction of G298A-β-subunit [29]) [29]

0.25 <1> (NADP$^+$, <1>, competitive inhibitor vs. NADPH [10]) [10]

0.3 <8> (NADP$^+$, <8>, uncompetitive inhibition with respect to glutamine [11]) [11]

0.48 <8> (methionine, <8>, competitive inhibitor with respect to α-keto-glutarate [11]) [11]

0.495 <7> (2'-phosphoadenosine-5'-diphospho-5'-β-D-ribose, <7>, for the NADPH:acceptor oxidoreductase activity of the β-subunit of the enzyme, substrate: iodonitrotetrazolium, cosubstrate: NADPH [26]) [26]

0.69 <1> (NADP$^+$, <1>, uncompetitive inhibitor vs. α-ketoglutarate [10]) [10]

0.76 <8> (DL-methionine DL-sulfoximine, <8>, noncompetitive inhibitor with respect to NADPH [11]) [11]

0.8 <1> (oxalylglycine, <1>, competitive inhibitor with respect to α-ketoglutarate [10]) [10]

0.86 <3> (p-chloromercuribenzoate) [19]

1.05 <7> (L-methionine, <7>, competitive inhibitor with respect to L-glutamine [8]) [8]

1.13 <1> (NADP$^+$, <1>, uncompetitive inhibitor vs. glutamine [10]) [10]

2.1 <1> (2'-adenylic acid, <1>, competitive inhibitor with respect to NADPH [10]) [10]

3 <8> (methionine, <8>, competitive inhibitor with respect to glutamine [11]) [11]

5.5 <8> (methionine, <8>, noncompetitive inhibitor with respect to NADPH [11]) [11]

17.1 <1> (2'-adenylic acid, <1>, uncompetitive inhibitor with respect to glutamine [10]) [10]

17.2 <1> (2'-adenylic acid, <1>, uncompetitive inhibitor with respect to α-keto-glutarate [10]) [10]

19.3 <1> (oxalylglycine, <1>, uncompetitive inhibitor with respect to glutamine [10]) [10]

20.6 <1> (oxalylglycine, <1>, uncompetitive inhibitor with respect to NADP [10]) [10]

25 <1> (glutamate, <1>, competitive inhibitor with respect to glutamine [10]) [10]

45 <1> (glutamate, <1>, competitive inhibitor with respect to α-ketoglutarate [10]) [10]

Additional information <3,7> [5,28,30]

pH-Optimum

6.5 <19> (<19>, optimum for both NH_3-dependent and L-glutamine-dependent reaction [27]) [27]

6.9 <17> (<17>, NADH-glutamine dependent activity [22]) [22]

7 <1> [10]

7.3-7.8 <15> [18]

7.5 <3, 6, 7> [5, 6, 9]

7.5-8.5 <5> (<5>, broad optimum [3]) [3]

7.6 <3> [1]

7.8 <2, 9, 17> (<9>, glutamine-dependent activity [12]; <17>, NADPH-glutamine dependent activity [22]; <2>, glutamine-dependent activity [23]) [12, 22, 23]

8 <2, 11, 12> (<2,11>, glutamine-dependent activity [14,23]) [14, 15, 23]

8.3 <16> [21]

8.4 <3> (<3>, for glutamate synthesis, NH_3-dependent glutamate synthase and apoenzyme [19]) [19]

8.5 <8> [11]

8.7 <9> (<9>, ammonia-dependent activity [12]) [12]

9 <2> (<2>, ammonia-dependent activity [23]) [23]

9.4 <3, 17> (<17>, NADPH-NH_3 dependent activity [22]; <3>, optimum for glutamate-dependent reduction of $NADP^+$, NH_3-dependent enzyme, apoglutamate synthase enzyme [19]) [19, 22]

pH-Range

6.4-9 <3> (<3>, at pH 6.4 and pH 9.0 about 50% of activity maximum [1]) [1]

6.5-8.7 <3> (<3>, at pH 6.5 and pH 8.7 about 50% of activity maximum [1]) [1]

7.4-9.1 <16> (<16>, at pH 7.4 and pH 9.1 about 50% of activity maximum [21]) [21]

8-9 <8> [11]

Temperature optimum (°C)

43 <12> [15]

75 <11> (<11>, glutamine-dependent reaction [14]) [14]

80 <19> (<19>, glutamine-dependent reaction [27]) [27]

90 <19> (<19>, NH$_3$-dependent reaction [27]) [27]

4 Enzyme Structure

Molecular weight

160000 <6, 11> (<6,11>, gel filtration [6,14]) [6, 14]

195000 <5> (<5>, sucrose density gradient centrifugation [3]) [3]

200000 <7> (<7>, approximate value, the $\alpha\beta$ protomeric form prevails at low protein concentration and at high ionic strength, gel filtration [30]) [30]

205000 <19> (<19>, gel filtration [27]) [27]

210000 <17> (<17>, sucrose density gradient sedimentation [22]) [22]

220000 <9> (<9>, sucrose density gradient sedimentation [12]) [12]

260000 <12> (<12>, gel filtration, in the absence of salt [15]) [15]

280000 <15> (<15>, gel filtration [18]) [18]

500000 <8> (<8>, gel filtration [11]) [11]

740000 <7> (<7>, gel filtration [9]) [9]

800000 <3, 7, 12> (<3>, sedimentation equilibrium, gel filtration [1]; <12>, gel filtration, presence of Cl$^-$ [25]; <7>, approximate value, the tetrameric form is found at high protein concentration and low ionic strength, gel filtration [30]) [1, 25, 30]

840000 <12, 14> (<12>, gel filtration, in the presence of 0.4 M NaCl [15]; <14>, sucrose density-gradient centrifugation [17]) [15, 17]

Additional information <12> (<12>, MW is salt-dependent [15]) [15]

Subunits

? <2, 12, 14, 15> (<2>, x * 175000 + x * 51500, SDS-PAGE [23]; <14>, x * 142000 + x * 55000, SDS-PAGE [17]; <15>, x * 72000 + x * 94000 [18]; <12>, x * 152000 + x * 53000, SDS-PAGE [25]) [17, 18, 23, 25]

dimer <5, 9, 17> (<5>, 1 * 145000 + 1 * 55000, SDS-PAGE [3]; <9>, 1 * 158000 + 1 * 54000, SDS-PAGE [12]; <17>, 1 * 160000 + 1 * 56000 [22]) [3, 12, 22]

octamer <3, 7> (<3>, 4 * 53000 + 4 * 135000, SDS-PAGE, urea polyacrylamide gel electrophoresis [1]; <7>, 4 * 135000 + 4 * 50000, SDS-PAGE [9]) [1, 9]

tetramer <19> (<19>, 4 * 53000, SDS-PAGE [27]) [27]

5 Isolation/Preparation/Mutation/Application

Localization

thylakoid membrane <12> (<12>, association with [15]) [15]

Purification

<1> (using ammonium sulfate fractionation, column chromatography on DEAE-cellulose, Sepharose 6B, hydroxyapatite, glutamate dehydrogenase antibody affinity chromatography and column chromatography on Sephacryl S-200 [10]) [10]

<2> (using sonication, heat treatment, protamine sulfate precipitation, ammonium sulfate precipitation, gel filtration on G-50, column chromatography on DEAE-Sephadex and gel filtration on Sephadex G-200 and Sepharose 6B [23]) [23]

<3> (using streptomycin sulfate treatment, ammonium sulfate fractionation, heat treatment, agarose gel filtration and DEAE-cellulose column chromatography [1,20]; using sulfate precipitation, gel filtration and column chromatography on 2',5'-ADP-Sepharose [4]) [1, 4, 20]

<5> (using protamine sulfate precipitation, ammonium sulfate precipitation, column chromatography on DE52, DEAE-Sephadex, hydroxyapatite, phenyl-Sepharose CL-4B and Bio-Gel filtration [3]) [3]

<6> (partial purification using column chromatography on DEAE-cellulose [6]) [6]

<7> (using heat treatment, ammonium sulfate precipitation, column chromatography on DEAE-Trisacryl, gel filtration and affinity chromatogray [9]; of the recombinant enzyme β-subunit, using ammonium sulfate fractionation, affinity chromatography on Reactive Red, ultrafiltration and column chromatography on Ultrogel AcA 54 [26]; of the α-subunit, using column chromatography on Q-Sepharose and Ultrogel AcA34 [28]; of the G298A-β-subunit, using column chromatography on Reactive Red or Amicon Red resins and gel filtration on Ultrogel AcA 34 [29]; of the recombinant enzyme from overproducing Escherichia coli cells, using ion exchange chromatography on Q-Sepharose, gel filtration on Sephacryl S-300 and affinity chromatography on 2',5' ADP-Sepharose 4B colum [30]) [9, 26, 28-30]

<8> (using ammonium sulfate fractionation, column chromatography on DEAE-cellulose, hydroxyapatite, Sepharose 6B and Sephadex G-200 [11]) [11]

<9> (using streptomycin sulfate treatment, ammonium sulfate precipitation, column chromatography on DEAE-Sephacel, Bio-Gel A 1.5 m, Red Sepharose CL6B and Sephadex G-25 [12]) [12]

<11> (using ammonium sulfate fractionation, gel filtration on Sephacryl S-300 and affinity chromatography on NADPH-Sepharose [14]) [14]

<12> (partial, using ultrasonic oscillation, ammonium sulfate precipitation, column chromatography on Sephacryl S300 and DE-52 cellulose [15]; using ammonium sulfate precipitation and column chromatography on DEAE-Sepharose, 2',5'-ADP-Sepharose and Sephacryl S-300 [25]) [15, 25]

<14> (using streptomycin sulfate treatment, ammonium sulfate precipitation, column chromatography on Ultrogel AcA 22 and DEAE-Sephadex A-50, ultrafiltration with an Amicon PM 30 membrane and chromatography on hydroxyapatite column [17]) [17]

<15> (using ammonium sulfate fractionation, column chromatography on DEAE-cellulose and Sephadex G-200 [18]) [18]

<17> [22]

<18> (using ammonium sulfate fractionation, column chromatography on DEAE-cellulose and Blue-Sepharose [24]) [24]
<19> (using heat treatment, anion-exchange column chromatography on Hi-Trap Q and cation exchange column chromatography on Resources S [27]) [27]

Cloning

<7> (production of the β-subunit in Escherichia coli [26]; production of the α- and β-subunits in Escherichia coli [28]; production of the G298A-β-subunit in Escherichia coli [29]; expression of the holoenzyme in Escherichia coli [30]; overproduction of the holoenzyme and of the α-subunit in Escherichia coli [32]) [26, 28-30, 32]
<19> (the gltA gene encoding the enzyme overexpressed in Escherichia coli [27]) [27]

Engineering

G298A <7> (<7>, mutant with an approximately 10fold decrease of the affinity of the enzyme for pyridine nucleotides with little or no effect on the rate of the enzyme reduction by NADPH, maintains the ability to bind NADPH and FAD, is catalytically active in the NADPH: iodonitrotetrazolium oxidoreductase reaction and has a monomeric state, mutation leads to production of insoluble protein under conditions that yield large amounts of soluble wild-type enzyme and to production of smaller amounts of enzyme [29]) [29]

6 Stability

pH-Stability

7.3-8.3 <7> [9]

Temperature stability

-20 <15> (<15>, complete loss of activity after 10 days [18]) [18]
-10 <12> (<12>, stable for at least 6 weeks [15]) [15]
0 <16> (<16>, stable for 6-8 h in crude extract [21]) [21]
4 <15> (<15>, unprotected enzyme loses 90% of activity in 48 hours, protected enzyme, addition of 5 mM EDTA and 5 mM 2-mercaptoethanol or 1 mM DTT retains 80 to 90% of the activity after 48 hours [18]) [18]
40 <16> (<16>, 10 min, 80% loss of activity, enzyme in crude extract [21]) [21]
45 <16> (<16>, 3 min, 80% loss of activity, enzyme in crude extract [21]) [21]
45 <8> (<8>, the enzyme retains its full activity after heating for 10 min in 0.1 M potassium phosphate buffer, pH 6.8 [11]) [11]
55 <8> (<8>, 10 min, in 0.1 M potassium phosphate buffer, pH 6.8, 70-80% loss of activity [11]) [11]
80 <11> (<11>, 10 min, 35% loss of activity [14]) [14]
100 <15> (<15>, 30 s, complete loss of activity [18]) [18]

Oxidation stability

<5>, presence of reducing agents is essential for stabilization [3]

General stability information

<3>, NADP$^+$ causes relatively small losses in activity compared with control values [1]

<3>, α-ketoglutarate, L-glutamine, L-glutamate and NADP when present together partial protect against enzyme inactivation [1]

<3>, incubation with L-glutamine or L-glutamate results in marked time- and temperature-dependent losses in activity [1]

<5>, 2 mM α-ketoglutarate, 10 mM β-mercaptoethanol and 10% glycerol when present together markedly enhance stability [3]

<5>, unprotected enzyme is unstable, losing 65% of its activity in 24 hours at 4°C [3]

<15>, DTT, 1 mM, stabilizes at 4°C [18]

<1, 3, 8>, α-ketoglutarate stabilizes [1, 10, 11]

<1, 8>, glutamine stabilizes [10, 11]

<3, 15>, 2-mercaptoethanol, stabilizes [1, 18]

<8, 15>, EDTA, stabilizes [11, 18]

Storage stability

<2>, 4°C, 200 mM potassium phosphate buffer, pH 7.6, 0.5 mM Na$_2$EDTA [23]

<3>, -80°C, 20 mM potassium-HEPES, pH 7.2, 2 mM α-ketoglutarate, 1 mM EDTA, stable for at least 6 months [1]

<3>, 4°C, 20 mM potassium-HEPES, pH 7.2, 2 mM α-ketoglutarate, 1 mM EDTA, 5 mM 2-mercaptoethanol, 20% loss of activity, after 1 month [1]

<5>, 4°C, unprotected enzyme loses 65% of activity after 24 h [3]

<15>, storing at -20°C, total loss of activity after 10 days [18]

References

[1] Miller, R.E.; Stadtman, E.R.: Glutamate synthase from Escherichia coli. An iron-sulfide flavoprotein. J. Biol. Chem., **247**, 7407-7419 (1972)

[2] Tempest, D.W.; Meers, J.L.: Synthesis of glutamate in Aerobacter aerogenes by a hitherto unknown route. Biochem. J., **117**, 405-407 (1970)

[3] Mei, B.; Jiao, R.: Purification and properties of glutamate synthase from Nocardia mediterranei. J. Bacteriol., **170**, 1940-1944 (1988)

[4] Schmidt, C.N.G.; Jervis, L.: Affinity purification of glutamate synthase from Escherichia coli. Anal. Biochem., **104**, 127-129 (1980)

[5] Rendina, A.R.; Orme-Johnson, W.H.: Glutamate synthase: on the kinetic mechanism of the enzyme from Escherichia coli W. Biochemistry, **17**, 5388-5393 (1978)

[6] Janssen, D.B.; op den Camp, H.J.M.; Leenen, P.J.M.; van der Drift, C.: The enzymes of the ammonia assimilation in Pseudomonas aeruginosa. Arch. Microbiol., **124**, 197-203 (1980)

[7] Vanoni, M.A.; Edmondson, D.E.; Zanetti, G.; Curti, B.: Characterization of the flavins and the iron-sulfur centers of glutamate synthase from Azospir-

illum brasilense by absorption, circular dichroism, and electron paramagnetic resonance spectroscopies. Biochemistry, **31**, 4613-4623 (1992)

[8] Vanoni, M.A.; Edmondson, D.E.; Rescigno, M.; Zanetti, G.; Curti, B.: Mechanistic studies on Azospirillum brasilense glutamate synthase. Biochemistry, **30**, 11478-11484 (1991)

[9] Ratti, S.; Curti, B.; Zanetti, G.; Galli, E.: Purification and characterization of glutamate synthase from Azospirillum brasilense. J. Bacteriol., **163**, 724-729 (1985)

[10] Sung, H.C.; Tachiki, T.; Kumagai, H.; Tochikura, T.: Properties of glutamate synthase from Brevibachterium flavum. J. Ferment. Technol., **62**, 569-575 (1984)

[11] Tachiki, T.; Sung, H.C.; Wakisaka, S.; Tochikura, T.: Purification and some properties of glutamate synthase from Gluconobacter suboxydans on glutamate as a nitrogen source. J. Ferment. Technol., **61**, 179-184 (1983)

[12] Schreier, H.J.; Bernlohr, R.W.: Purification and properties of glutamate synthase from Bacillus licheniformis. J. Bacteriol., **160**, 591-599 (1984)

[13] Suzuki, A.; Jacquot, J.P.; Gadal, P.: Glutamate synthase in rice roots. Studies on the electron donor specificity. Phytochemistry, **22**, 1543-1546 (1983)

[14] Schmidt, C.N.G.; Jervis, L.: Partial purification and characterization of glutamate synthase from a thermophilic Bacillus. J. Gen. Microbiol., **128**, 1713-1718 (1982)

[15] Yelton, M.M.; Yoch, D.C.: Nitrogen metabolism in Rhodospirillum rubrum: Characterization of glutamate synthase. J. Gen. Microbiol., **123**, 335-342 (1981)

[16] Nisbet, B.A.; Slaughter, J.C.: Glutamate dehydrogenase and glutamate synthase from the yeast Kluyveromyces fragilis: Variability in occurrence and properties. FEMS Microbiol. Lett., **7**, 319-321 (1980)

[17] Hemmilä, I.A.; Mäntsälä, P.I.: Purification and properties of glutamate synthase and glutamate dehydrogenase from Bacillus megaterium. Biochem. J., **173**, 45-52 (1978)

[18] Adachi, K.; Suzuki, I.: Purification and properties of glutamate synthase from Thiobacillus thioparus. J. Bacteriol., **129**, 1173-1182 (1977)

[19] Mäntsälä, P.; Zalkin, H.: Properties of apoglutamate synthase and comparison with glutamate dehydrogenase. J. Biol. Chem., **251**, 3300-3305 (1976)

[20] Mäntsälä, P.; Zalkin, H.: Glutamate synthase. Properties of the glutamine-dependent activity. J. Biol. Chem., **251**, 3294-3299 (1976)

[21] Hua, S.S.T.; Lichens, G.M.; Guirao, A.; Tsai, V.Y.: Biochemical properties of glutamate synthase of salt-tolerant Bradyrhizobium sp. strain WR1001. FEMS Microbiol. Lett., **37**, 209-213 (1986)

[22] Matsuoka, K.; Kimura, K.: Glutamate synthase from Bacillus subtilis PCI 219. J. Biochem., **99**, 1087-1100 (1986)

[23] Meister, A.: Glutamate synthase from Escherichia coli, Klebsiella aerogenes, and Saccharomyces cerevisiae. Methods Enzymol., **113**, 327-337 (1985)

[24] Rachim, M.A.; Nicholas, D.J.D.: Glutamine synthetase and glutamate synthase from Sclerotinia Schlerotiorum. Phytochemistry, **24**, 2541-2548 (1985)

[25] Carlberg, I.; Norlund, S.: Purification and partial characterization of gluta-
     mate synthase from Rhodospirillum rubrum grown under nitrogen-fixing
     conditions. Biochem. J., 279, 151-154 (1991)
[26] Vanoni, M.A.; Verzotti, E.; Zanetti, G.; Curti, B.: Properties of the recombi-
     nant β-subunit of glutamate synthase. Eur. J. Biochem., 236, 937-946 (1996)
[27] Jongsareejit, B.; Rahman, R.N.Z.A.; Fujiwara, S.; Imanaka, T.: Gene cloning,
     sequencing and enzymic properties of glutamate synthase from the hy-
     perthermophilic archaeon Pyrococcus sp. KOD1. Mol. Gen. Genet., 254,
     635-642 (1997)
[28] Vanoni, M.A.; Fischer, F.; Ravasio, S.; Verzotti, E.; Edmondson, D.E.; Hagen,
     W.R.; Zanetti, G.; Curti, B.: The recombinant α-subunit of glutamate
     synthase: spectroscopic and catalytic properties. Biochemistry, 37, 1828-
     1838 (1998)
[29] Morandi, P.; Valzasina, B.; Colombo, C.; Curti, B.; Vanoni, M.A.: Glutamate
     synthase: identification of the NADPH-binding site by site-directed muta-
     genesis. Biochemistry, 39, 727-735 (2000)
[30] Stabile, H.; Curti, B.; Vanoni, M.A.: Functional properties of recombinant
     Azospirillum brasilense glutamate synthase, a complex iron-sulfur flavo-
     protein. Eur. J. Biochem., 267, 2720-2730 (2000)
[31] Ravasio, S.; Curti, B.; Vanoni, M.A.: Determination of the midpoint poten-
     tial of the FAD and FMN flavin cofactors and of the 3Fe-4S cluster of gluta-
     mate synthase. Biochemistry, 40, 5533-5541 (2001)
[32] Ravasio, S.; Dossena, L.; Martin-Figueroa, E.; Florencio, F.J.; Mattevi, A.;
     Morandi, P.; Curti, B.; Vanoni, M.A.: Properties of the recombinant ferre-
     doxin-dependent glutamate synthase of Synechocystis PCC6803. Compari-
     son with the Azospirillum brasilense NADPH-dependent enzyme and its
     isolated α-subunit. Biochemistry, 41, 8120-8133 (2002)

Glutamate synthase (NADH) 1.4.1.14

1 Nomenclature

EC number
1.4.1.14

Systematic name
L-glutamate:NAD$^+$ oxidoreductase (transaminating)

Recommended name
glutamate synthase (NADH)

Synonyms
L-glutamate synthase (NADH)
L-glutamate synthetase
NADH-GOGAT
NADH-GltS
NADH-dependent glutamate synthase
NADH-glutamate synthase
NADH: GOGAT
glutamate (reduced nicotinamide adenine dinucleotide) synthase
glutamate synthase
glutamate synthase (NADH)
glutamate synthase (NADH-dependent)
glutamate synthetase
synthase, glutamate (reduced nicotinamide adenine dinucleotide)

CAS registry number
65589-88-0

2 Source Organism

<1> *Clostridium pasteurianum* (strain ATCC 6013 [5]) [5]
<2> *Saccharomyces cerevisiae* (strain X2180A [16]) [6, 16, 18, 24]
<3> *Neurospora crassa* (strains 74-A, 73-a, fl A, fl a, am 132, en(am)-2 C24 [7]; strain am-1 [12]) [7, 12]
<4> *Derxia gummosa* (obligate aerobic bacterium [8]) [8]
<5> *Chlamydomonas reinhardtii* (strain CCAP 11/32a [9]; strain 21 gr [15]) [9, 15]
<6> *Chlorobium vibrioforme f. thiosulfatophilum* (strain NCIB 8346 [10]) [10]
<7> *Euglena gracilis* (z [13]) [13]
<8> *Lupinus angustifolius* [1, 11, 14, 17]

<9> *Oryza sativa* [2, 20-22, 26]
<10> *Medicago sativa* [3, 24-26]
<11> *Phaseolus vulgaris* (2 isoenzymes: NADH-GOGAT I and NADH-GOGAT II [4]) [4]
<12> *Vicia faba* [19]
<13> *Spodoptera frugiperda* (Sf9 [23]) [23]
<14> *Bombyx mori* [23, 24]
<15> *Caenorhabditis elegans* (NADH-GltS or similar enzyme [24]) [24]
<16> *Plasmodium falciparum* (NADH-GltS or similar enzyme [24]) [24]
<17> *Nicotiana tabacum* (cv. Xanthi, transgenic tobacco that overexpress NADH-GOGAT from Medicago sativa, higher capacity to assimilate nitrogen due to a higher NADH-GOGAT activity [26]) [26]

3 Reaction and Specificity

Catalyzed reaction
2 L-glutamate + NAD^+ = L-glutamine + 2-oxoglutarate + NADH + H^+ (<8> kinetic mechanism [17])

Reaction type
oxidation
redox reaction
reduction
transamidation

Natural substrates and products
S L-glutamine + 2-oxoglutarate + NADH <1-3, 5, 6, 8-11, 13> (<8> role in overall ammonia assimilation system, involved in nitrogen metabolism [1]; <8,10> key enzyme in pathway of assimilation of symbiotically fixed N_2 into amino acids [3,14]; <10> predominant enzyme in N_2-fixing root nodules and roots, whereas in leaves and cotyledons Fd-GOGAT is the major form of GOGAT involved in nitrogen assimilation, NADH-GOGAT may be a rate-limiting step in NH_4^+ assimilation in root nodules [25]; <10> plays a critical role in the assimilation of symbiotically fixed nitrogen in nodules and during pollen development [26]; <1,5,9,11> ammonia assimilation [3-5,9,20]; <8> ammonia assimilation from fixed nitrogen, de novo synthesis of glutamate [11]; <10> root nodule ammonia assimilation, enzyme plays a central role in the functioning of effective root nodules [25]; <3,6> glutamine synthetase/glutamate synthase pathway for ammonia assimilation is favoured by growth at low concentrations of ammonia [10,12]; <6> key enzyme for glutamate production [10]; <9> role in synthesis of glutamate for growth and development in higher plants [2]; <1> regulatory role in biosynthesis pathways involving glutamate, early acceptor of fixed nitrogen [5]; <3> important role in glutamine degradation [7]; <13> involvement in glutamine metabolism, when glucose is in excess [23]; <2,9> glutamate biosynthesis [16,22]; <9> important role in the generation of glutamate for the assimilation of ammo-

nium ions via the glutamine synthetase reaction in the epidermis and exodermis of the root surface, role in the re-utilization of glutamine trans-
ported from phloem and xylem in young leaves and grains at the early
stage of ripening of rice plants [20,21]; <9> assimilation of exogeneously
supplied NH_4^+ is primarily via the cytosolic glutamine synthetase/plasti-
dial NADH-GOGAT cycle in specific regions of the epidermis and exodermis in roots [21]; <9> enzyme in developing organs, such as unexpanded
non-green leaves and developing grains, could be involved in the utilization of remobilized nitrogen [22]) (Reversibility: ir <2, 6, 8, 9, 13> [1, 10,
16, 20, 21, 23]; ? <1, 3, 5, 9-11> [2-5, 7, 9, 11, 12, 14, 22, 25, 26]) [1-5, 7, 9-
12, 14, 16, 20-23, 25, 26]

P glutamate + NAD^+ <1-3, 6, 8-11, 13> (<9> one of the two synthezised
glutamate molecules can be cycled back as substrate for glutamine synthetase reaction and the other can be used for many synthetic reactions
[20,21]) [3-5, 10-12, 14, 16, 20-23, 25, 26]

Substrates and products

S L-glutamine + 2-oxoglutarate + NADH <1-14, 17> (<8> absolute substrate specificity [1]; <1, 2, 4-7, 9> highly specific for its substrates [2, 5,
8, 10, 13, 15, 16]; <7, 8> absolute specificity for 2-oxoglutarate [11, 13];
<1, 2, 4-7, 9, 11, 13, 14> specific for NADH [2, 4-6, 8-10, 13, 15, 16, 18, 23];
<2> low activity when NH_3 is substituted for glutamine, NH_4Cl: 2.7% of
the activity with L-glutamine [6, 18]; <8> 2-oxoglutarate may bind first
and then L-glutamine, binding steps of NADH and 2-oxoglutarate are independent and random in order, or the mechanism may be ping-pong [1];
<8> NADH binds first before enzyme is capable to bind other substrates
[14]; <1, 8> 1 mol 2-oxoglutarate and 1 mol glutamine are consumed per
mol NADH oxidized [5, 17]; <8> mechanism involves compulsory, very
tightly binding of NADH as first substrate, followed by random-order
binding of glutamine and 2-oxoglutarate, NADH triggers a conformational
change, which then exposes the binding sites of the other substrates [17])
(Reversibility: ir <2, 6, 8, 9, 13, 14> [1, 2, 10, 16, 20, 23]; ? <1-5, 7-12, 17>
[3-9, 11-15, 17-19, 21, 22, 25, 26]) [1-26]

P L-glutamate + NAD^+ <1-3, 5-14, 17> [1-6, 10-26]

S L-glutamine + 2-oxoglutarate + reduced form of 3-acetylpyridine adenine
dinucleotide <8> (<8> reduced form of 3-acetylpyridine adenine dinucleotide is an alternative reductant to NADH, binds as tightly as NADH,
enzyme catalyzes NADH-dependent reduction of AcPdAD+ by a substi-
tuted-enzyme ping-pong mechanism [14]) (Reversibility: ? <8> [14]) [14]

P L-glutamine + oxidized form of 3-acetylpyridine adenine dinucleotide
<8> [14]

S Additional information <1, 2, 5, 8, 9, 11> (<9> no activity with L-asparagine, NH_4Cl or NADPH [2]; <8> no activity with L-asparagine, ammonium sulfate, 2-oxobutyrate, oxaloacetate [1]; <1, 5, 7, 11> no activity
with L-asparagine, NH_4Cl, pyruvate, oxaloacetate or NADPH [4, 5, 13,
15]; <5> no activity with hydroxylamine [15]; <2> no activity with L-
asparagine, D-glutamine, pyruvate, oxaloacetate or NADPH [16]; <1> no

activity with albizziin, 6-diazo-5-oxo-L-norvaline, azaserine, methionine sulfoximine, FADH or FMNH [5]; <2> no activity with 2-oxovalerate, 2-oxopimelate, 2-oxoadipate, oxaloacetate, D-glutamine, L- or D-asparagine, L- or D-homoglutamine, α-methyl-L-glutamine, DL-β-glutamine, L-isoglutamine, glutaramide or glutaramate [6,18]; <8> sodium dithionite is no alternative reductant [17]) [1, 2, 4-6, 14-17]

P ?

Inhibitors

2-hydroxyglutarate <8> (<8> D- and L-isomer, most potent competitive inhibitor [11]) [11]

2-oxoglutarate <8> [11]

3-acetylpyridine adenine dinucleotide <8> (<8> competitive inhibitor with respect to NADH [14]) [14]

6-diazo-5-oxo-L-norleucine <2, 5> (<2> 0.1 mM, 30 min, 0°C, pH 7.6, irreversible inhibition [6]; <5> 76% inhibition at 5 mM [15]) [6, 15]

ADP <4, 6> (<4> 10 mM, 84% inhibition [8]; <6> 20 mM, about 75% inhibition [10]) [8, 10]

AMP <4, 6> (<4> 10 mM, 61% inhibition [8]; <6> 20 mM, about 75% inhibition [10]) [8, 10]

ATP <4, 6> (<4> 10 mM, 37% inhibition [8]; <6> 5 mM, 75% inhibition [10]) [8, 10]

Co^{2+} <1> (<1> slight inhibition [5]) [5]

Cu^{2+} <1> (<1> Cu^{2+} and Fe^{2+} are most inhibitory metal ions [5]) [5]

D-glutamate <8> (<8> competitive inhibitor [11]) [11]

D-glutamine <2> (<2> slight inhibition [6]) [6, 18]

DL-ethionine sulfone <2> (<2> potent competitive inhibitor [6]) [6, 18]

DL-homocysteic acid <2> [18]

Fe^{2+} <1> (<1> Cu^{2+} and Fe^{2+} are most inhibitory metal ions [5]) [5]

L-2-amino-4-oxo-5-chloropentanoate <2> (<2> 1 mM, 30 min, 0°C, pH 7.6, irreversible inhibition [6]) [6]

L-albizziin <2> (<2> less inhibitory than L-homocysteine sulfonamide, methionine sulfone and ethionine sulfone [6]) [6, 18]

L-glutamate <1, 4, 8, 10> (<8> competitive with 2-oxoglutarate [1,11,17]; <8> glutamate binds 1.7times more tightly in the presence of NAD^+ [17]; <1> 4 mM, 20% inhibition, product inhibition [5]; <4> 10 mM, 24% inhibition [8]) [1, 3, 5, 8, 11, 17]

L-histidine <1> (<1> 25 mM, 20-30% inhibition [5]) [5]

L-homocysteine sulfonamide <2> (<2> very potent transition state inhibitor, competitive to L-glutamine [6,18]) [6, 18]

L-methionine sulfone <2> (<2> very potent competitive inhibitor [6]; <2> markedly inhibits [18]) [6, 18]

L-methionine-SR-sulfoxide <2> (<2> slight inhibition [6]) [6, 18]

L-methionine-SR-sulfoximine <2> (<2> potent irreversible inhibitor [6]) [6, 18]

Mg^{2+} <1> (<1> slight inhibition [5]) [5]

NAD$^+$ <6, 8, 10> (<6> 1 mM, 27% inhibition [10]; <8> competitive with NADH [1,17]) [1, 3, 10, 17]
NADH <8> (<8> substrate inhibition at high concentrations [14]) [14]
NADP$^+$ <6> (<6> 1 mM, 48% inhibition [10]) [10]
NADPH <4> (<4> 0.2 mM, 50% inhibition [8]) [8]
O-diazoacetyl-L-serine <2> (<2> 1 mM, 30 min, 0°C, pH 7.6, irreversible inhibition [6]) [6]
Zn^{2+} <1> (<1> slight inhibition [5]) [5]
acridine <5> (<5> inhibitor of flavoenzymes, 65% inhibition [15]) [15]
alanine <6> (<6> slight inhibition [10]) [10]
arginine <6> (<6> 20 mM, complete inhibition [10]) [10]
asparagine <1, 8> (<8> weak inhibition, competitive with 2-oxoglutarate [1]; <1> slight inhibition [5]) [1, 5]
aspartate <4, 8> (<4> 10 mM, 60% inhibition [8]; <8> weak inhibition, competitive with 2-oxoglutarate [1,11]) [1, 8, 11]
atebrin <6> [10]
azaserine <3-5, 7, 9-11, 13, 14> (<9> completely inhibited by 4 mM [2]; <10> glutamine antagonist, completely inhibited by 10 mM [3]; <11> completely inhibited by 0.4 mM [4]; <3> complete inhibition [12]; <7,13,14> completely inhibited by 1 mM, glutamine analog [13,23]; <5> 75% inhibition at 5 mM [15]) [2-4, 8, 9, 12, 13, 15, 23]
bathophenanthroline <5> (<5> 50% inhibition [15]) [15]
citrate <10> (<10> inhibited by high concentrations, 40 mM [3]) [3]
glutarate <8> (<8> most potent competitive inhibitor [11]) [11]
glycine <1, 6> (<1> slight inhibition [5]; <6> 48% inhibition at 10 mM, complete inhibition at 20 mM [10]) [5, 10]
leucine <6> (<6> 10 mM, 52% inhibition [10]) [10]
malate <6, 10> (<10> inhibited by high concentrations, 40 mM [3]; <6> 20 mM, 30% inhibition [10]) [3, 10]
methionine <1, 4, 6> (<1> 25 mM L-methionine, 20-30% inhibition [5]; <4> 10 mM, 47% inhibition [8]; <6> 10 mM, 72% inhibition [10]) [5, 8, 10]
oxaloacetate <8> (<8> weak inhibition, competitive with 2-oxoglutarate [1,11]) [1, 11]
p-hydroxymercuribenzoate <5> (<5> complete inhibition at 0.1-1 mM, at 0.0005 mM 33% inhibition or 75% inhibition after preincubation in presence of glutamine and 2-oxoglutarate, substrates induce a conformation, which makes essential sulfhydryl groups more accessible to reagent [15]) [15]
phenylalanine <6> (<6> 20 mM, complete inhibition [10]) [10]
proline <6> [10]
pyruvate <1, 6> (<1> slight inhibition, analogue of 2-oxoglutarate [5]; <6> 20 mM, 20% inhibition [10]) [5, 10]
serine <1, 6> (<1> 25 mM L-serine, 20-30% inhibition [5]; <6> 10 mM, 73% inhibition [10]) [5, 10]
sodium dithionite <1> (<1> strong inhibitory effect [5]) [5]
tryptophan <1, 6> (<1> 25 mM L-tryptophan, 20-30% inhibition [5]; <6> 29% inhibition at 10 mM, complete inhibition at 20 mM [10]) [5, 10]
valine <6> [10]

Additional information <1, 2, 4-7, 10> (<6> cumulative inhibition by combinations of amino acids and nucleotides, cumulative feedback inhibition [10]; <10> not inhibited by 2,2'-dipyridyl or N-ethylmaleimide [3]; <1> not inhibited by Mn^{2+} and Ca^{2+} [5]; <4> not inhibited by alanine, glycine, serine, lysine, proline, cysteine, valine, leucine, isoleucine, threonine, phenylalanine, histidine, tryptophan, asparagine, arginine, UTP, ITP, GTP [8]; <7> not inhibited by vitamin B_6-enzyme inhibitors [13]; <5> not inhibited by 1 mM cyanide or cyanate [15]; <2> not inhibited by the 20 common amino acids at a concentration of 20 mM [16]) [3, 5, 8, 10, 13, 15, 16]

Cofactors/prosthetic groups

3-acetylpyridine adenine dinucleotide <8> (<8> reduced form is an alternative reductant, binds as tightly as NADH [14]) [14]

FMN <8, 10> (<8> two flavin prosthetic groups per enzyme molecule, probably FMN [1]; <10> coding region important for binding of flavin mononucleotide is located on exon 16 [25]) [1, 25]

NADH <1-14, 17> (<8> NADH-dependent enzyme, very tight binding of NADH [1]; <1, 2, 4-7, 9, 11, 13, 14> specific for NADH as reductant [2, 4-6, 8-10, 13, 15, 16, 18, 23]; <13, 14> exclusively dependent on NADH as coenzyme [23]; <10> five conserved residues important for NADH binding are located in exon 20 [25]) [1-26]

flavin <2, 6, 8-10, 12> (<8, 9, 12> flavoprotein [2, 17, 19]; <6> flavin component [10]; <2, 10> iron-sulfur flavoprotein [24, 25]) [2, 10, 17, 19, 24, 25]

Additional information <1, 2, 5, 7, 9, 13> (<1, 2, 5, 7, 9, 13> no activity with NADPH [2, 5, 6, 9, 13, 16, 18, 23]; <1> absence of flavin and iron [5]) [2, 5, 6, 9, 13, 16, 18, 23]

Activating compounds

2-mercaptoethanol <8> (<8> 1%, presence of a reducing agent is essential for enzyme activity [1]) [1]

diphosphate <6> (<6> 5 mM, stimulates activity with 36% enhancement [10]) [10]

dithiothreitol <8> (<8> 0.5 mg per ml, presence of a reducing agent is essential for enzyme activity [1]) [1]

succinate <6> (<6> 50 mM, stimulates activity by 57% [10]) [10]

Additional information <7, 9> (<7> not activated by pyridoxal 5'-phosphate and pyridoxamine 5'-phosphate [13]; <9> supply of NH_4^+ at low concentrations increases the expression of the enzyme: supply with 1 mM NH_4^+ for 24 h increases the enzyme content in the epidermis and exodermis of the root surface [20,21]) [13, 20, 21]

Metals, ions

3Fe-4S-center <2, 10> (<2,10> iron-sulfur flavoprotein [24]; <10> flavoprotein containing an iron-sulfur cluster: 3Fe-4S cluster [25]; <12> enzyme contains non-heme iron and acid-labile sulfur [19]) [19, 24, 25]

Additional information <1> (<1> absence of flavin and iron [5]) [5]

Turnover number (min^{-1})

1080 <8> (3-acetylpyridine adenine dinucleotide, <8> reduced form as reductant, glutamate synthase reaction [14]) [14]

1920 <9> (L-glutamine) [2]

3060 <8> (3-acetylpyridine adenine dinucleotide, <8> NADH-dependent reduction of oxidized form of 3-acetylpyridine adenine dinucleotide [14]) [14]

4200 <8> (L-glutamine, <8> independent of pH 6.5-9.5 [1]) [1]

4200 <8> (NADH) [14]

Specific activity (U/mg)

2.66 <12> [19]

3.8 <1, 11> (<11> NADH-GOGAT I [4]) [4, 5]

4.56 <5> [15]

4.7 <6> [10]

4.76 <3> [12]

8.05 <11> (<11> NADH-GOGAT II [4]) [4]

9.8 <9> [2]

13.1 <10> [3]

16 <4> [8]

18 <8> [1]

41.7 <2> [18]

Additional information <3, 13, 14, 17> (<17> enzyme specific activity of transformants is significantly greater than that of control plants [26]) [7, 23, 26]

K$_m$-Value (mM)

0.00124 <8> (NADH) [17]

0.0013 <8> (NADH) [1, 14]

0.0014 <8> (NADH, <8> NADH-dependent reduction of oxidized form of 3-acetylpyridine adenine dinucleotide [14]) [14]

0.0026 <2> (NADH) [16]

0.003 <8> (3-acetylpyridine adenine dinucleotide, <8> reduced form [14]) [14]

0.003 <9> (NADH) [2]

0.0042 <10> (NADH) [3]

0.0052 <11> (NADH, <11> NADH-GOGAT II [4]) [4]

0.006 <8> (2-oxoglutarate, <8> with reduced form of 3-acetylpyridine adenine dinucleotide as reductant [14]) [14]

0.007 <5> (2-oxoglutarate) [9]

0.007 <2> (NADH) [6, 18]

0.0076 <5> (NADH) [15]

0.0096 <4> (NADH) [8]

0.013 <5, 6> (NADH) [9, 10]

0.014 <8> (3-acetylpyridine adenine dinucleotide, <8> NADH-dependent reduction of oxidized form of 3-acetylpyridine adenine dinucleotide [14]) [14]

0.014 <11> (NADH, <11> NADH-GOGAT I [4]) [4]

0.015 <7> (2-oxoglutarate) [13]

0.0182 <5> (2-oxoglutarate) [15]

0.02 <7> (NADH) [13]
0.022 <11> (2-oxoglutarate, <11> NADH-GOGAT I [4]) [4]
0.0238 <8> (2-oxoglutarate) [17]
0.024 <4, 8> (2-oxoglutarate) [8, 11, 14]
0.033 <10> (2-oxoglutarate) [3]
0.039 <8> (2-oxoglutarate) [1]
0.04 <2> (2-oxoglutarate) [6, 18]
0.076 <9> (2-oxoglutarate) [2]
0.087 <11> (2-oxoglutarate, <11> NADH-GOGAT II [4]) [4]
0.11 <1> (NADH) [5]
0.125 <8> (L-glutamine, <8> with reduced form of 3-acetylpyridine adenine dinucleotide as reductant [14]) [14]
0.14 <2> (2-oxoglutarate) [16]
0.16 <1> (L-glutamine) [5]
0.17 <1> (2-oxoglutarate) [5]
0.24 <11> (L-glutamine, <11> NADH-GOGAT II [4]) [4]
0.27 <6> (2-oxoglutarate) [10]
0.27 <4> (L-glutamine) [8]
0.28 <2> (L-glutamine) [6, 18]
0.3 <7> (L-glutamine) [13]
0.4 <8> (L-glutamine) [1]
0.466 <10> (L-glutamine) [3]
0.5 <8> (L-glutamine) [14]
0.5 <8> (L-glutamine) [17]
0.6 <5> (L-glutamine) [15]
0.769 <6> (L-glutamine) [10]
0.77 <11> (L-glutamine, <11> NADH-GOGAT I [4]) [4]
0.811 <9> (L-glutamine) [2]
0.9 <5> (L-glutamine) [9]
1 <2> (L-glutamine) [16]
Additional information <5> [15]

K_i-Value (mM)

0.001 <8> (3-acetylpyridine adenine dinucleotide) [14]
0.0036 <2> (L-homocysteine sulfonamide) [6]
0.005 <2> (L-methionine sulfone) [6]
0.01 <8> (NADH) [14]
0.015 <2> (DL-ethionine sulfone) [6]
0.1 <2> (L-albizziin) [6]
0.12 <8> (NAD$^+$) [1]
0.146 <2> (L-methionine-SR-sulfoximine) [6]
0.178 <2> (DL-homocysteic acid) [6]
0.25 <8> (L-2-hydroxyglutarate) [11]
0.28 <8> (D-2-hydroxyglutarate) [11]
0.3 <2> (D-glutamine) [6]
0.3 <2> (L-methionine-SR-sulfoxide) [6]
0.35 <8> (glutarate) [11]

0.6 <8> (D-glutamate) [11]
0.6 <8> (L-glutamate) [11]
0.7 <8> (L-glutamate) [1]
0.8 <4> (azaserine) [8]
1.1 <8> (glutamate) [17]
2.7 <8> (L-aspartate) [1]
2.7 <8> (L-aspartate) [11]
5 <8> (oxaloacetate) [1]
5 <8> (oxaloacetate) [11]
14 <8> (L-asparagine) [1]

pH-Optimum

7-7.5 <2> (<2> glutamine-dependent NADH oxidase activity [6,18]) [6, 18]
7.1-7.7 <2> [16]
7.2 <1> (<1> broad pH profile with optimum at pH 7.2 [5]) [5]
7.5 <4, 5, 7, 14> [8, 9, 13, 23]
7.5-8.5 <10> [3]
8 <11> (<11> NADH-GOGAT I, broad pH-optimum centering at pH 8 [4]) [4]
8.5 <8, 11> (<11> NADH-GOGAT II, narrow pH-optimum of 8.5 [4]) [1, 4]
9 <2> (<2> ammonia-dependent NADH oxidase activity, rate steadily increases from pH 6 to 9 [6,18]) [6, 18]

pH-Range

6-9 <2> (<2> at pH 6 and 9 about 50% of activity maximum [16]) [16]
7-9 <10> (<10> below and above activity declines rapidly [3]) [3]
7.2-8.2 <4> (<4> optimal pH-range with maximum at pH 7.5 [8]) [8]

Temperature optimum (°C)

25 <8> (<8> assay at [1,14,17]) [1, 14, 17]
27 <13> (<13> assay at [23]) [23]
30 <1, 5-7, 9, 11, 12, 14> (<11> NADH-GOGAT I/II [4]; <1,5,6,9,12,14> assay at [2,5,9,10,19,23]) [2, 4, 5, 9, 10, 13, 19, 23]
50 <1> [5]

4 Enzyme Structure

Molecular weight

194000 <9> (<9> gel filtration [2]) [2]
200000 <11> (<11> NADH-GOGAT II, gel filtration [4]) [4]
222000 <12> (<12> gel filtration [19]) [19]
229000 <10> (<10> calculated from the cDNA sequence, without N-terminal presequence of 101 amino acids [25]) [25]
235000 <8, 10> (<8> ultracentrifugation [1]; <10> gel filtration [3]) [1, 3]
236000 <12> (<12> native PAGE [19]) [19]
240000 <5, 11> (<5> sedimentation studies [9]; <11> NADH-GOGAT I, gel filtration [4]) [4, 9]

265000 <2> (<2> gel filtration [18]) [6, 18]
590000 <1> (<1> gel filtration [5]) [5]

Subunits

decamer <1> (<1> 2 * 91000 + 2 * 86000 + 2 * 68000 + 2 * 31000 + 2 * 17500, SDS-PAGE [5]) [5]

dimer <2> (<2> 1 * 169000 + 1 * 61000, SDS-PAGE in presence of 1% 2-mercaptoethanol [6,18]) [6, 18]

monomer <3, 8-12> (<8> 1 * 235000, SDS-PAGE [1]; <9> 1 * 196000, SDS-PAGE [2]; <10> 1 * 200000, SDS-PAGE [3]; <11> 1 * 200000, SDS-PAGE [4]; <3> 1 * 200000, SDS-PAGE [7]; <12> 1 * 195000, SDS-PAGE [19]) [1-4, 7, 19, 25]

Additional information <10> (<10> protein contains domains that correspond to both the large α-subunit with 46% identity and the small β-subunit with 38% identity of Escherichia coli NADPH-GOGAT, a 57 amino acid highly charged domain connects the two regions homologous to the prokaryotic subunits [25]) [25]

Posttranslational modification

proteolytic modification <2, 9, 10, 15, 16> (<2,10,15,16> enzyme is synthesized as proprotein with a N-terminal propeptide region [24]; <10> primary translation product is a 240 kDa protein, that contains a N-terminal 101 amino acid presequence resulting in a processed protein of 229 kDa [25,26]; <9> translational gene product has a 99 amino acid presequence at the N-terminal region [21]) [21, 24-26]

5 Isolation/Preparation/Mutation/Application

Source/tissue

cell culture <9> [2]

fat body <14> [23]

flower <10> (<10> enzyme activity is detectable in flowers [26]) [26]

leaf <9, 17> (<9> developing leaf blade [22]; <17> low concentrations [26]) [22, 26]

root <9, 17> (<9> roots of seedlings: root tips and area where the secondary roots are actively developing, enzyme in the central cylinder, apical meristem and the primordia of secondary roots of seedlings grown for 26 d in water, supply with 1 mM ammonium ions for 24 h increases the enzyme content in the epidermis and exodermis of the root surface [20]; <17> low concentrations, but 15-40% higher enzyme activity in the roots of transgenic plants than of control plants [26]) [20, 26]

root nodule <8, 10-12> (<10> high enzyme activity in nodules compared to roots and other organs [3]; <11> NADH-GOGAT II is twice as active as NADH-GOGAT I in mature nodules [4]; <10> enzyme activity increases markedly during development of effective root nodules [25, 26]) [1, 3, 4, 11, 14, 17, 19, 25, 26] root tip <9> (<9> shorter than 10 mm, supply with 1 mM ammonium ions for 24 h increases the enzyme content in the epidermis and exodermis cells of the root surface [21]) [21]

Additional information <10, 12> (<12> not found in crude soluble protein fractions from leaves or bacteroids [19]; <10> little or no enzyme activity and mRNA in leaves, stems, cotyledons and roots [25]) [19, 25]

Localization

chromatophore <6> [10]

cytoplasm <8> [1]

cytosol <7, 8> [13, 14]

plastid <9, 10> (<9> exclusively plastidial location, various forms of plastids in cells of epidermis and exodermis, in the cortex parenchyma and vascular parenchyma of root tips, epidermis and exodermis cells contain more plastids and much higher enzyme concentration than other root cells [21]; <10> major portion of enzyme activity resides in the plastidial fraction [25]) [21, 25]

Purification

<1> (58fold purification [5]) [5]

<2> (20fold partial purification [16]; 7500fold purification [6]) [6, 16, 18, 24]

<3> (150fold partial purification, strain am(132) [7]; 136fold purification [12]) [7, 12]

<4> (167fold purification [8]) [8]

<5> (partial purification [9]; 350fold purification [15]) [9, 15]

<6> (222fold purification [10]) [10]

<8> (500fold purification [1]) [1, 17]

<9> (815fold purification [2]) [2]

<10> (208fold purification [3]) [3, 24]

<11> (2 isoenzymes: NADH-GOGAT I with 877fold purification and NADH-GOGAT II with 631fold purification [4]) [4]

<12> (74fold partial purification [19]) [19]

Cloning

<9> (full-length cDNA clone and genomic clone for NADH-GOGAT gene [21]) [21, 26]

<9> (locus of structural gene encoding NADH-GOGAT is mapped on chromosome 1, detection of quantitative trait loci associated with the enzyme content in leaves [22]) [22]

<10> (isolation and characterization of NADH-GOGAT gene with 22 exons and 21 introns, isolation of 7.2 kbp cDNA encoding the complete enzyme [25,26]) [25, 26]

<17> (overexpression of chimeric NADH-GOGAT gene from Medicago sativa fused to the CaMV 35S promoter in transgenic Nicotiana tabacum [26]) [26]

<15, 16> (open reading frames encoding proteins similar to NADH-GltS are found during sequencing of genome [24]) [24]

<2, 10> (gene encoding NADH-GltS is cloned, N-terminal three quarters of the gene product is similar to the bacterial α-subunit from NADPH-GltS [24]) [24]

Engineering

Additional information <3> (<3> three en(am)-2 mutants, lacking enzyme activity, no growth on ammonia-limited conditions [7]) [7]

6 Stability

Temperature stability

35 <1> (<1> enzyme loses activity after incubation for 10 min [5]) [5]

45 <5, 7> (<7> rapid inactivation above [13]; <5> 10 min, 86% loss of activity [15]) [13, 15]

50 <1> (<1> half-life: 10 min, presence of 2-oxoglutarate and L-glutamine stabilizes against thermal denaturation, whereas preincubation with NADH or L-glutamate causes more heat sensitivity [5]) [5]

Additional information <5, 11> (<11> very sensitive to low temperatures [4]; <5> cold-labile enzyme [9]) [4, 9]

Oxidation stability

<5>, 4°C, a few days under air, stable [15]

<5>, sensitive to oxygen [9]

General stability information

<1>, 2-oxoglutarate, 2-mercaptoethanol and EDTA stabilize and have a synergystic effect [5]

<1>, sodium dithionite and glutamine stabilize [5]

<2>, 25% glycerol stabilizes [16]

<5>, high levels of sucrose stabilize [9]

<11>, dithiothreitol stabilizes [4]

<12>, instability of enzyme during purification process [19]

<10, 11>, highly labile enzyme activity [3, 4]

<2, 11>, EDTA stabilizes [4, 16]

<2, 5, 11>, 2-mercaptoethanol stabilizes [4, 9, 16, 18]

<8, 11>, phenylmethylsulphonyl fluoride is used as precaution against proteolysis [1, 4]

Storage stability

<2>, 25°C, 4°C or -20°C, relatively unstable [16]

<5>, 4°C, a few days under air, stable [15]

<8>, Tris buffer, overnight dialysis, total loss of activity [1]

<8>, phosphate buffer, per week, 20% loss of activity [1]

<10>, -80°C, 50% glycerol, 100 mM potassium phosphate, 3 mM DTT, stable for up to 1 month [3]

References

[1] Boland, M.J.; Benny, A.G.: Enzymes of nitrogen metabolism in legume nodules. Purification and properties of NADH-dependent glutamate synthase from lupin nodules. Eur. J. Biochem., **79**, 355-362 (1977)

[2] Hayakawa, T.; Yamaya, T.; Kamachi, K.; Ojima, K.: Purification, characterization, and immunological properties of NADH-dependent glutamate synthase from rice cell cultures. Plant Physiol., **98**, 1317-1322 (1992)

[3] Anderson, M.P.; Vance, C.P.; Heichel, G.H.; Miller, S.S.: Purification and characterization of NADH-glutamate synthase from alfalfa root nodules. Plant Physiol., **90**, 351-358 (1989)

[4] Chen, F.L.; Cullimore, J.V.: Two isoenzymes of NADH-dependent glutamate synthase in root nodules of Phaseolus vulgaris L.. Purification, properties and activity changes during nodule development. Plant Physiol., **88**, 1411-1417 (1988)

[5] Singhal, R.K.; Krishnan, I.S.; Dua, R.D.: Stabilization, purification, and characterization of glutamate synthase from Clostridium pasteurianum. Biochemistry, **28**, 7928-7935 (1989)

[6] Masters, D.S.; Meister, A.: Inhibition by homocysteine sulfonamide of glutamate synthase purified from Saccharomyces cerevisiae. J. Biol. Chem., **257**, 8711-8715 (1982)

[7] Romero, D.; Davila, G.: Genetic and biochemical identification of the glutamate synthase structural gene in Neurospora crassa. J. Bacteriol., **167**, 1043-1047 (1986)

[8] Wang, R.; Nicholas, D.J.D.: Some properties of glutamine synthetase and glutamate synthase from Derxia gummosa. Phytochemistry, **24**, 1133-1139 (1985)

[9] Cullimore, J.V.; Sims, A.P.: Occurrence of two forms of glutamate synthase in Chlamydomonas reinhardii. Phytochemistry, **20**, 597-600 (1981)

[10] Khanna, S.; Nicholas, D.J.D.: Some properties of glutamine synthetase and glutamate synthase from Chlorobium vibrioforme f. thiosulfatophilum. Arch. Microbiol., **134**, 98-103 (1983)

[11] Boland, M.J.; Court, C.B.: Glutamate synthase (NADH) from lupin nodules. Specificity of the 2-oxoglutarate site. Biochim. Biophys. Acta, **657**, 539-542 (1981)

[12] Hummelt, G.; Mora, J.: NADH-dependent glutamate synthase and nitrogen metabolism in Neurospora crassa. Biochem. Biophys. Res. Commun., **92**, 127-133 (1980)

[13] Miyatake, K.; Kitaoka, S.: NADH-dependent glutamate synthase in Euglena gracilis Z. Agric. Biol. Chem., **45**, 1727-1729 (1981)

[14] Boland, M.J.: NADH-dependent glutamate synthase from lupin nodules. Reactions with oxidised and reduced 3-acetylpyridine-adenine dinucleotide. Eur. J. Biochem., **115**, 485-489 (1981)

[15] Marquez, A.J.; Galvan, F.; Vega, J.M.: Purification and characterization of the NADH-glutamate synthase from Chlamydomonas reinhardii. Plant Sci. Lett., **34**, 305-314 (1984)

[16] Roon, R.J.; Even, H.I.; Larimore, F.: Glutamate synthase: Properties of the reduced nicotinamide adenine dinucleotide-dependent enzyme from Saccharomyces cerevisiae. J. Bacteriol., 118, 89-95 (1974)

[17] Boland, M.J.: Kinetic mechanism of NADH-dependent glutamate synthase from lupin nodules. Eur. J. Biochem., 99, 531-539 (1979)

[18] Meister, A.: Glutamate synthase from Escherichia coli, Klebsiella aerogenes, and Saccharomyces cerevisiae. Methods Enzymol., 113, 327-337 (1985)

[19] Cordovilla, M.P.; Perez, J.; Ligero, F.; Lluch, C.; Valpuesta, V.: Partial purification and characterization of NADH-glutamate synthase from faba bean (Vicia faba) root nodules. Plant Sci., 150, 121-128 (2000)

[20] Ishiyama, K.; Hayakawa, T.; Yamaya, T.: Expression of NADH-dependent glutamate synthase protein in the epidermis and exodermis of rice roots in response to the supply of ammonium ions. Planta, 204, 288-294 (1998)

[21] Hayakawa, T.; Hopkins, L.; Peat, L.J.; Yamaya, T.; Tobin, A.K.: Quantitative intercellular localization of NADH-dependent glutamate synthase protein in different types of root cells in rice plants. Plant Physiol., 119, 409-416 (1999)

[22] Obara, M.; Kajiura, M.; Fukuta, Y.; Yano, M.; Hayashi, M.; Yamaya, T.; Sato, T.: Mapping of QTLs associated with cytosolic glutamine synthetase and NADH-glutamate synthase in rice (Oriza sativa L.). J. Exp. Bot., 52, 1209-1217 (2001)

[23] Doverskog, M.; Jacobsson, U.; Chapman, B.E.; Kuchel, P.W.; Häggström, L.: Determination of NADH-dependent glutamate synthase (GOGAT) in Spodoptera frugiperda (Sf9) insect cells by a selective 1H/15N NMR in vitro assay. J. Biotechnol., 79, 87-97 (2000)

[24] Vanoni, M.A.; Curti, B.: Glutamate synthase: A complex iron-sulfur flavoprotein. Cell. Mol. Life Sci., 55, 617-638 (1999)

[25] Vance, C.P.; Miller, S.S.; Gregerson, D.L.; Samac, D.A.; Robinson, D.L.; Gantt, J.S.: Alfalfa NADH-dependent glutamate synthase: Structure of the gene and importance in symbiotic N_2 fixation. Plant J., 8, 345-358 (1995)

[26] Chichkova, S.; Arellano, J.; Vance, C.P.; Hernandez, G.: Transgenic tobacco plants that overexpress alfalfa NADH-glutamate synthase have higher carbon and nitrogen content. J. Exp. Bot., 52, 2079-2087 (2001)

Lysine dehydrogenase 1.4.1.15

1 Nomenclature

EC number
1.4.1.15

Systematic name
L-lysine:NAD$^+$ oxidoreductase (deaminating, cyclizing)

Recommended name
lysine dehydrogenase

Synonyms
L-lysine ε dehydrogenase <2> [2]

CAS registry number
68073-29-0

2 Source Organism

<-49> no activity in *Bos taurus* (liver homogenate [3]) [3]
<-48> no activity in *Oosporidium margaritiferum* (strain IFO 1208 [3]) [3]
<-47> no activity in *Endomycopsis burtonii* (strains IFO 1196 and 0844 [3]) [3]
<-46> no activity in *Trichosporon cutaneum* (strains IFO 0174 and 1198 [3]) [3]
<-45> no activity in *Rhodotorula glutinis* (strain IFO 0898 [3]) [3]
<-44> no activity in *Rhodotorula texensis var. minuta* (strain IFO 0412 [3]) [3]
<-43> no activity in *Rhodotorula rubra* (strains IFO 0001, 0003, 0870 and 0900 [3]) [3]
<-42> no activity in *Rhodotorula lactosa* (strain IFO 1006 [3]) [3]
<-41> no activity in *Rhodotorula marina* (strain IFO 0879 [3]) [3]
<-40> no activity in *Rhodotorula minuta* (strain IFO 0387 [3]) [3]
<-39> no activity in *Candida tropicalis* (strain IFO 1401 [3]) [3]
<-38> no activity in *Candida lypolytica* (strain IFO 1209 [3]) [3]
<-37> no activity in *Candida albicans* (strains IFO 0197 and 0601 [3]) [3]
<-36> no activity in *Torolupsis xylinus* (strain IFO 0454 [3]) [3]
<-35> no activity in *Torulopsis candida* (strains IFO 0380 and 0768 [3]) [3]
<-34> no activity in *Cryptococcus albidus* (strain IFO 0410 [3]) [3]
<-33> no activity in *Hansenula hansenii* (strain IFO 0794 [3]) [3]
<-32> no activity in *Hansenula holstii* (strain IFO 0980 [3]) [3]
<-31> no activity in *Hansenula canadenis* (strain IFO 0937 [3]) [3]
<-30> no activity in *Hansenula bimundalis* (strain IFO 1366 [3]) [3]
<-29> no activity in *Hansenula beijerinckii* (strain IFO 0981 [3]) [3]

<-28> no activity in *Hansenula sarturnus* (strain IFO 0125 and 0992 [3]) [3]
<-27> no activity in *Hansenula pettersonii* (strain IFO 1372 [3]) [3]
<-26> no activity in *Hansenula mrakii* (strain IFO 0895 [3]) [3]
<-25> no activity in *Hansenula anomala* (strain 0118 [3]) [3]
<-24> no activity in *Pichia polymorpha* (strain IFO 0195 [3]) [3]
<-23> no activity in *Kluyveromyces lactis* (strain IFO 1090 [3]) [3]
<-22> no activity in *Arthrobacter tumescens* (strain IFO 12960 [3]) [3]
<-21> no activity in *Brevibacterium divaricatum* (strain ICR 4100 [3]) [3]
<-20> no activity in *Pseudomonas aerofaciens* (strain IFO 3521 [3]) [3]
<-19> no activity in *Pseudomonas dacunhae* (strain ICR 3180 [3]) [3]
<-18> no activity in *Pseudomonas ovalis* (strain IFO 3738 [3]) [3]
<-17> no activity in *Pseudomonas marginalis* (strain IFO 3925 [3]) [3]
<-16> no activity in *Pseudomonas fluorescens* (strain IFO 3081 [3]) [3]
<-15> no activity in *Pseudomonas aeruginosa* (strain IFO 3080 [3]) [3]
<-14> no activity in *Sarcina subflava* (strain IFO 11992 [3]) [3]
<-13> no activity in *Staphylococcus aureus* (strain IFO 3060 [3]) [3]
<-12> no activity in *Corynebacterium glutamicum* (strain ATCC 13032 [3]) [3]
<-11> no activity in *Bacillus licheniformis* (strain IFO 12200 [3]) [3]
<-10> no activity in *Bacillus subtilis* (strain IFO 3037 [3]) [3]
 <-9> no activity in *Bacillus megaterium* (strain ICR 1340 [3]) [3]
 <-8> no activity in *Bacillus cereus* (strain IFO 3001 [3]) [3]
 <-7> no activity in *Achromobacter superficialis* (strain ICR 0890 [3]) [3]
 <-6> no activity in *Flavobacterium lutescens* (strain IFO 3014 [3]) [3]
 <-5> no activity in *Proteus vulgaris* (strain IFO 3167 [3]) [3]
 <-4> no activity in *Aerobacter aerogenes* (strain IFO 3320 [3]) [3]
 <-3> no activity in *Serratia marcescens* (strain IFO 3046 [3]) [3]
 <-2> no activity in *Erwinia aroidae* (strain IFO 3830 [3]) [3]
 <-1> no activity in *Escherichia coli* [3]
 <1> *Homo sapiens* [1]
 <2> *Agrobacterium tumefaciens* (strains ICR 1600 and IFO 3058, enzyme is induced by lysin [3]) [2, 3, 4, 6]
 <3> *Aerobacter aerogenes* (strains ICR 0220, IFO 3310 and IFO 12059 [3]) [3]
 <4> *Proteus morganii* (strain IFO 3848 [3]) [3]
 <5> *Alcaligenes faecalis* (strain IFO 3160 [3]) [3]
 <6> *Bacillus sphaericus* (strain IFO 3525 [3]) [3]
 <7> *Micrococcus flavus* (strain ICR 1820 [3]) [3]
 <8> *Micrococcus roseus* (strain IFO 3764 [3]) [3]
 <9> *Corynebacterium pseudodiphteriticum* (strain ICR 2210 [3]) [3]
 <10> *Hafnia alvei* (Bacterium cadaveris strain IFO 3731 [3]) [3]
 <11> *Pseudomonas fragi* (strain IFO 3458 [3]) [3]
 <12> *Brevibacterium ammoniagenes* (strain IFO 12071 [3]) [3]
 <13> *Arthrobacter atrocyneus* (strain IFO 12670 [3]) [3]
 <14> *Candida albicans* (strain SBUG 182 [5]) [5]

3 Reaction and Specificity

Catalyzed reaction
L-lysine + NAD$^+$ = 1,2-didehydropiperidine-2-carboxylate + NH$_3$ + NADH + H$^+$ (<2> the pro-R hydrogen at the prochiral C-6 carbon of L-lysine is specifically abstracted and transferred to the pro-R position at C-4 of NAD$^+$ [4])

Reaction type
deamination
oxidation
redox reaction
reduction

Natural substrates and products
S L-lysine + NAD$^+$ <1> (Reversibility: ? <1> [1]) [1]
P 1,2-didehydropiperidine-2-carboxylate + NH$_3$ + NADH <1> [1]
S L-lysine + NAD$^+$ + H$_2$O <2> (<2> ε amino group is oxidatively deaminated to α-aminoadipate δ-semialdehyde, which is spontaneously converted into 1,2-didehydropiperidine-6-carboxylate, first step of lysine degradation [2]) (Reversibility: ? <2> [2]) [2]
P 1,2-didehydropiperidine-6-carboxylate + NH$_3$ + NADH <2> [2]

Substrates and products
S 4-hydroxylysine + NADP$^+$ <14> (Reversibility: ? <14> [5]) [5]
P α-amino-γ-hydroxyadipate-semialdehyde + NADPH <14> [5]
S 5-hydroxylysine + NADP$^+$ <14> (Reversibility: ? <14> [5]) [5]
P α-amino-δ-hydroxyadipate-semialdehyde + NADPH <14> [5]
S L-lysine + NAD$^+$ <1> (Reversibility: ? <1> [1]) [1]
P 1,2-didehydropiperidine-2-carboxylate + NH$_3$ + NADH <1> [1]
S L-lysine + NAD$^+$ + H$_2$O <2-14> (<2> ε amino group is oxidatively deaminated to α-aminoadipate δ-semialdehyde, which is spontaneously converted into 1,2-didehydropiperidine-6-carboxylate [2]) (Reversibility: ? <2-14> [2-6]) [2-6]
P 1,2-didehydropiperidine-6-carboxylate + NH$_3$ + NADP <2-14> [2-6]
S S-(2-aminoethyl)-L-cysteine + NADP$^+$ + H$_2$O <14> (Reversibility: ? <14> [5]) [5]
P S-(2oxoethyl)-L-cysteine + NADPH + NH$_3$ <14> [5]

Inhibitors
5,5-dithiobis(2-nitrobenzoate) <2> (<2> 0.5 mM, 70% inhibition, 40% inhibition after preincubation with 10 mM L-lysine, 98.5% inhibition after preincubation with 1 mM NAD$^+$ [4]) [4]

Cofactors/prosthetic groups
NAD$^+$ <1, 2> (<2> NADP$^+$ can not replace NAD$^+$ [2,4]) [1, 2]

Activating compounds
5-aminovalerate <2> (<2> 5 mM, 1.65fold activation [4]) [4]
6-aminocaproate <2> (<2> 5 mM, 2.32fold activation [4]) [4]
7-aminoheptanoate <2> (<2> 5 mM, 2.3fold activation [4]) [4]

8-aminooctanoate <2> (<2> 5 mM, 2.26fold activation [4]) [4]
D-lysine <2> (<2> 5 mM, 1.58fold activation [4]) [4]
D-phenylalanine <2> (<2> 5 mM, 1.23fold activation [4]) [4]
D-tryptophan <2> (<2> 5 mM, 2.01fold activation [4]) [4]
DL-α-hydroxy-n-caproate <2> (<2> 5 mM, 1.85fold activation [4]) [4]
DL-homolysine <2> (<2> 5 mM, 1.8fold activation [4]) [4]
L-isoleucine <2> (<2> 5 mM, 1.72fold activation [4]) [4]
L-leucine <2> (<2> 5 mM, 1.69fold activation [4]) [4]
L-lysine <2> (<2> 5 mM, 2.29fold activation [4]) [4]
L-methionine <2> (<2> 5 mM, 1.54fold activation [4]) [4]
L-norleucine <2> (<2> 5 mM, 2.05fold activation [4]) [4]
L-norvaline <2> (<2> 5 mM, 1.71fold activation [4]) [4]
L-phenylalanine <2> (<2> 5 mM, 2.09fold activation [4]) [4]
L-phenylglycine <2> (<2> 5 mM, 1.74fold activation [4]) [4]
L-phenyllactate <2> (<2> 5 mM, 2.25fold activation [4]) [4]
L-tryptophan <2> (<2> 5 mM, 1.24fold activation [4]) [4]
L-tyrosine <2> (<2> 5 mM, 1.41fold activation [4]) [4]
L-valine <2> (<2> 5 mM, 1.47fold activation [4]) [4]
S-(2-aminoethyl)-L-cysteine <2> (<2> 5 mM, 2.32fold activation [4]) [4]

Specific activity (U/mg)

0.00004 <9> (<9> enzyme activity in crude extracts [3]) [3]
0.000059 <11> (<11> enzyme activity in crude extracts [3]) [3]
0.000069 <13> (<13> enzyme activity in crude extracts [3]) [3]
0.0001 <8> (<8> enzyme activity in crude extracts [3]) [3]
0.00017 <3> (<3> enzyme activity in crude extracts, strain IFO 3319 [3]) [3]
0.00021 <7> (<7> enzyme activity in crude extracts [3]) [3]
0.00026 <10> (<10> enzyme activity in crude extracts [3]) [3]
0.00067 <12> (<12> enzyme activity in crude extracts [3]) [3]
0.00069 <5> (<5> enzyme activity in crude extracts [3]) [3]
0.0016 <4> (<4> enzyme activity in crude extracts [3]) [3]
0.0029 <6> (<6> enzyme activity in crude extracts [3]) [3]
0.0052 <3> (<3> enzyme activity in crude extracts, strain ICR 0220 [3]) [3]
0.0144 <14> (<14> activity in cells grown on 5 mM lysine [5]) [5]
0.0392 <3> (<3> enzyme activity in crude extracts, strain IFO 12059 [3]) [3]
0.0696 <2> (<2> enzyme activity in crude extracts, strain IFO 3058 [3]) [3]
0.0739 <2> (<2> enzyme activity in crude extracts, strain ICR 1600 [3]) [3]
Additional information <1> (<1> 0.874 micromol/min/l homogenate, 3 months old baby girl with congenital lysine intolerance, in the absence of cysteine [1]) [1]
Additional information <1> (<1> 11.96 micromol/min/l homogenate, in the presence of cysteine [1]) [1]
Additional information <1> (<1> 2.70 micromol/min/l homogenate, 3 months old baby girl with congenital lysine intolerance, in the presence of cysteine [1]) [1]
Additional information <1> (<1> 6.24 micromol/min/l homogenate, in the absence of cysteine [1]) [1]

K$_m$-Value (mM)
 0.06 <2> (NAD$^+$) [4]
 1.85 <2> (deamino-NAD$^+$) [4]
 3.13 <2> (3-acetylpyridine-NAD$^+$) [4]

pH-Optimum
 9 <14> (<14> oxidative deamination of S-(β-aminoethyl)-cysteine, 5- and 4-hydroxylysine [5]) [5]
 9.3 <14> (<14> oxidative deamination of lysine [5]) [5]
 9.7 <2> [2]

Temperature optimum (°C)
 32 <14> (<14> with S-(2-aminoethyl)-cysteine and 5-hydroxylysine [5]) [5]
 37 <14> (<14> with lysine and 4-hydroxylysine [5]) [5]

4 Enzyme Structure

Subunits
 dimer <2> [4]

5 Isolation/Preparation/Mutation/Application

Source/tissue
 liver <1> [1]

Purification
 <2> [4]

Application
 analysis <2> (<2> amperometric biosensor for L-lysine based on immobilized enzyme [6]) [6]

6 Stability

Temperature stability
 30-50 <2> (<2> in the absence of lysine, approx. 50% activity after 10 min at 40°C [4]) [4]
 30-60 <2> (<2> in the presence of 5 mM L-lysine, approx. 50% activity after 10 min at 55°C [4]) [4]

References

[1] Buergi, W.; Richterich, R.; Colombo, J.P.: L-Lysine dehydrogenase deficiency in a patient with congenital lysine intolerance. Nature, **211**, 854-855 (1966)

[2] Misono, H.; Nagasaki, S.: Occurrence of L-lysine ε-dehydrogenase in Agrobacterium tumefaciens. J. Bacteriol., **150**, 398-401 (1982)

[3] Misono, H.; Nagasaki, S.: Distribution and physiological function of L-lysine ε-dehydrogenase. Agric. Biol. Chem., **47**, 631-633 (1983)

[4] Misono, H.: NAD$^+$-dependent lysine dehydrogenase from a plant-pathogenic bacterium, Agrobacterium tumefaciens. Vitamins (Japan), **65**, 1-12 (1991)

[5] Hammer, T.; Bode, R.: Enzymic production of α-aminoadipate-δ-semialdehyde and related compounds by lysine ε-dehydrogenase from Candida albicans. Zentralbl. Mikrobiol., **147**, 65-70 (1992)

[6] Dempsey, E.; Wang, J.; Wollenberger, U.; Ozsoz, M.; Smyth, M.R.: A lysine dehydrogenase-based electrode for biosensing of L-lysine. Biosens. Bioelectron., **7**, 323-327 (1992)

Diaminopimelate dehydrogenase

1 Nomenclature

EC number

1.4.1.16

Systematic name

meso-2,6-diaminoheptanedioate:NADP$^+$ oxidoreductase (deaminating)

Recommended name

diaminopimelate dehydrogenase

Synonyms

meso-α,ε-diaminopimelate dehydrogenase

meso-diaminopimelate dehydrogenase

CAS registry number

60894-21-5

2 Source Organism

<-7> no activity in *Nicotiana tabacum* [18]

<-6> no activity in *Glycine max* [18]

<-5> no activity in *Zea mays* [18]

<-4> no activity in *Chlamydomonas reinhardtii* [18]

<-3> no activity in *Staphylococcus aureus* [4]

<-2> no activity in *Klebsiella pneumoniae* (strains IFO 125059 and 12019 [4]) [4]

<-1> no activity in *Escherichia coli* (strains ICR 0010, K12, ICR 0050 and IFO 3301 [4]) [4]

<1> *Corynebacterium glutamicum* (strain RRL-5 [1]; strain AS019 [2]; strain ATCC 13032 [4]; Clostridium glutamicum [21]) [1-4, 9, 14, 16, 19, 20, 21, 22, 23]

<2> *Bacillus sphaericus* [4, 5, 7, 8, 10-15, 17, 23, 24]

<3> *Bacillus pasteurii* [8]

<4> *Bacillus globisporus* [8]

<5> *Bacillus macerans* [8]

<6> *Brevibacterium sp.* (strain ICR 7000 [6]) [4, 6, 14]

<7> *Brevibacterium protophormiae* [4]

<8> *Proteus morganii* [4]

<9> *Proteus rettgeri* [4]

<10> *Proteus vulgaris* [14]

<11> *Pseudomonas cruciviae* [4]

<12> *Pseudomonas fragi* [4]
<13> *Flavobacterium suaveolens* (low activity [4]) [4]
<14> *Micrococcus luteus* (low activity [4]) [4]
<15> *Alcaligenes viscolatus* (low activity [4]) [4]
<16> *Corynebacterium pseudodiphteriticum* (low activity [4]) [4]
<17> *Enterobacter cloacae* (low activity [4]) [4]
<18> *Brevibacterium ammoniagenes* (low activity [4]) [4]
<19> *Bacillus brevis* (low activity [4]) [4]
<20> *Hafnia alvei* (very low activity [4]) [4]
<21> *Arthrobacter tumescens* (very low activity [4]) [4]
<22> *Agrobacterium tumefaciens* (very low activity [4]) [4]
<23> *Pseudomonas putida* (very low activity [4]) [4]
<24> *Bacillus cereus* (very low activity [4]) [4]
<25> *Bacillus amyloliquefaciens* (very low activity [4]) [4]
<26> *Bacillus licheniformis* (very low activity [4]) [4]
<27> *Bacterium mycoides* (very low activity [4]) [4]
<28> *Achromobacter superficialis* (very low activity [4]) [4]
<29> *Achromobacter polymorph* (very low activity [4]) [4]
<30> *Micrococcus roseus* (very low activity [4]) [4]
<31> *Pseudomonas aureofaciens* (very low activity [4]) [4]
<32> *Pseudomonas alkanolytica* (very low activity [4]) [4]
<33> *Pseudomonas aeruginosa* (very low activity [4]) [4]
<34> *Erwinia aroideae* (low activity [14]) [14]
<35> *Proteus mirabilis* (low activity [14]) [14]
<36> *Sarcina subflava* (low activity [14]) [14]
<37> *Pseudomonas fluorescens* (low activity [14]) [14]

3 Reaction and Specificity

Catalyzed reaction

meso-2,6-diaminoheptanedioate + H_2O + $NADP^+$ = L-2-amino-6-oxoheptanedioate + NH_3 + NADPH + H^+ (<2> catalyzes the transfer of the pro-S hydrogen at C-4 of NADPH to the substrate, enzyme is B-stereospecific [12]; <1> transfer of the pro-R hydrogen, type A reductase [19]; <1> enzyme may exhibit a random order of addition of substrates [21]; <1> only amino acid dehydrogenase which stereospecifically oxidizes a D-stereocenter [22])

Reaction type

oxidation
redox reaction
reduction

Natural substrates and products

S L-2-amino-6-oxoheptanedioate + NH_3 + NADPH <1-12> (<1> involved in lysine biosynthesis [1]; <1> bypath of α,ε-diaminopimelic acid pathway [9]) (Reversibility: r <1-12> [1-15]) [1-15]
P meso-2,6-diaminoheptanedioate + H_2O + $NAD(P)^+$ <1-12> [1-15]

Substrates and products

S 4-methylene diaminopimelate + H_2O + $NADP^+$ <2> (<2> mixture of all possible stereoisomers, 4% of activity with meso-2,6-diaminopimelate [17]) (Reversibility: r <2> [17]) [17]

P 4-methyleneamino-6-oxopimelate + NH_3 + NADPH <2> [17]

S L-2-amino-6-oxoheptanedioate + NH_3 + NAD(P)H <1-37> (<1> diaminopimelate synthesis is thermodynamically favoured [22]) (Reversibility: r <1-37> [1-24]) [1-24]

P meso-2,6-diaminoheptanedioate + H_2O + $NAD(P)^+$ <1-37> [1-24]

S N-2-aminodiaminopimelate + H_2O + $NADP^+$ <2> (<2> mixture of all possible stereoisomers, 4% of activity with meso-2,6-diaminopimelate [17]) (Reversibility: r <2> [17]) [17]

P N-2-amino-6-oxopimelate + NH_3 + NADPH <2> [17]

S N-2-hydroxydiaminopimelate + H_2O + $NADP^+$ <2> (<2> mixture of all possible stereoisomers, 22% of activity with meso-2,6-diaminopimelate [17]) (Reversibility: r <2> [17]) [17]

P 2-hydroxyamino-6-oxopimelate + NH_3 + NADPH <2> [17]

Inhibitors

(2S,5S)-2-amino-3(3-carboxy-2-isoxazolin-5-yl)propanoic acid <1, 2> (<2> competitive inhibition [24]) [22, 24]

4,4'-dithiopyridine <2> (<2> complete inactivation [11]) [11]

5,5'-dithiobis(2-nitrobenzoate) <2> (<2> complete inactivation [11]) [11]

Cu^{2+} <1, 6> (<1> 0.01 mM, 82% inhibition [4]; <6> 0.01 mM, 68% inhibition [6]) [4, 6]

D-2,6-diaminopimelate <1, 2> (<1> 10 mM, 49% inhibition [4]) [4, 5, 12]

$HgCl_2$ <1, 2, 6> (<1> 0.01 mM, 92% inhibition [4]; <6> 0.01 mM, 100% inhibition [6]; <2> 0.01 mM, complete inhibition [12]) [4, 6, 11, 12]

L-2,6-diaminopimelate <1, 2> (<1> 10 mM, 27% inhibition [4]) [4, 5, 12]

L-2-amino-6-methylene-pimelate <2> (<2> competitive inhibition [23]) [23]

L-cysteine <1> (<1> 5 mM, 99% inhibition [4]) [4]

Li^+ <1> (<1> 1 mM, 32% inhibition [4]) [4]

N-ethylmaleimide <1, 2, 6> (<1> 1 mM, 39% inhibition [4]; <6> 1 mM, 52% inhibition [6]; <2> complete inactivation [11]) [4, 6, 11]

Ni^{2+} <1, 6> (<1> 1 mM, 36% inhibition [4]; <6> 1 mM, 21% inhibition [6]) [4, 6]

glyoxylate <2> (<2> 5 mM, 30% inhibition of reductive amination [12]) [12]

iodoacetate <1, 2, 6> (<1> 1 mM, 77% inhibition [4]; <6> 1 mM, 55% inhibition [6]; <2> complete inactivation [11]) [4, 6, 11]

meso-α,δ-diaminoadipate <2> [12]

meso-diaminopimelate <2> [5]

p-chloromercuribenzoate <1, 2> (<2> complete inactivation [11]; <2> 0.001 mM, complete inhibition [12]) [4, 11, 12]

thioglycolate <1> (<1> 5 mM, 62% inhibition [4]) [4]

Cofactors/prosthetic groups

NADH <1, 6> (<1,6> 3% activity of that with $NADP^+$ [4,6]) [4, 6]

NADPH <1-37> [1-24]

Specific activity (U/mg)

0.013 <2> (<2> activity in crude extracts at pH 8.0 [15]) [15]

0.03 <2> (<2> reductive amination of 2-amino-6-oxopimelate in crude extracts at pH 7.5 in the presence of 1.67 mM meso-diaminopimelate [5]) [5]

0.12 <5> (<5> activity in crude extracts [8]) [8]

0.13 <1> (strain AS019 transformed with plasmid harboring cloned gene [2]) [2]

0.15 <1> (<1> activity in crude extracts of cells grown on complex medium [16]) [16]

0.16 <1> (<1> activity in strain RRL-5 [1]) [1]

0.19 <4> (<4> activity in crude extracts [8]) [8]

0.237 <2> (<2> reductive amination of 2-amino-6-oxopimelate in crude extracts at pH 7.5 [10]) [10]

0.25 <2> (<2> reductive amination of 2-amino-6-oxopimelate in crude extracts at pH 7.5 [5]) [5]

0.287 <2> (<2> oxidative deamination of meso-diaminopimelate in crude extracts at pH 10.5 [10]) [10]

0.34 <2> (<2> activity in crude extracts at pH 10.5 [15]) [15]

0.45 <2> (<2> oxidative deamination of meso-diaminopimelate in crude extracts at pH 10.5 [5]) [5]

0.53 <3> (<3> activity in crude extracts [8]) [8]

2.22 <1> (<1> activity in strain RRL-5 transformed with plasmid harboring cloned gene [1]) [1]

128 <1> [20]

131 <6> [6]

131.6 <1> [4]

139 <2> [13]

K_m-Value (mM)

0.00083 <2> ($NADP^+$) [12, 13]

0.083-0.14 <1, 2, 6> ($NADP^+$) [4, 6, 12, 13]

0.13 <1> ($NADP^+$) [21]

0.13 <1> (NADPH) [4]

0.14 <6> ($NADP^+$) [6]

0.2 <2> (NADPH) [12]

0.21 <6> (L-2-amino-6-oxopimelate) [6]

0.23 <6> (NADPH) [6]

0.24 <2> (L-2-amino-6-oxopimelate) [12]

0.28 <1> (L-2-amino-6-oxopimelate) [4]

2 <6> (NAD^+) [6]

2-10 <1, 6> (NAD^+) [4, 6]

2.5 <2> (meso-2,6-diaminopimelate) [12, 13]

3.1 <1> (meso-diaminopimelate) [21]

6.25 <6> (meso-2,6-diaminopimelate) [6]

10 <1> (NAD^+) [4]

12.5 <2> (NH_3) [12]

36 <1> (NH_3) [4]

62.5 <6> (NH_3) [6]

K$_i$-Value (mM)

0.0042 <1> ((2S,5S)-2-amino-3(3-carboxy-2-isoxazolin-5-yl)propanoic acid, <1> reductive amination [22]) [22]

0.0042 <2> ((2S,5S)-2-amino-3(3-carboxy-2-isoxazolin-5-yl)propanoic acid, <2> competitive inhibition of L-2-amino-6-oxoheptanedioate [24]) [24]

0.005 <2> (L-2-amino-6-methylene-pimelate) [23]

0.023 <1> ((2S,5S)-2-amino-3(3-carboxy-2-isoxazolin-5-yl)propanoic acid, <1> oxidative deamination at pH 7.8 [22]) [22]

0.023 <2> ((2S,5S)-2-amino-3(3-carboxy-2-isoxazolin-5-yl)propanoic acid, <2> noncompetitive inhibition of meso-diaminopimelate [24]) [24]

2 <1> (D-2,6-diaminopimelate) [4]

3.12 <2> (L-α,ε-diaminopimelate) [12]

4 <2> (D-α,ε-diaminopimelate) [12]

4.16 <2> (meso-α,δ-diaminoadipate) [12]

8.5 <1> (L-2,6-diaminopimelate) [4]

pH-Optimum

7.5 <2> (<2> 2-amino-6-oxopimelate + NH$_3$ + NADPH [7,10,12]) [7, 10, 12]

7.9 <1> (<1> reductive amination of 2-amino-6-oxopimelate [4]) [4]

8.5 <6> (<6> 2-amino-6-oxopimelate + NH$_3$ + NADPH [6]) [6]

9 <1> (<1> oxidative deamination of meso-2,6-diaminopimelate [4]) [4]

10.5 <2, 6> (<2,6> meso-2,6-diaminopimelate + H$_2$O + NADP$^+$ [6, 7, 12, 13, 15]) [6, 7, 12, 13, 15]

pH-Range

4.9-10 <2> (<2> 2-amino-6-oxopimelate + NH$_3$ + NADPH [12]) [12]

6.8-11 <2> (<2> meso-2,6-diaminopimelate + H$_2$O + NADP$^+$ [12]) [12]

4 Enzyme Structure

Molecular weight

70000 <1, 6> (<1> gel filtration [4]; <6> gel filtration [6]) [4, 6, 21]

78500-80000 <2> (<2> gel filtration [12,13]; <2> sedimentation equilibrium centrifugation [12]) [12, 13]

Subunits

dimer <1, 2, 6> (<1> 2 * 35099, nucleotide sequencing [1]; <1> 2 * 39000, SDS-PAGE [4]; <6> 2 * 39000, SDS-PAGE [6]; <2> 2 * 41000, SDS-PAGE [12]; <1> 2 * 35198, electrospray MS, deduced from amino acid sequence [20]) [1, 4, 6, 12, 20, 21]

5 Isolation/Preparation/Mutation/Application

Purification

<1> [4, 19, 20, 21, 22]

<2> (ammonium sulfate, DEAE-cellulose, hydroxyapatite, Sephadex G-150 [12]) [12, 13]

<6> [6]

Crystallization

<1> (crystallization in the presence of NADP$^+$, only monoclinic crystals are suitable for X-ray diffraction, 2.2 A resolution [19]; <1> enzyme-NADP$^+$ complex, hanging drop vapor diffusion, crystals of maximum dimensions of 0.05 x 0.05 x 0.5 mm are observed in 1 M ammonium sulfate, 1.2 M lithium sulfate, 100 mM HEPES, pH 7.5, crystals of dimensions 0.3 x 0.05 x 0.5 mm are obtained from 13-17% polyethylen glycol 8000 in 100 mM sodium caco-dylate, pH 6.5 and 150-300 mM Mg(OAc)$_2$ [20]; <1> crystals of the ternary complex formed from the enzyme, NADP$^+$ and the inhibitor (2S,5S)-2-ami-no-3(3-carboxy-2-isoxazolin-5-yl)propanoic acid [22]; <1> crystals of the ternary complex formed from the enzyme, NADP$^+$ and the inhibitor L-2-ami-no-6-methylene-pimelate, 2.1 A resolution [23]) [19, 20, 22, 23]

Cloning

<1> (expression in Escherichia coli and Corynebacterim glutamicum [1]) [1-3, 19, 20, 21, 22]

6 Stability

pH-Stability

6.5-7 <1> [4]
7 <2> [12]
7-9 <6> (<6> at 40°C [6]) [6]

Temperature stability

45 <1, 2> (<1,2> unstable above [4,12]) [4, 12]
48 <6> (<6> 10 min, 10 mM potassium phosphate pH 7.4, unstable above [6]) [6]

Storage stability

<2>, -15°C, stable [7]
<2>, 4°C, 0.01% 2-mercaptoethanol, 20% glycerol, 1 month [12]
<1, 2>, -20°C, 0.01 M potassium phosphate buffer, pH 7.4, 0.01% 2-mercap-toethanol, 10% glycerol, 1 year [4, 12]

References

[1] Ishino, S.; Mizukami, T.; Yamaguchi, K.; Katsumata, R.; Araki, K.: Cloning and sequencing of the meso-diaminopimelate-D-dehydrogenase (ddh) gene of Corynebacterium glutamicum. Agric. Biol. Chem., 52, 2903-2909 (1988)

[2] Yeh, P.; Sicard, A.M.; Sinskey, A.J.: General organization of the genes specifically involved in the diaminopimelate-lysine biosynthetic pathway of Corynebacterium glutamicum. Mol. Gen. Genet., 212, 105-111 (1988)

[3] Ishino, S.; Mizukami, T.; Yamaguchi, K.; Katsumata, R.; Araki, K.: Nucleotide sequence of the meso-diaminopimelate D-dehydrogenase gene from Corynebacterium glutamicum. Nucleic Acids Res., 15, 3917 (1987)

[4] Misono, H.; Ogasawara, M.; Nagasaki, S.: Characterization of meso-diaminopimelate dehydrogenase from Corynebacterium glutamicum and its distribution in bacteria. Agric. Biol. Chem., **50**, 2729-2734 (1986)

[5] Bartlett, A.T.M.; White, P.J.: Regulation of the enzymes of lysine biosynthesis in Bacillus sphaericus NCTC 9602 during vegetative growth. J. Gen. Microbiol., **132**, 3169-3177 (1986)

[6] Misono, H.; Ogasawara, M.; Nagasaki, S.: Purification and properties of meso-diaminopimelate dehydrogenase from Brevibacterium sp.. Agric. Biol. Chem., **50**, 1329-1330 (1986)

[7] White, P.J.: 2,6-Diaminopimelate. Methods Enzym. Anal., 3rd Ed. (Bergmeyer, H.U., ed.), **8**, 377-383 (1985)

[8] Bartlett, A.T.M.; White, P.J.: Species of Bacillus that make a vegetative peptidoglycan containing lysine lack diaminopimelate epimerase but have diaminopimelate dehydrogenase. J. Gen. Microbiol., **131**, 2145-2152 (1985)

[9] Ishino, S.; Yamaguchi, K.; Shirahata, K.; Araki, K.: Involvement of meso-α,ε-diamino-pimelate D-dehydrogenase in lysine biosynthesis in Corynebacterium glutamicum. Agric. Biol. Chem., **48**, 2257-2260 (1984)

[10] White, P.J.: The essential role of diaminopimelate dehydrogenase in the biosynthesis of lysine by Bacillus sphaericus. J. Gen. Microbiol., **129**, 739-749 (1983)

[11] Misono, H.; Nagasaki, S.; Soda, K.: meso-α,ε-Diaminopimelate D-dehydrogenase: sulfhydryl group modification. Agric. Biol. Chem., **45**, 1455-1460 (1981)

[12] Misono, H.; Soda, K.: Properties of meso-α,ε-diaminopimelate D-dehydrogenase from Bacillus sphaericus. J. Biol. Chem., **255**, 10599-10605 (1980)

[13] Misono, H.; Soda, K.: Purification and properties of meso-α,ε-diaminopimelate D-dehydrogenase from Bacillus sphaericus. Agric. Biol. Chem., **44**, 227-229 (1980)

[14] Misono, H.; Togawa, H.; Soda, K.: meso-α,ε-Diaminopimelate D-dehydrogenase: distribution and the reaction product. J. Bacteriol., **137**, 22-27 (1979)

[15] Misono, H.; Togawa, H.; Yamamoto, T.; Soda, K.: Occurrence of meso-α,ε-diaminopimelate dehydrogenase in Bacillus sphaericus. Biochem. Biophys. Res. Commun., **72**, 89-93 (1976)

[16] Cremer, J.; Treptow, C.; Eggeling, L.; Sahm, H.: Regulation of enzymes of lysine biosynthesis in Corynebacterium glutamicum. J. Gen. Microbiol., **134**, 3221-3229 (1988)

[17] Lam, L.K.P.; Arnold, L.D.; Kalantar, T.H.; Kelland, J.G.; Lane-Bell, P.M.; Palcic, M.M.; Pickard, M.A.; Vederas, J.C.: Analogs of diaminopimelic acid as inhibitors of meso-diaminopimelate dehydrogenase and LL-diaminopimelate epimerase. J. Biol. Chem., **263**, 11814-11819 (1988)

[18] Chatterjee, S.P.; Singh, B.K.; Gilvarg, C.: Biosynthesis of lysine in plants: the putative role of meso-diaminopimelate dehydrogenase. Plant Mol. Biol., **26**, 285-290 (1994)

[19] Scapin, G.; Reddy, S.G.; Blanchard, J.S.: Three-dimensional structure of meso-diaminopimelic acid dehydrogenase from Corynebacterium glutamicum. Biochemistry, **35**, 13540-13551 (1996)

[20] Reddy, S.G.; Scapin, G.; Blanchard, J.S.: Expression, purification, and crystallization of meso-diaminopimelate dehydrogenase from Corynebacterium glutamicum. Proteins Struct. Funct. Genet., 25, 514-516 (1996)

[21] Wang, F.; Scapin, G.; Blanchard, J.S.; Angeletti, R.H.: Substrate binding and conformational changes of Corynebacterium glutamicum diaminopimelate dehydrogenase revealed by hydrogen/deuterium exchange and electrospray mass spectrometry. Protein Sci., 7, 293-299 (1998)

[22] Scapin, G.; Cirilli, M.; Reddy, S.G.; Gao, Y.; Vederas, J.C.; Blanchard, J.S.: Substrate and inhibitor binding sites in Corynebacterium glutamicum diaminopimelate dehydrogenase. Biochemistry, 37, 3278-3285 (1998)

[23] Cirilli, M.; Scapin, G.; Sutherland, A.; Vederas, J.C.; Blanchard, J.S.: The three-dimensional structure of the ternary complex of Corynebacterium glutamicum diaminopimelate dehydrogenase-NADPH-L-2-amino-6-methylene-pimelate. Protein Sci., 9, 2034-2037 (2000)

[24] Abbott, S.D.; Lane-Bell, P.; Sidhu, K.P.S.; Vederas, J.C.: Synthesis and testing of heterocyclic analogues of diaminopimelic acid (DAP) as inhibitors of DAP dehydrogenase and DAP epimerase. J. Am. Chem. Soc., 116, 6513-6520 (1994)

N-methylalanine dehydrogenase 1.4.1.17

1 Nomenclature

EC number
1.4.1.17

Systematic name
N-methyl-L-alanine:NADP⁺ oxidoreductase (demethylating, deaminating)

Recommended name
N-methylalanine dehydrogenase

CAS registry number
56379-51-2

2 Source Organism

<1> *Pseudomonas sp.* (MS ATCC 25262 [1]) [1]

3 Reaction and Specificity

Catalyzed reaction
N-methyl-L-alanine + H_2O + NADP⁺ = pyruvate + methylamine + NADPH + H⁺

Reaction type
oxidation
redox reaction
reduction

Natural substrates and products
S pyruvate + methylamine + NADPH <1> (<1> reverse reaction only to limited extend at very low rate [1]) (Reversibility: r <1> [1]) [1]
P N-methyl-L-alanine + H_2O + NADP⁺ <1> [1]

Substrates and products
S methylamine + α-ketobutyrate + NADPH <1> (<1> 14% of activity with pyruvate [1]) (Reversibility: ? <1> [1]) [1]
P glutamic acid + H_2O + NADP⁺
S methylamine + oxaloacetate + NADPH <1> (<1> 64% of activity with pyruvate [1]) (Reversibility: ? <1> [1]) [1]
P aspartic acid + H_2O + NADP⁺

S pyruvate + methylamine + NADPH <1> (<1> reverse reaction only to limited extend at very low rate [1]) (Reversibility: r <1> [1]) [1]

P N-methyl-L-alanine + H_2O + NADP$^+$ <1> [1]

Inhibitors

NADPH <1> (<1> substrate inhibition above 1 mM [1]) [1]

Cofactors/prosthetic groups

NADPH <1> [1]

Activating compounds

2-mercaptoethanol <1> (<1> required for full activity [1]) [1]

dithiothreitol <1> (<1> can replace 2-mercaptoethanol [1]) [1]

Specific activity (U/mg)

0.37 <1> (<1> purified enzyme [1]) [1]

K_m-Value (mM)

0.035 <1> (NADPH) [1]

15 <1> (pyruvate) [1]

75 <1> (methylamine) [1]

pH-Optimum

8.2-8.6 <1> [1]

4 Enzyme Structure

Molecular weight

36500 <1> (<1> SDS-PAGE [1]) [1]

77000 <1> (<1> gel filtration [1]) [1]

Subunits

dimer <1> (<1> 2 * 36500, SDS-PAGE, gel filtration [1]) [1]

5 Isolation/Preparation/Mutation/Application

Purification

<1> [1]

6 Stability

General stability information

<1>, unstable in the absence of 2-mercaptoethanol or dithiothreitol [1]

References

[1] Lin, M.C.M.; Wagner, C.: Purification and characterization of N-methylalanine dehydrogenase. J. Biol. Chem., **250**, 3746-3751 (1975)

Lysine 6-dehydrogenase

1 Nomenclature

EC number
1.4.1.18

Systematic name
L-lysine:NAD$^+$ 6-oxidoreductase (deaminating)

Recommended name
lysine 6-dehydrogenase

Synonyms
L-lysine ε-dehydrogenase

CAS registry number
89400-30-6

2 Source Organism

<-2> no activity in *Bos taurus* (liver [3]) [3]
<-1> no activity in *different yeasts* (more than 20 different yeast tested for activity [3]) [3]
<1> *Agrobacterium tumefaciens* (activity depends on growth conditions [3]) [1-3, 5-9]
<2> *Klebsiella pneumoniae* (Aerobacter aerogenes ICR 0220 and IFO 12059, traces of activity with strain IFO 3319 [3]) [3]
<3> *Proteus morganii* [3]
<4> *Alcaligenes faecalis* (traces of activity [3]) [3]
<5> *Bacillus sphaericus* [3]
<6> *Micrococcus sp.* (traces of activity in Micrococcus roseus and Micrococcus flavus [3]) [3]
<7> *Corynebacterium pseudodiphtheriticum* (traces of activity [3]) [3]
<8> *Hafnia alvei* (Bacterium cadaveris, traces of activity [3]) [3]
<9> *Pseudomonas fragi* (traces of activity [3]) [3]
<10> *Brevibacterium ammoniagenes* (low activity [3]) [3]
<11> *Arthrobacter atrocyaneus* (traces of activity [3]) [3]
<12> *Anabasis aphylla* [4]

3 Reaction and Specificity

Catalyzed reaction

L-lysine + NAD$^+$ = allysine + NH$_3$ + NADH + H$^+$

Reaction type

oxidation

redox reaction

reduction

Natural substrates and products

S L-lysine + NAD$^+$ <1-12> (<1> function in lysine catabolism [1,5]; <12> involved in biosynthesis of piperidine nucleus of alkaloids [4]) (Reversibility: ? <1-12> [1-4, 6-9]; <1> ir [5]) [1-9]

P α-aminoadipate δ-semialdehyde + NH$_3$ + NADH <1-12> (<1-12> product is spontaneosly converted into δ^1-piperideine-6-carboxylate [1-9]) [1-9]

Substrates and products

S L-lysine + NAD$^+$ <1-12> (<1> no activity with D-lysine [1]; <1> reaction A-stereospecific [5,9]) (Reversibility: ? <1-12> [1-4, 6-9]; ir <1> [5]) [1-9]

P α-aminoadipate δ-semialdehyde + NH$_3$ + NADH <1-12> (<1-12> product is spontaneosly converted into δ^1-piperideine-6-carboxylate [1-9]) [1-9]

S S-(β-aminoethyl)-L-cysteine + NAD$^+$ <1> (<1> 3% of the activity with L-lysine [1,6,7]) (Reversibility: ? <1> [1, 6]) [1, 6, 7]

P ?

Inhibitors

DTNB <1> (<1> almost complete inhibition in presence of NAD$^+$, 40% inhibition in presence of L-lysine [6]) [6]

HgCl$_2$ <1> (<1> complete inhibition at 0.001 mM after 10 min incubation [6]) [6]

N-ethylmaleimide <1> (<1> complete inhibition at 1 mM after 10 min incubation [6]) [6]

p-chloromercuribenzoate <1> (<1> complete inhibition at 0.01 mM after 10 min incubation, reversible with 2-mercaptoethanol or dithiothreitol [6]) [6]

Cofactors/prosthetic groups

3-acetylpyridine-NAD$^+$ <1> (<1> 16% of activity compared with NAD$^+$ [6]) [6]

NAD$^+$ <1-12> (<1> can not be replaced by NADP$^+$ [1]; <12> prosthetic group [4]) [1-9]

deamino-NAD$^+$ <1> (<1> 42% of activity compared with NAD$^+$ [6]) [6]

Activating compounds

6-aminocaproate <1> [7]

(DL-)homolysine <1> [7]

L-lysine <1> (<1> increased activity after 10 min preincubation [6]) [6]

Additional information <1> (<1> activiated by many amino acids [7]) [7]

Specific activity (U/mg)
4.54 <1> (<1> purified enzyme [1]) [1]
7.5 <1> (<1> preincubation in absence of L-lysine [6]) [6]
15.7 <1> (<1> preincubation in presence of L-lysine [6]) [6]

Km-Value (mM)
0.059 <1> (NAD⁺) [1, 6]
1.5 <1> (L-lysine) [1]
1.85 <1> (deamino-NAD⁺) [6]
3.13 <1> (3-acetylpyridine-NAD⁺) [6]

pH-Optimum
9.7 <1> [1, 2, 6]

4 Enzyme Structure

Molecular weight
39000 <1> (<1> SDS-PAGE [1]) [1]
70000 <1> (<1> gel filtration [1]) [1]
100000 <12> [4]
160000 <1> (<1> ultracentrifugation in presence of L-lysine [6]) [6]
165000 <1> (<1> gel filtration in presence of L-lysine [6]) [6]

Subunits
dimer <1> (<1> 2 * 39000, gel filtration, SDS-PAGE [1]; <1> after preincubation with NAD⁺ without L-lysine [6,7]) [1, 6, 7]
tetramer <1> (<1> gel filtration and centrifugation in presence of L-lysine [6,7]) [6, 7]

5 Isolation/Preparation/Mutation/Application

Source/tissue
branch <12> [4]

Purification
<1> [1, 2, 6-8]
<12> [4]

Application
analysis <1> (<1> highly specific determination of L-lysine concentration in serum [8]) [8]

6 Stability

pH-Stability
5-7.5 <1> (<1> most stable [7]) [7]

Temperature stability

40 <1> (<1> stable for 10 min in absence of L-lysine, unstable above 40°C
[6]) [6]

50 <1> (<1> stable for 10 min in presence of L-lysine, unstable above 50°C
[6]) [6]

General stability information

<1>, L-lysine stabilizes [6]

<1>, most stable after preincubation at 40°C, pH 5.5-7 for 10 min [1]

References

[1] Misono, H.; Uehigashi, H.; Morimoto, E.; Nagasaki, S.: Purification and
properties of L-lysine ε-deydrogenase from Agrobacterium tumefaciens. Agric. Biol. Chem., **49**, 2253-2255 (1985)

[2] Misono, H.; Nagasaki, S.: Occurrence of L-lysine ε dehydrogenase in Agrobacterium tumefaciens. J. Bacteriol., **150**, 398-401 (1982)

[3] Misono, H.; Nagasaki, S.: Distribution and physiological function of L-lysine
ε-dehydrogenase. Agric. Biol. Chem., **47**, 631-633 (1983)

[4] Esbolsev, E.O.; Klyschev, L.K.; Frantsev, A.P.: The enzymatic properties of
biosythesis of piperidine nucleus of alkaloids from lysine. F.E.C.S. (Int. Conf.
Chem. Biotechnol. Biol. Act. Nat. Prod., 3rd., Meeting Date 1985) VCH Wein-
heim, **5**, 60-64 (1987)

[5] Hashimoto, H.; Misono, H.; Nagata, S.; Nagasaki, S.: Stereospecificity of hydrogen transfer of the coenzyme catalyzed by L-lysine ε-dehydrogenase. Agric. Biol. Chem., **53**, 1175-1176 (1989)

[6] Misono, H.; Hashimoto, H.; Uehigashi, H.; Nagata, S.; Nagasaki, S.: Properties of L-lysine ε-dehydrogenase from Agrobacterium tumefaciens. J. Biochem., **105**, 1002-1008 (1989)

[7] Hashimoto, H.; Misono, H.; Nagata, S.; Nagasaki, S.: Activation of L-lysine ε-
dehydrogenase from Agrobacterium tumefaciens by several amino acids and
monocarboxylates. J. Biochem., **106**, 76-80 (1989)

[8] Hashimoto, H.; Misono, H.; Nagata, S.; Nagasaki, S.: Selective determination
of L-lysine with L-lysine ε dehydrogenase. Agric. Biol. Chem., **54**, 291-294
(1990)

[9] Misono, H.; Yoshimura, T.; Nagasaki, S.; Soda, K.: Stereospecific abstraction
of ε-pro-R-hydrogen of L-lysine by L-lysine ε-dehydrogenase from Agrobacterium tumefaciens. J. Biochem., **107**, 169-172 (1990)

Tryptophan dehydrogenase 1.4.1.19

1 Nomenclature

EC number
1.4.1.19

Systematic name
L-tryptophan:NAD(P)$^+$ oxidoreductase (deaminating)

Recommended name
tryptophan dehydrogenase

Synonyms
L-Trp-dehydrogenase
L-tryptophan dehydrogenase
NAD(P)-L-tryptophan dehydrogenase
TDH
dehydrogenase, tryptophan

CAS registry number
94047-13-9

2 Source Organism

<-1> no activity in *Brassica oleraceae* (kohlrabi [2]) [2, 4]
<1> *Nicotiana tabacum* (normal and crown-gall tumor tissue cultures, tissue transformed by Agrobacterium tumefaciens [1]) [1]
<2> *Pisum sativum* (pea [2,4,5]) [2, 4, 5]
<3> *Zea mays* (maize [2,4]) [2, 4]
<4> *Spinacia oleracea* (spinach [3]) [3]
<5> *Prosopis juliflora* [4]
<6> *Lycopersicon esculentum* (tomato [2,4]) [2, 4]

3 Reaction and Specificity

Catalyzed reaction
L-tryptophan + NAD(P)$^+$ = (indol-3-yl)pyruvate + NH_3 + NAD(P)H + H^+

Reaction type
amination
deamination
oxidation

redox reaction
reduction

Natural substrates and products

S L-tryptophan + NAD(P)$^+$ <1-3, 5, 6> (<-1,1-3,6> primary enzyme of indolylpyruvate pathway leading to indol-3-yl-acetic acid in plants, [1,2];
<2,3,5,6> metabolism of indole compounds in plants [4]) (Reversibility:
r <1-3, 5, 6> [1, 2, 4]) [1, 2, 4]

P indol-3-yl-pyruvate + NH$_3$ + NAD(P)H <1-3, 5, 6> [1, 2, 4]

Substrates and products

S L-tryptophan + NAD(P)$^+$ <1-3, 5, 6> (<2,3,6> amination proceeds more
intensively than deamination [2]; <2,3,5,6> no activity with D-tryptophan
[4]) (Reversibility: r <1-3, 5, 6> [1, 2, 4]) [1, 2, 4]

P indol-3-yl-pyruvate + NH$_3$ + NAD(P)H <1-3, 5, 6> [1, 2, 4]

Inhibitors

EDTA <2> (<2> reactivation by Ca^{2+} and Mn^{2+} [2]) [2]
Additional information <2> (<2> presence of a carbonyl group in the active
centre, less sensitive to SH-inhibitors than glutamate dehydrogenase [2]) [2]

Cofactors/prosthetic groups

NAD(P)$^+$ <1-6> [1, 2, 3, 4]
NAD(P)H <1-6> (<1> normal Nicotiana tabacum cells: higher specific activity with NADPH than with NADH, tumor cells of Nicotiana tabacum: no dif-
ference in specific activity with NADPH and NADH [1]; <4> effect of NADPH
compared to NADH is distinctly higher [3]) [1-4]

Metals, ions

Ca^{2+} <2, 4> (<4> optimal activation at 0.8 mM [3]) [2, 3]
Mn^{2+} <2> (<2> activation [2]) [2]
Additional information <2> (<2> absence of heavy metal ions in active cen-
tre [2]) [2]

Specific activity (U/mg)

0.014 <4> (<4> activity in organelles, cofactor NAD$^+$, deamination of L-tryptophan [3]) [3]
0.0155 <4> (<4> activity in organelles, cofactor NADP$^+$, deamination of L-
tryptophan [3]) [3]
0.0359 <4> (<4> activity in organelles, cofactor NADH, amination of L-tryptophan [3]) [3]
0.648 <4> (<4> activity in organelles, cofactor NADPH, amination of L-tryptophan [3]) [3]
0.87 <1> (<1> cofactor NADH, activity in normal BV-N strain [1]) [1]
1.22 <1> (<1> cofactor NADPH, activity in normal BV-N strain [1]) [1]
1.24 <1> (<1> cofactor NADH or NADPH, activity in tumor tobacco tissue
strains T-24 [1]) [1]
1.57 <1> (<1> cofactor NADH, activity in tumor tobacco tissue strain C-58
[1]) [1]

1.59 <1> (<1> cofactor NADPH, activity in tumor tobacco tissue strain C-58
[1]) [1]
6 <5> (<5> activity in shoots, cofactor NADH [4]) [4]
9 <5> (<5> activity in shoots, cofactor NADPH [4]) [4]
14 <2> (<2> activity in the apical part of the epicotyl in five-day-old etiolated seedlings 48 h after root excision [5]) [5]
18 <2> (<2> activity in the apical part of the epicotyl in five-day-old etiolated seedlings immidiately after root excision [5]) [5]
22 <5> (<5> activity in roots, cofactor NADH [4]) [4]
23 <2> (<2> activity in the apical part of the epicotyl in five-day-old etiolated seedlings 12 h after root excision [5]) [5]
28 <6> (<6> activity in shoots, cofactor NADH [4]) [4]
30 <5> (<5> activity in roots, cofactor NADPH [4]) [4]
39 <6> (<6> activity in shoots, cofactor NADPH [4]) [4]
45 <3> (<3> activity in shoots, cofactor NADH [4]) [4]
52 <2> (<2> activity in shoots, cofactor NADH [4]) [4]
64 <3> (<3> activity in shoots, cofactor NADPH [4]) [4]
70 <2> (<2> activity in shoots, cofactor NADPH [4]) [4]

K_m-Value (mM)

1.36 <1> (NADH, <1> tumor C-58 cell culture [1]) [1]
1.39 <1> (NADH, <1> tumor T-24 cell culture [1]) [1]
1.46 <1> (NADH, <1> normal cell culture [1]) [1]

pH-Optimum

8.5 <2> [2]

Temperature optimum (°C)

22 <1-3, 5, 6> (<1-3,5> assay at room temperature [1,4]) [1, 4]

5 Isolation/Preparation/Mutation/Application

Source/tissue

leaf <4> [3]
root <2, 3, 5, 6> [4]
seedling <2, 3, 6> [2, 5]
shoot <2, 3, 5, 6> [4]
tissue culture <1> (<1> normal and crown-gall tumor tissue cultures, tissue
transformed by Agrobacterium tumefaciens [1]) [1]

Localization

chloroplast <4> [3]
cytoplasm <4> [3]
Additional information <4> (<4> enzyme occurs in cytoplasm, chloroplast
and pellet of remaining organelles sedimenting at 97000 * g [3]) [3]

References

[1] El Bahr, M.K.; Kutacek, M.; Opatrny, Z.: L-tryptophan aminotransferase and L-tryptophan dehydrogenase, enzymes of IAA synthesis, in normal and tumorous tobacco tissue cultures. Biochem. Physiol. Pflanz., **182**, 213-222 (1987)

[2] Kutacek, M.: Auxin biosynthesis and its regulation on the molecular level. Biol. Plant., **27**, 145-153 (1985)

[3] Vackova, K.; Mehta, A.; Kutacek, M.: Tryptophan aminotransferase and tryptophan dehydrogenase activities in some cell compartments of spinach leaves: the effect of calcium ions on tryptophan dehydrogenase. Biol. Plant., **27**, 154-158 (1985)

[4] Ebeid, M.M.; Dimova, S.; Kutacek, M.: Substrate specificity of L-tryptophan dehydrogenase and its distribution in plants. Biol. Plant., **27**, 413-416 (1985)

[5] Tan Hoang Minh, Kutacek, M.; Sebanek, J.: Growth-correlative effect of the root on the apical part of the epicotyl in pea seedlings regarding the IAA content and L-tryptophan aminotransferase and L-tryptophan dehydrogenase activities. Biol. Plant., **26**, 342-348 (1984)

Phenylalanine dehydrogenase 1.4.1.20

1 Nomenclature

EC number
1.4.1.20

Systematic name
L-phenylalanine:NAD$^+$ oxidoreductase (deaminating)

Recommended name
phenylalanine dehydrogenase

Synonyms
L-phenylalanine dehydrogenase
PHD
PheDH
dehydrogenase, phenylalanine

CAS registry number
69403-12-9

2 Source Organism

<1> *Bacillus sphaericus* (SCRC-79a [1]; recombinant enzyme, expressed in Escherichia coli [6]) [1, 3, 6, 23]
<2> *Sporosarcina urea* [2, 3, 11, 14, 15]
<3> *Nocardia sp.* (239 [4,24]) [4, 24]
<4> *Rhodococcus maris* (K-18 [5]) [5]
<5> *Pediococcus acidilactici* [6]
<6> *Bacillus badius* [7, 14]
<7> *Brevibacterium sp.* [8, 10]
<8> *Rhodococcus sp.* (M4 [9,11,12]) [9, 11, 12, 13, 20, 21]
<9> *Thermoactinomyces intermedius* [16, 17, 18, 22]
<10> *Microbacterium sp.* (strain DM 86-1 [19]) [19]

3 Reaction and Specificity

Catalyzed reaction
L-phenylalanine + H_2O + NAD$^+$ = phenylpyruvate + NH_3 + NADH + H$^+$ (<4>, the reductive amination proceeds through a sequential ordered ternary-binary mechanism [5]; <8>, kinetic mechanism is ordered with NAD$^+$

binding prior to Phe and the products being released in the order of ammonia, phenylpyruvate, and NADH [20])

Reaction type
oxidative deamination
redox reaction
reductive amination

Natural substrates and products
S L-Phe + H_2O + NAD^+ <7> (<7>, the enzyme is involved in the degradation of Phe [8]) (Reversibility: ? <7> [8]) [8]
P phenylpyruvate + NH_3 + NADH <7> [8]

Substrates and products
S 2-oxo-4-methylpentanoate + NH_3 + NADH <6, 9> (<6>, 13% of the activity with phenylpyruvate [7]; <9>, 16% of the activity with phenylpyruvate [16]) (Reversibility: ? <6, 9> [7, 16, 18]) [7, 16, 18]
P L-Ile + NAD^+ + H_2O
S 2-oxo-4-methylthiobutanoate + NH_3 + NADH <6, 8, 9> (<6>, 16% of the activity with phenylpyruvate [7]; <8>, 33% of the activity with L-Phe [9]; <9>, 55% of the activity with phenylpyruvate [16]) (Reversibility: ? <6, 8, 9> [7, 9, 16, 18]) [7, 9, 16, 18]
P L-Met + NAD^+ + H_2O
S 2-oxobutanoate + NH_3 + NADH <9> (<9>, 5.5% of the activity with phenylpyruvate [16]) (Reversibility: ? <9> [16]) [16]
P 2-aminobutanoate + NAD^+ + H_2O
S 2-oxohexanoate + NH_3 + NADH <2, 3, 6, 9> (<3>, 240% of the activity with phenylpyruvate [4]; <6>, 31% of the activity with phenylpyruvate [7]; <9>, 130% of the activity with phenylpyruvate [16]) (Reversibility: ? <2, 3, 6, 9> [3, 4, 7, 16, 18]) [3, 4, 7, 16, 18]
P 2-aminohexanoate + H_2O + NAD^+
S 2-oxoisohexanoate + NH_3 + NADH <9> (<9>, 47% of the activity with phenylpyruvate [16]) (Reversibility: <9> [16, 18]) [16, 18]
P L-Leu + NAD^+ + H_2O
S 2-oxopentanoate + NH_3 + NADH <6, 9> (<6>, 12% of the activity with phenylpyruvate [7]; <9>, 37% of the activity with phenylpyruvate [16]) (Reversibility: ? <6, 9> [7, 16, 18]) [7, 16, 18]
P L-norvaline + NAD^+ + H_2O
S 4-hydroxyphenylpyruvate + NH_3 + NADH <1, 3, 6, 8, 9> (<3>, 28% of the activity with phenylpyruvate [4]; <6>, 53% of the activity with phenylpyruvate [7]; <8>, 5% of the activity with L-Phe [9]; <9>, 80% of the activity with phenylpyruvate [16]) (Reversibility: ? <1, 3, 6, 8, 9> [3, 4, 7, 9, 11, 16]) [3, 4, 7, 9, 11, 16]
P L-Tyr + NAD^+ + H_2O
S L-2-amino-n-butyric acid + H_2O + NAD^+ <6> (<6>, 1.0% of the activity with L-Phe [7]) (Reversibility: ? <6> [7]) [7]
P 2-oxobutanoate + NADH + NH_3

S L-2-aminohexanoic acid + H_2O + NAD^+ <6> (<6>, 19% of the activity with L-Phe [7]) (Reversibility: ? <6> [7]) [7]

P 2-oxohexanoate + NADH + NH_3

S L-Leu + H_2O + NAD^+ <3, 6, 9, 10> (<6>, 3.0% of the activity with L-Phe [7]) (Reversibility: ? <3, 6, 9, 10> [4, 7, 18, 19]) [4, 7, 18, 19]

P 3-methyl-2-oxopentanoic acid + NH_3 + NADH

S L-Met + H_2O + NAD^+ <6, 7, 8, 9, 10> (<6>, 8.0% of the activity with L-Phe [7]; <8>, 4% of the activity with L-Phe [9]; <9>, 2.2% of the activity with L-Phe [16]) (Reversibility: ? <6, 7, 8, 9, 10> [7, 9, 10, 16, 18, 19]) [7, 9, 10, 16, 18, 19]

P 2-oxo-4-methylthiobutanoate + NH_3 + NADH

S L-Phe + H_2O + NAD^+ <1-10> (<4>, the reductive amination proceeds through a sequential ordered ternary-binary mechanism [5]; <7,8>, the equilibrium favors L-Phe formation [10,12]) (Reversibility: r <1-10> [1-24]) [1-24]

P phenylpyruvate + NH_3 + NADH <1> [1]

S L-Trp + H_2O + NAD^+ <4, 6, 8> (<4>, 7.5% of the activity with L-Phe [5]; <6>, 4% of the activity with L-Phe [7]; <8>, 2% of the activity with L-Phe [9]) (Reversibility: ? <4, 6, 8> [5, 7, 9]) [5, 7, 9]

P indole-3-pyruvate + NH_3 + NADH

S L-Tyr + H_2O + NAD^+ <1, 6, 7, 8, 9, 10> (<6>, 9.0% of the activity with L-Phe [7]; <9>, 40% of the activity with L-Phe [16]) (Reversibility: ? <1, 6, 7, 8, 9, 10> [3, 7, 9, 10, 12, 16, 19]) [3, 7, 9, 10, 12, 16, 19]

P 4-hydroxyphenylpyruvate + NH_3 + NADH

S L-Val + H_2O + NAD^+ <6, 9, 10> (<6>, 4.0% of the activity with L-Phe [7]) (Reversibility: ? <6, 9, 10> [7, 18, 19]) [7, 18, 19]

P 3-methyl-2-oxobutanoate + NH_3 + NADH

S L-α-amino-β-phenylbutyrate + H_2O + NAD^+ <4> (<4>, 7.0% of the activity with L-Phe [5]) (Reversibility: ? <4> [5]) [5]

P 2-oxo-3-phenylbutyrate + NADH + NH_3

S L-ethionine + H_2O + NAD^+ <4, 6, 9> (<4>, 13.0% of the activity with L-Phe [5]; <6>, 7.0% of the activity with L-Phe [7]) (Reversibility: ? <4, 6, 9> [5, , 7, 18]) [5, 7, 18]

P 4-ethylthio-2-oxobutanoate + NH_3 + NADH

S L-norleucine + H_2O + NAD^+ <2, 4, 6, 9, 10> (<4>, 15.6% of the activity with L-Phe [5]; <6>, 19% of the activity with L-Phe [7]; <9>, 30% of the activity with L-Phe [16]) (Reversibility: ? <2, 4, 6, 9, 10> [3, 5, 7, 16, 19]) [3, 5, 7, 16, 19]

P 2-oxohexanoic acid + NADH + NH_3

S L-norvaline + H_2O + NAD^+ <6, 9, 10> (<6>, 5.0% of the activity with L-Phe [7]; <9>, 28% of the activity with L-Phe [16]) (Reversibility: ? <6, 9, 10> [7, 16, 19]) [7, 16, 19]

P 2-oxopentanoate + NH_3 + NADH

S L-phenylalanine + H_2O + 3-acetylpyridine-NAD^+ <4, 8> (<4>, at 241% of the activity with NAD^+ [5]; <8>, 350% of the activity with NAD^+ [20]) (Reversibility: ? <4, 8> [5, 20]) [5, 20]

P phenylpyruvate + NH_3 + acetylpyridine-NADH <4> [5]

S   L-phenylalanine + H_2O + 3-pyridinealdehyde-NAD$^+$ <4, 8> (<4>, at 9.2% of the activity with NAD$^+$ [5]; <8>, 55% of the activity with NAD$^+$ [20]) (Reversibility: ? <4, 8> [5, 20]) [5, 20]

P   phenylpyruvate + NH$_3$ + 3-pyridinealdehyde-NADH <4> [5]

S   L-phenylalanine + H_2O + NADP$^+$ <4> (<4>, at 6.6% of the activity with NAD$^+$ [5]) (Reversibility: ? <4> [5]) [5]

P   phenylpyruvate + NH$_3$ + NADPH <4> [5]

S   L-phenylalanine + H_2O + deamino-NAD$^+$ <4> (<4>, at 86% of the activity with NAD$^+$ [5]) (Reversibility: ? <4> [5]) [5]

P   phenylpyruvate + NH$_3$ + deamino-NADH <4> [5]

S   L-phenylalanine + H_2O + oxidized β-nicotinamide guanine dinucleotide <8> (<8>, at 86% of the activity with NAD$^+$ [20]) (Reversibility: ? <8> [20]) [20]

P   phenylpyruvate + NH$_3$ + reduced β-nicotinamide guanine dinucleotide <8> [20]

S   L-phenylalanine + H_2O + oxidized β-nicotinamide hypoxanthine dinucleotide <8> (<8>, at 86% of the activity with NAD$^+$ [20]) (Reversibility: ? <8> [20]) [20]

P   phenylpyruvate + NH$_3$ + reduced β-nicotinamide hypoxanthine dinucleotide <8> [20]

S   L-phenylalanine + H_2O + thio-NAD$^+$ <8> (<8>, at 86% of the activity with NAD$^+$ [20]) (Reversibility: ? <8> [20]) [20]

P   phenylpyruvate + NH$_3$ + thio-NADH <8> [20]

S   L-phenylalanine + H_2O + thionicotinamide-NAD$^+$ <4> (<4>, at 101% of the activity with NAD$^+$ [5]) (Reversibility: ? <4> [5]) [5]

P   phenylpyruvate + NH$_3$ + thionicotinamide-NADH <4> [5]

S   L-phenylalanine methyl ester + H_2O + NAD$^+$ <2, 6> (<6>, 38% of the activity with L-Phe [7]) (Reversibility: ? <2, 6> [3, 7]) [3, 7]

P   phenypyruvic acid methyl ester + NH$_3$ + NADH

S   L-phenylalaninol + H_2O + NAD$^+$ <6> (<6>, 9.4% of the activity with L-Phe [7]) (Reversibility: ? <6> [7]) [7]

P   1-hydroxyacetone + NADH + NH$_3$

S   alloisoleucine + H_2O + NAD$^+$ <6, 9> (<6>, 4.3% of the activity with L-Phe [7]; <9>, 26% of the activity with L-Phe [16]) (Reversibility: ? <6, 9> [7, 16]) [7, 16]

P   3-methyl-2-oxopentanoate + NADH + NH$_3$

S   indole-3-pyruvate + NH$_3$ + NADH <3> (<3>, 54% of the activity with phenylpyruvate [4]) (Reversibility: ? <3> [4]) [4]

P   L-Trp + NADH + H_2O

S   indolepyruvate + NH$_3$ + NADH <7, 8> (<8>, 3% of the activity with L-Phe [9]) (Reversibility: ? <7, 8> [9, 10]) [9, 10]

P   ?

S m-fluoro-DL-phenylalanine + H_2O + NAD$^+$ <4, 6, 9> (<4>, 7.7% of the activity with L-Phe [5]; <6>, 11% of the activity with L-Phe [7]; <9>, as effective as L-Phe [16]) (Reversibility: ? <4, 6, 9> [5, 7, 16]) [5, 7, 16]

P   3-(3-fluorophenyl)-2-oxopropionate + NADH + NH$_3$

S *o*-fluoro-DL-phenylalanine + H_2O + NAD^+ <6, 9> (<9>, 65% of the activity with L-Phe [16]; <6>, 2% of the activity with L-Phe [7]) (Reversibility: ? <6, 9> [7, 16]) [7, 16]

P 3-(2-fluorophenyl)-2-oxopropionate + NADH + NH_3

S *p*-amino-L-phenylalanine + H_2O + NAD^+ <9> (Reversibility: ? <9> [16]) [16]

P ?

S *p*-fluoro-DL-phenylalanine + H_2O + NAD^+ <4, 6, 9> (<4>, 8.3% of the activity with L-Phe [5]; <6>, 34% of the activity with L-Phe [7]; <9>, 118% of the activity with L-Phe [16]) (Reversibility: ? <4, 6, 9> [5, 7, 16]) [5, 7, 16]

P (4-fluorophenyl)-2-oxopropionate + NADH + NH_3

S phenylalaninamide + H_2O + NAD^+ <6> (<6>, 9.0% of the activity with L-Phe [7]) (Reversibility: ? <6> [7]) [7]

P 2-oxo-3-phenylpropionamide + NADH + NH_3

S phenylalanine hydroxamate + H_2O + NAD^+ <6> (<6>, 1.0% of the activity with L-Phe [7]) (Reversibility: ? <6> [7]) [7]

P 2-oxo-3-phenylpropionic acid hydroxamate + NADH + NH_3

S Additional information <9> (<9>, specific activity of the chimeric enzyme is 6% of that of the parental phenylalanine dehydrogenase and shows a broad substrate specificity in the oxidative deamination, like phenylalaine dehydrogenase. However, it acts much more effectively than phenylalanine dehydrogenase on Ile and Val. The parent enzyme and the chimeric enzyme belong to the pro-S specific dehydrogenase [16]) [16]

P ?

Inhibitors

3-phenylpropionate <4> (<4>, competitive with L-Phe [5]) [5]

4-hydroxyphenylethylamine <4> (<4>, competitive with L-Phe [5]) [5]

5,5'-dithio-bis(2-nitrobenzoic acid) <1, 2> (<1>, 0.17 mM: 26% inhibition [3]; <2>, 0.17 mM, complete inhibition [3]) [3]

$AgNO_3$ <1, 2> (<1,2>, 1 mM, complete inhibition [3]) [3]

Cu^{2+} <4> (<4>, slight [5]) [5]

D-Leu <1, 2> (<2>, 10 mM, 31% inhibition [3]; <1>, 10 mM, 17% inhibition [3]) [3]

D-Met <1, 2> (<2>, 10 mM, 25% inhibition [3]; <1>, 10 mM, 63% inhibition [3]) [3]

D-Phe <1, 2, 4, 8> (<2>, 10 mM, 77% inhibition [3]; <1>, 10 mM, 89% inhibition [3]; <4>, competitive with L-Phe [5]) [3, 5, 20]

D-Trp <1, 2> (<2>, 5 mM, 65% inhibition [3]; <1>, 5 mM, 83% inhibition [3]) [3]

D-Tyr <1, 4> (<1>, 1 mM, 29% inhibition [3]; <4>, competitive with L-Phe [5]) [3, 5]

D-Val <1> (<1>, 10 mM, 11% inhibition [3]) [3]

D-norleucine <1, 2> (<2>, 10 mM, 34% inhibition [3]; <1>, 10 mM, 52% inhibition [3]) [3]

HgCl$_2$ <1, 2, 4, 10> (<1>, 0.01 mM, 62% inhibition [3]; <2>, 0.01 M, complete inhibition [3]; <4>, 0.01 mM, complete inhibition [5]; <10>, 1.0 mM, 97% inhibition [19]) [3, 5, 19]
L-Cys <1> (<1>, 10 mM, 41% inhibition [3]) [3]
L-Ile <1> (<1>, 10 mM, 75% inhibition [3]) [3]
L-Leu <1, 2> (<2>, 10 mM, 39% inhibition [3]; <1>, 10 mM, 58% inhibition [3]) [3]
L-Phe <4, 8> (<4>, product inhibition [5]) [5, 9, 20]
L-Val <1> (<1>, 10 mM, 23% inhibition [3]) [3]
L-phenylglycine <4> (<4>, competitive with L-Phe [5]) [5]
L-phenyllactate <8> [20]
NAD$^+$ <4, 8> (<4>, product inhibition [5]) [5, 9]
NADH <8> [9, 20]
NEM <2> (<2>, 1 mM: 79% inhibition [3]) [3]
NH$_4^+$ <8> [20]
PCMB <1, 2, 4, 10> (<1,2>, 0.02 mM, complete inhibition [3]; <4>, 0.001 mM, complete inhibition [5]; <10>, 1.0 mM, 98% loss of activity [19]) [3, 5, 19]
α-methyl-DL-phenylalanine <4> [5]
methyl acetyl phosphate <9> (<9>, irreversible inactivation, simultaneous addition of substrate and coenzyme markedly protect from inactivation, the reagent can acetylate Lys69 and Lys81 [17]) [17]
phenylenediamine <4> [5]
phenylethylamine <4, 8> (<4>, competitive with L-Phe [5]) [5, 20]
phenylpropionate <8> [20]
phenylpyruvate <8> [9, 20]
trans-cinnamate <4, 8> (<4>, competitive with L-Phe [5]) [5, 20]

Cofactors/prosthetic groups

3-acetylpyridine-NAD$^+$ <4> (<4>, 241% of the activity with NAD$^+$ in the reaction with L-Phe [5]) [5]
3-pyridinealdehyde-NAD$^+$ <4> (<4>, 9.2% of the activity with NAD$^+$ in the reaction with L-Phe [5]) [5]
NAD$^+$ <1-10> [1-24]
NADH <1-10> [1-24]
deamino-NAD$^+$ <4> (<4>, 86% of the activity with NAD$^+$ in the reaction with L-Phe [5]) [5]
thionicotinamide-NAD$^+$ <4> (<4>, 101% of the activity with NAD$^+$ in the reaction with L-Phe [5]) [5]
Additional information <3, 8> (<3>, no activity with NADP$^+$ [4]; <8>, no activity with NADH [20]) [4, 20]

Turnover number (min^{-1})

6 <9> (L-ethionine, <9>, mutant enzyme F124M/V125S/H126I/A127I/A128Y/R129Q [18]) [18]
9 <9> (L-norvaline, <9>, mutant enzyme F124M/V125S/H126I/A127I/A128Y/R129Q [18]) [18]

9.6 <9> (L-Met, <9>, mutant enzyme F124M/V125S/H126I/A127I/A128Y/R129Q [18]) [18]

10.2 <9> (L-norleucine, <9>, mutant enzyme F124M/V125S/H126I/A127I/A128Y/R129Q [18]) [18]

12.6 <9> (L-Leu, <9>, mutant enzyme F124M/V125S/H126I/A127I/A128Y/R129Q [18]) [18]

14.4 <9> (L-Ile, <9>, wild-type enzyme [18]) [18]

16.8 <9> (L-Ile, <9>, mutant enzyme F124M/V125S/H126I/A127I/A128Y/R129Q [18]) [18]

35.4 <9> (L-norvaline, <9>, wild-type enzyme [18]) [18]

45 <9> (L-Phe, <9>, mutant enzyme F124M/V125S/H126I/A127I/A128Y/R129Q [18]) [18]

50.4 <9> (L-Met, <9>, wild-type enzyme [18]) [18]

66 <9> (L-ethionine, <9>, wild-type enzyme [18]) [18]

114 <9> (L-norleucine, <9>, wild-type enzyme [18]) [18]

216 <9> (L-Leu, <9>, wild-type enzyme [18]) [18]

1560 <9> (2-oxo-4-methylpentanoate, <9>, wild-type enzyme [18]) [18]

2220 <9> (2-oxo-4-methylthiobutanoate, <9>, wild-type enzyme [18]) [18]

2700 <9> (L-Phe, <9>, wild-type enzyme [18]) [18]

3420 <9> (2-oxoisohexanoate, <9>, wild-type enzyme [18]) [18]

4740 <9> (2-oxohexanoate, <9>, wild-type enzyme [18]) [18]

4920 <9> (2-oxopentanoate, <9>, wild-type enzyme [18]) [18]

5460 <9> (phenylpyruvate, <9>, wild-type enzyme [18]) [18]

Additional information <9> (<9>, turnover numbers of wild-type and mutant enzymes [17]) [17]

Specific activity (U/mg)

37.1 <10> [19]

65.2 <4> [5]

67.8 <6> [7]

87.9 <2> [2, 3]

111 <1> [3]

Additional information <9> [18]

K$_m$-Value (mM)

0.017 <9> (NADH, <9>, mutant enzyme K69A/K81A [17]) [17]

0.025 <1> (NADH) [3]

0.026 <9> (NADH, <9>, mutant enzyme K81A [17]) [17]

0.047 <7> (NADH) [10]

0.051 <9> (NADH, <9>, mutant enzyme K69A [17]) [17]

0.052 <9> (phenylpyruvate, <9>, mutant enzyme K81A [17]) [17]

0.057 <9> (NAD$^+$, <9>, mutant enzyme K81A [17]) [17]

0.06 <9> (L-Leu, <9>, chimeric enzyme [16]) [16]

0.06 <3> (phenylpyruvate) [4]

0.06 <9> (phenylpyruvate, <9>, wild-type enzyme [18]) [18]

0.065 <9> (phenylpyruvate, <9>, wild-type enzyme [17]) [17]

0.07 <10> (NADH) [19]

0.072 <2> (NADH) [3]
0.081 <9> (NAD⁺, <9>, mutant enzyme K69A/K81A [17]) [17]
0.083 <9> (NADH, <9>, wild-type enzyme [17]) [17]
0.09 <9> (L-Ile, <9>, wild-type enzyme [18]) [18]
0.09 <9> (L-Leu, <9>, wild-type enzyme [18]) [18]
0.09 <9> (L-Phe, <9>, mutant enzyme K81A [17]) [17]
0.1 <9, 10> (L-Phe, <9>, wild-type enzyme [17]) [17, 19]
0.11 <9> (L-Phe) [16, 18]
0.11 <7> (phenylpyruvate) [10]
0.12 <9> (L-Phe, <9>, chimeric enzyme [16]) [16]
0.125 <7> (NAD⁺) [10]
0.13 <8> (NADH) [9]
0.14 <2> (NAD⁺) [3]
0.16 <9> (NAD⁺, <9>, mutant enzyme K69A [17]) [17]
0.16 <2, 8> (phenylpyruvate) [3, 9]
0.17 <2, 9> (NAD⁺, <9>, wild-type enzyme [17]) [3, 16, 17]
0.18 <4> (deamino-NAD⁺) [5]
0.18 <4> (thionicotinamide-NAD⁺) [5]
0.2 <10> (NAD⁺) [19]
0.22 <9> (norleucine, <9>, wild-type enzyme [18]) [18]
0.23 <3> (NAD⁺) [4]
0.24 <7> (4-hydroxyphenylpyruvate) [10]
0.25 <4> (3-acetylpyridine-NAD⁺) [5]
0.25 <4> (NAD⁺) [5]
0.27 <9> (L-ethionine, <9>, wild-type enzyme [18]) [18]
0.27 <8> (NAD⁺) [9]
0.3 <10> (phenylpyruvate) [19]
0.34 <9> (L-Met, <9>, wild-type enzyme [18]) [18]
0.34 <2> (p-hydroxyphenylpyruvate) [3]
0.4 <7> (L-Phe) [10]
0.4 <2> (phenylpyruvate) [3]
0.43 <8> (L-Met) [9]
0.5 <4> (phenylpyruvate) [5]
0.53 <9> (L-norvaline, <9>, wild-type enzyme [18]) [18]
0.75 <3, 8> (L-Phe) [4, 9, 12]
1.1 <7> (L-Tyr) [10]
1.3 <4> (4-hydroxyphenylpyruvate) [5]
1.54 <10> (L-norleucine) [19]
1.6 <10> (L-Met) [19]
2.1 <8> (2-oxo-4-methylthiobutanoate) [9]
2.1 <9> (L-norleucine, <9>, mutant enzyme F124M/V125S/H126I/A127I/A128Y/R129Q [18]) [18]
2.4 <8> (p-hydroxyphenylpyruvate) [9]
2.44 <2> (2-oxohexanoate) [3]
2.5 <9> (2-oxohexanoate, <9>, wild-type enzyme [18]) [18]
2.8 <9> (2-oxo-4-methyl-thiobutanoate, <9>, wild-type enzyme [18]) [18]
3.1 <8> (L-Tyr) [9, 12]

3.5 <7> (L-Met) [10]
4 <9> (L-ethionine, <9>, mutant enzyme F124M/V125S/H126I/A127I/A128Y/ R129Q [18]) [18]
4.7 <9> (L-Leu, <9>, mutant enzyme F124M/V125S/H126I/A127I/A128Y/ R129Q [18]) [18]
7 <9> (phenylpyruvate, <9>, mutant enzyme K69A/K81A [17]) [17]
7.5 <9> (L-Met, <9>, mutant enzyme F124M/V125S/H126I/A127I/A128Y/ R129Q [18]) [18]
7.7 <8> (indolepyruvate) [9]
8 <7> (indolylpyruvate) [10]
8.3 <9> (phenylpyruvate, <9>, mutant enzyme K69A [17]) [17]
9.1 <9> (2-oxoisohexanoate, <9>, wild-type enzyme [18]) [18]
10 <9> (L-Ile, <9>, mutant enzyme F124M/V125S/H126I/A127I/A128Y/ R129Q [18]) [18]
10 <9> (L-norvaline, <9>, mutant enzyme F124M/V125S/H126I/A127I/ A128Y/R129Q [18]) [18]
11 <8> (L-Trp) [9]
13 <9> (2-oxo-4-methylpentanoate, <9>, wild-type enzyme [18]) [18]
20 <9> (L-Phe, <9>, mutant enzyme K69A/K81A [17]) [17]
23 <9> (L-Phe, <9>, mutant enzyme F124M/V125S/H126I/A127I/A128Y/ R129Q [18]) [18]
23 <9> (phenylpyruvate, <9>, mutant enzyme F124M/V125S/H126I/A127I/ A128Y/R129Q [18]) [18]
35 <9> (2-oxopentanoate, <9>, wild-type enzyme [18]) [18]
40 <8> (oxidized β-nicotinamide hypoxanthine dinucleotide) [20]
47 <9> (2-oxohexanoate, <9>, mutant enzyme F124M/V125S/H126I/A127I/ A128Y/R129Q [18]) [18]
50 <9> (2-oxo-4-methylpentanoate, <9>, mutant enzyme F124M/V125S/ H126I/A127I/A128Y/R129Q [18]) [18]
50 <8> (3-pyridinealdehyde-NAD$^+$) [20]
56 <9> (2-oxoisohexanoate, <9>, mutant enzyme F124M/V125S/H126I/ A127I/A128Y/R129Q [18]) [18]
60 <8> (thioNAD$^+$) [20]
78 <1> (NH$_3$) [3]
85 <2, 10> (NH$_3$) [3, 19]
87 <9> (2-oxo-4-methylthiobutanoate, <9>, mutant enzyme F124M/V125S/ H126I/A127I/A128Y/R129Q [18]) [18]
90 <8> (NAD$^+$) [20]
96 <3> (NH$_3$) [4]
107 <9> (2-oxopentanoate, <9>, mutant enzyme F124M/V125S/H126I/A127I/ A128Y/R129Q [18]) [18]
110 <8> (3-acetylpyridine-NAD$^+$) [20]
140 <9> (L-Phe, <9>, mutant enzyme K69A [17]) [17]
170 <8> (oxidized β-nicotinamide guanine dinucleotide) [20]
387 <8> (NH$_4^+$) [9]

K_i-Value (mM)

0.002 <8> (NADH) [9]
0.07 <8> (phenylpyruvate) [9]
0.14 <4> (D-Phe) [5]
1.27 <8> (NAD$^+$) [9]
1.52 <4> (phenylethylamine) [5]
1.58 <4> (D-Tyr) [5]
1.76 <4> (D-ethionine) [5]
2.6 <4> (3-phenylpropionate) [5]
4.36 <4> (phenylenediamine) [5]
7.34 <8> (phenylpyruvate) [9]
8.32 <4> (L-phenylglycine) [5]
12.7 <4> (trans-cinnamic acid) [5]
13.7 <4> (4-hydroxyphenylethylamine) [5]
17.85 <8> (Phe) [9]
20.8 <4> (α-methyl-DL-phenylalanine) [5]
Additional information <8> [20]

pH-Optimum

8.5 <7> (<7>, reductive amination of phenylpyruvate, crude extract [8]) [8]
9 <2, 9> (<2>, reductive amination of phenylpyruvate [3]; <9>, reductive amination of wilde-type enzyme and mutant enzyme F124M/V125S/H126I/A127I/A128Y/R129Q [18]) [3, 18]
9.3 <8> (<8>, reductive amination [9]) [9]
9.7-10.1 <9> (<9>, mutant enzyme F124M/V125S/H126I/A127I/A128Y/R129Q, oxidative deamination [18]) [18]
9.8 <4> (<4>, reductive amination of phenylpyruvate [5]) [5]
10 <3> [4]
10.3 <2> (<2>, reductive amination of phenylpyruvate [3]) [3]
10.5 <2, 7> (<2,7>, oxidative deamination of L-Phe [3,8]; <7>, crude extract [8]) [3, 8]
10.6 <9> (<9>, reductive amination of phenylpyruvate, chimeric enzyme [16]) [16]
10.7-11 <9> (<9>, oxidative deamination of L-Phe, chimeric enzyme [16]) [16]
10.8 <4> (<4>, oxidative deamination of L-Phe [5]) [5]
11 <9> (<9>, wild-type enzyme, oxidative deamination [18]) [18]
11.3 <2> (<2>, oxidative deamination of L-Phe [3]) [3]
12 <10> (<10>, oxidative deamination and reductive amination [19]) [19]

pH-Range

9.3-10.2 <3> (<3>, pH 9.3: about 40% of maximal activity, pH 10.2: about 35% of maximal activity [4]) [4]

Temperature optimum (°C)

40 <2> [3]
50 <1> [3]
53 <3> [4]
70 <10> [19]

Temperature range (°C)

20-70 <8> (<8>, 20°C: about 50% of maximal activity, 70°C: about 70% of maximal activity [9]) [9]

40-53 <3> (<3>, 40°C: about 65% of maximal activity, 53°C: optimum, no activity at 55°C [4]) [4]

4 Enzyme Structure

Molecular weight

42000 <3> (<3>, gel filtration [4]) [4]

55000 <9> (<9>, mutant enzyme F124M/V125S/H126I/A127I/A128Y/R129Q, monomeric enzyme form, gel filtration [18]) [18]

70000 <4> (<4>, gel filtration [5]) [5]

72000 <9> (<9>, chimeric enzyme consisting of the N-terminal domain of Thermoactinomyces intermedius phenylalanine dehydrogenase, containing the substrate-binding region and the C-terminal domain of leucine dehydrogenase from Bacillus stearothermophilus containing the NAD$^+$-binding region, gel filtration [16]) [16]

110000 <9> (<9>, mutant enzyme F124M/V125S/H126I/A127I/A128Y/R129Q, dimeric enzyme form, gel filtration [18]) [18]

150000 <8> (<8>, gel filtration [20]) [20]

290000 <2> (<2>, gel filtration [2]) [2]

305000 <2> (<2>, equilibrium sedimentation [3]) [3]

310000 <2> (<2>, gel filtration [3]) [3]

310000-340000 <1> (<1>, gel filtration [6]) [6]

330000 <10> (<10>, gel filtration [19]) [19]

340000 <1> (<1>, gel filtration, equilibrium sedimentation [3]) [3]

360000 <6> (<6>, gel filtration [7]) [7]

Subunits

? <1, 2> (<2>, x * 38000-39000, SDS-PAGE after treatment with 2-mercaptoethanol [2]; <2>, x * 39000, SDS-PAGE [2]; <1>, x * 41435, calculation from nucleotide sequence [1]) [1, 2]

dimer <4, 9> (<4>, 2 * 36000, SDS-PAGE [5]; <9>, 1 * 41000, the mutant enzyme F124M/V125S/H126I/A127I/A128Y/R129Q exists as a monomer or a dimer, which are in an equilibrium, SDS-PAGE [18]) [5, 18]

hexamer <9> (<9>, 6 * 41000, wild-type enzyme, SDS-PAGE [18]) [18]

monomer <3, 9> (<9>, 1 * 41000, the mutant enzyme F124M/V125S/H126I/A127I/A128Y/R129Q exists as a monomer or a dimer, which are in an equilibrium, SDS-PAGE [18]; <3>, 1 * 42000, SDS-PAGE [4]) [4, 18]

octamer <1, 6, 10> (<1>, 8 * 39000, SDS-PAGE [6]; <10>, 8 * 41000, SDS-PAGE [19]; <6>, 8 * 41000-42000, SDS-PAGE [7]) [6, 7, 19]

tetramer <8> (<8>, 4 * 39500, SDS-PAGE [20]) [20]

5 Isolation/Preparation/Mutation/Application

Localization

intracellular <2, 8> [11]

Purification

<1> [3, 6]

<2> [2, 3]

<3> (partial [4]) [4]

<4> [5]

<6> [7]

<8> [9, 21]

<9> (construction of a chimeric enzyme consisting of the N-terminal domain of Thermoactinomyces intermedius phenylalanine dehydrogenase, containing the substrate-binding region and the C-terminal domain of leucine dehydrogenase from Bacillus stearothermophilus containing the NAD$^+$-binding region [16]; mutant enzyme F124M/V125S/H126I/A127I/A128Y/R129Q [18]) [16, 18]

<10> [19]

Crystallization

<1> [3]

<2> [2, 3]

<3> (hanging-drop method of vapour diffusion using ammonium sulfate as the precipitant. Two crystal forms are obtained in the presence and absence of the enzyme substrates phenylpyruvate or Phe and its coenzyme NADH [24]) [24]

<8> (enzyme-NAD$^+$-phenylpyruvate complex and enzyme-NAD$^+$-β-phenyl-propionate complex [21]) [21]

Cloning

<1> (wild-type and mutant enzymes L307V, G124A and G124A/L307V expressed in Escherichia coli [23]) [1, 23]

<6> (expressed at high levels in Escherichia coli [7]) [7]

<8> (overexpression in Escherichia coli [21]) [21]

<9> (chimeric enzyme consisting of the N-terminal domain of Thermoactinomyces intermedius phenylalanine dehydrogenase, containing the substrate-binding region and the C-terminal domain of leucine dehydrogenase from Bacillus stearothermophilus containing the NAD$^+$-binding region, expression in Escherichia coli [16]) [16]

Engineering

F124M/V125S/H126I/A127I/A128Y/R129Q <9> (<9>, the catalytic efficiencies of the mutant enzyme with aliphatic amino acids and aliphatic keto acids as substrates are 0.5% to 2% of that of the wild-type enzyme. The efficiencies for L-Phe and phenylpyruvate decreases to 0.0008% and 0.035% of that of the wilde-type enzyme, respectively. Enzyme exists as monomeric or dimeric form, compared to wild-type enzyme which exists as hexameric enzyme form. Thermostability is lowered by mutation [18]) [18]

G124A <1> (<1>, mutant enzyme has lower activity towards L-Phe and enhanced activity towards almost all aliphatic amino acid substrates compared to the wild-type [23]) [23]

G124A/L307V <1> (<1>, mutant enzyme has lower activity towards L-Phe and enhanced activity towards almost all aliphatic amino acid substrates compared to the wild-type [23]) [23]

K173A <9> (<9>, 37°C, $t_{1/2}$ of the mutant enzyme is 60 min, compared to 48 min for the wild type enzyme, without addition of substrate or cofactor [17]) [17]

K69A <9> (<9>, 37°C, $t_{1/2}$ of the mutant enzyme is 50 min, compared to 48 min for the wild type enzyme, without addition of substrate or cofactor. K_m-value for L-Phe is 1400fold higher compared to wild type enzyme, K_m-value for phenylpyruvate is 128fold higher compared to wild type enzyme. Turnover number for deamination is 686fold lower than that of wild-type enzyme, turnover-number for amination is 43fold lower than that of wild-type enzyme [17]) [17]

K69A/K81A <9> (<9>, 37°C, $t_{1/2}$ of the mutant enzyme is 450 min, compared to 48 min for the wild type enzyme, without addition of substrate or cofactor. K_m-value for L-Phe is 200fold higher compared to wild type enzyme, K_m-value for phenylpyruvate is 108fold higher compared to wild type enzyme. Turnover number for deamination is 110fold lower than that of wild-type enzyme, turnover-number for amination is 61fold lower than that of wild-type enzyme [17]) [17]

K81A <9> (<9>, 37°C, $t_{1/2}$ of the mutant enzyme is 38 min, compared to 48 min for the wild type enzyme, without addition of substrate or cofactor. Turnover number for deamination is 440fold lower than that of wild-type enzyme, turnover-number for amination is 42fold lower than that of wild-type enzyme [17]) [17]

K89A <9> (<9>, 37°C, $t_{1/2}$ of the mutant enzyme is 75 min, compared to 48 min for the wild type enzyme, without addition of substrate or cofactor [17]) [17]

K90A <9> (<9>, 37°C, $t_{1/2}$ of the mutant enzyme is 80 min, compared to 48 min for the wild type enzyme, without addition of substrate or cofactor [17]) [17]

L307V <1> (<1>, mutant enzyme has lower activity towards L-Phe and enhanced activity towards almost all aliphatic amino acid substrates compared to the wild-type [23]) [23]

Additional information <9> (<9>, construction of a chimeric enzyme consisting of the N-terminal domain of Thermoactinomyces intermedius phenylalanine dehydrogenase, containing the substrate-binding region and the C-terminal domain of leucine dehydrogenase from Bacillus stearothermophilus containing the NAD⁺-binding region. The chimeric enzyme has a specific activity of 6% of that of the parental phenylalanine dehydrogenase and shows a broad substrate specificity in the oxidative deamination, like phenylalanine dehydrogenase. However, it acts much more effectively than phenylalanine dehydrogenase on Ile and Val [16]) [16]

Application

analysis <2, 6, 8> (<8>, monitoring of phenylketonuria, colorimetric method for the determination of plasma phenylalanine using L-phenylalanine dehydrogenase [12]; <2,6>, enzymatic cycling assays for the determination of L-Phe and phenylpyruvate. Assay 1 couples glutamine transferase K with L-phenylalanine dehydrogenase. Assay 2 combines phenylanine dehydrogenase, L-amino acid oxidase, and catalase [14]; <2>, high yield synthesis of L-amino acids in presence of formate dehydrogenase: L-Phe from phenylpyruvate, L-Tyr from 4-hydroxyphenylpyruvate, L-Trp from indole-3-pyruvate, L-Met from 2-oxo-4-methylthiobutanoate, L-Val from 2-oxoisopentanoate, L-Leu from 2-oxoisohexanoate [15]; <9>, the chimeric enzyme has a specific activity of 6% of that of the parental phenylalanine dehydrogenase [16]) [12, 14, 15, 16]

synthesis <5, 8, 9> (<5>, formation of L-Phe from phenylpyruvate [6]; <8>, formation of L-Phe from phenylpyruvate in presence of formate-dehydrogenase from Candida boidinii [13]; <8>, continous production of L-Phe in an enzyme-membrane-reactor [9]; <9>, synthesis of allysine (S)-2-amino-5-(1,3-dioxolan-2-yl)-pentanoic acid [22]) [6, 9, 13, 22]

6 Stability

pH-Stability

4 <1> (<1>, 30°C, 1 h, less than 20% loss of activity [3]) [3]
8-12.5 <10> (<10>, 30°C, 30 min, stable [19]) [19]
9 <1, 2> (<1,2>, 30°C, 1 h, stable [3]) [3]
11.3 <1> (<1>, 30°C, 1 h, stable [3]) [3]

Temperature stability

30 <2> (<2>, pH 9.0, stable [3]) [3]
37 <9> (<9>, $t_{1/2}$: 48 min, in absence of substrate or coenzyme. $t_{1/2}$: 2000 min, in presence of 10 mM L-Leu and 1.0 mM NAD^+ [17]) [17]
40 <2> (<2>, pH 9.0, 10 min, 25% loss of activity [3]) [3]
50 <2, 9> (<2>, pH 9.0, 10 min, complete inactivation [3]; <9>, 60 min, mutant enzyme F124M/V125S/H126I/A127I/A128Y/R129Q loses 50% of its activity [18]) [3, 18]
53 <3> (<3>, pH 6.0, 5 min, 50% loss of activity [4]; <3>, pH 9.5-10, 2 h, 50% loss of activity [4]; <3>, phenylpyruvate destabilizes the activity, 1 M sorbitol or 1 M glycerol stabilize [4]) [4]
54 <9> (<9>, 60 min, pH 7.0-9.5, chimeric enzyme is stable [16]) [16]
55 <1, 10> (<1>, pH 9.0, 10 min, stable [3]; <10>, 10 min, stable [19]) [3, 19]
58 <9> (<9>, pH 7.0-9.5, 60 min, chimeric enzyme, loss of activity above [16]) [16]
60 <1, 10> (<1>, pH 9.0, 10 min, most of the activity is lost [3]; <10>, pH 7.0, 10 min, stable [19]) [3, 19]

65-70 <9, 10> (<9>, 60 min, stable [16]; <10>, pH 7.0, 10 min, 50% loss of activity [19]) [16, 19]
70 <9> (<9>, 60 min, wild-type enzyme is stable [18]) [18]

General stability information

<4>, 20% glycerol, 0.25 M sodium malonate or 0.5 M sodium glutarate stabilize the enzyme during storage at 4°C [5]
<4>, unstable in absence of glycerol and salt [5]

Storage stability

<4>, 4°C, 0.25 M sodium malonate, stable for 1 month [5]
<1, 2>, 4°C, 0.01 M potassium phosphate buffer, pH 7.0, 0.1 mM EDTA, 5 mM 2-mercaptoethanol, stable for at least 2 years in crystalline form [3]

References

[1] Okazaki, N.; Hibino, Y.; Asano, Y.; Ohmori, M.; Numao, N.; Kondo, K.: Cloning and nucleotide sequencing of phenylalanine dehydrogenase gene of Bacillus sphaericus. Gene, **63**, 337-341 (1988)
[2] Asano, Y.; Nakazawa, A.: Crystallization of phenylalanine dehydrogenase from Sporosarcina urea. Agric. Biol. Chem., **49**, 3631-3632 (1985)
[3] Asano, Y.; Nakazawa, A.; Endo, K.: Novel phenylalanine dehydrogenases from Sporosarcina ureae and Bacillus sphaericus. Purification and characterization. J. Biol. Chem., **262**, 10346-10354 (1987)
[4] De Boer, L.; Van Rijssel, M.; Euverink, G.J.; Dijkhuizen, L.: Purification, characterization and regulation of a monomeric L-phenylalanine dehydrogenase from the facultative methylotroph Nocardia sp. 239. Arch. Microbiol., **153**, 12-18 (1989)
[5] Misono, H.; Yonezawa, J.; Nagata, S.; Nagasaki, S.: Purification and characterization of a dimeric phenylalanine dehydrogenase from Rhodococcus maris K-18. J. Bacteriol., **171**, 30-36 (1989)
[6] Asano, Y.; Endo, K.; Nakazawa, A.; Hibino, Y.; Okazaki, N.; Ohmori, M.; Numao, N.; Kondo, K.: Bacillus phenylalanine dehydrogenase produced in Escherichia coli - Its purification and application to L-phenylalanine synthesis. Agric. Biol. Chem., **51**, 2621-2623 (1987)
[7] Asano, Y.; Nakazawa, A.; Endo, K.; Hibino, Y.; Ohmori, M.; Numao, N.; Kondo, K.: Phenylalanine dehydrogenase of Bacillus badius. Purification, characterization and gene cloning [published erratum appears in Eur. J. Biochem. 1988 Jan 1;170(3):667]. Eur. J. Biochem., **168**, 153-159 (1987)
[8] Hummel, W.; Weiss, N.; Kula, M.R.: Isolation and characterization of a bacterium possessing L-phenylalanine dehydrogenase activity. Arch. Microbiol., **137**, 47-52 (1984)
[9] Hummel, W.; Schuette, H.; Schmidt, E.; Wandrey, C.; Kula, M.R.: Isolation of L-phenylalanine dehydrogenase from Rhodococcus sp. M4 and its application for the production of L-phenylalanine. Appl. Microbiol. Biotechnol., **26**, 409-416 (1987)

[10] Hummel, W.; Schuette, H.; Kula, M.R.: Enzymatic determination of L-phenylalanine and phenylpyruvate with L-phenylalanine dehydrogenase. Anal. Biochem., 170, 397-401 (1988)

[11] Campagna, R.; Bueckmann, A.F.: Comparison of the production of intracellular L-phenylalanine dehydrogenase by Rhodococcus species M4 and Sporosarcina urea at 50 liter scale. Appl. Microbiol. Biotechnol., 26, 417-421 (1987)

[12] Wendel, U.; Hummel, W.; Langenbeck, U.: Monitoring of phenylketonuria: a colorimetric method for the determination of plasma phenylalanine using L-phenylalanine dehydrogenase. Anal. Biochem., 180, 91-94 (1989)

[13] Hummel, W.; Schuette, H.; Schmidt, E.; Kula, M.R.: Neue Moeglichkeiten zur enzymatischen Herstellung von L-Phenylalanin. GBF Monogr. Ser. Volume date 1988, 11, 207-210 (1989)

[14] Cooper, A.J.L.; Leung, L.K.H.; Asano, Y.: Enzymatic cycling assay for phenylpyruvate. Anal. Biochem., 183, 210-214 (1989)

[15] Asano, Y.; Nakazawa, A.: High yield synthesis of L-amino acids by phenylalanine dehydrogenase from Sporosarcina urea. Agric. Biol. Chem., 51, 2035-2036 (1987)

[16] Kataoka, K.; Takada, H.; Tanizawa, K.; Yoshimura, T.; Esaki, N.; Ohshima, T.; Soda, K.: Construction and characterization of chimeric enzyme consisting of an amino-terminal domain of phenylalanine dehydrogenase and a carboxy-terminal domain of leucine dehydrogenase. J. Biochem., 116, 931-936 (1994)

[17] Kataoka, K.; Tanizawa, K.; Fukui, T.; Ueno, H.; Yoshimura, T.; Esaki, N.; Soda, K.: Identification of active site lysyl residues of phenylalanine dehydrogenase by chemical modification with methyl acetyl phosphate combined with site-directed mutagenesis. J. Biochem., 116, 1370-1376 (1994)

[18] Kataoka, K.; Takada, H.; Yoshimura, T.; Furuyoshi, S.; Esaki, N.; Ohshima, T.; Soda, K.: Site-directed mutagenesis of a hexapeptide segment involved in substrate recognition of phenylalanine dehydrogenase from Thermoactinomyces intermedius. J. Biochem., 114, 69-75 (1993)

[19] Asano, Y.; Tanetani, M.: Thermostable phenylalanine dehydrogenase from a mesophilic Microbacterium sp. strain DM 86-1. Arch. Microbiol., 169, 220-224 (1998)

[20] Brunhuber, N.M.W.; Thoden, J.B.; Blanchard, J.S.; Vanhooke, J.L.: Rhodococcus L-phenylalanine dehydrogenase: kinetics, mechanism, and structural basis for catalytic specifity. Biochemistry, 39, 9174-9187 (2000)

[21] Vanhooke, J.L.; Thoden, J.B.; Brunhuber, N.M.W.; Blanchard, J.S.; Holden, H.M.: Phenylalanine dehydrogenase from Rhodococcus sp. M4: High-resolution x-ray analyses of inhibitory ternary complexes reveal key features in the oxidative deamination mechanism. Biochemistry, 38, 2326-2339 (1999)

[22] Hanson, R.L.; Howell, J.M.; LaPorte, T.L.; Donovan, M.J.; Cazzulino, D.L.; Zannella, V.; Montana, M.A.; Nanduri, V.B.; Schwarz, S.R.; Eiring, R.F.; Durand, S.C.; Wasylyk, J.M.; Parker, W.L.; Liu, M.S.; Okuniewicz, F.J.; Chen, B.C.; Harris, J.C.; Natalic,: Synthesis of allysine ethylene acetal using phenylalanine dehydrogenase from Thermoactinomyces intermedius. Enzyme Microb. Technol., 26, 348-358 (2000)

[23] Seah, S.Y.; Britton, K.L.; Baker, P.J.; Rice, D.W.; Asano, Y.; Engel, P.C.: Alteration in relative activities of phenylalanine dehydrogenase towards different substrates by site-directed mutagenesis. FEBS Lett., **370**, 93-96. (1995)

[24] Pasquo, A.; Britton, K.L.; Baker, P.J.; Brearley, G.; Hinton, R.J.; Moir, A.J.G.; Stillman, T.J.; Rice, D.W.: Crystallization of NAD^+-dependent phenylalanine dehydrogenase from Nocardia sp239. Acta Crystallogr. Sect. D, **54**, 269-272 (1998)

Glycine dehydrogenase (cytochrome) 1.4.2.1

1 Nomenclature

EC number
1.4.2.1

Systematic name
glycine:ferricytochrome-c oxidoreductase (deaminating)

Recommended name
glycine dehydrogenase (cytochrome)

Synonyms
dehydrogenase, glycine (cytochrome)
glycine-cytochrome c reductase
reductase, glycine-cytochrome c

CAS registry number
9075-55-2

2 Source Organism

<1> *Nitrobacter agilis* [1]

3 Reaction and Specificity

Catalyzed reaction
glycine + H_2O + 2 ferricytochrome c = glyoxylate + NH_3 + 2 ferrocytochrome c

Reaction type
oxidation
oxidative deamination
redox reaction
reduction

Natural substrates and products
S glycine + ferricytochrome c <1> (Reversibility: ir <1> [1]) [1]
P glyoxylate + ferrocytochrome c + NH_3 <1> [1]

Substrates and products
S glycine + ferricytochrome c <1> (<1> highly specific [1]) (Reversibility: ir <1> [1]) [1]
P glyoxylate + ferrocytochrome c + NH_3 <1> [1]

Inhibitors

Cu^{2+} <1> (<1> 55% inhibition at 1 mM [1]) [1]

EDTA <1> (<1> 64% inhibition at 7.1 mM [1]) [1]

KCN <1> (<1> 47% inhibition at 7.1 mM [1]) [1]

KCl <1> (<1> 80% inhibition at 75 mM [1]) [1]

KNO_3 <1> (<1> 80% inhibition at 75 mM [1]) [1]

$MgCl_2$ <1> (<1> 80% inhibition at 40 mM [1]) [1]

NaCl <1> (<1> 80% inhibition at 75 mM [1]) [1]

Tris <1> (<1> buffer, 100 mM, pH 8.0 [1]) [1]

cytochrome c <1> (<1> above 0.15 mM [1]) [1]

diphosphate <1> (<1> buffer, 10 mM [1]) [1]

hydrazine <1> [1]

hydroxylamine <1> [1]

phosphate <1> (<1> buffer, 100 mM [1]) [1]

pyridoxal phosphate <1> (<1> up to 80% inhibition at 0.2-2 mM due to reaction with glycine [1]) [1]

pyruvate <1> (<1> 70% inhibition at 10 mM and 20% inhibition at 3 mM [1]) [1]

quinacrine hydrochloride <1> (<1> 0.3-1.0 mM [1]) [1]

sodium azide <1> (<1> 10% inhibition at 2 mM [1]) [1]

sodium barbital <1> (<1> buffer, 100 mM, pH 8.0 [1]) [1]

urea <1> (<1> 30% inhibition at 50 mM [1]) [1]

Specific activity (U/mg)

13.1 <1> [1]

K_m-Value (mM)

0.016 <1> (cytochrome c) [1]

0.8 <1> (glycine) [1]

pH-Optimum

6.1-6.9 <1> (<1> in potassium phosphate or diphosphate buffer [1]) [1]

4 Enzyme Structure

Molecular weight

69000 <1> (<1> gel filtration, sucrose density gradient centrifugation [1]) [1]

5 Isolation/Preparation/Mutation/Application

Purification

<1> [1]

6 Stability

Temperature stability

4 <1> (<1> 4-10 weeks, 30-50% loss of activity [1]) [1]
70 <1> (<1> 5 min: 75% loss of activity, 2 min: 50% loss of activity [1]) [1]
80 <1> (<1> 2 min: 100% loss of activity [1]) [1]

Storage stability

<1>, 4°C, 4-10 weeks, 30-50% loss of activity [1]

References

[1] Sanders, H.K.; Becker, G.E.; Nason, A.: Glycine-cytochrome c reductase from Nitrobacter agilis. J. Biol. Chem., **247**, 2015-2025 (1972)

D-Aspartate oxidase

1 Nomenclature

EC number
1.4.3.1

Systematic name
D-aspartate:oxygen oxidoreductase (deaminating)

Recommended name
D-aspartate oxidase

Synonyms
D-aspartic oxidase
DASOX
DASPO <1> [1, 3, 16]
DDO <3> [5, 8]
aspartic oxidase

CAS registry number
9029-20-3

2 Source Organism

<1> *Bos taurus* [1-4, 7, 13-21]
<2> *Rattus norvegicus* (Sprague-Dawley [2]) [2, 12]
<3> *Homo sapiens* [2, 5, 8]
<4> *Xenopus laevis* [4]
<5> *Cyprinus carpio* [6]
<6> *Carassius auratus langsdorfii* [6]
<7> *Oncorhynchus mykiss* [6]
<8> *Seriola quinqueradiata* [6]
<9> *Pagrus major* [6]
<10> *Cryptococcus humicola* (strain UJ1 [9]) [9]
<11> *Octopus vulgaris* [10, 12, 13]
<12> *Fusarium sacchari* (strain elongantum Y-105 [11]) [11]
<13> *Mus musculus* [12]
<14> *Sus scrofa* [22, 24]
<15> *Oryctolagus cuniculus* [23, 24]
<16> *Ovis aries* [24]

3 Reaction and Specificity

Catalyzed reaction
D-aspartate + H_2O + O_2 = oxaloacetate + NH_3 + H_2O_2

Reaction type
oxidation
redox reaction
reduction

Natural substrates and products

S D-aspartate + H_2O + O_2 <1-16> (<4> intermediate product is iminoaspartate [4]; <3> this enzyme is proposed to have a role in the inactivation of the synaptically released D-aspartate [5]) (Reversibility: ? <1-16> [1-24]) [1-24]

P oxaloacetate + NH_3 + H_2O_2

S meso-2,3-diaminosuccinate + H_2O + O_2 <1, 5-8> (Reversibility: ? <1, 5-8> [6, 21]) [6, 21]

P pyrazine 2,5-dicarboxylic acid + pyrazine 2,6-dicarboxylic acid + H_2O_2 + NH_3 <1, 5-8> [6, 21]

Substrates and products

S D-α-aminoadipic acid + H_2O + O_2 <1> (Reversibility: ? <1> [3]) [3]

P 2-oxoadipic acid + NH_3 + H_2O_2

S D-asparagine + H_2O + O_2 <11> (Reversibility: ? <11> [10]) [10]

P 2-oxosuccinamic acid + H_2O_2 + NH_3

S D-aspartate + H_2O + O_2 <1-16> (<4> intermediate product is iminoaspartate [4]; <15> ferricyanide and 2,6-dichlorophenolindophenol can also act as electron acceptors instead of O_2 [23]) (Reversibility: ? <1-16> [1-24]) [1-24]

P oxaloacetate + NH_3 + H_2O_2 <1-16> [1-24]

S D-aspartate dimethylester + H_2O + O_2 <1, 11> (Reversibility: ? <1, 11> [13]) [13]

P oxaloacetic acid dimethylester + H_2O_2 + NH_3

S D-aspartic acid-β-hydroxamate + H_2O + O_2 <1> (Reversibility: ? <1> [3]) [3]

P 4-(hydroxyamino)-2,4-dioxobutanoic acid + NH_3 + H_2O_2

S D-glutamate + H_2O + O_2 <1-3, 5-12, 14, 15> (<15> ferricyanide and 2,6-dichlorophenolindophenol can also act as electron acceptors instead of O_2 [23]) (Reversibility: ? <1-3, 5-12, 14, 15> [1-3, 6, 8-11, 13, 16, 19, 22, 23]) [1-3, 6, 8-11, 13, 16, 19, 22, 23]

P 2-oxoglutarate + NH_3 + H_2O_2

S D-glutamine + H_2O + O_2 <11> (Reversibility: ? <11> [10]) [10]

P 2-oxoglutaric acid + H_2O_2 + NH_3

S D-homocysteic acid + H_2O + O_2 <1> (Reversibility: ? <1> [3]) [3]

P 2-oxo-4-sulfobutanoic acid + NH_3 + H_2O_2

S D-proline + H_2O + O_2 <1, 11> (Reversibility: ? <1, 11> [1, 10, 15, 16]) [1, 10, 15, 16]

P 2-oxopentanoic acid + NH$_3$ + H$_2$O$_2$
S D-thiazolidine-2-carboxylate + H$_2$O + O$_2$ <1> (Reversibility: ? <1> [15, 16, 19]) [15, 16, 19]
P (ethylthio)oxoacetic acid + H$_2$O$_2$ + NH$_3$
S DL-2-amino-3-phosphonopropanoic acid + H$_2$O + O$_2$ <1> (Reversibility: ? <1> [3]) [3]
P 2-oxo-3-phosphonopropanoic acid + NH$_3$ + H$_2$O$_2$
S DL-cysteic acid + H$_2$O + O$_2$ <1> (Reversibility: ? <1> [3]) [3]
P 2-oxo-3-sulfopropionic acid + NH$_3$ + H$_2$O$_2$
S N-methyl-D-aspartate + H$_2$O + O$_2$ <1-3, 5-12> (Reversibility: ? <1-3, 5-12> [1-3, 6, 8-11, 13]) [1-3, 6, 8-11, 13]
P oxaloacetate + CH$_3$NH$_2$ + H$_2$O$_2$ <1-3, 5-12> [1-3, 6, 8-11, 13]
S cis-2,3-piperidine dicarboxylic acid + H$_2$O + O$_2$ <1> (Reversibility: ? <1> [3]) [3]
P ?
S glycyl-D-aspartic acid + H$_2$O + O$_2$ <1> (Reversibility: ? <1> [3]) [3]
P ?
S *meso*-2,3-diaminosuccinate + H$_2$O + O$_2$ <1, 5-8> (Reversibility: ? <1, 5-8> [6, 21]) [6, 21]
P pyrazine 2,5-dicarboxylic acid + pyrazine 2,6-dicarboxylic acid + H$_2$O$_2$ + NH$_3$ <1, 5-8> [6, 21]
S Additional information <1> (<1> chemical modification with phenyl-glyoxal results in irreversible loss of activity towards dicarboxylic D-amino acids, paralleled with a transient appearance of activity versus mono-carboxylic ones [14-16]) [14-16]
P ?

Inhibitors

2-oxoglutaric acid <15> (<15> at 6.67 mM 64% inhibition [23]) [23]
D-malate <10> (<10> competitive inhibitor, at 20 mM 34% inhibition [9]; <15> 90% inhibition at 0.33 mM [23]) [9, 23]
D-tartaric acid <15> (<15> at 6.67 mM 77% inhibition [23]) [23]
KCN <14> [22]
L-leucine <15> (<15> 6% inhibition at 6.67 mM [23]) [23]
L-malic acid <15> (<15> at 6.67 mM 79% inhibition [23]) [23]
L-tartaric acid <15> (<15> at 6.67 mM 24% inhibition [23]) [23]
L-tartrate <1, 4> (<4> at 10 mM [4]; <1> competitive inhibitor [16]) [1, 3, 4, 16]
L-valine <15> (<15> 6% inhibition at 6.67 mM [23]) [23]
adipic acid <15> (<15> at 6.67 mM 15% inhibition [23]) [23]
benzoate <5, 6> (<5> 43% inhibition at 10 mM [6]; <6> 31% inhibition at 10 mM [6]) [6]
citric acid <15> (<15> at 6.67 mM 29% inhibition [23]) [23]
fumaric acid <15> (<15> at 6.67 mM 78% inhibition [23]) [23]
glutaric acid <15> (<15> at 6.67 mM 62% inhibition [23]) [23]
malonate <10, 15> (<10> competitive inhibitor, at 20 mM 53% inhibition [9]; <15> at 0.66 mM 78% inhibition [23]) [9, 23]

meso-tartrate <5-10, 15> (<5> 53% inhibition at 40 mM [6]; <6> 56% inhibition at 40 mM [6]; <7,9> 34% inhibition at 10 mM [6]; <8> 70% inhibition at 10 mM [6]; <10> 9% inhibition at 20 mM [9]) [6, 9, 24]
oxalic acid <15> (<15> at 6.67 mM 40% inhibition [23]) [23]
oxaloacetic acid <15> (<15> at 6.67 mM 90% inhibition [23]) [23]
succinic acid <15> (<15> at 6.67 mM 66% inhibition [23]) [23]

Cofactors/prosthetic groups

FAD <1-7, 9-15> (<1, 11> 1 mol flavin per mol protein [10, 15, 17]) [2-17, 20-22, 24]

Turnover number (min^{-1})

11 <1> (D-aspartate, <1> 50 mM potassium phosphate buffer, 0.3 mM EDTA, pH 7.4, 4°C [15]) [15, 17]
43.5 <1> (D-aspartate, <1> 50 mM potassium phosphate buffer, 0.3 mM EDTA, pH 7.4, 25°C [17]) [17]

Specific activity (U/mg)

0.00021 <7> (<7> from kidney [6]) [6]
0.00048 <5> (<5> from liver [6]) [6]
0.00053 <8> (<8> from brain [6]) [6]
0.00055 <8> (<8> from liver [6]) [6]
0.0006 <6> (<6> from kidney [6]) [6]
0.00062 <8> (<8> from kidney [6]) [6]
0.00076 <5> (<5> from kidney [6]) [6]
0.00119 <6> (<6> from liver [6]) [6]
0.00119 <9> (<9> from kidney [6]) [6]
0.00144 <9> (<9> from liver [6]) [6]
0.00233 <7> (<7> from liver [6]) [6]
0.0035 <1> [20]
0.4 <14> [22]
2.4 <15> (<15> ferricyanide as electron acceptor [23]) [23]
25.3 <1> [18]
29.1 <1> [3]
76.1 <10> [9]

K$_m$-Value (mM)

0.17 <1> (O$_2$, <1> at 4°C [15]) [15]
0.2 <1> (N-methyl-D-aspartate) [1]
0.46 <1> (O$_2$, <1> at 25°C [17]) [17]
0.59 <1> (D-glutamate, <1> at pH 8.3 [19]) [19]
0.86 <15> (D-aspartate, <15> O$_2$ as acceptor [23]) [23]
0.9 <1> (D-proline) [1]
1.5 <1> (N-methyl-D-aspartate) [3]
1.6 <1> (D-aspartate, <1> at 25°C [17]) [17]
1.7 <1> (DL-cysteic acid) [3]
2.1 <1> (D-glutamate, <1> at pH 7.4 [19]) [19]
2.2 <1> (D-aspartate, <1> at 4°C [15]) [15]
2.5 <1> (D-aspartate, <1> at pH 8.3 [19]) [19]

2.7 <1, 3> (D-aspartate) [1, 8]
3 <1> (D-aspartic acid-β-hydroxamate) [3]
3.7 <1, 10> (D-aspartate) [3, 9]
4.3 <11> (D-aspartate) [10]
4.5 <1> (D-aspartate) [13]
4.5 <1> (cis-2,3-piperidine dicarboxylic acid) [3]
4.8 <1> (*meso*-2,3-diaminosuccinate) [21]
5.2 <15> (D-aspartate, <15> ferricyanide as acceptor [23]) [23]
5.6 <1> (D-glutamate) [3]
5.7 <1> (D-aspartate, <1> at pH 7.4 [19]) [19]
5.7 <15> (D-glutamate, <15> ferricyanide as acceptor [23]) [23]
5.8 <11> (D-aspartate) [13]
6.2 <11> (D-glutamate) [13]
6.8 <3> (N-methyl-D-aspartate) [8]
7.7 <1> (D-aspartate) [21]
8.8 <1> (D-glutamate) [1]
9.7 <1, 11> (D-glutamate) [10, 13]
9.8 <1> (D-homocysteic acid) [3]
15.2 <1> (DL-amino-3-phosphopropanoic acid) [3]
16.2 <1> (D-α-aminoadipic acid) [3]
28 <10> (N-methyl-D-aspartate) [9]
42 <11> (N-methyl-D-aspartate) [13]
47 <1> (N-methyl-D-aspartate) [13]
47.7 <11> (N-methyl-D-aspartate) [10]
53 <11> (D-asparagine) [13]
60 <1> (D-asparagine) [13]
143 <1> (glycyl-D-aspartic acid) [3]
150 <11> (D-aspartate dimethylester) [13]
152 <10> (D-glutamate) [9]
166 <15> (D-glutamate, <15> O_2 as acceptor [23]) [23]
240 <11> (D-asparagine) [10]
440 <11> (D-proline) [10]
540 <1> (D-aspartate dimethylester) [13]
600 <11> (D-glutamine) [13]
815 <1, 11> (D-glutamine) [10, 13]

K_i-Value (mM)

0.011 <15> (D-malate) [23]
0.06 <15> (malonate) [23]
5.4 <10> (malonate) [9]
13.3 <10> (D-malate) [9]
61 <10> (*meso*-tatrate) [9]

pH-Optimum

7.5 <10, 12> [9, 11]
8.7 <14> [22]
9.5 <15> [23]

Additional information <15> (<15> optimum depends on the order of addition of the reactands [23]) [23]

pH-Range

4.5-11 <15> (<15> largely destroyed below or over this range [23]) [23]

6.5-9.5 <11> [10]

Temperature optimum (°C)

33-37 <10> [9]

40 <12> [11]

4 Enzyme Structure

Molecular weight

30000 <4> (<4> native PAGE [4]) [4]

37000 <3, 11> (<3> recombinant protein, SDS-PAGE [8]; <11> SDS-PAGE [10]) [8, 10]

37660 <1> (<1> apoprotein, amino acid analysis [3]) [3]

38000 <1> (<1> recombinant protein, SDS-PAGE [3,15]) [3, 15]

39000 <1> (<1> containing 1 molecule of FAD, SDS-PAGE [18]) [18]

40000 <10> (<10> SDS-PAGE [9]) [9]

Subunits

homotetramer <10> (<10> gel filtration [9]) [9]

monomer <1> (<1> 1 * 37659, amino acid analysis [3]) [3]

monomer <1> (<1> 1 * 39000, SDS-PAGE [17]) [17]

5 Isolation/Preparation/Mutation/Application

Source/tissue

brain <2, 3, 8, 13> (<3> mainly localized in neurons [2]; <3> two forms encoded named DDO-1 and DDO-2, DDO-1 is characterized [8]) [2, 6, 8, 12]

egg <4> [4]

embryo <4> (<4> 504 hours old [4]) [4]

hepatopancreas <11> [13]

kidney <1, 5-9, 13, 15> (<1> tubule [2]; <1> cortex [7,14,18]) [1-3, 6, 7, 12-21, 23, 24]

liver <5-9, 13, 15> [6, 12, 24]

muscle <13> [12]

ovary <4> [4]

thyroid gland <14> [22]

Localization

mitochondrion <14> [22]

peroxisome <1-3> (<3> its C-terminus has a peroxisome targeting signal [5]) [2, 5, 15]

Purification

<1> (expressed in Escherichia coli, mixture of 80-91% active FAD-containing and 9-20% inactive OH-FAD containing enzyme [3,4,15]) [3, 4, 15, 18]
<3> (recombinant [8]) [8]
<4> [4]
<10> [9]
<14> [22]
<15> [23]

Cloning

<1> (expressed in Escherichia coli GI724 cells [3,7]) [3, 7]
<3> (expressed in Escherichia coli [5]; expressed in Escherichia coli BL21 [8]) [5, 8]

Application

molecular biology <3> (<3> This enzyme is proposed to have a role in the inactivation of the synaptically released D-aspartate. Its C-terminus has a peroxisome targeting signal. [5]) [5]

6 Stability

pH-Stability

6-8.5 <15> (<15> fairly stable [23]) [23]
9 <12> (<12> stable [11]) [11]

Temperature stability

50 <12> (<12> stable [11]) [11]

Storage stability

<15>, -10°C, deionized water, several days, 100% activity [24]

References

[1] Sacchi, S.; Lorenzi, S.; Molla, G.; Pilone, M.S.; Rossetti, C.; Pollegioni, L.: Engineering the substrate specificity of D-amino-acid oxidase. J. Biol. Chem., **277**, 27510-27516 (2002)
[2] Zaar, K.; Kost, H.P.; Schad, A.; Volkl, A.; Baumgart, E.; Fahimi, H.D.: Cellular and subcellular distribution of D-aspartate oxidase in human and rat brain. J. Comp. Neurol., **450**, 272-282. (2002)
[3] Negri, A.; Tedeschi, G.; Ceciliani, F.; Ronchi, S.: Purification of beef kidney D-aspartate oxidase overexpressed in Escherichia coli and characterization of its redox potentials and oxidative activity towards agonists and antagonists of excitatory amino acid receptors. Biochim. Biophys. Acta, **1431**, 212-222 (1999)
[4] Tedeschi, G.; Negri, A.; Bernardini, G.; Oungre, E.; Ceciliani, F.; Ronchi, S.: D-Aspartate oxidase is present in ovaries, eggs and embryos but not in testis of Xenopus laevis. Comp. Biochem. Physiol. B, **124**, 489-494 (1999)

[5] Amery, L.; Brees, C.; Baes, M.; Setoyama, C.; Miura, R.; Mannaerts, G.P.; Van Veldhoven, P.P.: C-terminal tripeptide Ser-Asn-Leu (SNL) of human D-aspartate oxidase is a functional peroxisome-targeting signal. Biochem. J., **336**, 367-371 (1998)

[6] Kera, Y.; Hasegawa, S.; Watanabe, T.; Segawa, H.; Yamada, R.H.: D-Aspartate oxidase and free acidic D-amino acids in fish tissues. Comp. Biochem. Physiol. B, **119**, 95-100 (1998)

[7] Simonic, T.; Duga, S.; Negri, A.; Tedeschi, G.; Malcovati, M.; Tenchini, M.L.; Ronchi, S.: cDNA cloning and expression of the flavoprotein D-aspartate oxidase from bovine kidney cortex. Biochem. J., **322**, 729-735 (1997)

[8] Setoyama, C.; Miura, R.: Structural and functional characterization of the human brain D-aspartate oxidase. J. Biochem., **121**, 798-803 (1997)

[9] Yamada, R.h.; Ujiie, H.; Kera, Y.; Nakase, T.; Kitagawa, K.; Imasaka, T.; Arimoto, K.; Takahashi, M.; Matsumura, Y.: Purification and properties of D-aspartate oxidase from Cryptococcus humicolus UJ1. Biochim. Biophys. Acta, **1294**, 153-158 (1996)

[10] Tedeschi, G.; Negri, A.; Ceciliani, F.; Ronchi, S.; Vetere, A.; DÁniello, G.; DÁniello, A.: Properties of the flavoenzyme D-aspartate oxidase from Octopus vulgaris. Biochim. Biophys. Acta, **1207**, 217-222 (1994)

[11] Wakayama, M.; Nakashima, S.; Sakai, K.; Moriguchi, M.: Isolation, enzyme production and characterization of D-aspartate oxidase from Fusarium sacchari var. elongatum Y-105. J. Ferment. Bioeng., **78**, 377-379 (1994)

[12] DÁniello, A.; DÓnofrio, G.; Pischetola, M.; DÁniello, G.; Vetere, A.; Petrucelli, L.; Fisher, G.H.: Evidence for the in vivo biological activity of D-amino acid oxidase and D-aspartate oxidase. Effects of D-amino acids. J. Biol. Chem., **268**, 26941-26949 (1993)

[13] DÁniello, A.; Vetere, A.; Petrucelli, L.: Further study on the specificity of D-amino acid oxidase and of D-aspartate oxidase and time course for complete oxidation of D-amino acids. Comp. Biochem. Physiol. B, **105**, 731-734 (1993)

[14] Tedeschi, G.; Negri, A.; Ceciliani, F.; Biondi, P.A.; Secchi, C.; Ronchi, S.: Chemical modification of functional arginyl residues in beef kidney D-aspartate oxidase. Eur. J. Biochem., **205**, 127-132 (1992)

[15] Negri, A.; Tedeschi, G.; Ceciliani, F.; Simonic, T.: Structural studies of beef kidney D-aspartate oxidase. Flavins and Flavoproteins (Proc. Int. Symp., 10th, Meeting Date 1990, Curti, B., Ronchi S., Zanetti, G., eds.) de Gruyter, Berlin, New York, 179-187 (1990)

[16] Tedeschi, G.; Negri, A.; Biondi, P.A.; Secchi, C.; Ronchi, S.: Modification of substrate specificity of D-aspartate oxidase chemically modified phenylglyoxal. Flavins and Flavoproteins (Proc. Int. Symp., 10th, Meeting Date 1990, Curti, B., Ronchi S., Zanetti, G., eds.) de Gruyter, Berlin, New York, 189-192 (1990)

[17] Negri, A.; Massey, V.; Williams, C.H.; Schopfer, L.M.: The kinetic mechanism of beef kidney D-aspartate oxidase. J. Biol. Chem., **263**, 13557-13563 (1988)

[18] Negri, A.; Massey, V.; Williams, C.H.: D-Aspartate oxidase from beef kidney. Purification and properties. J. Biol. Chem., **262**, 10026-10034 (1987)

[19] Burns, C.L.; Main, D.E.; Buckthal, D.J.; Hamilton, G.A.: Thiazolidine-2-carboxylate derivatives formed from glyoxylate and L-cysteine or L-cysteinylglycine as possible physiological substrates for D-aspartate oxidase. Biochem. Biophys. Res. Commun., 125, 1039-1045 (1984)

[20] Nasu, S.; Wicks, F.D.; Gholson, R.K.: The mammalian enzyme which replaces B protein of E. coli quinolinate synthetase is D-aspartate oxidase. Biochim. Biophys. Acta, 704, 240-252 (1982)

[21] Rinaldi, A.; Pellegrini, M.; Crifo, C.; de Marco, C.: Oxidation of *meso*-diaminosuccinic acid, a possible natural substrate for D-aspartate oxidase. Eur. J. Biochem., 117, 635-638 (1981)

[22] Jaroszewicz, L.: D-Asparatate oxidase in the thyroid gland. Enzyme, 20, 80-89 (1975)

[23] Dixon, M.; Kenworthy, P.: D-Aspartate oxidase of kidney. Biochim. Biophys. Acta, 146, 54-76 (1967)

[24] Still, J.L.; Sperling, E.: On the prosthetic group of the D-aspartic oxidase. J. Biol. Chem., 182, 585-589 (1950)

1 Nomenclature

EC number
1.4.3.2

Systematic name
L-amino-acid:oxygen oxidoreductase (deaminating)

Recommended name
L-amino-acid oxidase

Synonyms
L-amino acid oxidase
L-amino acid:O_2 oxidoreductase
L-aminooxidase
LAAO
LAO
aromatic L-amino acid oxidase
ophio-amino-acid oxidase

CAS registry number
9000-89-9

2 Source Organism

<-18> no activity in *Alcaligenes faecalis* [3]
<-17> no activity in *Acinetobacter anitratum* [3]
<-16> no activity in *Pseudomonas aeruginosa* [3]
<-15> no activity in *Bacillus subtilis* [3]
<-14> no activity in *Mycobacterium phlei* [3]
<-13> no activity in *Serratia marcescens* [3]
<-12> no activity in *Enterobacter aerogenes* [3]
<-11> no activity in *Klebsiella pneumoniae* [3]
<-10> no activity in *Shigella flexneri* [3]
 <-9> no activity in *Neisseria catatthalis* [3]
 <-8> no activity in *Diplococcus pneumoniae* [3]
 <-7> no activity in *Streptococcus viridans* [3]
 <-6> no activity in *Micrococcus tetragenus* [3]
 <-5> no activity in *Staphylococcus epidermidis* [3]
 <-4> no activity in *Staphylococcus aureus* [3]
 <-3> no activity in *Bacillus subtilis (NCIB 3610)* [19]

<-2> no activity in *Escherichia coli (NCIB 9483)* [19]
<-1> no activity in *Pseudomonas putida T1* [19]
<1> *Proteus mirabilis* [21]
<2> *Rattus norvegicus* [1, 4, 29, 31, 33, 37]
<3> *Agkistrodon piscivorus piscivorus* (cottonmouth moccasin [1]) [1, 2]
<4> *Crotalus adamanteus* (eastern diamondback rattlesnake, multiple electro-
      phoretic components [27]) [1, 2, 5, 9, 11-14, 17, 18, 25-28, 30, 32, 34, 35,
      39]
<5> *Proteus sp.* (30 different strains [3]) [3]
<6> *Providencia sp.* (10 different strains [3]; PCM 1298 [15]) [3, 15]
<7> *Ophiophagus hannah* (king cobra [7]) [7, 60, 61]
<8> *Trimeresurus mucrosquamatus* (Taiwan habu snake [8]) [8]
<9> *snake* [6, 16, 47, 48]
<10> *Crotalus terrificus terrificus* [10]
<11> *Proteus rettgeri* [10, 20]
<12> *Corynebacterium sp.* (A20 [19]) [19]
<13> *Neurospora crassa* (inducible, expression in cells derepressed for nitrogen
      in presence of an amino acid, intracellular and extracellular location [38])
      [22, 23, 38, 57]
<14> *Meleagris gallopavo* (turkey [24]) [24]
<15> *Gallus gallus* [36]
<16> *Anacystis nidulans* (Synechococcus leopoliensis (blue-green algae) [40-
      45]) [40-45]
<17> *Synechococcus sp.* [46]
<18> *Chlamydomonas reinhardtii* [49]
<19> *Calloselasma rhodostoma* [50, 51, 53, 54]
<20> *Mus musculus* (milk [52]) [52]
<21> *Eristocophis macmahoni* [55]
<22> *Agkistrodon contortrix laticinctus* (<22> venom [56]) [56]
<23> *Bothrops sp.* (venom [58]) [58]
<24> *Bacillus carotarum* (2Pfa [59]) [59]

3 Reaction and Specificity

Catalyzed reaction
      an L-amino acid + H_2O + O_2 = a 2-oxo acid + NH_3 + H_2O_2 (<2-4,17> me-
      chanism [1, 2, 13, 23, 25, 30, 46]; <4> allosteric effects, activation energy
      [35]; <13> ping-pong mechanism with binary complexes, derived from ki-
      netic data [23])

Reaction type
      oxidative deamination
      redox reaction

Natural substrates and products
      S L-amino acid + H_2O + O_2 <4, 12, 16, 17, 18, 20> (<12> utilization of
      amino acids as nitrogen source [19]; <16> first step in reaction sequence

of formation of phenylacetic acid from L-phenylalanine [44]; <16> functional in photosynthetic and respiratory activities as well as in L-arginine degradation [41, 45]; <16,17> involved in the water-splitting reaction of photosystem II [40, 42, 46]; <4> bactericidal action of L-amino acid oxidase, [26]; <20> responsible for killing bacteria in the mammary gland [52]; <18> operates as scavanger of ammonium from extracellular amino acids [49]; <9> biological functions of snake venom enzymes [16]) (Reversibility: ? <4, 12, 16-18, 20> [19, 26, 44, 45, 46, 49, 52]) [16, 19, 26, 44, 45, 46, 49, 52]

P 2-oxo acid + NH_3 + H_2O_2

Substrates and products

S L-alanine + H_2O + O_2 <8, 12> (<8> not [8]) (Reversibility: ? <12> [19]) [8, 19]

P 2-oxopropanoic acid + NH_3 + H_2O_2

S L-albizziin + H_2O + O_2 <4> (<4> L-α-amino-β-ureidopropionic acid [39]) (Reversibility: ? <4> [39]) [39]

P α-keto-β-ureidopropionic acid + NH_3 + H_2O_2 (<4> product: corresponding cyclic lactam [39])

S L-α-aminobutyric acid + H_2O + O_2 <5, 6, 12> (<5, 6> low reaction rate [3]) (Reversibility: ? <5, 6, 12> [3, 19]) [3, 19]

P α-ketobutyrate + NH_3 + H_2O_2

S L-amino acid + H_2O + O_2 <1-24> (<2> other electron acceptors: methylene blue [1]; <18> all amino acids except cysteine [49]) (Reversibility: ? <1-24> [1-52, 55-60]) [1-52, 55-60]

P 2-oxo acid + NH_3 + H_2O_2

S L-arginine + H_2O + O_2 <5, 6, 8, 12, 14, 16, 24> (<5, 6, 8> low reaction rate [3, 8]) (Reversibility: ? <5, 6, 8, 12, 14, 16, 24> [3, 8, 19, 24, 43, 59]) [3, 8, 19, 24, 43, 59]

P 2-oxo-5-guanidinovaleric acid + NH_3 + H_2O_2

S L-asparagine + H_2O + O_2 <8> (<8> very low activity [8]) (Reversibility: ? <8> [8]) [8]

P α-ketosuccinamic acid + NH_3 + H_2O_2

S L-aspartic acid + H_2O + O_2 <4, 8> (<4, 8> low activity [5, 8]; <2,12> not [4, 19]) (Reversibility: ? <4, 8> [5, 8]) [5, 8]

P oxaloacetate + NH_3 + H_2O_2

S L-cysteine + H_2O + O_2 <8, 12> (<8> not [8]) (Reversibility: ? <12> [19]) [8, 19]

P 2-oxo-3-mercaptopropanoic acid + NH_3 + H_2O_2

S L-cystine + H_2O + O_2 <8> (<8> very low activity [8]) (Reversibility: ? <8> [8]) [8]

P ? + NH_3 + H_2O_2

S L-glutamic acid + H_2O + O_2 <4, 8> (<4, 8> low activity [5, 8]; <2> not [4]) (Reversibility: ? <4, 8> [5, 8]) [5, 8]

P 2-oxoglutarate + NH_3 + H_2O_2

S L-glutamine + H_2O + O_2 <8> (<8> very low activity [8]) (Reversibility: ? <8> [8]) [8]

P 2-oxoglutarate + NH_3 + H_2O_2

S L-histidine + H_2O + O_2 <5, 6, 8, 12, 14, 16> (<5, 6, 8> low reaction rate [3,8]) (Reversibility: ? <5, 6, 8, 12, 14, 16> [3, 8, 19, 24, 43]) [3, 8, 19, 24, 43]

P 3-(1H-imidazol-4-yl)-2-oxopropionic acid + NH_3 + H_2O_2

S L-homocysteine + H_2O + O_2 <9> (Reversibility: ? <9> [47]) [47]

P 4-thio-2-ketobutyric acid + NH_3 + H_2O_2

S L-isoleucine + H_2O + O_2 <5, 6, 8, 12> (<5, 6, 8> low reaction rate [3, 8]) (Reversibility: ? <5, 6, 8, 12> [3, 8, 19]) [3, 8, 19]

P α-keto-β-methyl-pentanoic acid + NH_3 + H_2O_2 (i.e. 2-keto-3-methylvalerate)

S L-lactic acid + H_2O + O_2 <2> (Reversibility: ? <2> [1, 4, 37]) [1, 4, 37]

P pyruvate + NH_3 + H_2O_2

S L-leucine + H_2O + O_2 <2, 4-6, 8, 12, 14, 19, 21, 23, 24> (<22> no reaction with D-leucine [56]; <4-6, 8, 23> high reaction rate [3, 5, 8, 58]) (Reversibility: ? <2, 4-6, 8, 12, 14, 21, 23, 24> [3, 5, 8, 16, 19, 24, 37, 50, 55, 57-59]) [3, 5, 8, 19, 24, 37, 50, 55, 57-59]

P 4-methyl-2-oxo-pentanoic acid + NH_3 + H_2O_2

S L-lysine + H_2O + O_2 <4, 12, 14, 16, 24> (<4> low activity [5]; <2, 8, 17> not [4, 8, 46]) (Reversibility: ? <4, 12, 14, 16, 24> [5, 19, 24, 43, 59]) [5, 19, 24, 43, 59]

P 6-amino-2-oxohexanoic acid + NH_3 + H_2O_2

S L-methionine + H_2O + O_2 <2, 4-6, 8, 12, 23, 24> (<4-6, 8, 23> high reaction rate [3, 5, 8, 58]) (Reversibility: ? <2, 4-6, 8, 12, 23, 24> [3, 5, 8, 19, 37, 58, 59]) [3, 5, 8, 19, 37, 58, 59]

P 4-methylsulfanyl-2-oxobutanoate + NH_3 + H_2O_2

S L-norleucine + H_2O + O_2 <5, 6> (<5> high reaction rate [3]) (Reversibility: ? <5, 6> [3]) [3]

P α-ketocaproate + NH_3 + H_2O_2 (i.e. 2-oxo-n-hexanoate)

S L-norvaline + H_2O + O_2 <5, 6> (high reaction rate [3]) (Reversibility: ? <5, 6> [3]) [3]

P 2-oxo-valeric acid + NH_3 + NADH + NH_3 + H_2O_2 (i.e. 2-oxopentanoic acid or α-ketovalerate)

S L-ornithine + H_2O + O_2 <8, 12, 14, 16> (<8> not [8]) (Reversibility: ? <12, 14, 16> [19, 24, 43]) [8, 19, 24, 43]

P 2-oxo-5-aminopentanoate + NH_3 + H_2O_2

S L-phenylalanine + H_2O + O_2 <2, 4-6, 8, 12, 14, 23, 24> (<4-6, 8, 23> high reaction rate [3, 5, 8, 58]) (Reversibility: ? <2, 4-6, 8, 12, 14, 23, 24> [3, 5, 8, 16, 19, 24, 37, 57-59]) [3, 5, 8, 19, 24, 37, 57-59]

P phenylpyruvate + NH_3 + H_2O_2

S L-propargylglycine + H_2O + O_2 <4> (Reversibility: ? <4> [18]) [18]

P 2-oxopent-4-ynoic acid + NH_3 + H_2O_2

S L-serine + H_2O + O_2 <4, 12> (<4> low activity [5]; <2,8> not [4, 8]) (Reversibility: ? <4, 12> [5, 19]) [5, 19]

P 2-oxo-3-hydroxypropionic acid + NH_3 + H_2O_2

S L-threonine + H_2O + O_2 <4> (<4> low activity [5]) (Reversibility: ? <4> [5]) [5]

P 2-oxo-3-hydroxybutanoic acid + NH_3 + H_2O_2

S L-tryptophan + H_2O + O_2 <2, 4-6, 8, 12, 14, 23> (<4-6,23> high reaction rate [3,5,58]; <8> low reaction rate [8]) (Reversibility: ? <2, 4-6, 8, 12, 14, 23> [3, 5, 8, 19, 24, 37, 58]) [3, 5, 8, 19, 24, 37, 58]

P ? + NH_3 + H_2O_2

S L-tyrosine + H_2O + O_2 <4-6, 8, 12, 14> (<-1, 4-6, 8> high reaction rate [3, 5, 8]) (Reversibility: ? <4-6, 8, 12, 14> [3, 5, 8, 19, 24]) [3, 5, 8, 19, 24]

P p-hydroxyphenylpyruvate + NH_3 + H_2O_2

S L-valine + H_2O + O_2 <8> (<8> very low activity [8]) (Reversibility: ? <8> [8]) [8]

P 2-oxoisovalerate + NH_3 + H_2O_2

S O-carbamoyl-L-serine + H_2O + O_2 <4, 5, 6> (Reversibility: ? <4, 5, 6> [3, 39]) [3, 39]

P 3-O-carbamoyl-2-keto-propionic acid + NH_3 + H_2O_2 (<4> product: corresponding cyclic lactam [39])

S S-adenosyl-L-homocysteine + H_2O + O_2 <2> (Reversibility: ? <2> [31]) [31]

P 4-thioadenosyl-2-oxobutyrate + NH_3 + H_2O_2

S S-carbamoyl-L-cysteine + H_2O + O_2 <4, 5, 6> (Reversibility: ? <4, 5, 6> [3, 39]) [3, 39]

P 3-thiocarbamoyl-2-keto-propionic acid + NH_3 + H_2O_2 (<4> product: corresponding cyclic lactam [39])

S α-hydroxyisocaproic acid + H_2O + O_2 <2> (Reversibility: ? <2> [37]) [37]

P 2-oxo-4-methylpentanoic acid + NH_3 + H_2O_2

S α-hydroxyvaleric acid + H_2O + O_2 <2> (Reversibility: ? <2> [37]) [37]

P 2-oxopentanoic acid + NH_3 + H_2O_2

S glycine + H_2O + O_2 <2, 4, 8, 12> (<4> low activity [5]; <2, 8, 12> not, [4, 8, 19, 37]) (Reversibility: ? <2, 4, 8, 12> [4, 58, 19, 37]) [4, 5, 8, 19, 37]

P glyoxylate + NH_3 + H_2O_2 (i.e. oxoacetic acid)

S hydroxyisovaleric acid + H_2O + O_2 <2> (Reversibility: ? <2> [37]) [37]

P ? + NH_3 + H_2O_2

S mandelic acid + H_2O + O_2 <2> (Reversibility: ? <2> [37]) [37]

P phenylpyruvate + NH_3 + H_2O_2

S thyroxine + H_2O + O_2 <2> (Reversibility: ? <2> [4]) [4]

P 3-[4-(4-hydroxy-3,5-diiodophenoxy)-3,5-diiodophenylpyruvate] + NH_3 + H_2O_2

S triiodothyronine + H_2O + O_2 <2> (Reversibility: ? <2> [4]) [4]

P 3-[4-(4-hydroxy-3-iodophenoxy)-3,5-diiodophenyl]-2-oxoproprionoc acid + NH_3 + H_2O_2

S Additional information <1-4, 7, 8-12, 20-22> (<11> 2 L-amino acid oxidases with different specificity: 1. oxidative deamination of aromatic carboxylic, sulfur containing imino and β-hydroxy-L-amino acids, no activity to basic L-amino acids or L-citrulline, 2. oxidative deamination of L-arginine, L-histidine, L-ornithine, L-citrulline and L-lysine, with no affinity for any of the other L-amino acids tested [20]; <1> 2 distinct L-amino acid oxidases with different specificity, 1. aliphatic or aromatic L-amino acids with non-polar side chains, 2. L-amino acids with positively charged side chains [21]; <20> reacts in order Phe > Met,Tyr > Cys,Leu > His > other amino acids [52]; <2-4> not: D-amino acids, N,N-dimethylleucine, α-aminoisobutyric acid [1]; <2> not: α-hydroxyisobutyrate [1, 4]; <2, 8, 12> not: L-threonine [4, 8, 19]; <8,12> not: proline [8, 19]; <9> overview on assays [6]; <10, 11> investigation of oxidative deamination of sulfur amino acids [10]; <4> S-carbamoyl-and S-thiocarbamoyl derivatives of L-cysteine [14]; <2-4> proline and N-methyl derivatives of leucine, methionine, homocysteine and S-benzylhomocysteine also oxidized [1]; <4> has bactericidal activity [26]; <21, 22> shows apoptosis-inducing activity in cell cultures [55, 56]; <9> enzymatic properties of snake venom enzymes, mechanisms of enzyme-induced platelet aggregation and apoptosis [16]; <7> induces human platelet aggregation [60]; <7> cytotoxic to different cell lines, resulting in loss of ability in attachment and inhibition of cell proliferation [61]) [1, 4, 6, 8, 10, 14, 19-21, 26, 52, 55, 56, 60, 61]

P ?

Inhibitors

2-naphthol <16> [43]

Ba^{2+} <16> [43]

Ca^{2+} <16> (<16> $CaCl_2$, $Ca(OOCCH_3)_2$ [40]; <16> inhibition can be relieved by L-arginine [41]) [40, 41, 43]

Cd^{2+} <8, 16> (<8> $CdCl_2$ [8]) [8, 43]

Co^{2+} <16> [43]

Cu^{2+} <14, 16> (<14> in presence of the activator Mn^{2+} [24]) [24, 43]

$CuSO_4$ <2> [1]

EDTA <18> (<18> 10 mM, 90% inhibition [49]) [49]

HgCl <5, 6, 8> [3, 8]

K^+ <16> (<16> KCl, $K(OOCH_3)$ [40]) [40]

KCN <5, 6, 8, 18> (<18> 10 mM, complete inhibition [49]) [3, 8, 49]

La^{3+} <16> (<16> LaCl3, $La(OOCCH_3)_3$ [40]) [40]

Mg^{2+} <16> (<16> EDTA, ATP and ADP, but not AMP can overcome inhibition [43]) [43]

Mn^{2+} <16> [40, 43]

$MnCl_2$ <8> [8]

NH_4^+ <2-4, 16> (<2-4> inhibition of amino acid oxidase activity, no inhibition of L-hydroxy acid oxidase activity [1]) [1, 40]

Na^+ <16> (<16> sodium salts, order of effectiveness: SCN^- > NO_3^- > Cl^-, Br^- > I^- > F^- > $HCOO^-$ > CH_3COO- [40]) [40]

NaF <18> (<18> 10 mM, 85% inhibition [49]) [49]

Ni^{2+} <16> [43]

Sr^{2+} <16> [43]

Zn^{2+} <16> [40, 43]

$ZnCl_2$ <8> [8]

anthranilate <19> (<19> competitive [50]) [50]

aromatic carboxylates <4> (<4> competitive [34]) [34]

atebrine <5, 6> [3]

bathophenanthroline disulfonic acid <16> [43]

benzenearsonic acid <2> [37]
benzoate <4> (<4> and derivatives [35]) [34, 35]
benzoic acid <2-4, 8> [1, 8]
butanedione <9> (<9> good substrates but not poor substrates protect against inactivation [48]) [48]
chloropromazine <16> (<16> inhibition can be relieved by L-arginine [41]) [41]
hydroxylamine <18> (<18> 5 mM, 75% inhibition [49]) [49]
iodoacetamide <8> [8]
iodoacetic acid <2-4> [1]
m-aminobenzoate <4> [34]
m-chlorobenzoate <4> [34]
m-chlorobenzoate <4> [34]
m-fluorobenzoate <4> [34]
m-hydroxybenzoate <4> [34]
m-nitrobenzoate <4> [34]
mandelate <4> [34]
nordihydroguariaretic acid <16> [43]
o-aminobenzoate <4> [34]
o-chlorobenzoate <4> [34]
o-fluorobenzoate <4> [34]
o-hydroxybenzoate <4> [34]
o-mercaptobenzoate <4> [34]
o-nitrobenzoate <4> [34]
o-phenanthroline <16> [43]
orthanilic acid <4> [34]
p-aminobenzoic acid <8> [8]
p-chloromercuribenzoate <2, 8> [8, 37]
quinine sulfate <5, 6> [3]
sodium azide <5, 6, 16> (<5, 6> slight [3]) [3, 43]
thiosemicarbazide <18> (<18> 5 mM, 40% inhibition [49]) [49]
vinylglycine <4> [18]
Additional information <4, 7, 9, 12, 15, 16> (<4, 7, 12> substrate inhibition [5, 7, 19, 26, 35]; <4> good substrates, not poor substrates are inhibitors [5]; <4> insensitive to KCN at 1 mM [26]; <15> naturally occuring inhibitors [36]; <16> inhibition by cations increases in alkaline pH, inhibition by anions increases in acidic pH [40]; <9> not: 5,5-dithiobis-(2-nitrobenzoic acid), trinitrobenzene sulfonic acid [48]) [5, 7, 19, 35, 36, 40, 48]

Cofactors/prosthetic groups

FAD <2-4, 7, 13, 16, 18, 19, 23, 24> (<3, 4, 7, 24> 2 mol of FAD per mol of enzyme [1, 2, 5, 7, 9, 59]; <13> 4 mol of FAD per mol of enzyme [22]; <16> 1 mol of FAD per mol of enzyme [43]; <18> non-covalently bound [49]) [1, 2, 5, 7, 9, 22, 43, 49, 50, 58, 59]
FMN <2, 8, 14, 19> (<2> 6 mol of FMN per mol of enzyme [4, 33]; <8,19> 2 mol of FMN per mol of enzyme [8, 54]; <14> FMN firmly bound to enzyme [24]) [4, 8, 24, 33, 37, 54]

Activating compounds

$CaCl_2$ <18> (<18> 2 mM, 50% activation [49]) [49]

Metals, ions

Cl^- <4> (<4> favors reaction [5]) [5]

Mn^{2+} <14> (<14> powerful activator [24]; <8> $MnCl_2$ inhibits [8]) [24]

Additional information <14> (<14> magnesium, iron and molybdene are no activators [24]) [24]

Turnover number (min^{-1})

378 <2> (L-leucine) [4]

1560 <2> (L-lactic acid) [4]

5196 <7> (L-leucine) [7]

Additional information <4> (<4> mol amino acid/min x mol of enzyme bound FAD: 11000 L-arginine, 280 L-phenylalanine, 40 L-valine, 600 L-leucine [13]) [13]

Specific activity (U/mg)

1.4 <18> (<18> glycine [49]) [49]

1.45 <13> [22]

1.98 <2> [37]

4.4 <18> (<18> L-threonine [49]) [49]

4.9 <18> (<18> L-aspartate [49]) [49]

6 <18> (<18> L-glutamate [49]) [49]

12.5 <18> (<18> L-histidine [49]) [49]

14.4 <18> (<18> L-valine [49]) [49]

14.8 <18> (<18> L-tyrosine [49]) [49]

15.5 <18> (<18> L-tryptophan [49]) [49]

16.6 <18> (<18> L-arginine [49]) [49]

17 <18> (<18> L-isoleucine [49]) [49]

17.4 <18> (<18> S-adenosy-L-cysteine [49]) [49]

17.8 <18> (<18> L-alanine [49]) [49]

18.4 <18> (<18> L-asparagine [49]) [49]

18.8 <18> (<18> L-leucine [49]) [49]

18.9 <18> (<18> L-lysine [49]) [49]

20.4 <18> (<18> L-cystine [49]) [49]

21.4 <18> (<18> L-ornithine [49]) [49]

24.2 <18> (<18> L-serine [49]) [49]

24.3 <18> (<18> L-methionine [49]) [49]

25.2 <18> (<18> S-adenosyl-L-homocysteine [49]) [49]

26.5 <18> (<18> L-ethionine [49]) [49]

26.5 <18> (<18> L-phenylalanine [49]) [49]

30.5 <18> (<18> L-methionine sulfoximine [49]) [49]

39.5 <18> (<18> L-glutamine [49]) [49]

120 <16> [43]

159 <22> (<22> L-alanine [56]) [56]

272 <22> (<22> L-histidine [56]) [56]

282 <22> (<22> L-tryptophan [56]) [56]

564 <22> (<22> L-leucine [56]) [56]
590 <22> (<22> L-arginine [56]) [56]
751 <22> (<22> L-valine [56]) [56]
772 <22> (<22> L-methionine [56]) [56]
776 <22> (<22> L-phenylalanine [56]) [56]
868 <22> (<22> L-isoleucine [56]) [56]
Additional information <2, 4, 9, 14, 23> (<2,9> enzyme assays [6,16,31]; <23> activity in 19 different genus Bothrops venoms [58]) [4-6, 16, 24, 31, 58]

K_m-Value (mM)

0.011 <12> (L-phenylalanine) [19]
0.037 <24> (L-arginine) [59]
0.04 <12> (L-lysine) [19]
0.05 <19> (L-phenylalanine) [54]
0.05 <12> (cysteine) [19]
0.08 <19> (L-tryptophan) [54]
0.13 <19> (L-norleucine) [54]
0.14 <7> (L-leucine) [7]
0.16 <13> (L-phenylalanine) [23]
0.24 <19> (L-methionine) [54]
0.259 <24> (L-asparagine) [59]
0.27 <11> (L-arginine, <11> fraction II [20]) [20]
0.34 <24> (L-glutamine) [59]
0.35 <10> (L-methionine) [10]
0.56 <19> (L-leucine) [50]
0.63 <19> (L-leucine) [54]
0.72 <10> (L-homocystine) [10]
0.8 <19> (L-isoleucine) [54]
0.91 <19> (L-norvaline) [54]
1 <15> (L-leucine) [36]
1 <11> (L-tyrosine, <11> fraction I [20]) [20]
1.17 <8> (L-leucine) [8]
1.2 <13> (O_2) [23]
1.3 <10, 11> (L-homocysteine) [10]
1.6 <14> (L-lysine) [24]
1.8 <11> (L-histidine, <11> fraction II [20]) [20]
1.9 <10> (S-adenosyl-L-homocysteine) [10]
2 <19> (L-arginine) [54]
2.2 <14> (L-tyrosine, L-tryptophan) [24]
2.2 <10, 18> (S-ribosyl-L-homocysteine) [10, 49]
2.5 <19> (L-asparagine) [54]
2.5 <19> (L-histidine) [54]
2.6 <11> (L-leucine, <11> fraction I [20]) [20]
2.9 <14> (L-arginine) [24]
3.1 <11> (L-phenylalanine, <11> fraction I [20]) [20]
3.1 <10> (djenkolic acid) [10]

3.5 <14> (L-phenylalanine) [24]
4.2 <11> (L-tryptophan, <11> fraction I [20]) [20]
4.5 <10> (L-cysteine) [10]
5 <19> (L-aminobutyric acid) [54]
5 <10> (L-cystine) [10]
5.3 <11> (L-cysteine) [10]
6.2 <14> (L-histidine) [24]
6.2 <14> (L-ornithine) [24]
8 <19> (L-glutamine) [54]
9.6 <11> (L-methionine) [10, 20]
10 <19> (L-lysine) [54]
10 <14> (L-methionine) [24]
12.5 <19> (L-alanine) [54]
12.5 <11> (L-homocystine) [10]
13 <11> (S-ribosyl-L-homocysteine) [10]
13.3 <19> (L-ornithine) [54]
14 <11> (S-adenosylhomocysteine) [20]
14 <11> (S-adenosylmethionine) [10]
14.2 <11> (L-ornithine, <11> fraction II [20]) [20]
21.3 <11> (L-citrulline, <11> fraction II [20]) [20]
23.2 <11> (L-lysine, <11> fraction II [20]) [20]
24.8 <2> (S-adenosyl-L-homocysteine) [31]
45 <11> (methionine sulfoxide) [10]
71 <11> (L-isoleucine, <11> fraction I [20]) [20]
83 <11> (djenkolic acid) [10]
100 <11> (S-adenosyl-L-homocysteine sulfoxide) [10]
Additional information <4> (<4> effect of pH on K_m [28]) [28]

K_i-Value (mM)

0.15 <4> (*m*-chlorobenzoate) [34]
0.34 <4> (*m*-nitrobenzoate) [34]
0.37 <4> (*m*-fluorobenzoate) [34]
0.42 <4> (*o*-aminobenzoate) [34]
0.7 <4> (*p*-chlorobenzoate) [34]
0.86 <4> (*o*-nitrobenzoate) [34]
1 <4> (*o*-chlorobenzoate) [34]
1 <4> (*o*-mercaptobenzoate) [34]
1.2 <4> (orthanilic acid) [34]
1.3 <4> (*p*-fluorobenzoate) [34]
1.4 <4> (*o*-flourobenzoate) [34]
1.4 <4> (*p*-aminobenzoate) [34]
1.6 <4> (*m*-aminobenzoate) [34]
1.7 <4> (benzoate) [34]
2.1 <4> (*m*-hydroxybenzoate) [34]
2.2 <4> (*o*-hydroxybenzoate) [34]
2.3 <4> (*p*-hydroxybenzoate) [34]
2.3 <4> (*p*-nitrobenzoate) [34]

11 <4> (D-mandelate) [34]
11 <4> (L-mandelate) [34]

pH-Optimum

6.5-7 <4> [26]
6.6 <15> (<15> L-leucine [36]) [36]
7 <8> (<8> L-Ile, L-Leu, L-Cys, L-Met, L-Phe, L-Tyr [8]) [8]
7-7.6 <5, 6> [3]
7.2-7.5 <2-4, 10> (<10> 2 optima, one between pH 7.2 and 7.5 and the other
above pH 8.5, L-cysteine, S-adenosyl-L-homocysteine [10]) [1, 10]
7.4-7.8 <11> (<11> S-ribosyl-L-homocysteine, S-adenosyl-L-homocysteine,
S-adenosyl-L-methionine, L-methionine [10]) [10]
7.5 <4> (<4> L-leucine [5]) [5, 9]
8 <2, 8> (<2> L-lactic acid [4]; <8> L-His [8]) [4, 8]
8-8.4 <10> (<10> homocysteine, cysteine [10]) [10]
8-8.5 <24> (<24> L-phenylalanine [59]) [59]
8.5 <7, 10> (<10> 2 optima, one between pH 7.2 and 7.5 and the other above
pH 8.5, L-cysteine, S-adenosylhomocysteine, with S-adenosyl-homocysteine
and djenkolic acid increase of activity beyond pH 8.5 [10]) [7, 10]
8.5-8.8 <11> (<11> djenkolic acid [10]) [10]
8.7 <2> (<2> L-leucine [4]) [4]
8.8-9.2 <2> (<2> S-adenosyl-L-homocysteine [31]) [31]
9 <18> (<18> broad optimum [49]) [49]
9 <8> (<8> L-Trp [8]) [8]
9.5 <13> (<13> L-phenylalanine [22]) [22]

pH-Range

3.5-12 <13> (<13> half maximal activity at pH 3.5 and 12 [22]) [22]
4.5-10.5 <24> [59]

Temperature optimum (°C)

49 <13> (<13> L-phenylalanine [22]) [22]
50 <24> [59]

4 Enzyme Structure

Molecular weight

12000 <22> (<22> gel filtration [56]) [56]
59000 <21> (<21> matrix-assisted laser desorption/ionization mass spectroscopy [55]) [55]
98000 <16> (<16> gel filtration [43]) [43]
102000-115000 <24> (<24> gel filtration, native PAGE [59]) [59]
113000 <20> (<20> gel filtration [52]) [52]
128000-153000 <4> (<4> approach to equilibrium method [9]) [9]
130000 <4> [5, 9]
130000-140000 <4, 12> (<4> sedimentation equilibrium [32]; <12> gel filtration [19]) [19, 32]

132000 <19> (<19> gel filtration [54]) [54]
135000 <7> (<7> gel filtration [60]) [60]
138000 <2> (<2> calculation from diffusion constant, sedimentation coefficient and FAD content [1]) [1]
140000 <7, 8> (<7> gel filtration [7]; <8> gel filtration [8]) [7, 8]
150000 <3> (<3> approach to equilibrium method [1]) [1]
150000 <7> (<7> gel filtration [61]) [61]
300000 <13> (<13> disc gel electrophoresis, linear acrylamide concentration [22]) [22]
310000-314000 <2> (<2> high speed equilibrium method, gel filtration [4,33]) [4, 33]
900000-1000000 <18> (<18> form Mα, gel filtration, sedimentation equilibrium [49]) [49]
1200000-1300000 <18> (<18> form Mβ, gel filtration, sedimentation equilibrium [49]) [49]

Subunits

dimer <4, 7, 8, 16, 19, 20, 22, 24> (<16> 2 * 49000, SDS-PAGE [43]; 2 * 68000, <7> SDS-PAGE [7]; <8> 2 * 70000, SDS-PAGE [8]; <4> 2 * 70000, sedimentation equilibrium in presence of 6 M guanidine hydrochloride, 2 types of polypeptide chains [32]; <19> 2 * 66000, SDS-PAGE [54]; <20> 2 * 60000, SDS-PAGE [52]; <22> 2 * 60000, SDS-PAGE [56]; <24> 2 * 54000, SDS-PAGE [59]; <7> 2 * 65000-70000, SDS-PAGE [60, 61]) [7, 8, 32, 43, 52, 54, 56, 59, 60, 61]
oligomer <18> (<18> Mα, ? * 66000, Mβ, ? * 66000 + ? * 135000, latter polypeptide not required for amino acid oxidase activity [49];) [49]
Additional information <9> (<9> overview on structural properties [16]) [16]

Posttranslational modification

glycoprotein <4, 7, 18, 19> (<4> 2-5% carbohydrate, including sialic acid [5]; <7> 3.8% carbohydrate [7]; <18> both polypeptide chains are glycosylated [49]; <19> glycan is a bis-sialylated, biantennary, core-fucosylated dodecasaccharide [51]) [5, 7, 32, 49, 51, 53, 54, 61]

5 Isolation/Preparation/Mutation/Application

Source/tissue

blood <4> [27]
culture filtrate <13> [22]
kidney <2> [1, 4, 29, 31, 33, 37]
liver <2, 14, 15> [1, 24, 31, 36]
mycelium <13> [22, 23]
venom <3, 4, 7, 8, 10, 11> [1, 5, 6, 7, 8, 9, 10, 16, 17, 27, 32, 34, 39, 47, 48]

Localization

cell envelope <1> [21]
cytoplasmic membrane <5, 6> (<5,6> constituent part of [3]) [3]

membrane <16> (<16> enzyme is part of photosystem II particles [42]; <16> photosynthetic membranes [44]) [42, 44]
mitochondrion <2, 14> [24, 37]
periplasm <18> [49]
peroxisome <2> [4, 29]
soluble <2, 12, 13> [19, 22, 29, 37]

Purification

<2> [1, 4, 31, 37]
<3> [1]
<4> [1, 5, 9]
<7> [60, 61]
<7> [7]
<8> [8]
<13> [22]
<14> [24]
<16> [43]
<18> [49]
<19> [50, 54]
<21> [55]
<22> [56]
<24> [59]

Crystallization

<19> (in presence of citrate and o-aminobenzoate, comparison with D-amino acid oxidases [53]) [53]
<22> [56]
<2, 4> [1, 4, 5, 9, 37]

Cloning

<19> [50]
<20> [52]

Engineering

Additional information <13> (<13> gln-1bR8, mutant altered in the regulation of L-amino acid oxidase [57]) [57]

6 Stability

pH-Stability

5 <8> (<8> 4°C, 24 h, complete inactivation [8]) [8]
7 <4, 19> (<4> reversible inactivation: at pH near neutrality change into inactive configuration, regaining of active configuration on lowering the pH, monovalent anions, substrate and substrate analogs prevent inactivation [17]; <19> at pH above neutrality reversible inactivation [50]) [17, 50]
7-8 <8> (<8> 4°C, 24 h, stable [8]) [8]
9 <4, 7> (<7> 25°C, 1 h stable [7]; <4> unstable at [12]; <7> 25°C, 5 min, 47% loss of activity, complete loss of activity after 30 min [7]) [7, 12]

10 <8> (<8> 4°C, 24 h, complete inactivation [8]) [8]
Additional information <4> (<4> protection by acetate (high concentration) and some aliphatic and aromatic monocarboxylic acids at pH 9.0 [12]; <4> slow reversible and temperature-dependent transition to an inactive form under alkaline conditions [5]) [5, 12]

Temperature stability

4 <7> (<7> 3 months, no loss of activity [7]) [7]
25 <7> (<7> 1 month, no loss of activity [7]) [7]
37 <7> (<7> 5 days, no loss of activity, 14 days, 20% loss of activity [7]) [7]
45 <2> (<2> 5 min, stable [37]) [37]
50 <8> (<8> 10 min, pH 6.8, stable up to [8]) [8]
51 <2> (<2> 5 min, inactivation above 51°C [37]) [37]
65 <8> (<8> 10 min, 54% loss of activity [8]) [8]
70 <3> (<3> rapid loss of activity [1]) [1]
71 <2> (<2> 5 min, complete inactivation [37]) [37]
73 <4> (<4> no loss of activity after 5 min, heat-stable in presence of 0.01 mM L-leucine [9]) [9]
75 <8> (<8> 10 min, 97% loss of activity [8]) [8]
Additional information <4, 19> (<4> reduced form of the enzyme is much more stable to heat than oxidized form, substrates: e.g. L-Leu, L-Phe, L-Met protect from heat denaturation, no protection by D-leucine and L-lysine [9]; <4> labile to heating at 60°C [26]; <19> freezing causes reversible inactivation [50]) [9, 26, 50]

General stability information

<4>, acetate at high concentrations increases pH-stability at pH 9 [12]
<4>, aliphatic and aromatic monocarboxylic acids protect at pH 9.0 [12]
<4>, freezing, inactivation between -5°C and -60°C, maximal inactivation at -20°C, rate of inactivation is dependent on pH of storage, in most cases complete reactivation by heating at pH 5, inactivation not prevented by monovalent cations [11, 17]
<4>, freezing, quick freezing with dry ice/acetone mixture and storage at -15°C for 60 h, enzyme loses 33% of activity, complete reactivation by heating at pH 5, inactivation not prevented by monovalent cations [11, 17]
<6>, stability of enzyme in immobilized whole cells [15]
<7>, freezing, quick freezing with dry ice/acetone mixture and storage at -15°C for 60 h, enzyme is stable [7]
<8>, freezing, irreversible inactivation [8]
<8>, irreversible inactivation by lyophilization [8]
<12>, 2-mercaptoethanol stabilizes [19]
<12>, KCl stabilizes [19]

Storage stability

<2>, 0-3°C, dialyzed enzyme, 2 weeks, 10% loss of activity [4]
<4>, 0-5°C, crystalline suspension in water, pH 7.2, stable for several months [5]
<7>, 4°C, pH 7.4, 20 mM Tris-HCl, 3 months [7]

<8>, 4°C, 40% v/v glycerol, 3 months [8]
<12>, 4°C, 66 mM KCl, 10 mM 2-mercaptoethanol, 1 month, 5-10% loss of activity [19]
<13>, 4°C, purified enzyme stable for weeks [22]
<24>, -20°C, 4 months, no loss of activity [59]
<24>, 4°C, 4 months, no loss of activity [59]
<5, 6>, 0°C, several weeks [3]

References

[1] Meister, A.; Wellner, D.: Flavoprotein amino acid oxidases. The Enzymes, 2nd Ed (Boyer, P.D., Lardy, H., Myrbäck, K., eds.), 7, 609-648 (1963)

[2] Bright, H.J.; Porter, D.J.T.: Flavoprotein oxidases. The Enzymes, 3rd Ed. (Boyer, P.D., ed.), 12, 421-505 (1975)

[3] Cioaca, C.; Ivanof, A.: Bacterial amino acid oxidases. I. L-amino acid oxidase and its distribution in bacteria. Arch. Roum. Pathol. Exp. Microbiol., 33, 211-222 (1974)

[4] Nakano, M.; Danowski, T.S.: L-Amino acid oxidase (rat kidney). Methods Enzymol., 17B, 601-605 (1971)

[5] Weller, D.: L-Amino acid oxidase (snake venom). Methods Enzymol., 17B, 597-600 (1971)

[6] Weller, D.; Lichtenberg, L.A.: Assay of amino acid oxidase. Methods Enzymol., 17B, 593-596 (1971)

[7] Tan, N.H.; Saifuddin, M.N.: Isolation and characterization of an unusual form of L-amino acid oxidase from King cobra (Ophiophagus hannah) venom. Biochem. Int., 19, 937-944 (1989)

[8] Ueda, M.; Chang, C.C.; Ohno, M.: Purification and characterization of L-amino acid oxidase from the venom of Trimeresurus mucrosquamatus (Taiwan habu snake). Toxicon, 26, 695-706 (1988)

[9] Wellner, D.; Meister, A.: Crystalline L-amino acid oxidase of Crotalus adamanteus. J. Biol. Chem., 235, 2013-2018 (1960)

[10] Chen, S.S.; Walgate, J.H.; Duerre, J.A.: Oxidative deamination of sulfur amino acids by bacterial and snake venom L-amino acid oxidase. Arch. Biochem. Biophys., 146, 51-63 (1971)

[11] Curti, B.; Massey, V.; Zmudka, M.: Inactivation of snake venom L-amino acid oxidase by freezing. J. Biol. Chem., 243, 2306-2314 (1968)

[12] Paik, W.K.; Kim, S.: Studies on the stability of L-amino-acid oxidase of snake venom. Biochim. Biophys. Acta, 139, 49-55 (1967)

[13] Massey, V.; Curti, B.: On the reaction mechanism of Crotalus adamanteus L-amino acid oxidase. J. Biol. Chem., 242, 1259-1264 (1967)

[14] Kimura, T.; Esaki, N.; Tanaka, H.; Soda, K.: Action of S-carbamoyl and S-thiocarbamoyl derivatives of L-cysteine on L-amino acid oxidase. Agric. Biol. Chem., 48, 3157-3159 (1984)

[15] Szwajcer, E.; Brodelius, P.; Mosbach, K.: Production of α-keto acids: 2. Immobilized whole cells of Providencia sp. PCM 1298 containing L-amino acid oxidase. Enzyme Microb. Technol., 4, 409-413 (1982)

[16] Du, X.Y.; Clemetson, K.J.: Snake venom L-amino acid oxidases. Toxicon, 40, 659-665 (2002)

[17] Couer, C.J.; Edmondson, D.F.; Singer. T.P.: Reversible inactivation of L-amino acid oxidase. Properties of the three conformational forms. J. Biol. Chem., 252, 8035-8039 (1977)

[18] Marcotte, P.; Walsh, C.: Vinylglycine and proparglyglycine: complementary suicide substrates for L-amino acid oxidase and D-amino acid oxidase. Biochemistry, 15, 3070-3076 (1976)

[19] Coudert, M.; Vandecasteele, J.P.: Charcterization and physiological function of a soluble L-amino acid oxidase in Corynebacterium. Arch. Microbiol., 102, 151-153 (1975)

[20] Duerre, J.A.; Chakrabarty, S.: L-Amino acid oxidases of Proteus rettgeri. J. Bacteriol., 121, 656-663 (1975)

[21] Pelmont, J.; Arlaud, G.; Rossat, A.M.: L-Amino acid oxidases of Proteus mirabilis: general properties. Biochimie, 54, 1359-1374 (1972)

[22] Aurich, H.; Luppa, D.; Schucker, G.: Purification and properties of L-amino acid oxidase from neurospora. Acta Biol. Med. Ger., 28, 209-220 (1972)

[23] Luppa, D.; Aurich, H.: Kinetic studies on the reaction mechanism of L-amino acid oxidase from Neurospora crassa. Acta Biol. Med. Ger., 27, 839-850 (1971)

[24] Mizon, J.; Biserte, G.; Boulanger, P.: Properties of turkey (Meleagris gallopavo L.) liver L-amino acid oxidase. Biochim. Biophys. Acta, 212, 33-42 (1970)

[25] Page, D.S.; Vanetten, R.L.: L-Amino acid oxidase. II. Deuterium isotope effects and the action mechanism for the reduction of L-amino acid oxidase by L-leucine. Biochim. Biophys. Acta, 227, 16-31 (1971)

[26] Skarnes, R.C.: L-Amino-acid oxidase, a bactericidal system. Nature, 225, 1072-1073 (1970)

[27] Hayes, M.B.; Wellner, D.: Microheterogeneity of L-amino acid oxidase. Separation of multiple components by polyacrylamide gel electrofocusing. J. Biol. Chem., 244, 6636-6644 (1969)

[28] Page, D.S.; Van Etten, R.L.: L-Amino-acid oxidase. I. Effect of pH. Biochim. Biophys. Acta, 191, 38-45 (1969)

[29] Nakano, M.; Saga, M.; Tsutsumi, Y.: Distribution and immunochemical properties of rat kidney L-amino-acid oxidase, with a note on peroxisomes. Biochim. Biophys. Acta, 185, 19-30 (1969)

[30] Porter, D.J.T.; Bright, H.J.: Location of hydrogen transfer steps in the mechanism of reduction of L-amino acid oxidase. Biochem. Biophys. Res. Commun., 36, 209-214 (1969)

[31] Miller, C.H.; Duerre, J.A.: Oxidative deamination of S-adenosyl-L-homocysteine by rat kidney L-amino acid oxidase. J. Biol. Chem., 244, 4273-4276 (1969)

[32] De Kok, A.; Rawitch, A.B.: Studies on L-amino acid oxidase. II. Dissociation and characterization of its subunits. Biochemistry, 8, 1405-1411 (1969)

[33] Nakano, M.; Tarutani, O.; Danowski, T.S.: Molecular weight of mammalian L-amino-acid oxidase from rat kidney. Biochim. Biophys. Acta, 168, 156-157 (1968)

[34] De Kok, A.; Veeger, C.: Studies on L-amino-acid oxidase. I. Effects of pH and competitive inhibitors. Biochim. Biophys. Acta, 167, 35-47 (1968)

[35] Koster, J.K.; Veeger, C.: The relation between temperature-inducible allosteric effects and the activation energies of amino-acid oxidases. Biochim. Biophys. Acta, 167, 48-63 (1968)

[36] Shinwari, M.A.; Falconer, I.R.: Naturally occuring inhibition and activation of avian liver L-amino acid oxidase. Biochem. J., 104, 538-548 (1967)

[37] Nakano, M.; Tsutsumi, Y.; Danowski, T.S.: Crystalline L-amino-acid oxidase from the soluble fraction of rat-kidney cells. Biochim. Biophys. Acta, 139, 40-48 (1967)

[38] Sikora, L.; Marzluf, G.A.: Regulation of L-amino acid oxidase and of D-amino acid oxidase in Neurospora crassa. Mol. Gen. Genet., 186, 33-39 (1982)

[39] Cooper, A.J.L.; Meister, A.: Action of liver glutamine transaminase and L-amino acid oxidase on several glutamine analogs. Preparation and properties of the 4-S, O, and NH analogs of α-ketoglutaramic acid. J. Biol. Chem., 248, 8499-8505 (1973)

[40] Pistorius, E.K.: Further evidence for a functional relationship between L-amino acid oxidase activity and photosynthetic oxygen evolution in Anacystis nidulans. Effect of chloride on the two reactions. Z. Naturforsch. C, 40, 806-813 (1985)

[41] Pistorius, E.K.: Effects of Mn^{2+}, Ca^{2+} and chlorpromazine on photosystem II of Anacystis nidulans. An attempt to establish a functional relationship of amino acid oxidase to photosystem II. Eur. J. Biochem., 135, 217-222 (1983)

[42] Pistorius, E.K.; Voss, H.: Presence of an amino acid oxidase in photosystem II of Anacystis nidulans. Eur. J. Biochem., 126, 203-209 (1982)

[43] Pistorius, E.K.; Voss, H.: Some properties of a basic L-amino-acid oxidase from Anacystis nidulans. Biochim. Biophys. Acta, 611, 227-240 (1980)

[44] Löffelhardt, W.: The biosynthesis of phenylacetic acids in the bluegreen alga Anacystis nidulans: Evidence for the involvement of a thylakoid-bound L-amino acid oxidase. Z. Naturforsch. C, 32, 345-350 (1977)

[45] Pistorius, E.K.; Kertsch, R.; Faby, S.: Investigations about various possible functions of the L-amino acid oxidase in the cyanobacterium Anacystis nidulans. Z. Naturforsch. C, 44, 370-377 (1989)

[46] Meyer, R.; Pistorius, E.K.: Some properties of photosystem II preparations from the cyanobacterium Synechococcus sp. - presence of an L-amino acid oxidase in photosystem II complexes from Synechococcus sp.. Biochim. Biophys. Acta, 893, 426-433 (1987)

[47] Cooper, A.J.L.; Meister, A.: Enzymatic oxidation of L-homocysteine. Arch. Biochem. Biophys., 239, 556-566 (1985)

[48] Christman, M.F.; Cardenas, J.M.: Essential arginine residues occur in or near the catalytic site of L-amino acid oxidase. Experientia, 38, 537-538 (1982)

[49] Vallon, O.; Bulte, L.; Kuras, R.; Olive, J.; Wollman, F.A.: Extensive accumulation of an extracellular L-amino acid oxidase during gametogenesis of Chlamydomonas reinhardtii. Eur. J. Biochem., 215, 351-360 (1993)

[50] Macheroux, P.; Seth, O.; Bollschweiler, C.; Schwarz, M.; Kurfurst, M.; Au, L.C.; Ghisla, S.: L-Amino-acid oxidase from the Malayan pit viper Calloselasma rhodostoma: comparative sequence analysis and characterization of active and inactive forms of the enzyme. Eur. J. Biochem., 268, 1679-1686 (2001)

[51] Geyer, A.; Fitzpatrick, T.B.; Pawelek, P.D.; Kitzing, K.; Vrielink, A.; Ghisla, S.; Macheroux, P.: Structure and characterization of the glycan moiety of L-amino-acid oxidase from the Malayan pit viper Calloselasma rhodostoma. Eur. J. Biochem., 268, 4044-4053 (2001)

[52] Sun, Y.; Nonobe, E.; Kobayashi, Y.; Kuraishi, T.; Aoki, F.; Yamamoto, K.; Sakai, S.: Characterization and expression of L-amino acid oxidase of mouse milk. J. Biol. Chem., 277, 19080-19086 (2002)

[53] Pawelek, P.D.; Cheah, J.; Coulombe, R.; Macheroux, P.; Ghisla, S.; Vrielink, A.: The structure of L-amino acid oxidase reveals the substrate trajectory into an enantiomerically conserved active site. EMBO J., 19, 4204-4215 (2000)

[54] Ponnudurai, G.; Chung, M.C.; Tan, N.H.: Purification and properties of the L-amino acid oxidase from Malayan pit viper (Calloselasma rhodostoma) venom. Arch. Biochem. Biophys., 313, 373-378 (1994)

[55] Ali, S.A.; Stoeva, S.; Abbasi, A.; Alam, J.M.; Kayed, R.; Faigle, M.; Neumeister, B.; Voelter, W.: Isolation, structural, and functional characterization of an apoptosis-inducing L-amino acid oxidase from leaf-nosed viper (Eristocophis macmahoni) snake venom. Arch. Biochem. Biophys., 384, 216-226 (2000)

[56] Souza, D.H.F.; Eugenio, L.M.; Fletcher, J.E.; Jiang, M.S.; Garratt, R.C.; Oliva, G.; Selistre-de-Araujo, H.S.: Isolation and structural characterization of a cytotoxic L-amino acid oxidase from Agkistrodon contortrix laticinctus snake venom: Preliminary crystallographic data. Arch. Biochem. Biophys., 368, 285-290 (1999)

[57] Calderon, J.; Olvera, L.; Martinez, L.M.; Davila, G.: A Neurospora crassa mutant altered in the regulation of L-amino acid oxidase. Microbiology, 143, 1969-1974 (1997)

[58] Pessatti, M.; Fontana, J.D.; Furtado, M.F.; Guimaraes, M.F.; Zanette, L.R.; Costa, W.T.; Baron, M.: Screening of Bothrops snake venoms for L-amino acid oxidase activity. Appl. Biochem. Biotechnol., 51-52, 197-210 (1995)

[59] Brearley, G.M.; Price, C.P.; Atkinson, T.; Hammond, P.M.: Purification and partial characterization of a broad-range L-amino acid oxidase from Bacillus carotarum 2Pfa isolated from soil. Appl. Microbiol. Biotechnol., 41, 670-676 (1994)

[60] Li, Z.Y.; Yu, T.F.; Lian, C.Y.: Purification and characterization of L-amino acid oxidase from king cobra (Ophiophagus hannah) venom and its effects on human platelet aggregation. Toxicon, 32, 1349-1358 (1994)

[61] Ahn, M.Y.; Lee, B.M.; Kim, Y.S.: Characterization and cytotoxicity of L-amino acid oxidase from the venom of king cobra (Ophiophagus hannah). Int. J. Biochem. Cell Biol., 29, 911-919 (1997)

D-Amino-acid oxidase   1.4.3.3

1 Nomenclature

EC number
   1.4.3.3

Systematic name
   D-amino-acid:oxygen oxidoreductase (deaminating)

Recommended name
   D-amino-acid oxidase

Synonyms
   D-aminoacid oxidase
   DAAO
   DAMOX
   DAO
   L-amino acid:O_2 oxidoreductase
   ophio-amino-acid oxidase
   oxidase, D-amino acid

CAS registry number
   9000-88-8

2 Source Organism

<1> *Sus scrofa* [1-5, 9-12, 16, 17, 19, 23-25, 28, 34, 36, 39]
<2> *Rhodotorula gracilis* [6, 7, 18, 26, 29, 31, 32, 35, 36]
<3> *Candida tropicalis* [8]
<4> *Trigonopsis variabilis* [13, 20, 21, 30-32, 36, 37, 40, 22]
<5> *Neurospora crassa* [14]
<6> *Chlorella vulgaris* [15]
<7> *Candida boidinii* [27]
<8> *Homo sapiens* [33, 36]
<9> *Achatina achatina* [38]
<10> *Arion ater* [38]
<11> *Mus musculus* [38]

3 Reaction and Specificity

Catalyzed reaction

a D-amino acid + H_2O + O_2 = a 2-oxo acid + NH_3 + H_2O_2 (<2> ping pong bi
bi mechanism [7])

Reaction type

oxidation
oxidative deamination
redox reaction
reduction

Natural substrates and products

S D-2-aminobutyrate + H_2O + O_2 <1> (Reversibility: ? <1> [2]) [2]
P 2-oxobutyrate + NH_3 + H_2O_2
S D-2-aminobutyrate + H_2O + O_2 <1> (Reversibility: ? <1> [23]) [23]
P 2-oxobutyrate + NH_3 + H_2O_2
S D-alanine + H_2O + O_2 <1-10> (Reversibility: ? <1-10> [2, 4, 7, 8, 14, 15,
       18, 21, 23, 26, 27, 33, 34, 35, 38, 39, 22]) [2, 4, 7, 8, 14, 15, 18, 21, 23, 26, 27,
       33, 34, 35, 38, 39, 22]
P pyruvic acid + NH_3 + H_2O_2
S D-α-aminoadipate + H_2O + O_2 <4> (Reversibility: ? <4> [21]) [21]
P 2-oxoadipate + NH_3 + H_2O_2
S D-arginine + H_2O + O_2 <2, 4> (Reversibility: ? <4> [21, 35]) [21, 35]
P 5-guanidino-2-oxopentanoic acid + NH_3 + H_2O_2
S D-asparagine + H_2O + O_2 <2, 6, 7> (Reversibility: ? <2, 6, 7> [15, 26, 27])
       [15, 26, 27]
P 2-oxosuccinamic acid + NH_3 + H_2O_2
S D-aspartic acid + H_2O + O_2 <1> (<1> weak activity [4]) (Reversibility: ?
       <1> [4]) [4]
P oxaloacetate + NH_3 + H_2O_2
S D-citrulline + H_2O + O_2 <4> (Reversibility: ? <4> [21]) [21]
P 2-oxo-5-ureidopentanoic acid + NH_3 + H_2O_2
S D-cysteine + H_2O + O_2 <2, 3> (Reversibility: ? <2, 3> [7, 8, 26]) [7, 8, 26]
P 2-oxo-3-thiopropionic acid + NH_3 + H_2O_2
S D-ethionine + H_2O + O_2 <4, 5> (Reversibility: ? <4, 5> [14, 21]) [14, 21]
P 4-ethylsulfanyl-2-oxobutyric acid + NH_3 + H_2O_2
S D-glutamic acid + H_2O + O_2 <6> (Reversibility: ? <6> [15]) [15]
P α-ketoglutarate + NH_3 + H_2O_2
S D-glutamine + H_2O + O_2 <2, 3, 6> (Reversibility: ? <2, 3, 6> [8, 15, 26])
       [8, 15, 26]
P 2-oxoglutaramate + NH_3 + H_2O_2
S D-histidine + H_2O + O_2 <1, 3, 4, 6> (Reversibility: ? <1, 3, 4, 6> [2, 8, 13,
       15, 21]) [2, 8, 13, 15, 21]
P 3-(1H-imidazol-4-yl)-2-oxopentanoate + NH_3 + H_2O_2
S D-isoleucine + H_2O + O_2 <1-4, 6> (Reversibility: ? <1-4, 6> [4, 8, 15, 21,
       26]) [4, 8, 15, 21, 26]

P 3-methyl-2-oxopentanoic acid + NH_3 + H_2O_2
S D-leucine + H_2O + O_2 <1-7> (Reversibility: ? <1-7> [4, 7, 8, 13-15, 18, 21, 26, 27, 39]) [4, 7, 8, 13-15, 18, 21, 26, 27, 39]
P 4-methyl-2-oxopentanoic acid + NH_3 + H_2O_2
S D-methionine + H_2O + O_2 <1, 2, 4-7, 9, 10> (Reversibility: ? <1, 2, 4-7, 9, 10> [2, 4, 7, 13-15, 18, 21, 23, 26, 27, 34, 35, 38, 39, 22]) [2, 4, 7, 13-15, 18, 21, 23, 26, 27, 34, 35, 38, 39, 22]
P 4-methylthio-2-oxobutanoic acid + NH_3 + H_2O_2
S D-norleucine + H_2O + O_2 <1, 2, 5> (Reversibility: ? <1, 2, 5> [2, 7, 14, 23]) [2, 7, 14, 23]
P 2-oxohexanoate + NH_3 + H_2O_2
S D-norvaline + H_2O + O_2 <1, 2, 5> (Reversibility: ? <1, 2, 5> [2, 7, 14, 23, 35]) [2, 7, 14, 23, 35]
P 2-oxopentanoate + NH_3 + H_2O_2
S D-ornithine + H_2O + O_2 <1> (Reversibility: ? <1> [2]) [2]
P 5-amino-2-oxopentanoate + NH_3 + H_2O_2
S D-phenylalanine + H_2O + O_2 <1, 2, 4-7, 9, 10> (Reversibility: ? <1, 2, 4-7, 9, 10> [4, 7, 14, 15, 21, 26, 27, 34, 35, 38, 39, 22]) [4, 7, 14, 15, 21, 26, 27, 34, 35, 38, 39, 22]
P phenylpyruvic acid + NH_3 + H_2O_2
S D-proline + H_2O + O_2 <1-3, 6, 7, 9, 10> (Reversibility: ? <1-3, 6, 7, 9, 10> [2, 7, 8, 12, 15, 18, 23, 26, 27, 34, 35, 38]) [2, 7, 8, 12, 15, 18, 23, 26, 27, 34, 35, 38]
P 2-oxopentanoic acid + NH_3 + H_2O_2
S D-serine + H_2O + O_2 <1-4, 6, 7> (Reversibility: ? <1-4, 6, 7> [7, 8, 15, 18, 21, 26-28, 22]) [7, 8, 15, 18, 21, 26-28, 22]
P 2-oxo-3-hydroxypropionic acid + NH_3 + H_2O_2
S D-threonine + H_2O + O_2 <2, 6, 7> (Reversibility: ? <2, 6, 7> [15, 26, 27]) [15, 26, 27]
P 2-oxo-3-hydroxybutyric acid + NH_3 + H_2O_2
S D-tryptophan + H_2O + O_2 <1, 2, 4> (Reversibility: ? <1, 2, 4> [4, 7, 13, 21, 26, 35, 39]) [4, 7, 13, 21, 26, 35, 39]
P indol-3-pyruvic acid + NH_3 + H_2O_2
S D-tyrosine + H_2O + O_2 <1, 2, 4, 6> (Reversibility: ? <1, 2, 4, 6> [4, 15, 18, 21, 39]) [4, 15, 18, 21, 39]
P 4-hydroxyphenylpyruvic acid + NH_3 + H_2O_2
S D-valine + H_2O + O_2 <1, 2, 4-7> (Reversibility: ? <1, 2, 4-7> [2, 4, 7, 14, 15, 21, 23, 26, 27, 35, 39, 22]) [2, 4, 7, 14, 15, 21, 23, 26, 27, 35, 39, 22]
P α-ketoisovaleric acid + NH_3 + H_2O_2
S cephalosporin C + H_2O + O_2 <2, 4> (Reversibility: ? <2, 4> [20, 31, 36, 37, 22]) [20, 31, 36, 37, 22]
P 7-(5-oxoadipoamido)cephalosporanic acid + NH_3 + H_2O_2
S ε-N-benzoyl-D-lysine + H_2O + O_2 <1> (Reversibility: ? <1> [4]) [4]
P 6-benzylamino-2-oxohexanoate + NH_3 + H_2O_2
S glycine + H_2O + O_2 <1, 6> (<5> not [14]) (Reversibility: ? <1, 6> [12, 15]) [12, 15]
P formic acid + NH_3 + H_2O_2

S phenylglycine + H_2O + O_2 <1, 2, 4> (Reversibility: ? <1, 2, 4> [21, 26, 39])
[21, 26, 39]

P benzoylformic acid + NH_3 + H_2O_2

S thiazolidine-2-carboxylic acid + H_2O + O_2 <1, 2> (Reversibility: ? <2> [7,
26]) [2, 7, 26]

P ? + NH_3 + H_2O_2

Substrates and products

S D-2-aminobutyrate + H_2O + O_2 <1> (Reversibility: ? <1> [2]) [2]

P 2-oxobutyrate + NH_3 + H_2O_2

S D-2-aminobutyrate + H_2O + O_2 <1> (Reversibility: ? <1> [23]) [23]

P 2-oxobutyrate + NH_3 + H_2O_2

S D-alanine + H_2O + O_2 <1-10> (Reversibility: ? <1-10> [2, 4, 7, 8, 14, 15,
18, 21, 23, 26, 27, 33, 34, 35, 38, 39, 22]) [2, 4, 7, 8, 14, 15, 18, 21, 23, 26, 27,
33, 34, 35, 38, 39, 22]

P pyruvic acid + NH_3 + H_2O_2

S D-α-aminoadipate + H_2O + O_2 <4> (Reversibility: ? <4> [21]) [21]

P 2-oxoadipate + NH_3 + H_2O_2

S D-arginine + H_2O + O_2 <2, 4> (Reversibility: ? <4> [21, 35]) [21, 35]

P 5-guanidino-2-oxopentanoic acid + NH_3 + H_2O_2

S D-asparagine + H_2O + O_2 <2, 6, 7> (Reversibility: ? <2, 6, 7> [15, 26, 27])
[15, 26, 27]

P 2-oxosuccinamic acid + NH_3 + H_2O_2

S D-aspartic acid + H_2O + O_2 <1> (<1> weak activity [4]) (Reversibility: ?
<1> [4]) [4]

P oxaloacetate + NH_3 + H_2O_2

S D-citrulline + H_2O + O_2 <4> (Reversibility: ? <4> [21]) [21]

P 2-oxo-5-ureidopentanoic acid + NH_3 + H_2O_2

S D-cysteine + H_2O + O_2 <2, 3> (Reversibility: ? <2, 3> [7, 8, 26]) [7, 8, 26]

P 2-oxo-3-thiopropionic acid + NH_3 + H_2O_2

S D-ethionine + H_2O + O_2 <4, 5> (Reversibility: ? <4, 5> [14, 21]) [14, 21]

P 4-ethylsulfanyl-2-oxobutyric acid + NH_3 + H_2O_2

S D-glutamic acid + H_2O + O_2 <6> (Reversibility: ? <6> [15]) [15]

P α-ketoglutarate + NH_3 + H_2O_2

S D-glutamine + H_2O + O_2 <2, 3, 6> (Reversibility: ? <2, 3, 6> [8, 15, 26])
[8, 15, 26]

P 2-oxoglutaramate + NH_3 + H_2O_2

S D-histidine + H_2O + O_2 <1, 3, 4, 6> (Reversibility: ? <1, 3, 4, 6> [2, 8, 13,
15, 21]) [2, 8, 13, 15, 21]

P 3-(1H-imidazol-4-yl)-2-oxopentanoate + NH_3 + H_2O_2

S D-isoleucine + H_2O + O_2 <1-4, 6> (Reversibility: ? <1-4, 6> [4, 8, 15, 21,
26]) [4, 8, 15, 21, 26]

P 3-methyl-2-oxopentanoic acid + NH_3 + H_2O_2

S D-leucine + H_2O + O_2 <1-7> (Reversibility: ? <1-7> [4, 7, 8, 13-15, 18, 21,
26, 27, 39]) [4, 7, 8, 13-15, 18, 21, 26, 27, 39]

P 4-methyl-2-oxopentanoic acid + NH_3 + H_2O_2

S D-methionine + H_2O + O_2 <1, 2, 4-7, 9, 10> (Reversibility: ? <1, 2, 4-7, 9, 10> [2, 4, 7, 13-15, 18, 21, 23, 26, 27, 34, 35, 38, 39, 22]) [2, 4, 7, 13-15, 18, 21, 23, 26, 27, 34, 35, 38, 39, 22]

P 4-methylthio-2-oxobutanoic acid + NH_3 + H_2O_2

S D-norleucine + H_2O + O_2 <1, 2, 5> (Reversibility: ? <1, 2, 5> [2, 7, 14, 23]) [2, 7, 14, 23]

P 2-oxohexanoate + NH_3 + H_2O_2

S D-norvaline + H_2O + O_2 <1, 2, 5> (Reversibility: ? <1, 2, 5> [2, 7, 14, 23, 35]) [2, 7, 14, 23, 35]

P 2-oxopentanoate + NH_3 + H_2O_2

S D-ornithine + H_2O + O_2 <1> (Reversibility: ? <1> [2]) [2]

P 5-amino-2-oxopentanoate + NH_3 + H_2O_2

S D-phenylalanine + H_2O + O_2 <1, 2, 4-7, 9, 10> (Reversibility: ? <1, 2, 4-7, 9, 10> [4, 7, 14, 15, 21, 26, 27, 34, 35, 38, 39, 22]) [4, 7, 14, 15, 21, 26, 27, 34, 35, 38, 39, 22]

P phenylpyruvic acid + NH_3 + H_2O_2

S D-proline + H_2O + O_2 <1-3, 6, 7, 9, 10> (Reversibility: ? <1-3, 6, 7, 9, 10> [2, 7, 8, 12, 15, 18, 23, 26, 27, 34, 35, 38]) [2, 7, 8, 12, 15, 18, 23, 26, 27, 34, 35, 38]

P 2-oxopentanoic acid + NH_3 + H_2O_2

S D-serine + H_2O + O_2 <1-4, 6, 7> (Reversibility: ? <1-4, 6, 7> [7, 8, 15, 18, 21, 26-28, 22]) [7, 8, 15, 18, 21, 26-28, 22]

P 2-oxo-3-hydroxypropionic acid + NH_3 + H_2O_2

S D-threonine + H_2O + O_2 <2, 6, 7> (Reversibility: ? <2, 6, 7> [15, 26, 27]) [15, 26, 27]

P 2-oxo-3-hydroxybutyric acid + NH_3 + H_2O_2

S D-tryptophan + H_2O + O_2 <1, 2, 4> (Reversibility: ? <1, 2, 4> [4, 7, 13, 21, 26, 35, 39]) [4, 7, 13, 21, 26, 35, 39]

P indol-3-pyruvic acid + NH_3 + H_2O_2

S D-tyrosine + H_2O + O_2 <1, 2, 4, 6> (Reversibility: ? <1, 2, 4, 6> [4, 15, 18, 21, 39]) [4, 15, 18, 21, 39]

P 4-hydroxyphenylpyruvic acid + NH_3 + H_2O_2

S D-valine + H_2O + O_2 <1, 2, 4-7> (Reversibility: ? <1, 2, 4-7> [2, 4, 7, 14, 15, 21, 23, 26, 27, 35, 39, 22]) [2, 4, 7, 14, 15, 21, 23, 26, 27, 35, 39, 22]

P α-ketoisovaleric acid + NH_3 + H_2O_2

S cephalosporin C + H_2O + O_2 <2, 4> (Reversibility: ? <2, 4> [20, 31, 36, 37, 22]) [20, 31, 36, 37, 22]

P 7-(5-oxoadipoamido)cephalosporanic acid + NH_3 + H_2O_2

S ε-N-benzoyl-D-lysine + H_2O + O_2 <1> (Reversibility: ? <1> [4]) [4]

P 6-benzylamino-2-oxohexanoate + NH_3 + H_2O_2

S glycine + H_2O + O_2 <1, 6> (<5> not [14]) (Reversibility: ? <1, 6> [12, 15]) [12, 15]

P formic acid + NH_3 + H_2O_2

S phenylglycine + H_2O + O_2 <1, 2, 4> (Reversibility: ? <1, 2, 4> [21, 26, 39]) [21, 26, 39]

P benzoylformic acid + NH_3 + H_2O_2

S thiazolidine-2-carboxylic acid + H_2O + O_2 <1, 2> (Reversibility: ? <2> [7, 26]) [2, 7, 26]

P ? + NH_3 + H_2O_2

Inhibitors

1,3-butadien 1-carboxylic acid <2> (<2> D-phenylglycine oxidation [7]) [7]
2-hydroxy carboxylic acids <1> [2]
2-hydroxybenzoic acid <2> (<2> D-phenylglycine oxidation [7]) [7]
2-hydroxybutyric acid <2> (<2> D-phenylglycine oxidation [7]) [7]
2-ketobutyric acid <2> (<2> D-phenylglycine oxidation [7]) [7]
2-oxo carboxylic acids <1> [2]
2-oxobutyrate <1> (<1> D-alanine oxidation [2]) [2]
3-bromobenzoic acid <2> (<2> D-phenylglycine oxidation [7]) [7]
3-chlorobenzoic acid <2> (<2> D-phenylglycine oxidation [7]) [7]
3-hydroxybenzoic acid <2> (<2> D-phenylglycine oxidation [7]) [7]
3-methylbenzoic acid <2> (<2> D-phenylglycine oxidation [7]) [7]
3-nitrobenzoic acid <2> (<2> D-phenylglycine oxidation [7]) [7]
4-hydroxybenzoic acid <2> (<2> D-phenylglycine oxidation [7]) [7]
5,5'-dithiobis-nitrobenzoate <4> [21]
5-methylpyrazole-3-carboxylate <1> [17]
5-methylthiophene-2-carboxylate <1> [17]
ADP <1> [17]
ADPribose <1> [17]
AMP <1> [17, 23]
Ag^+ <7> (<7> 1 mM [27]) [27]
D-2-hydroxy-3-methylvalerate <1> (<1> D-alanine oxidation [2]) [2]
D-lactate <1> (<1> slight, D-alanine oxidation [2]) [2]
Δ^1-piperidine 2-carboxylate <1> (D-lysine oxidation) [12]
DL-2-hydroxybutyrate <1> (<1> D-alanine oxidation [2]) [2]
DL-2-hydroxyoctanoate <1> (<1> D-alanine oxidation [2]) [2]
H_2O_2 <1> [17]
Hg^{2+} <5, 7> (<7> 1 mM [27]) [14, 27]
L-histidine <6> (<6> D-histidine oxidation [15]) [15]
L-leucine <1> (<1> D-alanine oxidation [2]) [2]
L-methionine <1> (<1> D-alanine oxidation [2]) [2]
L-norvaline <1> (<1> D-alanine oxidation [2]) [2]
NADH <1> [17]
NADPH <1> [17]
UMP <1> [23]
Zn^{2+} <5> [14]
acetaldehyde <1> [17]
acetylsalicylate <1> [17]
adenosine 5'-monophosphate <1> [12]
aniline <1> (<1> slight, D-alanine oxidation [2]) [2, 12]
anthranilate <4> [40]
benzoic acid <2, 4, 9> (<2> D-phenylglycine oxidation [7]; <4> not [21]; <4> competitive inhibitor [40]) [7, 21, 38, 40]

chloramphenicol <1> [12]
chlorpromazine <1> [12]
chlortetracycline <1> [12]
crotonic acid <2> (<2> D-phenylglycine oxidation [7]) [7]
cysteine <1> [17]
dCMP <1> [23]
dephospho-CoA <1> [17]
ethacrynic acid <1> [17]
flufenamic acid <1> [17]
formaldehyde <1> [17]
furosemide <1> [17]
glycyl-DL-norvaline <1> (<1> slight, D-alanine oxidation [2]) [2]
glycyl-DL-phenylalanine <1> (<1> slight, D-alanine oxidation [2]) [2]
glycyl-DL-tryptophan <1> (<1> slight, D-alanine oxidation [2]) [2]
glycyl-DL-valine <1> (<1> slight, D-alanine oxidation [2]) [2]
glyoxylate <4> [21]
histidyl-histidine <1> (<1> slight, D-alanine oxidation [2]) [2]
indomethacin <1> [17]
iodoacetamide <7> (<7> 1 mM [27]) [27]
kojic acid <2> (<2> D-phenylglycine oxidation [7]) [7]
malonate <9> [38]
mefenamic acid <1> [17]
mersalyl <1> [17]
nicotinate <1> [17]
nicotinic acid <2> (<2> D-phenylglycine oxidation [7]) [7]
p-aminosalicylic acid <1> [12]
p-chloromercuribenzoate <1, 4-7> (<7> 0.1 mM [27]) [1, 12-15, 27]
penicillamine <1> [17]
penicillin <1> [12]
phenylbutazone <1> [17]
phenylmercuriacetate <4> [21]
picolinic acid <2> (<2> D-phenylglycine oxidation [7]) [7]
pyrrole-2-carboxylate <1> [17]
pyrrolidine <1> (<1> slight, D-alanine oxidation [2]) [2]
pyruvate <1> (<1> slight, D-alanine oxidation [2]) [2]
riboflavin 5'-monophosphate <1> [12]
salicylate <1> [17]
straight-chain fatty acids <1> [2]
streptomycin <1> [12]
succinate semialdehyde <1> [17]
trans-2-pentenoic acid <2> (<2> D-phenylglycine oxidation [7]) [7]
tropolone <1> [17]

Cofactors/prosthetic groups

FAD <1-4, 6-8> (<1-4, 6, 8> flavoprotein [1, 6-8, 11, 13, 15, 21, 23, 25, 26, 28, 33, 36]; <1, 2, 4> 1 FAD per monomer of MW 38000-39000 Da [6, 25, 30, 40]; <2, 4> tightly bound [13, 26]) [1, 6-8, 11, 13, 15, 21, 23, 25-33, 36, 40]

Metals, ions

Fe <4> (<4> 2 identical subunits, each carrying probably 1 molecule of iron [20]) [20]

Turnover number (min^{-1})

0.01 <4> (D-alanine, <4> mutant D206A [30]) [30]
0.06 <4> (D-alanine, <4> mutant D206S [30]) [30]
0.14 <4> (D-alanine, <4> mutant D206E [30]) [30]
0.34 <4> (D-alanine, <4> wild-type [30]) [30]
3 <2> (D-alanine, <2> mutant R285A [29]) [29, 36]
48 <2> (D-alanine, <2> mutant R285K [29]) [29, 36]
54 <1> (D-valine, <1> mutant T317A [28]) [28]
55.2 <1> (D-alanine, <1> mutant T317A [28]) [28]
90 <8> (D-alanine, <8> mutant G281C [33]) [33]
108 <1> (D-serine, <1> mutant T317A [28]) [28]
144 <1> (D-phenylalanine, <1> mutant T317A [28]) [28]
150 <1> (D-valine, <1> wild-type [28]) [28]
156 <8> (D-alanine, <8> FAD-S-mutant enzyme [33]) [33]
162 <1> (D-serine, <1> wild-type [28]) [28]
252 <2> (D-alanine, <2> mutant Y223S [36]) [36]
300 <1> (D-alanine, <1> mutant Y228F [34]) [34]
360 <2> (D-valine, <2> D isotope [32]) [32]
426 <1> (D-phenylalanine, <1> mutant Y228F [34]) [34]
440 <4> (D-alanine, <4> D isotope [32]) [32]
510 <1> (D-alanine, <1> wild-type [28]) [28]
520 <4> (D-valine, <4> D isotope [32]) [32]
582 <1> (D-proline, <1> mutant T317A [28]) [28]
600 <1, 8> (D-alanine, <8> wild-type [33]) [33, 36]
696 <1> (D-alanine, <1> mutant Y224F [34]) [34]
702 <1> (D-proline, <1> mutant Y228F [34]) [34]
762 <1> (D-alanine, <1> wild-type [34]) [34]
1020 <1> (D-phenylalanine, <1> wild-type [28]) [28]
1260 <1> (D-proline, <1> wild-type [28]) [28]
1730 <2> (D-valine) [32]
2500 <4> (D-valine) [32]
2850 <1> (D-methionine, <1> mutant Y228F [34]) [34]
2928 <1> (D-phenylalanine, <1> mutant Y224F [34]) [34]
3150 <4> (D-alanine) [32]
3354 <1> (D-phenylalanine, <1> wild-type [34]) [34]
4250 <2> (D-alanine, <2> D isotope [32]) [32]
5022 <1> (D-proline, <1> mutant Y224F [34]) [34]
5232 <1> (D-proline, <1> wild-type [34]) [34]
5520 <1> (D-methionine, <1> mutant Y224F [34]) [34]
6000 <4> (cephalosporin C) [40]
8580 <1> (D-methionine, <1> wild-type [34]) [34]
12600 <2> (D-alanine, <2> mutant Y223F [36]) [36]
18000 <4> (D-alanine) [40]

20700 <2> (D-alanine) [32, 36]
21000 <2> (D-alanine, <2> wild-type [29]) [29, 36]
31370 <2> (D-valine) [7]
43250 <2> (D-alanine) [7]
44200 <2> (cephalosporin C) [36]

Specific activity (U/mg)

0.15 <1> [34]
0.5 <5> [14]
1.04 <4> (<4> D206A [30]) [30]
5 <3> [8]
6.41 <4> (<4> D206S [30]) [30]
8.8 <1> [12]
16.71 <4> (<4> D206E [30]) [30]
18.61 <4> (<4> wild-type [30]) [30]
57.8 <2> [26]
86.6 <4> [40]
139 <7> [27]
175 <2> [6]
244 <6> [15]

K_m-Value (mM)

0.00045 <1> (FAD) [23]
0.015 <1> (2,6-dichlorophenolindophenol, <1> with D-alanine [2,23]) [2, 23]
0.06 <2> (O_2, <2> R285K [29]) [29]
0.06 <4> (phenylalanine) [21]
0.063 <1> (D-norvaline) [2]
0.067 <1> (methylene blue, <1> with D-alanine [2,23]) [2, 23]
0.08 <4> (D-leucine) [21]
0.08 <2> (O_2, <2> R285A [29]) [29]
0.17 <4> (D-tryptophan) [21]
0.18 <1, 2> (O_2, <2> with D-methionine, [7]; <1> with D-alanine [2]) [2, 7, 23]
0.2 <4> (D-phenylalanine) [22]
0.23 <1> (D-leucine) [39]
0.24 <5> (D-methionine) [14]
0.27 <4> (D-methionine) [21]
0.28 <4, 5> (D-leucine) [13, 14]
0.29 <4> (D-methionine) [22]
0.29 <2> (D-phenylalanine) [7]
0.3 <2> (D-tryptophan) [7]
0.38 <4> (D-phenylglycine) [22]
0.39 <2> (D-norleucine) [7]
0.4 <1> (D-norleucine) [2]
0.46 <1> (D-2-aminobutyrate) [2]
0.5 <2> (D-leucine) [7]
0.67 <1> (D-valine, <1> wild-type [28]) [28]
0.7 <1> (D-tryptophan, <1> immobilized enzyme [16]) [16]

0.76 <4> (D-methionine) [20]
0.81 <1> (D-methionine) [39]
0.83 <2> (D-alanine) [35]
0.83 <1, 2> (D-methionine) [2, 7]
0.83 <4> (cephalosporin C) [22]
0.89 <2> (D-norvaline) [7]
0.95 <4> (O_2, <4> cephalosporin C [40]) [40]
1 <1> (D-alanine, <1> T317A [28]) [28]
1 <8> (D-alanine, <8> wild-type [33]) [33]
1 <4> (ethionine) [21]
1.02 <4> (O_2, <4> D-alanine [40]) [40]
1.1 <1> (D-alanine, <1> wild-type [28]) [28]
1.1 <1> (D-proline, <1> wild-type [28]) [28]
1.1 <1> (O_2) [17]
1.2 <1> (D-proline, <1> T317A [28]) [28]
1.2 <1, 4> (D-tryptophan, <1> free enzyme [16]) [16, 21]
1.29 <2> (D-alanine, <2> 38.3 kDa form [35]) [35]
1.4 <1> (D-phenylalanine, <1> wild-type [28]) [28]
1.4 <1> (D-tryptophan) [39]
1.4 <1> (D-valine) [2, 39]
1.5 <1> (D-phenylalanine) [39]
1.7 <1> (D-proline) [2]
1.8 <1> (D-alanine) [2]
1.99 <1> (thiazolidine-2-carboxylate) [2]
2.3 <1> (D-alanine, <1> free enzyme [16]; <1> wild-type [34]) [16, 34]
2.3 <2> (O_2, <2> wild-type [29]) [29]
2.6 <2> (D-alanine, <2> wild-type [29]) [29]
2.8 <1> (D-alanine, <1> Y224F [34]) [34]
3 <4> (D-valine) [22]
3.5 <1> (D-valine, <1> free enzyme [16]) [16]
3.95 <7> (D-valine) [27]
4.28 <7> (D-alanine) [27]
4.3 <1> (D-valine, <1> T317A [28]) [28]
4.63 <4> (D-alanine, <4> wild-type [30]) [30]
4.8 <1> (serine, <1> wild-type [28]) [28]
5.8 <1> (tyrosine) [39]
6.5 <4> (D-alanine) [22]
6.7 <4> (citrulline) [21]
7 <1> (D-phenylglycine) [39]
7.5 <1> (D-phenylalanine, <1> Y224F [34]) [34]
8 <8> (D-alanine, <8> G281C [33]) [33]
8 <4> (arginine) [21]
9.1 <1> (D-phenylalanine, <1> T317A [28]) [28]
9.1 <1> (D-proline, <1> Y224F [34]) [34]
9.36 <4> (D-alanine, <4> D206E [30]) [30]
9.5 <1> (D-proline, <1> wild-type [34]) [34]
10 <4> (D-phenylalanine) [20]

10.9 <7> (D-leucine) [27]
12.2 <1> (D-proline, <1> Y228F [34]) [34]
13 <8> (D-alanine, <8> FAD-S-mutant enzyme [33]) [33]
13 <4> (cephalosporin C) [20]
13.75 <2> (D-serine) [7]
15.5 <2> (D-proline, <2> 38.3 kDa form [35]) [35]
15.6 <1> (D-methionine, <1> Y224F [34]) [34]
17.3 <4> (D-serine) [22]
18 <1> (D-alanine) [39]
18.12 <4> (D-alanine, <4> D206A [30]) [30]
18.3 <1> (D-phenylalanine, <1> Y228F [34]) [34]
18.89 <2> (D-valine) [7]
20 <4> (D-valine) [21]
20.52 <4> (D-alanine, <4> D206S [30]) [30]
20.8 <1> (D-serine, <1> T317A [28]) [28]
21.41 <2> (D-cysteine) [7]
21.47 <2> (D-proline) [7]
21.5 <2> (D-proline) [35]
23 <4> (cephalosporin C) [40]
25 <4> (D-serine) [21]
27 <1> (D-alanine, <1> Y228F [34]) [34]
27.4 <3> (D-alanine) [8]
27.4 <7> (D-methionine) [27]
33.7 <7> (D-serine) [27]
43 <4> (D-alanine) [40]
50 <1> (D-methionine, <1> Y228F [34]) [34]
55.9 <1> (D-phenylalanine, <1> wild-type [34]) [34]
76 <4> (D-alanine) [20]
143 <1> (D-methionine, <1> wild-type [34]) [34]
310 <2> (D-alanine, <2> R285A [29]) [29]
800 <2> (D-alanine, <2> R285K [29]) [29]

pH-Optimum

7.5 <4> [40]
7.5-8 <4> [22]
8-8.5 <2> [7]
8-9 <1> (<1> D-alanine, immobilized enzyme [4]) [4]
8.5 <3> [8]
8.5-9.5 <1> (<1> D-alanine, free enzyme [4]) [4]
9-9.2 <5> (<5> D-methionine [14]) [14]

pH-Range

5-7 <4> [40]
6.8-8.5 <2, 4> [31]
7.3-10.3 <1> (<1> less than 50% of maximal activity above and below, D-alanine, immobilized enzyme [4]) [4]
7.7-10.5 <1> (<1> less than 50% of maximal activity above and below, D-alanine, free enzyme [4]) [4]

Temperature optimum (°C)

20-30 <5> [14]
40-45 <2> [7]
45 <1> (<1> D-alanine, free and immobilized enzyme [4]) [4]
55 <4> [40]

Temperature range (°C)

27-60 <1> (<1> less than 50% of maximal activity above and below, D-alanine [4]) [4]
30-50 <2> (<2> less than 50% of maximal activity above and below [7]) [7]

4 Enzyme Structure

Molecular weight

38000 <1> (<1> sedimentation equilibrium [25]) [25]
39000 <3> (<3> gel filtration) [8]
43000 <7> (<7> gel filtration [27]) [27]
79000 <2> (<2> gel filtration [26]) [26, 35, 36]
86000 <4> (<4> disc gel electrophoresis [20]) [20]
112000-115000 <1> (<1> sedimentation and diffusion data [12]) [12]
118000-125000 <5> (<5> PAGE, gel filtration [14]) [14]
170000 <4> (<4> PAGE [22]) [22]
320000 <4> (<4> PAGE [22]) [22]
570000 <4> (<4> gel filtration [22]) [22]

Subunits

? <2, 4> (<2> x * 38200, SDS-PAGE [6]; <4> x * 39000, SDS-PAGE [21]) [6, 21]
dimer <2, 4> (<4> α_2, 2 * 40000, SDS-PAGE [20,36,40]; <2> 2 * 37000, SDS-PAGE [26]) [20, 26, 30, 34, 36, 40]
monomer <1, 3, 7> (<3> 1 * 39000, SDS-PAGE [8,36]; <1> 1 * 38000, SDS-PAGE [25]; <7> 1 * 41000, SDS-PAGE [27]) [8, 25, 27, 36]

Posttranslational modification

glycoprotein <4> (<4> contains 7% w/v covalently bound carbohydrate [21]) [21]
Additional information <4> (<4> non-glycosylated [40]) [40]

5 Isolation/Preparation/Mutation/Application

Source/tissue

body wall <10> [38]
crop <10> [38]
digestive gland <9, 10> [38]
intestine <10> [38]
kidney <1, 8, 10, 11> [1-5, 9, 12, 16, 17, 19, 23, 25, 28, 33, 34, 36, 38]

liver <8> [36]
mycelium <5> [14]
ovotestis <10> [38]
salivary gland <10> [38]
stomach <10> [38]

Localization

mitochondrion <5> (<5> matrix [14]) [14]
peroxisome <3, 7, 9-11> [8, 27, 38]

Purification

<1> (large scale purification of apo-D-aminoacid oxidase [5]; affinity chromatography on Cibacron Blue Sepharose [24]) [5, 12, 24, 25, 34]
<2> [6, 26]
<3> [8]
<4> [13, 20, 21, 40]
<5> (partial [14]) [14]
<6> (partial [15]) [15]
<8> [33]
<9, 10> [38]

Renaturation

<1> (renaturation studies of free and immobilized enzyme [19]) [19]

Crystallization

<1> [12]

Cloning

<1> (expression in Escherichia coli [9]) [9, 34]
<2> (expression in Escherichia coli [29]) [29, 36]
<4> (expression in Escherichia coli [30]) [30]
<4> (expression in Kluyveromyces lactis [36]) [36]
<4> (expression in Saccharomyces cerevisiae [36]) [36]
<7> (overexpression in Candida boidinii strain aod1 Δ [27]) [27]
<8> (expression in Escherichia coli [33]) [33]

Engineering

D206A <4> (<4> decreased activity with D-amino acids [30]) [30]
D206E <4> (<4> decreased activity with D-amino acids [30]) [30]
D206G <4> (<4> no activity with D-amino acids [30]) [30]
D206L <4> (<4> no activity with D-amino acids [30]) [30]
D206N <4> (<4> no activity with D-amino acids [30]) [30]
D206S <4> (<4> decreased activity with D-amino acids [30]) [30]
G281C <8> (<8> decreased activity with D-amino acids [33]) [33]
G313A <1> (<1> decreased activities to various D-amino acids [28]) [28]
R285A <2> (<2> decreased activity with D-amino acids [29]) [29]
R285D <2> (<2> decreased activity with D-amino acids [29]) [29]
R285K <2> (<2> decreased activity with D-amino acids [29]) [29]
R285Q <2> (<2> decreased activity with D-amino acids [29]) [29]
T317A <1> (<1> decreased activity to FAD [28]) [28]

Y223F <4> (<4> slower substrate binding than the wild-type [36]) [36]
Y223F <4> (<4> slower substrate binding than the wild-type [36]) [36]
Y224F <1> (<1> turnover numbers similar to wild-type [34]) [34]
Y228F <1> (<1> turnover numbers lower than turnover numbers of wild-type [34]) [34]

Application

medicine <2, 4> (<2,4> human gene therapy [31,36]) [31, 36]
synthesis <2, 4> (<2,4> oxidation of cephalosporin C in the two-step formation of 7-aminocephalosporanic acid [31,36]) [31, 36]

6 Stability

pH-Stability

7.3-7.9 <2> (<2> 30 min stable [7]) [7]
8 <2> (<2> unstable above [26]) [26]

Temperature stability

4 <4> (<4> 24 h, 50% loss of activity [21]) [21]
45 <6> (<6> 10 min, 35% loss of activity [15]) [15]
50 <6> (<6> 10 min, 57% loss of activity [15]) [15]
52 <4, 6> (<4> pH 5.2, addition of benzoic acid or m-toluic acid, 7 min stable [13]; <6> 10 min, 76% loss of activity [15]) [13, 15]
60 <1, 6> (<1> 10 min, about 55% loss of activity, free and immobilized enzyme [4]; <6> 10 min, 93% loss of activity [15]) [4, 15]
65 <6> (<6> 10 min, complete loss of activity [15]) [15]

General stability information

<1>, chymotrypsin, 10% w/w, no effect [10]
<1>, trypsin, 10% w/w, inactivation and proteolysis under nondenaturing conditions [10]
<2>, cetylpyridinium bromide stabilizes [26]
<2>, changes in ionic strength, I: 16-82 mM, have only moderate effect on enzyme stability, stability maximum is 50 mM [7]
<2>, glycerol, 10%, stabilizes [7, 26]
<3>, glycerol, 50% stabilizes [8]
<4>, benzoic acid stabilizes [13]
<4>, m-toluic acid stabilizes [13]
<4>, stable to guanidine [40]
<2, 4>, 2-mercaptoethanol stabilizes [7, 40]

Storage stability

<1>, 4°C, pH 6.5, 3.2 M ammonium sulfate, crystallized, several months [23]
<2>, -20°C, several months [29]
<2>, -80°C, 2-3 months, stable [26]
<4>, frozen, 50% loss of activity after 7 days [21]

References

[1] Dixon, M.; Kleppe, K.: D-Amino acid oxidase I. Dissociation and recombination of the holoenzyme. Biochim. Biophys. Acta, **96**, 357-367 (1965)

[2] Dixon, M.; Kleppe, K.: D-Amino acid oxidase II. specificity, competitive inhibition and reaction sequence. Biochim. Biophys. Acta, **96**, 368-382 (1965)

[3] Dixon, M.; Kleppe, K.: D-Amino acid oxidase II. Effect of pH. Biochim. Biophys. Acta, **96**, 383-389 (1965)

[4] Tosa, T.; Sano, R.; Chibata, I.: Immobilized D-amino acid oxidase. Preparation, some enzymatic properties, and potential uses. Agric. Biol. Chem., **38**, 1529-1534 (1974)

[5] Harbron, S.; Fisher, M.; Rabin, B.R.: Large scale preparation and purification of apo-D-aminoacid oxidase for use in novel amplification assays. Biotechnol. Tech., **6**, 55-60 (1992)

[6] Simonetta, M.P.; Pollegioni, L.; Casalin, P.; Curti, B.; Ronchi, S.: Properties of D-amino-acid oxidase from Rhodotorula gracilis. Eur. J. Biochem., **180**, 199-204 (1989)

[7] Pollegioni, L.; Falbo, A.; Pilone, M.S.: Specificity and kinetics of Rhodotorula gracilis D-amino acid oxidase. Biochim. Biophys. Acta, **1120**, 11-16 (1992)

[8] Yoshizawa, M.; Ueda, M.; Mozaffar, S.; Tanaka, A.: Some properties of peroxisome-associated D-amino acid oxidase from Candida tropicalis. Agric. Biol. Chem., **50**, 2637-2638 (1986)

[9] Watanabe, F.; Fukui, K.; Momoi, K.; Miyake, Y.: Expression of normal and abnormal porcine kidney D-amino acid oxidase in Escherichia coli: purification and characterization of the enzymes. Biochem. Biophys. Res. Commun., **165**, 1422-1427 (1989)

[10] Tarelli, G.T.; Vanoni, M.A.; Negri, A.; Curti, B.: Characterization of a fully active N-terminal 37-kDa polypeptide obtained by limited tryptic cleavage of pig kidney D-amino acid oxidase. J. Biol. Chem., **265**, 21242-21246 (1990)

[11] Bright, H.J.; Porter, D.J.T.: Flavoprotein oxidases. The Enzymes, 3rd Ed. (Boyer, P.D., ed.), **12**, 421-505 (1975)

[12] Yagi, K.: D-Amino acid oxidase and its complexes (hog kidney). Methods Enzymol., **17B**, 608-622 (1971)

[13] Berg, C.P.; Rodden, F.A.: Purification of D-amino oxidase from Trigonopsis variabilis. Anal. Biochem., **71**, 214-222 (1976)

[14] Rosenfeld, M.G.; Leiter, E.H.: Isolation and characterization of a mitochondrial D-amino acid oxidase from Neurospora crassa. Can. J. Biochem., **55**, 66-74 (1977)

[15] Pistorius, E.K.; Voss, H.: A D-amino acid oxidase from Chlorella vulgaris. Biochim. Biophys. Acta, **481**, 395-406 (1977)

[16] Parkin, K.; Hultin, H.O.: Immobilization and characterization of D-amino acid oxidase. Biotechnol. Bioeng., **21**, 939-953 (1979)

[17] Hamilton, G.A.; Buckthal, D.J.: The inhibition of mammalian D-amino acid oxidase by metabolites and drugs. Inferences concerning physiological function. Bioorg. Chem., **11**, 350-370 (1982)

[18] Simonetta, M.P.; Vanoni, M.A.; Curti, B.: D-Amino acid oxidase activity in the yeast Rhodotorula gracilis. FEMS Microbiol. Lett., 15, 27-31 (1982)

[19] Carrera, G.; Pasta, P.; Curti, B.: Renaturation studies of free and immobilized D-amino-acid oxidase. Biochim. Biophys. Acta, 745, 181-188 (1983)

[20] Szwajcer, E.; Mosbach, K.: Isolation and partial characterization of a D-amino acid oxidase active against cephalosporin. c from the yeast Trigonopsis variabilis. Biotechnol. Lett., 7, 1-7 (1985)

[21] Kubicek-Pranz, E.M.; Röhr, M.: D-Amino acid oxidase from the yeast Trigonopsis variabilis. J. Appl. Biochem., 7, 104-113 (1985)

[22] Schraeder, T.; Andreesen, J.R.: Properties and chemical modification of D-amino acid oxidase from Trigonopsis variabilis. Arch. Microbiol., 165, 41-47 (1996)

[23] Bergmeyer, H.U.: D-Aminosäure-Oxydase. Methods Enzym. Anal., 3rd Ed. (Bergmeyer, H.U., ed.), 1, 460-461 (1974)

[24] Leonil, J.; Langrene, S.; Sicsic, S.; Le Goffic, F.: Purification of D-amino acid oxidase apoenzyme by affinity chromatography on Cibacron Blue Sepharose. J. Chromatogr., 347, 316-319 (1985)

[25] Tu, S.C.; Edelstein, S.J.; McCormick, D.B.: A modified purification method and properties of pure porcine D-amino acid oxidase. Arch. Biochem. Biophys., 159, 889-896 (1973)

[26] Simonetta, M.P.; Vanoni, M.A.; Casalin, P.: Purification and properties of D-amino-acid oxidase, an inducible flavoenzyme from Rhodotorula gracilis. Biochim. Biophys. Acta, 914, 136-142 (1987)

[27] Yurimoto, H.; Hasegawa, T.; Sakai, Y.; Kato, N.: Characterization and high-level production of D-amino acid oxidase in Candida boidinii. Biosci. Biotechnol. Biochem., 65, 627-633 (2001)

[28] Setoyama, C.; Nishina,Y.; Tamaoki, H.; Mizutani, H.; Miyahara, I.; Hirotsu, K.; Shiga, K.; Miura, R.: Effects of hydrogen bond in association with flavin and substrate in flavoenzyme D-amino acid oxidase. The catalytic and structural roles of Gly313 and Thr317. J. Biochem., 131, 59-69 (2002)

[29] Molla, G.; Porrini, D.; Job, V.; Motteran, L.; Vegezzi, C.; Campaner, S.; Pilone, M.S.; Pollegioni, L.: Role of arginine 285 in the active site of Rhodotorula gracilis D-amino acid oxidase. A site-directed mutagenesis study. J. Biol. Chem., 275, 24715-24721 (2000)

[30] Ju, S.S.; Lin, L.L.; Wang, W.C.; Hsu, W.H.: A conserved aspartate is essential for FAD binding and catalysis in the D-amino acid oxidase from Trigonopsis variabilis. FEBS Lett., 436, 119-122 (1998)

[31] Pollegioni, L.; Porrini, D.; Molla, G.; Pilone, M.S.: Redox potentials and their pH dependence of D-amino-acid oxidase of Rhodotorula gracilis and Trigonopsis variabilis. Eur. J. Biochem., 267, 6624-6632 (2000)

[32] Pollegioni, L.; Langkau, B.; Tischer, W.; Ghisla, S.; Pilone, M.S.: Kinetic mechanism of D-amino acid oxidases from Rhodotorula gracilis and Trigonopsis variabilis. J. Biol. Chem., 268, 13850-13857 (1993)

[33] Raibekas, A.A.; Fukui, K.; Massey, V.: Design and properties of human D-amino acid oxidase with covalently attached flavin. Proc. Natl. Acad. Sci. USA, 97, 3089-3093 (2000)

[34] Pollegioni, L.; Fukui, K.; Massey, V.: Studies on the kinetic mechanism of pig kidney D-amino acid oxidase by site-directed mutagenesis of tyrosine 224 and tyrosine 228. J. Biol. Chem., **269**, 31666-31673 (1994)

[35] Pollegioni, L.; Ceciliani, F.; Curti, B.; Ronchi, S.; Pilone, M.S.: Studies on the structural and functional aspects of Rhodotorula gracilis D-amino acid oxidase by limited trypsinolysis. Biochem. J., **310**, 577-583. (1995)

[36] Pilone, M.S.: D-Amino acid oxidase: new findings. Cell. Mol. Life Sci., **57**, 1732-1747 (2000)

[37] Vikartovska-Welwardova, A.; Michalkova, E.; Gemeiner, P.; Welward, L.: Stabilization of D-amino-acid oxidase from Trigonopsis variabilis by manganese dioxide. Folia Microbiol., **44**, 380-384 (1999)

[38] Parveen, Z.; Large, A.; Grewal, N.; Lata, N.; Cancio, I.; Cajaraville, M.P.; Perry, C.J.; Connock, M.J.: D-Aspartate oxidase and D-amino acid oxidase are localized in the peroxisomes of terrestrial gastropods. Eur. J. Cell Biol., **80**, 651-660 (2001)

[39] Moreno, J.A.; Montes, F.J.; Catalan, J.; Galan, M.A.: Inhibition of D-amino acid oxidase by α-keto acids analogs of amino acids. Enzyme Microb. Technol., **18**, 379-382. (1996)

[40] Pollegioni, L.; Buto, S.; Tischer, W.; Ghisla, S.; Pilone, M.S.: Characterization of D-amino acid oxidase from Trigonopsis variabilis. Biochem. Mol. Biol. Int., **31**, 709-717 (1993)

Amine oxidase (flavin-containing)   1.4.3.4

1 Nomenclature

EC number
1.4.3.4

Systematic name
amine:oxygen oxidoreductase (deaminating) (flavin-containing)

Recommended name
amine oxidase (flavin-containing)

Synonyms
MAO
epinephrine oxidase
monoamine oxidase
monoamine:O_2 oxidoreductase (deaminating)
serotonin deaminase
tyraminase
adrenaline oxidase
tyramine oxidase
Additional information <3, 4, 5> (<3, 4, 5> MAO-A and MAO-B are 2 different isoenzymic forms which differ in substrate specificity, distribution among tissues and their structures [15, 21, 23, 27]; <3, 4, 5> isoforms differ in inhibitor sensitivity [21, 23, 25, 27]; <4> the N-terminal region of the two isoenzymes is not involved in the different specificity of the two isoenzymes for substrates and inhibitors [35]; <4> MAO is a novel type of disulfide oxidoreductase [6]) [6, 15, 21, 23, 27, 33, 35]

CAS registry number
9001-66-5

2 Source Organism

<1> *Aspergillus niger* (two amine oxidases isolated: the first one corresponds to the copper-dependent enzyme, the second one is a evolutionary prototype form of MAO-A and B [37]) [14, 37]
<2> *Bos taurus* (MAO-B [41]) [1, 4, 8, 16, 24, 26, 29, 31, 33, 34, 36]
<3> *Sus scrofa* [2, 10, 12, 23, 30]
<4> *Homo sapiens* [3-5, 6, 8, 13, 17-19, 21, 25, 28, 32, 35, 39, 41]
<5> *Rattus norvegicus* (MAO-A [9]) [7, 9, 11, 12, 15, 18, 20, 22, 27, 38, 40]

3 Reaction and Specificity

Catalyzed reaction

$RCH_2NH_2 + H_2O + O_2 = RCHO + NH_3 + H_2O_2$ (<2,4> mechanism of monoamine oxidase A and B [8]; <4> mechanism of monoamine oxidase A [28]; <1, 3, 5> ping-pong kinetic mechanism [12, 14]; <2, 4> mechanism [28, 29])

Reaction type

deamination
oxidation
redox reaction
reduction

Natural substrates and products

S $RCH_2NH_2 + H_2O + O_2$ <2, 4, 5> (<2, 5> responsible for the catabolism of various biogenic amine neurotransmitters as well as for the metabolism of certain exogenous amines [4, 27, 36]; <2, 4> important functions in the metabolism of biogenic amines in the central nervous system and peripheral tissues [21, 26, 41]) (Reversibility: ? <4, 5> [9, 21, 25-28, 36, 41]) [21, 25-28, 36, 41]

P $RCHO + NH_3 + H_2O_2$ <2, 4, 5> [21, 26, 27]

Substrates and products

S 1,4-benzyl-1-cyclopropyl-1,2,3,6-tetrahydropyridine <2> (Reversibility: ? <2> [36]) [36]

P ?

S 1-methyl-4-phenyl-1,2,3,6-tetrahydropyridine + H_2O + O_2 <2, 4> (Reversibility: <2, 4> [8, 28]) [8, 28]

P ?

S 2-phenylethylamine + H_2O + O_2 <1-5> (<2, 4, 5> MAO-B selective substrate [19, 27, 32, 33]) (Reversibility: ? <1-5> [5, 6, 8, 9, 15, 19, 20, 22, 23, 25, 27, 29, 30, 32, 33, 37]) [5, 6, 8, 9, 15, 19, 20, 22, 23, 25, 27, 29, 30, 32, 33, 37]

P 2-phenylethanal + NH_3 + H_2O_2 <1-5> [5, 6, 8, 9, 15, 19, 20, 22, 23, 25, 29, 30, 32, 33, 37]

S 3-phenylpropylamine + H_2O + O_2 <3> (Reversibility: ? <3> [30]) [30]

P 3-phenylpropanal + NH_3 + H_2O_2 <3> [30]

S $RCH_2NH_2 + H_2O + O_2$ <1-5> (<2> oxidizes primary, secondary and tertiary amines [1, 9, 27]; <2, 3> α-methylbenzylamine [26, 30]; <3> o-, m-, p-chlorobenzylamine [30]; <2, 4, 5> 5-hydroxytryptamine, MAO-A selective substrate [1, 15, 18, 19, 25, 32, 33, 35]; <1,5> norepinephrine [22, 37]; <5> octopamine [22]; <1> no activity with octopamine [37]; <3> amylamine [2, 10, 30]; <3> isoamylamine [10]; <3> 3-hydroxytyramine [10]; <3,5> phenylethylhydrazine [10, 12]; <1> n-butylamine [14]; <2> benzylhydrazine [26]; <5> 4-hydroxy-3-methoxy-2-phenylethylamine, nonselective substrate, 4-methoxy-2-phenylethylamine, 5-methoxytryptamine [15]; <2, 5> specificity [1, 15, 26, 27]; <1, 2, 3> no activity with histamine [1, 2, 10, 37]; <3> no activity with spermine [10]; <3> no activity with spermidine [2, 10]) (Reversibility: r <3, 5> [12]; ? <1, 2-5> [1, 5, 6, 8-11, 13-23, 25-30, 32, 33, 35-37, 40]) [1, 5, 6, 8-23, 25-30, 32, 33, 35-37, 40]

P RCHO + NH_3 + H_2O_2 <2, 3> [1, 2, 5, 12, 14, 26, 27, 37]

S benzylamine + H_2O + O_2 <1-5> (<2, 5> MAO-B selective substrate [27, 33]; <4> analogs [6]) (Reversibility: ? <1-5> [6, 10, 12-15, 17, 18, 22, 23, 25, 26, 29, 30, 32, 33, 37]) [6, 10, 12-15, 17, 18, 22, 23, 25-27, 29, 30, 32, 33, 37]

P benzaldehyde + NH_3 + H_2O_2 <1-5> [6, 10, 12-15, 17, 18, 22, 23, 25-27, 29, 30, 32, 33, 37]

S dopamine + H_2O + O_2 <1, 4, 5> (<2> nonselective substrate [33]) (Reversibility: ? <1, 4, 5> [5, 6, 13, 15, 19, 22, 37]) [5, 6, 13, 15, 19, 22, 37]

P (3,4-dihydroxyphenyl)acetaldehyde + NH_3 + H_2O_2 <1, 4, 5> [5, 6, 13, 15, 19, 22, 37]

S ethylamine + H_2O + O_2 <1> (Reversibility: ? <1> [14]) [14]

P ethanal + NH_3 + H_2O_2 <1> [14]

S kynuramine + H_2O + O_2 <2, 3, 4> (<5> nonselective substrate [27]) (Reversibility: ? <2, 3, 4> [1, 6, 8, 10, 21, 27, 28, 40]) [1, 6, 8, 10, 21, 27, 28, 40]

P 3-(2-aminophenyl)-3-oxopropanal + NH_3 + H_2O_2 <2, 3, 4> [1, 6, 8, 10, 21, 27, 28, 40]

S serotonin + H_2O + O_2 <1, 3, 5> (Reversibility: ? <1, 3, 5> [5, 9, 10, 11, 20, 22, 23, 37, 40]) [5, 9, 10, 11, 20, 22, 23, 37, 40]

P (5-hydroxy-1H-indol-3-yl)acetaldehyde + NH_3 + H_2O_2 <1, 3, 5> [5, 9, 10, 11, 20, 22, 23, 37, 40]

S tryptamine + H_2O + O_2 <2, 3, 4> (Reversibility: ? <2, 3, 4> [1, 10, 13, 15, 22, 33]) [1, 10, 13, 15, 22, 33]

P 1H-indol-3-yl-acetaldehyde + NH_3 + H_2O_2 <2, 3, 4> [1, 10, 13, 15, 22, 33]

S tyramine + H_2O + O_2 <2-5> (<2,4,5> nonselective substrate [19,27,33]) (Reversibility: ? <2-5> [1, 9, 10, 11, 13, 15, 18-20, 22, 25, 27, 33, 40]) [1, 9, 10, 11, 13, 15, 18-20, 22, 25, 27, 33, 40]

P (4-hydroxyphenyl)acetaldehyde + NH_3 + H_2O_2 <2-5> [1, 9, 10, 11, 13, 15, 18-20, 22, 25, 27, 33, 40]

Inhibitors

(aminomethyl)trimethylsilane <-1> [4]

1,10 phenanthroline <2> (<2> weak inhibition [1]) [1]

1-O-n-octyl-β-D-glucopyranoside <4> (<4> at concentrations well below the critical micelle concentration [19]) [19]

2,2'-dipyridyl disulfide <4> [21]

2-hydroxyquinoline <2> (<2> reversible by dialysis [1]) [1]

4-benzyl-1-cyclopropyl-1,2,3,6-tetrahydropyridine <2> [36]

4-cyanophenol <2, 5> (<2> weak inhibition [1]) [1, 15]

5-(aminomethyl)-3-aryl-2-oxazolidinones <4> [4]

8-hydroxyquinoline <2> (<2> reversible by the addition of Zn^{2+}, Ni^{2+}, Co^{2+} [1]) [1]

N-(2-aminoethyl)-aryl-carboxamide <-1> [5]

Ro 19-6327 <4> [35]

Ro 41-1049 <4> [35]

α-naphthol <2> (<2> reversible by dialysis [1]) [1, 26] amphetamine <2, 4> (<4> a competitive inhibitor of MAO-A [21]) [6, 8, 21, 28, 41]

benzyl cyanide <5> (<5> reversible inhibitor [15]) [15]

β-naphthol <2> (<2> reversible by dialysis [1]) [1, 26]

clorgyline <1, 4, 5> (<3,5> MAO-A highly sensitive and MAO-B less sensitive to inhibition [15,22,23]; <4> irreversible inhibitor [19]; <5> poor inhibition [22]) [5, 6, 7, 11, 13, 15, 20-22, 25, 33, 37, 40]

d-amphetamine <2, 4> [8, 41]

deoxycholate <2> [24]

deprenyl <1, 4, 5> (<5> MAO-B sensitive [22,23]) [19, 21, 22, 23, 25, 33, 35-37]

harmaline <2, 4> [26, 35]

lazabemide <4> [35]

pargyline <1, 4, 5> (<4,5> MAO-B sensitive [4,22]; <5> irreversible inhibitor [40]) [4, 11, 22, 25, 37, 38, 40]

phenylethylhydrazine <3, 5> [12]

phenylhydrazine <2> [26]

semicarbazide <2> [26]

sodium diethyldithiocarbamate <2> (<2> weak inhibition [1]) [1]

tetrahydropyridines <2> [36]

tranylcypromine <5> [11]

Additional information <5> (<5> inhibitors with a substitution at the 5-position in 2-[N-(2-propynyl) aminomethyl]-1-methyl indole [7]; <5> overview: selective inhibitors of form A and B [27]) [7, 27]

Cofactors/prosthetic groups

flavin <1-5> (<2-5> flavoprotein [3,10,17,21,24,27]; <2> 1 mol of flavin: per 100000 g of protein [1]; <3> 1 mol FAD per 114000 g of protein [2]; <3> 1 mol FAD per 115000 g of protein [10]; <4> 1 flavin per subunit [21]; <5> FAD covalently attached to each subunit of the apoprotein [27]; <2> contains 2 subunits: only one of which possesses covalently linked flavin [33]; <2> only one of two subunits contains 8-α-cysteinyl FAD [34]; <4> covalent coupling of FAD to MAO occurs specifically at the -SH-groups of cysteine [35]; <1> noncovalently bound flavin [37]; <5> two forms of the enzyme: a catalytically active form with covalently bound FAD and an inactive form [38]; <4> FAD is positioned in MAO-B through noncovalent binding at Glu34 and Tyr44 and covalent linkage at Cys397 [39]; <5> FAD is covalent linkage at Cys406 in MAO-A and is not essential for the catalytically activity [40]; <4> speculation that, when an amine oxidized, electrons pass from the amine to the disulfide and then to the flavin [41]; <4> 1 mol FAD per mol of MAO B is covalently bound by an 8-α-S-cysteinyl 397 linkage [4]; <4> MAO-A contain 1 mol of 8-α-S-cysteinyl 406 FAD per mol [5]; <5> tyrosine residues near Cys 406 may form a pocket to facilitate FAD incorporation and a stable conformation, probably through interactions among the aromatic rings of the tyrosine residues and FAD [9]) [1-3, 5, 7, 9, 10, 17, 21, 22, 24, 27, 33-35, 37-41]

Metals, ions

Cu^{2+} <2, 3> (<2> 0.00014-0.00015 mg copper per mg of protein, nonessential for activity [1]; <3> 1 gatom of copper per 59000 g of protein [2]; <3> negligible amounts in the purified enzym [10]) [1, 2, 10]

Turnover number (min⁻¹)

9.6 <4> (1-methyl-4-phenyl-1,2,3,6-tetrahydropyridine, <4> MAO-A [8]) [8]
10.8 <4> (1-methyl-4-phenyl-1,2,3,6-tetrahydropyridine) [28]
114 <4> (kynuramine, <4> MAO-A [8]) [8, 28]
125 <4> (kynuramine, <4> MAO-A [21]) [21]
318 <4> (tyramine) [28]
420-470 <2> (benzylamine) [33]
600 <4> (benzylamine, <4> recombinant MAO-B [4]) [4]
618 <1> (benzylamine) [14]
666 <1> (ethylamine) [14]
1080 <1> (n-butylamine) [14]
1092-1248 <2> (phenylethylamine, <2> MAO-B [8]) [8]
1250 <2> (phenylethylamine) [29]
Additional information <4> (<4> turnover number of MAO-A mutants [6]) [6]

Specific activity (U/mg)

0.0125 <4> [18]
0.0161 <5> [18]
0.8 <5> [22]
Additional information <2-5> [1, 4, 6, 10, 17, 23, 26, 27, 33, 34]

K_m-Value (mM)

0.006 <3> (*o*-chlorobenzylamine) [30]
0.0159 <5> (5-methoxytryptamine) [15]
0.0185 <5> (tryptamine) [15]
0.0208 <5> (2-phenylethylamine) [15]
0.022 <5> (phenylethylamine) [12]
0.0238 <5> (4-methoxy-2-phenylethylamine) [15]
0.04 <2, 4> (1-methyl-4-phenyl-1,2,3,6-tetrahydropyridine, <4> MAO-A [8]) [8]
0.04 <3> (*m*-chlorobenzylamine) [30]
0.048 <3> (phenylethylamine) [12]
0.066 <4> (tryptamine) [13]
0.07 <2> (kynuramine) [1]
0.079 <4> (tyramine) [13]
0.087 <5> (benzylamine, <5> recombinant enzyme [38]) [38]
0.091 <4> (benzylamine) [13]
0.095 <3> (*p*-chlorobenzylamine) [30]
0.11 <4> (kynuramine, <4> MAO-A [8]) [8]
0.111 <4> (dopamine) [13]
0.12 <2> (benzylamine) [1]
0.12 <5> (benzylamine, <5> rat liver enzyme [38]) [38]

0.12 <4> (kynuramine) [28]
0.17 <3> (*m*-methylbenzylamine) [30]
0.187 <5> (5-hydroxytryptamine) [15]
0.19 <2> (2-phenylethylamine) [29]
0.19 <4> (5-hydroxytryptamine) [35]
0.19-0.25 <2> (phenylethylamine, <2> MAO-B [8]) [8]
0.245 <5> (benzylamine) [15]
0.282 <5> (tyramine) [15]
0.33 <4> (O_2) [4]
0.405 <5> (dopamine) [15]
0.414 <2> (4-benzyl-1-cyclopropyl-1,2,3,6-tetrahydropyridine) [36]
0.475 <5> (4-hydroxy-3-methoxy-2-phenylethylamine) [15]
0.48-0.51 <2> (benzylamine, <2> MAO-B [8]) [8]
0.5 <4> (benzylamine) [4]
0.51 <2> (benzylamine) [29]
0.56 <1> (benzylamine) [14]
1.7 <4> (benzylamine hydrobromide) [17]
2.36 <1, 3> (n-butylamine) [14, 30]
2.8 <3> (3-phenylpropylamine) [30]
3.6 <3> (β-phenylethylamine) [30]
5 <3> (α-methylbenzylamine) [30]
20 <3> (amylamine) [30]
20 <1> (ethylamine) [14]
Additional information <2-5> (<3,4> K_m of membrane-bound and Triton X-100-solubilized enzyme [10,21,32]; <4,5> K_m of membrane-bound and 1-O-n-octyl-β-D-glucopyranoside-treated enzyme [19,20]; <5> K_m of mitochondrial outer membrane and purified enzyme [22]; <2,3> effect of phospholipids on K_m [23,24]; <4,5> K_m of wild-type MAO-A and B and of their chimera [35,40]; <4> comparison of K_m of MAO expressed in Saccharomyces cerevisiae and Pichia pastoris [5]; <4> comparison of MAO-A mutants using kynuramine, benzylamine and *para*-substituted benzylamine analogues as substrates [6]; <5> comparison of K_m of mutant enzymes using serotonin, phenylethylamine, tyramine, tryptamine as substrates [9]) [4, 5, 9, 10, 15, 19-26, 30, 32, 35, 40]

K_i-Value (mM)

0.02 <4> (amphetamine) [21]
0.023 <5> (phenylethylhydrazine) [12]
0.042 <3> (phenylethylhydrazine) [12]
0.1 <2> (β-naphthol) [1]
0.182 <2> (1-methyl-4-phenyl-1,2,3,6-tetrahydropyridine) [36]
0.2 <2> (α-naphthol) [1]
0.25 <2> (2-hydroxyquinoline) [1]
0.65 <2> (8-hydroxyquinoline) [1]

pH-Optimum

7.75 <4> (<4> soluble enzyme [32]) [32]
8 <2> (<2> purified enzyme [24]) [24]

8-9 <3> (<3> purified enzyme [23]) [23]
8-10 <2> (<2> phosphatidylcholine-treated enzyme [24]) [24]
8.5 <2> [1]
8.7 <3, 4> (<3> tyramine as substrate [10]) [10, 17]
9.2 <3> (<3> benzylamine as substrate [10]) [10]
9.5 <5> [15]
Additional information <4> (<4> pH-value for optimal deamination by a membrane-bound preparation of enzyme was dependent of the concentration of the amine employed [32]) [32]

pH-Range
5.5-11.8 <4> (<4> no activity below and above [17]) [17]
7-11 <3> (<3> about 50% of activity maximum at pH 7 and 11 [10]) [10]
7.5-10 <4> (<4> about 50% of activity maximum at pH 7.5 and 10 [17]) [17]
Additional information <5> (<5> lability increased at pH-values greater than 10 [15]) [15]

Temperature optimum (°C)
30 <4, 5> (<4,5> assay at [18]) [18]
37 <2> (<2> assay at [1]) [1]

Temperature range (°C)
Additional information <2> (<2> temperature-dependencies of enzyme activity of purified and lipid-treated enzyme [24]) [24]

4 Enzyme Structure

Molecular weight
102000 <3> (<3> gel filtration [2]) [2]
115000 <3> (<3> gel filtration, lowest molecular weight determined, variations due to different states of aggregation [10]) [10]
120000 <3> (<3> calculation per FAD molecule [2]) [2]
146000 <2> (<2> calculation on basis of flavin content [33]) [33]
215000 <1> (<1> gel filtration [37]) [37]

Subunits
? <3-5> (<4> x * 64000, SDS-PAGE [17]; <5> x * 60000, SDS-PAGE [20]; <4> x * 60000-64000, SDS-PAGE [21]; <5> x * 65000, SDS-PAGE [22]; <3> x * 58000, SDS-PAGE [23]; <4> x * 60000, SDS-PAGE, x * 59474, electrospray mass spectroscopy, including one covalent FAD, MAO-B [4]; <4> x * 59163, electrospray mass spectroscopy, including one covalent FAD, MAO-B [4]; <4> x * 60512, electrospray mass spectroscopy, including one covalent FAD, MAO-A [5]) [4, 5, 17, 20-23]
dimer <2> (<2> 2 * 62000, SDS-PAGE in presence of mercaptoethanol [33]; <2> 2 * 52000, SDS-PAGE, gel filtration chromatography with 6 M guanidine HCl as solvent [34]) [33, 34]
tetramer <1> (<1> 4 * 55000, SDS-PAGE [37]) [37]

Posttranslational modification

Additional information <3, 4> (<3> enzyme is bound to the mitochondria by acidic phospholipids, soluble enzyme can form complexes with such phospholipids [10]; <4> lipid binding can cause significant alterations in properties of the enzyme [13]; <3> lipid environment is essential for enzymatic activity, lipid microenvironment may regulate the mode of action of the enzyme in the mitochondrial outer membrane [23]; <2> the lipid enviroment appears to be an indispensable prerequisite for proper enzyme activity [24]) [10, 13, 23, 24]

5 Isolation/Preparation/Mutation/Application

Source/tissue

blood <4> (<4> predominantly B form [25]) [25]
brain <3, 4, 5> [2, 12, 13, 19, 32]
granulocyte <4> (<4> predominantly B form [25]) [25]
kidney <2> [1, 26]
liver <2-5> [4, 5, 6, 7, 8, 10-12, 15-18, 20, 22-24, 27, 29, 31, 33, 34, 41]
lymphocyte <4> (<4> predominantly B form [25]) [25]
mycelium <1> [37]
placenta <4> [21, 28]
plasma <3> [30]
spleen <3> [23]

Localization

mitochondrial outer membrane <4, 5> [4, 9, 15, 20, 22, 24, 39]
mitochondrion <2-5> (<2> intra- and intermitochondrial membrane [26]) [1, 2, 7, 10, 11, 16, 18, 19, 21, 23, 26-29, 33, 34, 36, 38, 40]

Purification

<1> [37]
<2> [1, 24, 26, 33, 34]
<3> [10, 23]
<4> [4, 17, 18, 21]
<5> [18, 20, 22, 27]

Cloning

<4> (MAO-A and MAO-B are products of two different genes [3]; expression in a human embryonic kidney cell line and expression in mammalian cells [35]; expression in mammalian COS-7 cells [38]; expression in Pichia pastoris [4,5]; expression in Saccharomyces cerevisiae [5,6]; advantage of the expression in Pichia pastoris compared with the expression in Saccharomyces cerevisiae: a higher level of homogeneity of the isolated enzyme [5]) [3-5, 6, 35, 38]

<5> (expression in Esherichia coli, two forms of the enzyme: a catalytically active form exhibits similar properties as rat liver enzyme and an inactive form [38]; expression in Saccharomyces cerevisiae YSA-1C [40]; expression in Saccharomyces cerevisiae BJ 2168 [9]) [38, 40]

Engineering

C397A <4> (<4> expressed protein catalytically inactive [35]) [35]

C397H <4> (<4> expressed protein catalytically inactive [35]) [35]

C406A <5> (<5> MAO-A, K_m for serotonin and tyramine not altered [40]) [40]

Y402A <5> (<5> decrease in activity not significant, decrease in the FAD incorporation [9]) [9]

Y402F <5> (<5> decrease in activity not significant [9]) [9]

Y407A <5> (<5> no activity with serotonin, phenylethylamine, tyramine, low activity with tryptamine, decrease in the FAD incorporation [9]) [9]

Y407F <4> (<4> K_m for kynuramine not altered, higher K_m for benzylamine analogue [6]) [6]

Y407F <5> (<5> little changes in K_m for all substrates [9]) [9]

Y410A <5> (<5> decrease in activity not significant [9]) [9]

Y444F <4> (<4> mutation alters the substrate specificity, that depends on the size of the aromatic ring and on the length of the alkyl side chain of various substrate analogues [6]) [6]

Additional information <4> (<4> MAO-A chimeric form containing the N-terminus of MAO-B, the first 36 acid sequence, do not significantly differ in their affinity for 5-hydroxytryptamine and phenylethylamine, MAO-B chimeric form containing the N-terminus of MAO-A, the first 45 acid sequence, but kinetic properties could not be detemined [35]) [35]

Application

medicine <4> (<4> ethiology of schizophrenia and affective disorders e.g. Parkinsons disease and epilepsy [25]) [25, 9]

6 Stability

Temperature stability

15 <4> (<4> below, enzyme expressed in Saccharomyces cerevisiae, 60 min stable [5]) [5]

20 <4> (<4> enzyme expressed in Pichia pastoris, 60 min stable [5]) [5]

20 <4> (<4> enzyme expressed in Saccharomyces cerevisiae, 60 min 15% loss of activity [5]) [5]

25 <4> (<4> enzyme expressed in Pichia pastoris, 60 min, 8% loss of activity [5]) [5]

25 <4> (<4> enzyme expressed in Saccharomyces cerevisiae, 60 min 50% loss of activity [5]) [5]

30 <4> (<4> enzyme expressed in Pichia pastoris, 60 min, 60% loss of activity [5]) [5]

30 <4> (<4> enzyme expressed in Saccharomyces cerevisiae, 60 min 80% loss of activity [5]) [5]

Oxidation stability

<2>, treatment with chemical agents (e.g. oxidizing agents) induce appearance of ability to catalyze hydrolytic deamination of AMP, ADP and ATP [16]

<5>, under certain experimental conditions e.g. preincubation with Cu^{2+}, treatment with oxidized oleic acid reversible qualitative alterations in catalytic properties occur, e.g. decrease in monoamine oxidase activity and appearance of ability to deaminate some nitrogenous compounds such as histamine, putrescine, cadaverine, lysine, AMP [11]

Organic solvent stability

2-mercaptoethanol <2> (<2> enzyme very stable in the presence of mercaptoethanol and low concentrations of detergents such as Triton X-100 or cholate [34]) [34]

Triton X-100 <2, 4> (<2> disintegrates the lipid-enzyme cluster to the smallest active units [24]; <4> solubilization of MAO-A and MAO-B alters the energies of activation and K_m for a number of substrates [32]) [24, 32, 34]

deoxycholate <2> (<2> optimal extraction [26]) [24, 26]

Additional information <3, 4> (<3> when enzyme liberated by methyl ethyl ketone extraction K_m about 3times lower than that for enzyme bound to the mitochondria or solubilzed by Triton X-100 [10]; <4> 1-O-n-octyl-β-D-glucopyranoside, solubilizes enzyme without dramatically altering the lipid microenviroment, a state analogous to the native species [19]) [10, 19, 20]

Storage stability

<1>, -20°C, stable for several weeks [37]

<4>, considerable stability to handling and storage, MAO-B [4]

<4>, frozen ammonium sulfate precipitate dissolved in 0.1 M phosphate buffer, pH 7.4, 50% loss of activity after 2 months [17]

<5>, 30°C, 8 h, 30% of initial activity of Ala-mutated enzyme, in the presence of 0.005 mM FAD 70% of activity [40]

<5>, 30°C, 8 h, no decrease in activity of wild-type enzyme [40]

References

[1] Erwin, V.G.; Hellerman, L.: Mitochondrial monoamine oxidase. I. Purification and characterization of the bovine kidney enzyme. J. Biol. Chem., **242**, 4230-4238 (1967)

[2] Tipton, K.F.: The prosthetic groups of pig brain mitochondrial monoamine oxidase. Biochim. Biophys. Acta, **159**, 451-459 (1968)

[3] Powell, J.F.: Molecular biological studies of monoamine oxidase: structure and function. Biochem. Soc. Trans., **19**, 199-201 (1991)

[4] Newton-Vinson, P.; Hubalek, F.; Edmondson, D.E.: High-level expression of human liver monoamine oxidase B in Pichia pastoris. Protein Expr. Purif., **20**, 334-345 (2000)

[5] Li, M.; Hubalek, F.; Newton-Vinson, P.; Edmondson, D.E.: High-level expression of human liver monoamine oxidase A in Pichia pastoris: comparison with the enzyme expressed in Saccharomyces cerevisiae. Protein Expr. Purif., **24**, 152-162 (2002)

[6] Nandigama, R.K.; Miller, J.R.; Edmondson, D.E.: Loss of serotonin oxidation as a component of the altered substrate specificity in the Y444F mutant of recombinant human liver MAO A. Biochemistry, **40**, 14839-14846 (2001)

[7] Balsa, D.; Fernandez-Alverez, E.; Tipton, K.F.; Unzeta, M.: Monoamine oxidase inhibitory potencies and selectivities of 2-[N-(2-propynyl)-aminomethyl]-1-methyl indole derivatives. Biochem. Soc. Trans., **19**, 215-218 (1991)

[8] Ramsay, R.R.; Singer, T.P.: The kinetic mechanisms of monoamine oxidases A and B. Biochem. Soc. Trans., **19**, 219-223 (1991)

[9] Ma, J.; Ito, A.: Tyrosine residues near the FAD binding site are critical for FAD binding and for the maintenance of the stable and active conformation of rat monoamine oxidase A. J. Biochem., **131**, 107-111 (2002)

[10] Oreland, L.: Purification and properties of pig liver mitochondrial monoamine oxidase. Arch. Biochem. Biophys., **146**, 410-412 (1971)

[11] Veryovkina, L.V.; Samed, M.M.A.; Gorkin, V.Z.: Mitochondrial monoamine oxidase of rat liver: reversible qualitative alterations in catalytic properties. Biochim. Biophys. Acta, **258**, 56-70 (1972)

[12] Tipton, K.F.; Spires, I.P.C.: The kinetics of phenethylhydrazine oxidation by monoamine oxidase. Biochem. J., **125**, 521-524 (1971)

[13] Tipton, M.F.; Houslay, M.D.; Garrett, N.J.: Rapid screening assay for revertants derived from MSV-transformed cells. Nature, **246**, 213-214 (1973)

[14] Suzuki, H.; Ogura, Y.; Yamada, H.: Kinetic studies on the amine oxidase reaction. J. Biochem., **72**, 703-712 (1972)

[15] Houslay, M.D.; Tipton, K.F.: A kinetic evaluation of monoamine oxidase activity in rat liver mitochondrial outer membranes. Biochem. J., **139**, 645-652 (1974)

[16] Akopyan, Z.I.; Kulygina, A.A.; Terzeman, I.I.; Gorkin, V.Z.: Induced appearance of adenylate-deaminating activity in highly purified bovine liver mitochondrial monoamine oxidase. Biochim. Biophys. Acta, **289**, 44-56 (1972)

[17] Norstrand, I.F.; Glantz, M.D.: Purification and properties of human liver monoamine oxidase. Arch. Biochem. Biophys., **158**, 1-11 (1973)

[18] Dennick, R.G.; Mayer, R.J.: Purification and immunochemical characterization of monoamine oxidase from rat and human liver. Biochem. J., **161**, 167-174 (1977)

[19] Pearce, L.B.; Roth, J.A.: Human brain monoamine oxidase: solubilization and kinetics of inhibition by octylglucoside. Arch. Biochem. Biophys., **224**, 464-472 (1983)

[20] Stadt, M.A.; Banks, P.A.; Kobes, R.D.: Purification of rat liver monoamine oxidase by octyl glucoside extraction and reconstitution. Arch. Biochem. Biophys., **214**, 223-230 (1982)

[21] Weyler, W.; Salach, J.I.: Purification and properties of mitochondrial monoamine oxidase type A from human placenta. J. Biol. Chem., **260**, 13199-13207 (1985)

[22] McCauley, R.: Properties of a monoamine oxidase from rat liver mitochondrial outer membranes. Arch. Biochem. Biophys., **189**, 8-13 (1978)

[23] Inagaki, T.; Rao, N.A.; Yagi, K.: Modulation by phospholipids of the activity of monoamine oxidase purified from pig liver. J. Biochem., 100, 597-603 (1986)

[24] Pohl, B.; Schmidt, W.: Comparative studies of purified and reconstituted monoamine oxidase from bovine liver mitochondria. Biochim. Biophys. Acta, 731, 338-345 (1983)

[25] Balsa, M.D.; Gomez, N.; Unzeta, M.: Characterization of monoamine oxidase activity present in human granulocytes and lymphocytes. Biochim. Biophys. Acta, 992, 140-144 (1989)

[26] Dugal, B.S.: Localization, purification and substrate specificity of monoamine oxidase. Biochim. Biophys. Acta, 480, 56-69 (1977)

[27] Youdim, M.B.H.; Tenne, M.: Assay and purification of liver monoamine oxidase. Methods Enzymol., 142, 617-627 (1987)

[28] Ramsay, R.R.: Kinetic mechanism of monoamine oxidase A. Biochemistry, 30, 4624-4629 (1991)

[29] Husain, M.; Edmondson, D.E.; Singer, T.P.: Kinetic studies on the catalytic mechanism of liver monoamine oxidase. Biochemistry, 21, 595-600 (1982)

[30] Lindström, A.; Olsson, B.; Petterson, G.; Szymanska, J.: Kinetics of the interaction between pig-plasma benzylamine oxidase and various monoamines. Eur. J. Biochem., 47, 99-105 (1974)

[31] Weyler, W.; Salach, J.I.: Iron content and spectral properties of highly purified bovine liver monoamine oxidase. Arch. Biochem. Biophys., 212, 147-153 (1981)

[32] Roth, J.A.; Eddy, B.J.: Kinetic properties of membrane-bound and Triton X-100-solubilized human brain monoamine oxidase. Arch. Biochem. Biophys., 205, 260-266 (1980)

[33] Salach, J.I.: Monoamine oxidase from beef liver mitochondria: simplified isolation procedure, properties, and determination of its cysteinyl flavin content. Arch. Biochem. Biophys., 192, 128-137 (1979)

[34] Minamiura, N.; Yasunobu, K.T.: Bovine liver monoamine oxidase. A modified purification procedure and preliminary evidence for two subunits and one FAD. Arch. Biochem. Biophys., 189, 481-489 (1978)

[35] Gottowik, J.; Cesura, A.M.; Malherbe, P.; Lang, G.; Da Prada, M.: Characterization of wild-type and mutant forms of human monoamine oxidase A and B expressed in a mammalian cell line. FEBS Lett., 317, 152-156 (1993)

[36] Kuttab, S.; Kalgutkar, A.; Castagnoli, N., Jr.: Mechanistic studies on the monoamine oxidase B catalyzed oxidation of 1,4-disubstituted tetrahydropyridines. Chem. Res. Toxicol., 7, 740-744 (1994)

[37] Schilling, B.; Lerch, K.: Amine oxidases from Aspergillus niger: identification of a novel flavin-dependent enzyme. Biochim. Biophys. Acta, 1243, 529-537 (1995)

[38] Hirashiki, I.; Ogata, F.; Ito, A.: Rat monoamine oxidase B expressed in Escherichia coli has a covalently-bound FAD. Biochem. Mol. Biol. Int., 37, 39-44 (1995)

[39] Zhou, B.P.; Lewis, D.A.; Kwan, S.W.; Kirksey, T.J.; Abell, C.W.: Mutagenesis at a highly conserved tyrosine in monoamine oxidase B affects FAD incorporation and catalytic activity. Biochemistry, 34, 9526-9531 (1995)

[40] Hiro, I.; Tsugeno, Y.; Hirashiki, I.; Ogata, F.; Ito, A.: Characterization of rat monoamine oxidase A with noncovalently-bound FAD expressed in yeast cells. J. Biochem., **120**, 759-765 (1996)

[41] Sablin, S.O.; Ramsay, R.R.: Monoamine oxidase contains a redox-active disulfide. J. Biol. Chem., **273**, 14074-14076 (1998)

Pyridoxamine-phosphate oxidase 1.4.3.5

1 Nomenclature

EC number
1.4.3.5

Systematic name
pyridoxamine-5'-phosphate:oxygen oxidoreductase (deaminating)

Recommended name
pyridoxamine-phosphate oxidase

Synonyms
FprA protein
PMP oxidase
PNP/PMP oxidase
PNPOx
oxidase, pyridoxamine phosphate
pyridoxamine 5'-phosphate oxidase
pyridoxamine phosphate oxidase
pyridoxaminephosphate oxidase (EC 1.4.3.5: deaminating)
pyridoxine (pyridoxamine) 5'-phosphate oxidase
pyridoxine (pyridoxamine) phosphate oxidase

CAS registry number
9029-21-4

2 Source Organism

<-9> no activity in *Nocardia sp.* (CBS 2 [10]) [10]
<-8> no activity in *Arthrobacter globiformis* [10]
<-7> no activity in *Arthrobacter sp.* (SuC 3 [10]) [10]
<-6> no activity in *Micrococcus lysodeicticus* [10]
<-5> no activity in *Bacillus pasteurii* [10]
<-4> no activity in *Bacillus polymyxa* [10]
<-3> no activity in *bacillus circulans* [10]
<-2> no activity in *Bacillus cereus* [10]
<-1> no activity in *Bacillus subtilis* [10]
<1> *Raphanus sativus* (Japanese radish [5]) [5]
<2> *Brassica juncea* [5]
<3> *Brassica rapa* [5]
<4> *Daucus carota* [5]

<5> *Spinacia oleracea* [5]
<6> *Lactuca sativa* [5]
<7> *Chrysanthemum coronarium* [5]
<8> *Medicago sativa* [5]
<9> *Phaseolus mungo* [5]
<10> *Oryza sativa* [5]
<11> *Avena sativa* [5]
<12> *Glycine max* [5]
<13> *Phaseolus radiatus* [5]
<14> *Escherichia coli* (B, K12 [10]) [10, 37, 38, 39, 40]
<15> *Salmonella typhimurium* [10]
<16> *Proteus mirabilis* [10]
<17> *Serratia marcescens* [10]
<18> *Flavobacterium sp.* (CB6 [10]) [10]
<19> *Pseudomonas aureofaciens* [10]
<20> *Pseudomonas fluorescens* [10]
<21> *Pseudomonas putida* [10]
<22> *Pseudomonas sp.* (CBS 4 [10]) [10]
<23> *Xanthomonas sp.* [10]
<24> *Agrobacterium tumefaciens* [10]
<25> *Rhodospirillum rubrum* [10]
<26> *Chloridazon-degrading bacteria* [10]
<27> *Triticum aestivum* (wheat, properties of the enzyme induced by light may differ from those of a constitutive enzyme [31]; isozymes: E1 and E2 [15]) [1, 5, 15, 26, 31]
<28> *Oryctolagus cuniculus* [2-4, 6, 12, 14, 16-19, 21-25, 29, 35]
<29> *Rattus norvegicus* (no activity in Morris hepatoma 7777 cells [11]) [7, 11, 13, 21, 32]
<30> *Saccharomyces cerevisiae* (2 forms: FI and FII [30]; strains FL100 and AL114 [36]) [8, 20, 30, 33, 34, 36]
<31> *Sus scrofa* [9]
<32> *Homo sapiens* (overview [2]) [2, 27, 28]
<33> *Micrococcus radiodurans* [10]

3 Reaction and Specificity

Catalyzed reaction

pyridoxamine 5'-phosphate + H_2O + O_2 = pyridoxal 5'-phosphate + NH_3 + H_2O_2 (<28> kinetic mechanims via either a binary or a ternary complex mechanism, depending on nature of substrate, ternary complex mechanism with pyridoxamine 5'-phosphate [12]; <28> enzyme is not stereospecific and catalyzes removal of either pro-R or pro-S hydrogen from 4-methylene of pyridoxamine 5' phosphate [6]; <14> enzyme is specific for removal of the pro-R hydrogen atom from the prochiral C4' carbon atom of pyridoxamine 5'-phosphate, hydride ion mechanism is suggested [40])

Reaction type
  deamination
  oxidation
  redox reaction
  reduction

Natural substrates and products
  S   pyridoxamine 5'-phosphate + H_2O + O_2 <1-13, 27, 28, 32> (<1-13,27>
         may be the regulatory enzyme at the final step of pyridoxal 5'-phosphate
         biosynthesis [5]; <32> enzyme plays a role in regulation of vitamin B_6
         metabolism in erythrocytes [28]; <28> in conjugation with pyridoxal ki-
         nase the enzyme is responsible for the formation of the coenzyme pyri-
         doxal 5'-phosphate, from the B_6 vitamers pyridoxine and pyridoxamine
         [29]) (Reversibility: ? <1-13, 27, 28, 32> [5, 28, 29]) [5, 28, 29]
  P   pyridoxal 5'-phosphate + NH_3 + H_2O_2 <1-13, 27, 28, 32> [5, 28, 29]

Substrates and products
  S   5'-homopyridoxine-phosphate + H_2O + O_2 <28> (Reversibility: ? <28>
         [25]) [25]
  P   homopyridoxal 5'-phosphate + H_2O_2 <28> [25]
  S   5'-phospho-pyridoxal-O-carboxymethyloxime + H_2O + O_2 <28> (Rever-
         sibility: ? <28> [25]) [25]
  P   O-carboxymethylpyridoxal 5'-phosphate + hydroxylamine + H_2O_2 <28>
         [25]
  S   5'-phospho-pyridoxaloxime + H_2O + O_2 <28> (Reversibility: ? <28> [25])
         [25]
  P   pyridoxal 5'-phosphate + hydroxylamine + H_2O_2 <28> [25]
  S   5'-phospho-pyridoxyl-2-aminobutyrate + H_2O + O_2 <28> (<28> 110% of
         activity with pyridoxamine 5'-phosphate [24]) (Reversibility: ? <28> [24])
         [24]
  P   pyridoxal 5'-phosphate + 2-aminobutyrate + H_2O_2 <28> [24]
  S   5'-phospho-pyridoxyl-3-alanine + H_2O + O_2 <28, 31> (<28> 85% of ac-
         tivity with pyridoxamine 5'-phosphate [24]) (Reversibility: ? <28, 31> [9,
         24]) [9, 24]
  P   pyridoxal 5'-phosphate + alanine + H_2O_2 <28, 31> [9, 24]
  S   5'-phospho-pyridoxyl-3-aminobenzoate + H_2O + O_2 <31> (Reversibility:
         ? <31> [9]) [9]
  P   pyridoxal 5'-phosphate + 3-aminobenzoate + H_2O_2 <31> [9]
  S   5'-phospho-pyridoxyl-4-aminobenzoate + H_2O + O_2 <31> (Reversibility:
         ? <31> [9]) [9]
  P   pyridoxal 5'-phosphate + 4-aminobenzoate + H_2O_2 <31> [9]
  S   5'-phospho-pyridoxyl-4-aminobutyrate + H_2O + O_2 <31> (Reversibility: ?
         <31> [9]) [9]
  P   pyridoxal 5'-phosphate + 4-aminobutyrate + H_2O_2 <31> [9]
  S   5'-phospho-pyridoxyl-4-aminonitrobenzene + H_2O + O_2 <31> (Reversi-
         bility: ? <31> [9]) [9]
  P   pyridoxal 5'-phosphate + 4-aminonitrobenzene + H_2O_2 <31> [9]

S 5'-phospho-pyridoxyl-4-aminophenol + H_2O + O_2 <31> (Reversibility: ? <31> [9]) [9]

P pyridoxal 5'-phosphate + 4-aminophenol + H_2O_2 <31> [9]

S 5'-phospho-pyridoxyl-5-aminovalerate + H_2O + O_2 <31> (Reversibility: ? <31> [9]) [9]

P pyridoxal 5'-phosphate + 5-aminovalerate + H_2O_2 <31> [9]

S 5'-phospho-pyridoxyl-D-2-aminobutyrate + H_2O + O_2 <28> (<28> 120% of activity with pyridoxamine 5'-phosphate [24]) (Reversibility: ? <28> [24]) [24]

P pyridoxal 5'-phosphate + D-2-aminobutyrate + H_2O_2 <28> [24]

S 5'-phospho-pyridoxyl-D-alanine + H_2O + O_2 <28> (<28> 130% of activity with pyridoxamine 5'-phosphate [24]) (Reversibility: ? <28> [24]) [24]

P pyridoxal 5'-phosphate + D-alanine + H_2O_2 <28> [24]

S 5'-phospho-pyridoxyl-D-leucine + H_2O + O_2 <28> (<28> 120% of activity with pyridoxamine 5'-phosphate [24]) (Reversibility: ? <28> [24]) [24]

P pyridoxal 5'-phosphate + D-leucine + H_2O_2 <28> [24]

S 5'-phospho-pyridoxyl-D-tyrosine + H_2O + O_2 <28> (<28> 55% of activity with pyridoxamine 5'-phosphate [24]) (Reversibility: ? <28> [24]) [24]

P pyridoxal 5'-phosphate + D-tyrosine + H_2O_2 <28> [24]

S 5'-phospho-pyridoxyl-L-2-aminobutyrate + H_2O + O_2 <28> (Reversibility: ? <28> [24]) [24]

P pyridoxal 5'-phosphate + L-2-aminobutyrate + H_2O_2 <28> [24]

S 5'-phospho-pyridoxyl-L-alanine + H_2O + O_2 <28> (<28> 140% of activity with pyridoxamine 5'-phosphate [24]) (Reversibility: ? <28> [24]) [24]

P pyridoxal 5'-phosphate + L-alanine + H_2O_2 <28> [24]

S 5'-phospho-pyridoxyl-L-leucine + H_2O + O_2 <28> (<28> 86% of activity with pyridoxamine 5'-phosphate [24]) (Reversibility: ? <28> [24]) [24]

P pyridoxal 5'-phosphate + L-leucine + H_2O_2 <28> [24]

S 5'-phospho-pyridoxyl-L-phenylalanine + H_2O + O_2 <28> (<28> 54% of activity with pyridoxamine 5'-phosphate [24]) (Reversibility: ? <28> [24]) [24]

P pyridoxal 5'-phosphate + L-phenylalanine + H_2O_2 <28> [24]

S 5'-phospho-pyridoxyl-L-serine + H_2O + O_2 <28> (<28> 39% of activity with pyridoxamine 5'-phosphate [24]) (Reversibility: ? <28> [24]) [24]

P pyridoxal 5'-phosphate + L-serine + H_2O_2 <28> [24]

S 5'-phospho-pyridoxyl-L-tryptophan + H_2O + O_2 <28> (<28> 18% of activity with pyridoxamine 5'-phosphate [24]) (Reversibility: ? <28> [24]) [24]

P pyridoxal 5'-phosphate + L-tryptophan + H_2O_2 <28> [24]

S 5'-phospho-pyridoxyl-L-tyrosine + H_2O + O_2 <28> (<28> 57% of activity with pyridoxamine 5'-phosphate [24]) (Reversibility: ? <28> [24]) [24]

P pyridoxal 5'-phosphate + L-tyrosine + H_2O_2 <28> [24]

S 5'-phospho-pyridoxyl-aniline + H_2O + O_2 <31> (Reversibility: ? <31> [9]) [9]

P pyridoxal 5'-phosphate + aniline + H_2O_2 <31> [9]

S 5'-phospho-pyridoxyl-benzylamine + H_2O + O_2 <28> (<28> 120% of activity with pyridoxamine 5'-phosphate [24]) (Reversibility: ? <28> [24]) [24]

P pyridoxal 5'-phosphate + benzylamine + H_2O_2 <28> [24]

S 5'-phospho-pyridoxyl-glycine + H_2O + O_2 <28> (<28> 110% of activity with pyridoxamine 5'-phosphate [24]) (Reversibility: ? <28> [24]) [24, 25]

P pyridoxal 5'-phosphate + glycine + H_2O_2 <28> [24]

S N-(5'-phospho-4'-pyridoxyl)-N'-(1-naphthyl)ethylenediamine + H_2O + O_2 <28> (Reversibility: ? <28> [17]) [17]

P pyridoxal 5'-phosphate + N-(1-naphthyl)-ethylenediamine + H_2O_2 <28> [17]

S N-(5'-phospho-4'-pyridoxyl)-N'-(1-naphthyl-5-sulfonate)-ethylenediamine + H_2O + O_2 <28> (Reversibility: ? <28> [17]) [17]

P pyridoxal 5'-phosphate + N-(1-naphthyl-5-sulfonate)-ethylenediamine + H_2O_2 <28> [17]

S N-(5'-phospho-4'-pyridoxyl)-N'-(9-acridyl)ethylenediamine + H_2O + O_2 <28> (Reversibility: ? <28> [17]) [17]

P pyridoxal 5'-phosphate + N-(9-acridyl)ethylenediamine + H_2O_2 <28> [17]

S N-(5'-phospho-4'-pyridoxyl)-N'-(dansyl)ethylenediamine + H_2O + O_2 <28> (Reversibility: ? <28> [17]) [17]

P pyridoxal 5'-phosphate + N-(dansyl)ethylenediamine + H_2O_2 <28> [17]

S N-(5'-phospho-4'-pyridoxyl)-N'-(dansyl)pentylenediamine + H_2O + O_2 <28> (Reversibility: ? <28> [17]) [17]

P pyridoxal 5'-phosphate + N-(dansyl)pentylenediamine + H_2O_2 <28> [17]

S N-phosphopyridoxyl tryptamine + H_2O + O_2 <29> (Reversibility: ? <29> [13]) [13]

P pyridoxal 5'-phosphate + tryptamine + H_2O_2 <29> [13]

S N^{10}-(5'-phosphopyridoxyl)-1,10-diaminodecane + H_2O + O_2 <28> (Reversibility: ? <28> [24]) [24]

P pyridoxal 5'-phosphate + 1,10-diaminodecane + H_2O_2 <28> [24]

S N^2-acetyl-N^6-(5'-phosphopyridoxyl)-L-lysine + H_2O + O_2 <28> (Reversibility: ? <28> [16]) [16]

P pyridoxal 5'-phosphate + N^2-acetyl-L-lysine + H_2O_2 <28> [16]

S α-N-(5'-phospho-4'-pyridoxyl)-ornithine + O_2 + H_2O <28> (Reversibility: ? <28> [25]) [25]

P pyridoxal 5'-phosphate + α-ornithine + H_2O_2 <28> [25]

S ε-pyridoxyllysine 5'-phosphate + H_2O + O_2 <28> (Reversibility: ? <28> [16]) [16]

P pyridoxal 5'-phosphate + lysine + H_2O_2 <28> [16]

S pyridoxamine 5'-phosphate + H_2O + O_2 <1-33> (<28> enzyme is not stereospecific, catalyzes equally well removal of either pro-R or pro-S-hydrogen from the 4-methylene of pyridoxamine 5'-phosphate [6]; <27> isozymes E I uses only pyridoxine 5'-phosphate as substrate [15]; <27> E II uses pyridoxine 5'-phosphate and pyridoxamine 5'-phosphate [15]; <28> no oxidation of the unphosphorylated analogs of pyridoxamine phosphate, pyridoxine phosphate and N-(phosphopyridoxyl)amines [25];

<32> 6 to 7fold more active as with pyridoxamine 5'-phosphate [2]) (Reversibility: ? <1-33> [1-40]) [1-40]
P pyridoxal 5'-phosphate + NH_3 + H_2O_2 <1-33> [1-40]
S pyridoxine 5'-phosphate + H_2O + O_2 <28, 30, 31, 32> (Reversibility: ? <28, 30, 31, 32> [2, 8, 9, 12, 25, 29]) [2, 8, 9, 12, 25, 29]
P pyridoxal 5'-phosphate + H_2O_2 <28, 30, 31, 32> [2, 8, 9, 12, 25, 29]

Inhibitors

2,3-butanedione <28> (<28> 10 mM, complete inhibition after 50 min in borate buffer, inhibition is fully reversible upon removal of borate [4]) [4]
2,4-pentandione <28> (<28> 0.3 mM, approx. 90% inhibition of holoenzyme [4]) [4]
4-chloromercuribenzoate <28> (<28> 0.1 mM, 90-95% inhibition [14]) [14]
4-deoxypyridoxine 5'-phosphate <14, 28> (<28> 0.1 mM, 42% inhibition [14]; <14> competitive inhibition [37]) [14, 29, 37]
4-pyridoxic acid phosphate <28> (<28> 0.1 mM, 33.5% inhibition [14]) [14]
Cd^{2+} <27, 30> (<27> 0.5 mM, isozyme E I: insensitive, isozyme E II: 44% inhibition [15]; <30> 2.85 mM, 82% inhibition of pyridoxamine 5'-phosphate oxidation, 27% inhibition of pyridoxine oxidation [34]) [15, 34]
Cu^{2+} <28, 30> (<28> 1 mM, 90-95% inhibition [14]; <30> 2.85 mM, 38% inhibition of pyridoxamine 5'-phosphate oxidation, 23% inhibition of pyridoxine oxidation [34]) [14, 34]
Fe^{2+} <30> (<30> 2.85 mM, 44% inhibition of pyridoxamine 5'-phosphate oxidation, 81% inhibition of pyridoxine oxidation [34]) [34]
Fe^{3+} <30> (<30> 2.85 mM, 26% inhibition of pyridoxamine 5'-phosphate oxidation, 50% inhibition of pyridoxine oxidation [34]) [34]
Hg^{2+} <27, 28, 30> (<27> 0.5 mM, isozyme E I: insensitive, isozyme E II: 54% inhibition [15]; <28> 1 mM, 90-95% inhibition [14]; <30> 2.85 mM, 30% inhibition of pyridoxamine 5'-phosphate oxidation, 2% inhibition of pyridoxine oxidation [34]) [14, 15, 34]
Zn^{2+} <27, 30> (<27> 0.5 mM, isozyme E I: insensitive, isozyme E II: 54% inhibition [15]; <30> 2.85 mM, 92% inhibition of pyridoxamine 5'-phosphate oxidation, 60% inhibition of pyridoxine oxidation [34]) [15, 34]
ammonium sulfate <27> (<27> 1.9 M, E II: activation at short incubation time, E I: complete inactivation after 10 h incubation [15]) [15]
citrate-phosphate buffer <27> [15]
diethyldicarbonate <28> (<28> 1.2 mM, complete inactivation after 10 min, activity decrease of 60% can be restored by a 20 min incubation with 900 mM hydroxylamine, pyridoxal 5'-phosphate oxime protects from inactivation, inactivation is due to modification of histidine residues [18]) [18]
phenylglyoxal <28> (<28> pyridoxal 5'-phosphate protects [4]) [4]
pyridoxal 5'-phosphate <14, 28, 29, 31, 32> (<28,29> competitive inhibition [21]; <14> competitive inhibition [37]) [2, 9, 14, 21, 29, 37]
pyridoxaloxime 5'-phosphate <28, 31> (<31> competitive inhibition [9]; <28> 0.2 mM, 80% inhibition [14]) [9, 14, 18, 29]
pyridoxamine 5'-phosphate <29> (<29> substrate inhibition above 0.0006 mM [7]) [7]

pyridoxine 5'-phosphate <28> (<28> substrate inhibition [12]) [12]
Additional information <28, 29> (<28> no substrate inhibition with pyridoxamine 5'-phosphate [12]; <29> no inhibition by tryptophan metabolites [7])
[7, 12]

Cofactors/prosthetic groups

FAD <28> (<28> about 0.1% as active as FMN [14]; <28> inactive [25]; <28>
bound poorly to apoenzyme [29]) [14, 29]
FMN <14, 28, 30, 31> (<28, 31> flavoprotein [3, 9, 14, 23, 25]; <28,31> binds
1 mol of FMN per mol of dimer [9, 29]; <30> 1 mol of FMN per mol of
enzyme [33]; <28> analogs modified at position 2, 3, 7, 8, 8α and at the
ribityl side chain are all bound by the apoenzyme but less tightly than FMN
[29]; <14> 2 molecules of FMN bound per homodimer [39]) [3, 9, 14, 23, 25,
29, 33, 39]
Additional information <28> (<28> specific for flavin-phosphate [23]) [23]

Activating compounds

2-aminobutanoate <28> (<28> activation [14]) [14]
3-hydroxyanthranilate <29> (<29> 145% increase in pyridoxine 5-phosphate
oxidase activity, 28% increase in pyridoxamine 5-phosphate oxidase activity
[7]) [7]
3-hydroxykynurenine <29> (<29> 125% increase in pyridoxine 5-phosphate
oxidase activity, 20% increase in pyridoxamine 5-phosphate oxidase activity
[7]) [7]
5-hydroxyindolacetate <29> (<29> 125% increase in pyridoxine 5-phosphate
oxidase activity, 9% increase in pyridoxamine 5-phosphate oxidase activity
[7]) [7]
5-hydroxytrpamine <29> (<29> 20% increase pyridoxamine 5-phosphate
oxidase activity [7]) [7]
Tris <28> (<28> activation [14]) [14]
alanine <28> (<28> activation [14]) [14]
cysteine <28> (<28> activation [14]) [14]
guanidine hydrochloride <28> (<28> 500 mM, activation at low concentrations and brief periods of exposure, inactivation at higher concentrations
and longer periods of exposure [19]) [19]
tryptamine <29> (<29> 27% increase in pyridoxamine 5-phosphate oxidase
activity [7]) [7]
urea <28> (<28> 2.0-2.5 M, rapid and reversible activation at low concentrations and brief periods of exposure, inactivation at higher concentrations and
longer periods of exposure [19]) [19]
valine <28> (<28> activation [14]) [14]
xanthurenate <29> (<29> 21% increase in pyridoxine 5-phosphate oxidase
activity [7]) [7]

Metals, ions

Additional information <30> (<30> no requirement for bivalent cations [34])
[34]

Turnover number (min⁻¹)

0.24 <31> (5'-phospho-pyridoxyl-*p*-aminophenol) [9]

0.48 <14> (pyridoxine 5'-phosphate, <14> R197M mutant enzyme [40]) [40]

0.54 <31> (pyridoxamine 5'-phosphate) [9]

0.6 <31> (5'-phospho-pyridoxyl-aniline) [9]

1.8 <14> (pyridoxine 5'-phosphate, <14> H199N mutant enzyme [40]) [40]

1.8 <14> (pyridoxine 5'-phosphate, <14> R197M mutant enzyme [40]) [40]

2.1 <31> (5'-phospho-pyridoxyl-5-aminovalerate) [9]

2.4 <31> (5'-phospho-pyridoxyl-3-alanine) [9]

2.4 <31> (5'-phospho-pyridoxyl-4-aminobutyrate) [9]

3.6 <14> (pyridoxine 5'-phosphate, <14> D49A mutant enzyme [40]) [40]

5.4 <31> (pyridoxine 5'-phosphate) [9]

6 <31> (5'-phospho-pyridoxyl-*m*-aminobenzoate) [9]

6 <31> (5'-phospho-pyridoxyl-*p*-aminonitrobenzene) [9]

6.2 <28, 30> (pyridoxamine 5'-phosphate) [8, 12]

7.8 <14> (pyridoxine 5'-phosphate) [40]

8.4 <14> (pyridoxine 5'-phosphate, <14> H199A mutant enzyme [40]) [40]

8.4 <14> (pyridoxine 5'-phosphate, <14> R14M mutant enzyme [40]) [40]

9.6 <14> (pyridoxine 5'-phosphate, <14> R14E mutant enzyme [40]) [40]

12 <14> (pyridoxine 5'-phosphate, <14> recombinant enzyme, in phosphate and HEPES buffer at pH 7.6 [39]) [39]

18 <14> (pyridoxine 5'-phosphate, <14> recombinant enzyme, in Tris buffer at pH 7.6 [39]) [39]

36 <14> (pyridoxine 5'-phosphate, <14> Y17F mutant enzyme [40]) [40]

42 <28> (pyridoxine 5'-phosphate) [12]

42 <30> (pyridoxine 5'-phosphate) [8]

45.6 <14> (pyridoxine 5'-phosphate, <14> recombinant enzyme [37]) [37]

103.2 <14> (pyridoxamine 5'-phosphate, <14> recombinant enzyme [37]) [37]

1680 <14> (pyridoxamine 5'-phosphate, <14> recombinant enzyme after cleavage from fusion protein [38]) [38]

1980 <14> (pyridoxal 5'-phosphate, <14> recombinant enzyme after cleavage from fusion protein [38]) [38]

Specific activity (U/mg)

0.00000062 <13> (<13> pyridoxine 5'-phosphate activity in seedlings [5]) [5]

0.000001 <6> (<6> pyridoxine 5'-phosphate activity in seedlings [5]) [5]

0.0000012 <4> (<4> pyridoxine 5'-phosphate activity in root extracts [5]) [5]

0.0000013 <7, 9> (<7,9> pyridoxine 5'-phosphate activity in seedlings [5]) [5]

0.0000015 <8> (<8> pyridoxine 5'-phosphate activity in seedlings [5]) [5]

0.0000043 <30> (<30> activity in strain FL100 [36]) [36]

0.000012 <12> (<12> pyridoxine 5'-phosphate activity in seedlings [5]) [5]

0.0000178 <1> (<1> pyridoxine 5'-phosphate activity in root extracts [5]) [5]

0.000019 <5> (<5> pyridoxine 5'-phosphate activity in leaf extracts [5]) [5]

0.000022 <25> (<25> activity in crude extracts [10]) [10]

0.000025 <26> (<26> activity in crude extracts [10]) [10]

0.000028 <30> (<30> activity in strain Al114 [36]) [36]

0.000035 <3> (<3> pyridoxine 5'-phosphate activity in extracts of seedlings [5]) [5]

0.0000366 <1> (<1> pyridoxine 5'-phosphate activity in mabikina extracts [5]) [5]

0.00004 <18> (<18> activity in crude extracts [10]) [10]

0.000042 <15> (<15> activity in crude extracts [10]) [10]

0.000065 <17> (<17> activity in crude extracts [10]) [10]

0.000065 <27> (<27> activity in suspension of cultered cells, substrate pyridoxamine 5'-phosphate [1]) [1]

0.000067 <14> (<14> activity in crude extracts, strain B [10]) [10]

0.000068 <24> (<24> activity in crude extracts [10]) [10]

0.000072 <27> (<27> activity in callus, substrate pyridoxamine 5'-phosphate [1]) [1]

0.000073 <10> (<10> pyridoxine 5'-phosphate activity in seedlings [5]) [5]

0.000073 <16> (<16> activity in crude extracts [10]) [10]

0.0000745 <2> (<2> pyridoxine 5'-phosphate activity in extracts of seedlings [5]) [5]

0.00008 <33> (<33> activity in crude extracts [10]) [10]

0.000084 <11> (<11> pyridoxine 5'-phosphate activity in seedlings [5]) [5]

0.000085 <23> (<23> activity in crude extracts [10]) [10]

0.000097 <27> (<27> activity in cell suspension cultures, substrate pyridoxine 5'-phosphate [1]) [1]

0.0001 <27> (<27> activity in callus, substrate pyridoxine 5'-phosphate [1]) [1]

0.00012 <27> (<27> activity in seedlings, substrate pyridoxamine 5'-phosphate [1]) [1]

0.0005 <19> (<19> activity in crude extracts [10]) [10]

0.0005 <27> (<27> activity in seedlings, substrate pyridoxine 5'-phosphate [1]) [1]

0.000667 <21> (<21> activity in crude extracts [10]) [10]

0.0007 <20> (<20> activity in crude extracts [10]) [10]

0.000717 <22> (<22> activity in crude extracts, cells grown on minimal medium [10]) [10]

0.0012 <27> (<27> partialy purified enzyme [26]) [26]

0.00188 <27> (<27> isozyme EII [15]) [15]

0.00367 <27> (<27> isozyme EI [15]) [15]

0.0417 <30> (<30> activity of apoenzyme after dialysis of holoenzyme against potassium acetate buffer pH 4.0 and potassium phosphate buffer pH 7.0 for 24 h each [33]) [33]

0.056 <30> (<30> substrate pyridoxine 5'-phosphate, enzyme species FI which is assumed to be an artefact of the purification procedure [30]) [30]

0.09 <31> [9]

0.107 <30> (<30> substrate pyridoxine 5'-phosphate, enzyme species FII [30]) [30]

0.138 <29> [7]

0.15 <29> [11]

0.156 <28> [24]

0.293 <14> (<14> recombinant enzyme [39]) [39]
0.3 <28> [3]
0.41 <28> [29]
0.432 <28> [14]
0.435 <30> [33]
0.462 <30> [34]
1.37 <14> (<14> recombinant enzyme [37]) [37]
1.708 <29> [32]
Additional information <27> (<27> 0.2 nmol/min/g fresh seedlings, activity in six-day old seedlings [31]) [31]
Additional information <27> (<27> 0.63 nmol/min/g fresh seedlings, activity in six-day-old seedlings after illumination with red light [31]) [31]
Additional information <32> (<32> 25 nmol/g hemoglobin/h, activity in haemolysate [28]) [28]
Additional information <32> (<32> 3.3 nmol/g hemoglobin/h, subject with a slow rate of pyridoxine conversion [27]) [27]
Additional information <32> (<32> 7.8 nmol/g hemoglobin/h, subject with a slow rate of pyridoxine conversion given 24 mg/day oral riboflavin for 2 weeks [27]) [27]

K_m-Value (mM)
0.000016 <30> (FMN) [33]
0.000031 <28> (FMN) [14]
0.0003 <14> (pyridoxine 5'-phosphate) [40]
0.0003 <14> (pyridoxine 5'-phosphate, <14> H199N mutant enzyme [40]) [40]
0.00065 <29> (N-phosphopyridoxyl tryptamine) [13]
0.00092 <29> (pyridoxine 5'-phosphate) [7]
0.001 <14> (pyridoxine 5'-phosphate, <14> D49A and Y17F mutant enzyme [40]) [40]
0.0012 <30> (pyridoxine 5'-phosphate, <30> immobilized enzyme [20]) [20]
0.0013 <32> (pyridoxamine 5'-phosphate, <32> measured in hemolysate [28]) [28]
0.002 <14> (pyridoxine 5'-phosphate) [39]
0.002 <14> (pyridoxine 5'-phosphate, <14> R14E mutant enzyme [40]) [40]
0.002 <14> (pyridoxine 5'-phosphate, <14> recombinant enzyme [37]) [37]
0.0026 <14> (pyridoxine 5'-phosphate, <14> R14M mutant enzyme [40]) [40]
0.0027 <30> (pyridoxine 5'-phosphate, <30> soluble enzyme [20]) [20]
0.0031 <28> (pyridoxine 5'-phosphate) [14]
0.0036 <28> (pyridoxamine 5'-phosphate) [12]
0.004 <29> (pyridoxine 5'-phosphate, <29> enzyme activity in liver homogenate [21]) [21]
0.0059 <28> (5'-homopyridoxine-phosphate) [25]
0.0059 <28> (5'-methylpyridoxine-phosphate) [25]
0.008 <28> (N-(5'-phospho-4'-pyridoxyl)-N'-(1-naphthyl)ethylenediamine) [17]

0.008 <29> (pyridoxamine 5'-phosphate, <29> enzyme activity in liver homogenate [21]) [21]

0.0082 <28> (pyridoxine 5'-phosphate) [12]

0.009 <30> (pyridoxine 5'-phosphate, <30> enzyme species FI which is assumed to be an artefact of the purification procedure [30]) [30]

0.01 <31> (5'-phospho-pyridoxyl-3-aminobenzoate) [9]

0.01 <31> (5'-phospho-pyridoxyl-4-aminobenzoate) [9]

0.01 <31> (5'-phospho-pyridoxyl-aniline) [9]

0.01 <28> (pyridoxamine 5'-phosphate) [25]

0.01 <29> (pyridoxamine 5'-phosphate) [7]

0.01 <30> (pyridoxamine 5'-phosphate, <30> immobilized enzyme [20]) [20]

0.011 <28> (pyridoxamine 5'-phosphate, <28> enzyme activity in liver homogenate [21]) [21]

0.0112 <28> (pyridoxamine 5'-phosphate) [19]

0.013 <28> (pyridoxamine 5'-phosphate) [16]

0.013 <31> (pyridoxine 5'-phosphate) [9]

0.015 <28> (pyridoxamine 5'-phosphate, <28> enzyme activity in cytosol [21]) [21]

0.016 <28> (pyridoxine 5'-phosphate, <28> pure enzyme in 200 mM potassium phosphate pH 7.0 [21]) [21]

0.017 <28> (pyridoxine 5'-phosphate, <28> enzyme activity in cytosol [21]) [21]

0.017 <28> (pyridoxine 5'-phosphate, <28> enzyme activity in liver homogenate [21]) [21]

0.018 <30> (pyridoxamine 5'-phosphate, <30> soluble enzyme [20]) [20]

0.0187 <28> (pyridoxamine 5'-phosphate, <28> in the presence of 1 M urea [19]) [19]

0.02 <28> (pyridoxine 5'-phosphate) [16]

0.021 <28> (5'-phospho-pyridoxaloxime) [25]

0.021 <28> (pyridoxal phosphate oxime) [25]

0.022 <29> (pyridoxine 5'-phosphate, <29> enzyme activity in liver homogenate in the presence of 0.005 mM pyridoxal 5'-phosphate [21]) [21]

0.0238 <28> (pyridoxine 5'-phosphate) [19]

0.024 <28> (pyridoxamine 5'-phosphate, <28> pure enzyme in 200 mM potassium phosphate pH 7.0 [21]) [21]

0.025 <28> (5'-phospho-pyridoxal-O-carboxymethyloxime) [25]

0.025 <28> (pyridoxal phosphate-carboxymethyloxime ester) [25]

0.025 <30> (pyridoxamine 5'-phosphate, <30> enzyme species FI which is assumed to be an artefact of the purification procedure [30]) [30]

0.03 <28> (5'-phospho-pyridoxyl-5-aminovalerate) [25]

0.03 <28, 31> (phospho-pyridoxyl-4-aminobutyrate) [25, 9]

0.03 <31> (pyridoxine 5'-phosphate) [9]

0.031 <28> (5'-methylpyridoxine-phosphate) [25]

0.031 <28> (5'-methylpyridoxine-phosphate) [25]

0.031 <28> (N-(5'-phospho-4'-pyridoxyl)-L-tyrosine) [25]

0.033 <28> (N-(5'-phospho-4'-pyridoxyl)-L-benzylamine) [25]

0.035 <31> (phospho-pyridoxyl-3-alanine) [9]

0.039 <29> (pyridoxamine 5'-phosphate, <29> enzyme activity in liver homogenate in the presence of 0.005 mM pyridoxal 5'-phosphate [21]) [21]

0.0417 <28> (pyridoxamine 5'-phosphate, <28> in the presence of 2 M urea [19]) [19]

0.065 <28> (N^2-acetyl-N^6-(5'-phospho-pyridoxyl)L-lysine) [16]

0.0667 <28> (pyridoxine 5'-phosphate, <28> in the presence of 2 M urea [19]) [19]

0.068 <28> (N-(5'-phospho-4'-pyridoxyl)-glycine) [25]

0.07 <14> (pyridoxine 5'-phosphate, <14> H199A mutant enzyme [40]) [40]

0.075 <28> (N-(5'-phospho-4'-pyridoxyl)-L-leucine) [25]

0.085 <28> (O_2, <28> + pyridoxamine 5'-phosphate [12]) [12]

0.09 <14> (pyridoxine 5'-phosphate, <14> R197M mutant enzyme [40]) [40]

0.091 <28> (N-(5'-phospho-4'-pyridoxyl)-L-α-aminobutyrate) [25]

0.095 <28> (N-(5'-phospho-4'-pyridoxyl)-L-phenylalanine) [25]

0.1 <31> (pyridoxamine-5'-phosphate) [9]

0.105 <14> (pyridoxamine 5'-phosphate, <14> recombinant enzyme [37]) [37]

0.11 <28> (N-(5'-phospho-4'-pyridoxyl)-β-alanine) [25]

0.12 <28> (N-(5'-phospho-4'-pyridoxyl)-L-tryptophan) [25]

0.125 <28> (N-(5'-phospho-4'-pyridoxyl)-D-leucine) [25]

0.13 <28> (N-(5'-phospho-4'-pyridoxyl)-L-serine) [25]

0.14 <28> (pyridoxamine 5'-phosphate) [14]

0.182 <28> (O_2, <28> cosubstrate pyridoxine 5'-phosphate [12]) [12]

0.2 <28> (α-N-(5'-phospho-4'-pyridoxyl)-lysine) [25]

0.22 <28> (N-(5'-phospho-4'-pyridoxyl)-L-alanine) [25]

0.29 <28> (N-(5'-phospho-4'-pyridoxyl)-D-α-aminobutyrate) [25]

0.368 <14> (pyridoxal 5'-phosphate, <14> recombinant enzyme after cleavage from fusion protein [38]) [38]

0.4 <28> (N^{10}-(5'-phospho-4'-pyridoxyl)-1,10-diaminodecane) [25]

0.4 <14> (pyridoxamine 5'-phosphate, <14> recombinant enzyme after cleavage from fusion protein [38]) [38]

0.53 <28> (α-N-(5'-phospho-4'-pyridoxyl)-ornithine) [25]

0.77 <28> (N-(5'-phospho-4'-pyridoxyl)-D-alanine) [25]

1.6 <28> (N-(5'-phospho-4'-pyridoxyl)-D-tyrosine) [25]

2.4 <14> (pyridoxine 5'-phosphate, <14> R197M mutant enzyme [40]) [40]

K_i-Value (mM)

0.0018 <28> (4'-deoxypyridoxine-5'-phosphate) [29]

0.002 <28> (pyridoxal 5'-phosphate oxime) [29]

0.003 <28> (pyridoxal 5'-phosphate, <28> pure enzyme in 200 mM potassium phosphate pH 7.0 or enzyme activity in liver homogenate, substrate pyridoxamine 5' phosphate or pyridoxine 5'-phosphate [21]) [21,29,35]

0.008 <14> (pyridoxal 5'-phosphate) [37,38]

0.05 <28> (pyridoxine 5'-phosphate) [12]

0.105 <14> (4'-deoxypyridoxine-5'-phosphate) [37]

pH-Optimum

7.5 <30> (<30> pyridoxine 5'-phosphate, immobilized enzyme [20]) [20]

8 <30, 32> (<30> pyridoxine 5'-phosphate, soluble and immobilized enzyme [8]; <30> pyridoxine 5'-phosphate, [20, 34]; <30> soluble enzyme [20]) [8, 20, 28, 34]

8.3 <14> (<14> recombinant enzyme after cleavage from fusion protein [38]) [38]

8.5 <14> (<14> pyridoxine phosphate oxidase activity, slight optimum [37]) [37]

8.5-10 <14> (<14> pyridoxamine phosphate oxidase activity remains nearly constant between pH 8.5 and 10.0 [37]) [37]

9 <30> (<30> pyridoxamine 5'-phosphate, soluble enzyme [8]; <30> pyridoxamine 5'-phosphate, soluble and immobilized enzyme [20]; <30> pyridoxamine 5'-phosphate [34]; <32> in hemolysate [28]) [8, 20, 28, 34]

9-10 <28> (<28> pyridoxine 5'-phosphate, pyridoxamine 5'-phosphate [14]) [14]

9.5 <27, 30> (<30> immobilized enzyme, pyridoxamine 5'-phosphate [8]) [8, 15]

pH-Range

6-10 <14> (<14> pyridoxine phosphate oxidase activity [37]) [37]

6.5-10 <14> (<14> pyridoxamine phosphate oxidase activity is absent at pH 6.0, rises sharply from pH 6.0 to 8.5 and remains constantly high between pH 8.5 and 10.0 [37]) [37]

Temperature optimum (°C)

36 <30> (<30> immobilized enzyme [8]) [8]

40 <27> (<27> isozymes EI and EII, rapid decrease above [15]) [15]

44 <30> (<30> soluble enzyme [8]) [8]

45 <30> (<30> soluble enzyme [20]) [20, 34]

50 <14> [37]

55 <30> (<30> immobilized enzyme [20]) [20]

Temperature range (°C)

25-50 <30> (<30> immobilized enzyme at 25°C: about 60% of maximal activity, soluble enzyme approx. 40%, immobilized enzyme at 50°C: about 30% of maximal activity, soluble enzyme approx. 20% [8]) [8]

30-55 <30> (<30> at 30°C: about 50% of maximal activity, at 55°C: about 15% of maximal activity [20]) [20]

38-60 <30> (<30> at 38°C: about 55% of maximal activity, at 60°C: about 20% of maximal activity [20]) [20]

4 Enzyme Structure

Molecular weight

25500 <14> (<14> recombinant enzyme after cleavage from fusion protein [38]) [38]

51000 <14> (<14> gel filtration [37]) [37, 39]

55000 <28, 30> (<30> analytical polyacrylamide disc gel electrophoresis [33]; <28> gel filtration [25]) [25, 33]

60000 <31> (<31> polyacrylamide gradient gel electrophoresis [9]) [9]

Subunits

? <29> (<29> x * 25000-28000, assumed to be a dimer, SDS-PAGE [11]) [11]

dimer <14, 28, 30, 31> (<31> 2 * 30000, drastic denaturation conditions: 0.1% SDS and 2-mercaptoethanol, SDS-PAGE [9]; <28> 2 * 27000, SDS-PAGE [25,29]; <30> 1 * 27000 + 1 * 25000, SDS-PAGE [33]; <30> 1 * 26762, deduced from nucleotide sequence [36]; <14> 2 * 28000, SDS-PAGE [37]; <14> 2 * 28000, SDS-PAGE [39]) [9, 25, 29, 33, 36, 37, 39]

monomer <14> (<14> 1 * 25500, recombinant enzyme after cleavage from fusion protein, gel filtration [38]) [38]

5 Isolation/Preparation/Mutation/Application

Source/tissue

brain <29, 31> [9, 32]

callus <9, 12, 13> [5]

cell suspension culture <27> [1]

erythrocyte <32> [27, 28]

grain <27> [5]

liver <28, 29> [3, 7, 14, 16, 18, 19, 21, 22, 24, 25]

plant <1-13, 27> [5]

plumule <27> [5]

radicle <27> [5]

seedling <2, 3, 6-11, 27> [5, 15, 26]

Localization

Additional information <28> (<28> entirely present in 105000 * g supernatant fluid [22]) [22]

Purification

<14> (recombinant enzyme, DEAE-cellulose, hydroxyapatite, CM-cellulose, Bio-Gel [37]) [37, 39]

<27> (partial [26]) [26]

<28> (partial [14]; acid treatment, ammonium sulfate, DEAE-Sephadex, Sephadex G-100, phosphopyridoxal-Sepharose [3]) [3, 14, 24, 25, 29, 35]

<29> (affinity chromatography [32]) [7, 32, 11]

<30> (partial [8,20,30,33]; affinity chromatography [34]) [8, 20, 30, 33, 34]

<31> [9]

Crystallization

<14> (enzyme in complex with pyridoxal 5'-phosphate, space group C_2, 2.07 A resolution [40]) [40]

Cloning

<14> (expression in Escherichia coli [37]; expression in Escherichia coli as maltose binding protein fusion [38]; wild-type and mutant enzyme [40]) [37, 38, 39, 40]

<30> (expression in Escherichia coli [36]) [36]

Engineering

D49A <14> (<14> decrease in affinity for pyridoxine 5'-phosphate [40]) [40]

H199A <14> (<14> decrease in affinity for pyridoxine 5'-phosphate [40]) [40]

H199N <14> (<14> little decrease in pyridoxine 5'-phosphate turnover [40]) [40]

R14E <14> (<14> decrease in affinity for pyridoxine 5'-phosphate [40]) [40]

R14M <14> (<14> decrease in affinity for pyridoxine 5'-phosphate [40]) [40]

R197E <14> (<14> strong decrease in affinity for pyridoxine 5'-phosphate [40]) [40]

R197M <14> (<14> catalyzes removal of the proS hydrogen atom from (4'R)-^{3}H-pyridoxamine 5'-phosphate, decrease in affinity for pyridoxine 5'-phosphate [40]) [40]

Y17F <14> (<14> decrease in affinity for pyridoxine 5'-phosphate [40]) [40]

Application

analysis <29> (<29> pyridoxamine phosphate oxidase assay applicable for measuring activity in erythrocytes [13]) [13]

synthesis <30> (<30> continuous production of pyridoxal 5'-phosphate with enzyme immobilized to iodo- and bromoacetyl polysaccharides [8]) [8]

6 Stability

pH-Stability

6.1 <27> (<27> 37°C, 30 min, isozyme E I: 85% loss of activity, isozyme E II: 50% loss of activity [15]) [15]

Temperature stability

5 <30> (<30> pH 7.0, immobilized enzyme stable for 25 days, free enzyme: rapid loss of activity [8]) [8]

24 <30> (<30> pH 7.0, immobilized enzyme: 50% loss of activity after 10 days, free enzyme: 50% loss of activity after 2-3 days [8]) [8]

25 <30> (<30> half-life: 4 days immobilized enzyme, less than 1 day free enzyme [20]) [20]

45 <30> (<30> above 45°C, loss of activity after 30 min [34]) [34]

65 <14> (<14> no loss in activity after incubation at temperatures below 50°C, 30% to 40% activity loss after 15 min at 65°C [37]) [37]

General stability information

<28>, bovine serum albumin stabilizes enzyme in dilute solution [29]

<28>, repeated thawing and refreezing: denaturation [25, 29]

<30>, immobilized enzyme is more stable than purified free enzyme against heat and pH change [8]

Storage stability

<14>, -75°C, 15% glycerol, several months, no loss in activity [37]

<14>, 4°C, several weeks, no loss of activity [37]

<28>, -20°C, 18 months, over 80% of activity remains, apo- and holoenzyme [25]

<28>, -20°C, 4 years stable, shell-frozen apo- and holoenzyme [29]

<28>, -80°C [3]

<30>, approx. 25°C, immobilized enzyme, 200 mM K_3PO_4, pH 7.0, longer than 1 week [20]

<30>, approx. 5°C, immobilized enzyme, 200 mM K_3PO_4, pH 7.0, longer than 1 month [20]

References

[1] Miyata, H.; Iwamoto, A.; Kojima, Y.; Furuhashi, K.; Tsuge, H.: Effects of culture medium on pyridoxine (pyridoxamine) phosphate oxidase activity in wheat callus. Agric. Biol. Chem., **52**, 343-348 (1988)

[2] Merrill, A. H.; Wang, E.: Highly sensitive methods for assaying the enzymes of vitamin B6 metabolism. Methods Enzymol., **122**, 110-115 (1986)

[3] Bowers-Komro, D.M.; Hagen, T.M.; McCormick, D.B.: Modified purification of pyridoxamine (pyridoxine) 5-phosphate oxidase from rabbit liver by 5-phosphopyridoxyl affinity chromatography. Methods Enzymol., **122**, 116-120 (1986)

[4] Choi, J.D.; McCormick, D.B.: Roles of arginyl residues in pyridoxamine-5-phosphate oxidase from rabbit liver. Biochemistry, **20**, 5722-5728 (1981)

[5] Tsuge, H.; Iwamoto, A.; Yamamoto, I.; Miyata, H.; Hattori, K.; Ohashi, K.: Distribution of pyridoxaminephosphate oxidase activity in various plants and calluses. Agric. Biol. Chem., **50**, 289-296 (1986)

[6] Bowers-Komro, D.M.; McCormick, D.B.: Pyridoxamine-5-phosphate oxidase exhibits no specificity in prochiral hydrogen abstraction from substrate. J. Biol. Chem., **260**, 9580-9582 (1985)

[7] Takeuchi, F.; Tsubouchi, R.; Shibata, Y.: Effect of tryptophan metabolites on the activities of rat liver pyridoxal kinase and pyridoxamine 5-phosphate oxidase in vitro. Biochem. J., **227**, 537-544 (1985)

[8] Tsuge, H.; Okada, T.: Immobilization of yeast pyridoxaminephosphate oxidase to halogenacetyl polysaccharides. Biotechnol. Bioeng., **26**, 412-418 (1984)

[9] Churchich, J.E.: Brain pyridoxine-5-phosphate oxidase. A dimeric enzyme containing one FMN site. Eur. J. Biochem., **138**, 327-332 (1984)

[10] Pflug, W.; Lingens, F.: Occurrence of pydixaminephosphate oxidase and pyridoxal kinase in Gram-negative and Gram-positive bacteria. Hoppe-Seyler's Z. Physiol. Chem., **364**, 1627-1630 (1983)

[11] Nutter, L.M.; Meisler, N.T.; Thanassi, J.W.: Absence of pyridoxine- (pyridox-amine-) 5-phosphate oxidase in Morris hepatoma 7777. Biochemistry, **22**, 1599-1604 (1983)

[12] Choi, J.D.; Bowers-Komro, M.; Davis, M.D.; Edmondson, D.E.; McCormick, D.B.: Kinetic properties of pyridoxamine (pyridoxine)-5-phosphate oxidase from rabbit liver. J. Biol. Chem., **258**, 840-845 (1983)

[13] Langham, L.; Garber, B.M.; Roe, D.A.; Kazarinoff, M. N.: A radiometric as-say of pyridoxamine (pyridoxine) 5-phosphate oxidase. Anal. Biochem., **125**, 329-334 (1982)

[14] Wada, H.; Snell, E.E.: The enzymatic oxidation of pyridoxine and pyridox-amine phosphates. J. Biol. Chem., **236**, 2089-2095 (1961)

[15] Tsuge, H.; Kuroda, Y.; Iwamoto, A.; Ohashi, K.: Partial purification and property of pyridoxine (pyridoxamine)-5-phosphate oxidase isozymes from wheat seedlings. Arch. Biochem. Biophys., **217**, 479-484 (1982)

[16] Gregory III, J.F.: Effects of ε-pyridoxyllysine and related compounds on li-ver and brain pyridoxal kinase and liver pyridoxamine (pyridoxine) 5-phosphate oxidase. J. Biol. Chem., **255**, 2355-2359 (1980)

[17] DePecol, M.E.; McCormick, D.B.: Syntheses, properties, and use of fluores-cent N-(5-phospho-4-pyridoxyl)amines in assay of pyridoxamine (pyridox-ine) 5-phosphate oxidase. Anal. Biochem., **101**, 435-441 (1980)

[18] Horiike, K.; Tsuge, H.; McCormick, D.B.: Evidence for an essential histidyl residue at the active site of pyridoxamine (pyridoxine)-5-phosphate oxi-dase from rabbit liver. J. Biol. Chem., **254**, 6638-6643 (1979)

[19] Horiike, K.; Merrill, A.H.; McCormick, D.B.: Activation and inactivation of rabbit liver pyridoxamine (pyridoxine) 5-phosphate oxidase activity by urea and other solutes. Arch. Biochem. Biophys., **195**, 325-335 (1979)

[20] Tsuge, H.; Sen-Maru, K.; Ohashi, K.: Immobilization of yeast pyridoxamine (pyridoxine) 5-phosphate oxidase by organomercurial agarose. FEBS Lett., **93**, 331-334 (1978)

[21] Merrill, A.H.; Horiike, K.; McCormick, D.B.: Evidence for the regulation of pyridoxal 5-phosphate formation in liver by pyridoxamine (pyridoxine) 5 -phosphate oxidase. Biochem. Biophys. Res. Commun., **83**, 984-990 (1978)

[22] Guerrieri, P.: Intracellular localization of pyridoxamine-5-phosphate oxi-dase in rabbit liver. FEBS Lett., **41**, 11-13 (1974)

[23] Kazarinoff, M.N.; McCormick, D.B.: Specificity of pyridoxine (pyridoxa-mine) 5-phosphate oxidase for flavin-phosphates. Biochim. Biophys. Acta, **359**, 282-287 (1974)

[24] Kazarinoff, M.N.; McCormick, D.B.: N-(5-phospho-4-pyridoxyl)amines as substrates for pyridoxine (pyridoxamine) 5-phosphate oxidase. Biochem. Biophys. Res. Commun., **52**, 440-446 (1973)

[25] Kazarinoff, M.N.; McCormick, D.B.: Rabbit liver pyridoxamine (pyridox-ine) 5-phosphate oxidase. Purification and properties. J. Biol. Chem., **250**, 3436-3442 (1975)

[26] Tsuge, H.; Watanabe, K.; Ohashi, K.: Evidence for a pyridoxine (pyridoxa-mine) 5'-phosphate oxidase from wheat seedlings. FEBS Lett., **88**, 205 207 (1978)

[27] Clements, J.E.; Anderson, B.B.: Glutathione reductase activity and pyridoxine (pyridoxamine) phosphate oxidase activity in the red cell. Biochim. Biophys. Acta, **632**, 159-163 (1980)

[28] Clements, J.E.; Anderson, B.B.: Pyridoxine (pyridoxamine) phosphate oxidase activity in the red cell. Biochim. Biophys. Acta, **613**, 401-409 (1980)

[29] Merrill, A.H.; Kazarinoff, M.N.; Tsuge, H.; Horiike, K.; McCormick, D.B.: Pyridoxamine (pyridoxine) 5-phosphate oxidase from rabbit liver. Methods Enzymol., **62**, 568-574 (1979)

[30] Tsuge, H.; Itoh, K.; Ozeki, K.; Ohashi, K.: Separation and characterization of two pyridoxamine (pyridoxine) 5'-phosphate oxidases from baker's yeast. Agric. Biol. Chem., **46**, 1075-1077 (1982)

[31] Tsuge, H.; Otani, K.; Ishido, T.; Ohashi, K.: Induction of plant pyridoxine (pyridoxamine) 5'-phosphate oxidase activity by light. Agric. Biol. Chem., **45**, 1725-1726 (1981)

[32] Cash, C.D.; Maitre, M.; Rumigny, J.F.; Mandel, P.: Rapid purification by affinity chromatography of rat brain pyridoxal kinase and pyridoxamine-5-phosphate oxidase. Biochem. Biophys. Res. Commun., **96**, 1755-1760 (1980)

[33] Tsuge, H.; Ozeki, K.; Ohashi, K.: Molecular and enzymatic properties of pyridoxamine (pyridoxine) 5'-phosphate oxidase from baker's yeast. Agric. Biol. Chem., **44**, 2329-2335 (1980)

[34] Tsuge, H.; Ozeki, K.; Sen-Maru, K.; Ohashi, K.: Purification and properties of pyridoxamine (pyridoxine) 5'-phosphate oxidase from baker's yeast. Agric. Biol. Chem., **43**, 1801-1807 (1979)

[35] Merrill, A.H.; Korytnyk, W.; Horiike, K.; McCormick, D.B.: Spectroscopic studies of complexes between pyridoxamine (pyridoxine)-5'-phosphate oxidase and pyridoxyl 5'-phosphate compounds differing at position 4'. Biochim. Biophys. Acta, **626**, 57-63 (1980)

[36] Loubbardi, A.; Marcireau, C.; Karst, F.; Guilloton, M.: Sterol uptake induced by an impairment of pyridoxal phosphate synthesis in Saccharomyces cerevisiae: cloning and sequencing of the PDX3 gene encoding pyridoxine (pyridoxamine) phosphate oxidase. J. Bacteriol., **177**, 1817-1823 (1995)

[37] Zhao, G.; Winkler, M.E.: Kinetic limitation and cellular amount of pyridoxine (pyridoxamine) 5'-phosphate oxidase of Escherichia coli K-12. J. Bacteriol., **177**, 883-891 (1995)

[38] Notheis, C.; Drewke, C.; Leistner, E.: Purification and characterization of the pyridoxol-5'-phosphate:oxygen oxidoreductase (deaminating) from Escherichia coli. Biochim. Biophys. Acta, **1247**, 265-271 (1995)

[39] Di Salvo, M.; Yang, E.; Zhao, G.; Winkler, M.E.; Schirch, V.: Expression, purification, and characterization of recombinant Escherichia coli pyridoxine 5'-phosphate oxidase. Protein Expr. Purif., **13**, 349-356 (1998)

[40] Di Salvo, M.L.; Ko, T.P.; Musayev, F.N.; Raboni, S.; Schirch, V.; Safo, M.K.: Active site structure and stereospecificity of Escherichia coli pyridoxine-5'-phosphate oxidase. J. Mol. Biol., **315**, 385-397 (2002)

Amine oxidase (copper-containing)

1 Nomenclature

EC number
1.4.3.6

Systematic name
amine:oxygen oxidoreductase (deaminating) (copper-containing)

Recommended name
amine oxidase (copper-containing)

Synonyms
2-phenylethylamine oxidase
ABP
amiloride-binding protein
amine oxidase
amine oxidase [copper-containing]
BOLAO
copper amine oxidase
Cu-Amine oxidase
DAO
Diamine oxidase
HPAO
MAOXI
MAOXII
methylamine oxidase
monamine oxidase
PK-DAO
RAO
SAO
serum amine oxidase
tyramine oxidase
VP97
amine oxidase (pyridoxal containing)
amine oxygen oxidoreductase
benzylamine oxidase
diamine:O_2 oxidoreductase (deaminating)
diamino oxhydrase
diaminooxidase
histaminase
histamine deaminase
histamine oxidase

monoamine oxidase
semicarbazid-sensitiv amine oxidase

CAS registry number
   9001-53-0

2 Source Organism

<1>   *Rattus norvegicus* [15, 16]
<2>   *Oryctolagus cuniculus* [16]
<3>   *Mus musculus* [16]
<4>   *Cavia porcellus* [16]
<5>   *Pisum sativum* [6, 7, 15, 18, 22, 55, 56, 57]
<6>   *Cicer arietinum* [18, 57]
<7>   *Lens esculenta* [19, 40, 44, 57, 58]
<8>   *Oryza sativa* (rice [20]) [20]
<9>   *Trifolium subterraneum* (clover [21]) [21]
<10> *Euphorbia characias* [29, 57, 58]
<11> *Bos taurus* [1, 2, 14, 15, 23, 30, 32, 33, 57]
<12> *Sus scrofa* [3-5, 8-14, 15, 16, 26-28, 30, 31, 33, 49, 57, 58]
<13> *Homo sapiens* [16, 17, 24, 25, 32, 41, 57]
<14> *Canis familiaris* [16]
<15> *Lathyrus cicera* [34]
<16> *Phaseolus vulgaris* [34]
<17> *Vicia faba* [35]
<18> *Arachis hypogaea* (groundnut [36]) [36]
<19> *Onobrychis viciifolia* [37, 48]
<20> *Seriola quinqueradiata* [39]
<21> *Glycine max* [38, 42]
<22> *Aspergillus niger* [33, 45, 47]
<23> *plants* (overview [32]) [32, 57]
<24> *bacteria* (overview [32]) [32]
<25> *Lens culinaris* [18]
<26> *Hansenula polymorpha* (1A2V) [43, 50, 54, 57]
<27> *Triticum aestivum* [46]
<28> *Aspergillus niger* [47]
<29> *Klebsiella oxytoca* [51]
<30> *Arabidopsis thaliana* [52]
<31> *Lathyrus odoratus* (sweet pea [53]) [53]
<32> *Lathyrus sativus* (grass pea [53]) [53, 55]
<33> *Arthrobacter globiformis* [56]

3 Reaction and Specificity

Catalyzed reaction

$RCH_2NH_2 + H_2O + O_2 = RCHO + NH_3 + H_2O_2$ (<12> mechanism [11]; <5> ping-pong mechanism is suggested [7]; <26> proposed mechanism, crystal structure [50]; <26> role of copper in enzyme activity [54]; <5, 6, 7, 10, 11, 12, 13, 23, 26> catalytic mechanism [57])

Reaction type

oxidation
oxidative deamination
redox reaction
reduction

Natural substrates and products

S $RCH_2NH_2 + H_2O + O_2$ <1-4, 12-14, 23, 24> (< 1-4,12-14> enzyme plays a protective role against histamine in diseases such as ischaemic bowel syndrome, mesenteric infarction and ulcerative colitis [16]; <23, 24> overview: possible functions [32]) (Reversibility: ? <1-4, 12-14, 23, 24> [16, 32]) [16, 32]

P $RCHO + NH_3 + H_2O_2$ <1-4, 12-14, 23, 24> [16, 32]

Substrates and products

S   (8-arginine)-vasopressin + H_2O + O_2 <13> (Reversibility: ? <13> [25]) [25]

P   ?

S   (8-lysine)-vasopressin + H_2O + O_2 <13> (Reversibility: ? <13> [25]) [25]

P   ?

S   (E,Z)-1,4-diamino-2-butene + H_2O + O_2 <19, 22, 28, 32, 31> (Reversibility: ? <19, 22, 28, 31, 32> [37, 47, 53]) [37, 47, 53]

P   4-aminobuten-2-al + NH_3 + H_2O_2 <19, 22, 28, 31, 32> [37, 47, 53]

S   1,10-diaminodecane + H_2O + O_2 <10, 13> (Reversibility: ? <10, 13> [16, 29]) [16, 29]

P   10-decanal + NH_3 + H_2O_2 <10, 13> [16, 29]

S   1,2-diaminoethane + H_2O + O_2 <13> (Reversibility: ? <13> [16]) [16]

P   2-aminoethanal + NH_3 + H_2O_2 <13> [16]

S   1,3-diaminopropane + H_2O + O_2 <13, 20, 27> (<13> 109% of activity with putrescine [41]) (Reversibility: ? <13, 20, 27> [16, 39, 46]) [16, 39, 41, 46]

P   3-aminopropanal + NH_3 + H_2O_2 <13, 20, 27> [16, 39, 41, 46]

S   1,4-diaminobutane + H_2O + O_2 <6-8, 10, 12, 13, 15-17, 19, 20, 21, 22, 27, 28, 31, 32> (<6> trivial name putrescine [18]) (Reversibility: ? <6-8, 10, 12, 13, 15-17, 19, 20, 21, 22, 27, 28, 31, 32> [14, 16-20, 29, 34, 35, 37, 38, 39, 46, 47, 53]) [14, 16-20, 29, 34, 35, 37, 38, 39, 41, 46, 47, 48, 53]

P   4-aminobutanal + NH_3 + H_2O_2 <6-8, 10, 12, 13, 15-17, 19, 20, 21, 22, 27, 28, 31, 32> [14, 16-20, 29, 34, 35, 37, 38, 39, 41, 46, 47, 48, 53]

S   1,4-methylhistamine + H_2O + O_2 <22> (Reversibility: ? <22> [45]) [45]

P   ?

S 1,5-diaminopentane + H_2O + O_2 <6-8, 10, 12, 13, 15-17, 19, 20, 22, 27, 28, 31, 32> (<6> trivial name cadaverine [18]; <8> 60% of activity with putrescine [20]; <13> most active [41]) (Reversibility: ? <6-8, 10, 12, 13, 15-17, 19, 20, 22, 27, 28, 31, 32> [14, 16, 18-20, 29, 34, 35, 39, 46, 47, 48, 53]) [14, 16, 18-20, 29, 34, 35, 39, 41, 46, 47, 48, 53]

P 5-aminopentanal + NH_3 + H_2O_2 <6-8, 10, 12, 13, 15-17, 19, 20, 22, 27, 28, 31, 32> [14, 16, 18-20, 29, 34, 35, 39, 41, 46, 47, 48, 53]

S 1,6-diaminohexane + H_2O + O_2 <7, 10, 13, 15, 16, 20, 31, 32> (Reversibility: ? <7, 10, 13, 15, 16, 20, 31, 32> [16, 19, 29, 34, 53]) [16, 19, 29, 34, 39, 41, 53]

P 6-aminohexanal + NH_3 + H_2O_2 <7, 10, 13, 15, 16, 20, 31, 32> [16, 19, 29, 34, 39, 41, 53]

S 1,7-diaminoheptane + H_2O + O_2 <10, 13, 15, 16, 20> (Reversibility: ? <10, 13, 15, 16, 20> [16, 29, 34, 39]) [16, 29, 34, 39, 41]

P 7-aminoheptanal + NH_3 + H_2O_2 <10, 13, 15, 15, 20> [16, 29, 34, 39, 41]

S 1,8-diaminooctane + H_2O + O_2 <13, 20> (Reversibility: ? <13, 20> [16, 39]) [16, 39]

P 8-aminooctanal + NH_3 + H_2O_2 <13, 20> [16, 39]

S 1,9-diaminononane + H_2O + O_2 <13> (Reversibility: ? <13> [16]) [16]

P 9-aminononanal + NH_3 + H_2O_2 <13> [16]

S 2-(3,4-dihydroxyphenyl)-ethylamine + H_2O + O_2 <1, 5, 11, 22, 12> (<11> trivial name dopamine, nonstereospecific hydrogen abstraction at C-1 and C-2, proposed mechanism [23]; <5,11> abstraction of pro-S hydrogen from the α-carbon [15]; <1,5,12> stereospecific abstraction of pro-S hydrogen from the α-carbon atom [15]) (Reversibility: ? <1, 5, 11, 22, 12> [15, 23, 45]) [15, 23, 45]

P 3,4-dihydroxyphenylacetaldehyde + NH_3 + H_2O_2 <1, 5, 11, 22, 12> [15, 23, 45]

S 2-butyne-1,4-diamine + H_2O + O_2 <32> (Reversibility: ? <32> [55]) [55]

P 4-amino-2-butynal + H_2O_2 + NH_3 <32> [55]

S 2-hydroxyputrescine + H_2O + O_2 <32, 31> (Reversibility: ? <32, 31> [53]) [53]

P 2-hydroxy-4-aminobutanal + H_2O_2 + NH_3 <32, 31> [53]

S 2-methylhistamine + H_2O + O_2 <13> (Reversibility: ? <13> [16]) [16]

P ?

S 2-phenylethylamine + H_2O + O_2 <17, 22, 29> (Reversibility: ? <17, 22, 29> [35, 45, 51]) [35, 45, 51]

P 2-phenylethanal + NH_3 + H_2O_2 <17, 22, 29> [35, 45, 51]

S 3-hydroxycadaverine + H_2O + O_2 <32, 31> (Reversibility: ? <32, 31> [53]) [53]

P 3-hydroxy-4-aminopentanal + H_2O_2 + NH_3 <32, 31> [53]

S 5-methylhistamine + H_2O + O_2 <13> (Reversibility: ? <13> [16]) [16]

P ?

S N^{π}-methylhistamine + H_2O + O_2 <13> (Reversibility: ? <13> [16]) [16]

P ?

S N^r-methylhistamine + H_2O + O_2 <13> (Reversibility: ? <13> [16]) [16]

P ?

S N-acetyl-1,4-diaminobutane + H_2O + O_2 <13> (<13> 8.6% of activity with putrescine [41]) (Reversibility: ? <13> [41]) [41]

P N-acetyl-aminobutanal + acetylamine + H_2O_2 <13> [41]

S N-acetyl-1,5-diaminopentane + H_2O + O_2 <13> (<13> 3.8% of activity with putrescine [41]) (Reversibility: ? <13> [41]) [41]

P N-acetyl-aminopentanal + acetylamine + H_2O_2 <13> [41]

S N^1-acetyl-spermidine + H_2O + O_2 <13> (<13> 8.6% of activity with putrescine [41]) (Reversibility: ? <13> [41]) [41]

P ?

S RCH_2NH_2 + H_2O + O_2 <1-33> (<6> no activity with 1,3-diaminopropane [18]; <6> no activity with ornithine [18]; <6> no activity with 4-amino-butyric acid [18]; <23,24> overview [32]; <12> treatment of diamine oxidase with reducing agents induces ability to catalyze oxidative deamination of substrates of monoamine oxidase EC 1.4.3.4 [13]; <13> short-chain aliphatic diamines are deaminated with highest reaction velocity [16]; <10> only aliphatic diamines from C_4 to C_{10} and cystamine are oxidized, enzymatic activity decreases sharply with increasing chain length of diamines [29]; <10, 13, 19> no activity with benzylamine [17, 29, 37]; <29> no activity with benzylamine, methylamine, cadaverine or putrescine [51]; <10> no activity with p-dimethylaminomethylbenzylamine [29]; <6, 7, 10> no activity with histamine [18, 19, 29]; <10> no activity with spermidine [29]; <32, 31> very low activity with: 2-phenylethyla-mine, tryptamine, histamine, N^1-naphthylethylenediamine, dopamine, spermine, benzylamine or homoveratrylamine [53]) (Reversibility: ? <1-33> [1-58]) [1-58]

P RCHO + NH_3 + H_2O_2 <1-33> [1-58]

S agmatine <13, 15-17, 20, 31, 32> (<13> 14% of activity with putrescine [41]) (Reversibility: ? <13, 15-17, 20, 31, 32> [34, 35, 39, 41, 53]) [34, 35, 39, 41, 53]

P ?

S aminobutane + H_2O + O_2 <22, 28, 29> (<29> 65% of activity with phe-nylethylamine, tyramine or tryptamine [51]) (Reversibility: ? <22, 28, 29> [45, 47, 51]) [45, 47, 51]

P butanal + NH_3 + H_2O_2 <22, 28, 29> [45, 47, 51]

S aminoheptan + H_2O + O_2 <11> (Reversibility: ? <11> [1]) [1]

P heptanal + NH_3 + H_2O_2 <11> [1]

S aminohexane + H_2O + O_2 <22, 28> (Reversibility: ? <22, 28> [45, 47]) [45, 47]

P hexanal + NH_3 + H_2O_2 <22, 28> [45, 47]

S aminopentane + H_2O + O_2 <22> (Reversibility: ? <22> [45]) [45]

P pentanal + NH_3 + H_2O_2 <22> [45]

S aminopropane + H_2O + O_2 <22> (Reversibility: ? <22> [45]) [45]

P propanal + NH_3 + H_2O_2 <22> [45]

S benzylamine + H_2O + O_2 <5, 11, 12, 17, 22, 28> (Reversibility: ? <5, 11, 12, 17, 22, 28> [2, 4, 6, 35, 45, 47]) [2, 4, 6, 35, 45, 47]

P benzaldehyde + NH_3 + H_2O_2 <5, 11, 12, 17, 22, 28> [2, 4, 6, 35, 45, 47]

S collagen + H_2O + O_2 <13> (Reversibility: ? <13> [25]) [25]
P ?
S cystamine + H_2O + O_2 <32, 31> (Reversibility: ? <32, 31> [53]) [53]
P 2-mercaptoethanal + H_2O_2 + NH_3 <32, 31> [53]
S histamine + H_2O + O_2 <12, 13, 17, 20, 22, 28> (<13> 28% of activity with putrescine [41]) (Reversibility: ? <12, 13, 17, 20, 22, 28> [14, 16, 17, 35, 39, 45, 47]) [14, 16, 17, 35, 39, 41, 45, 47]
P 4-imidazolylethanal + NH_3 + H_2O_2 <12, 13, 17, 20, 22, 28> [14, 16, 17, 18, 19, 29, 35, 39, 41, 45, 47]
S lysine + H_2O + O_2 <17> (Reversibility: ? <17> [35]) [35]
P 2-aminohexanoic acid + NH_3 + H_2O_2 <17> [35]
S methylamine + H_2O + O_2 <26> (Reversibility: ? <26> [54]) [54]
P methanal + NH_3 + H_2O_2 <26> [54]
S norepinephrine + H_2O + O_2 <22> (Reversibility: ? <22> [45]) [45]
P ?
S octopamine + H_2O + O_2 <22> (Reversibility: ? <22> [45]) [45]
P ?
S *p*-dimethylaminomethylbenzylamine + H_2O + O_2 <7, 12, 13, 15, 16, 22, 28, 31, 32> (Reversibility: ? <7, 12, 13, 15, 16, 22, 28, 31, 32> [11, 19, 25, 34, 47, 53]) [11, 19, 25, 34, 47, 53]
P *p*-dimethylaminomethylbenzaldehyde + NH_3 + H_2O_2 <7, 12, 13, 15, 16, 22, 28, 31, 32> [11, 19, 25, 29, 34, 47, 53]
S propene-1,3-diamine + H_2O + O_2 <32> (Reversibility: ? <32> [55]) [55]
P ?
S serotonin + H_2O + O_2 <22> (Reversibility: ? <22> [45]) [45]
P ?
S spermidine + H_2O + O_2 <6, 7, 11, 13, 15-17, 20, 21, 31, 32> (<6> 15% of activity with putrescine [18]) (Reversibility: ? <6, 7, 11, 13, 15-17, 20, 21, 31, 32> [1, 18, 19, 34, 35, 38, 39, 41, 53]) [1, 18, 19, 34, 35, 38, 39, 41, 53]
P ?
S spermine + H_2O + O_2 <6, 11, 15-17, 21> (<6> low activity [18]; <10> no reaction [29]) (Reversibility: ? <6, 11, 15-17, 21> [1, 18, 34, 35, 38]) [1, 18, 29, 34, 35, 38]
P ?
S tropocollagen + H_2O + O_2 <13> (Reversibility: ? <13> [25]) [25]
P ?
S tryptamine <17, 22, 29> (Reversibility: ? <17, 22, 29> [35, 45, 51]) [35, 45, 51]
P ?
S tyramine + H_2O + O_2 <17, 22, 28, 29, 31, 32> (Reversibility: ? <17, 22, 28, 29, 31, 32> [35, 45, 47, 51, 53]) [35, 45, 47, 51, 53]
P 4-hydroxyphenylethanal + NH_3 + H_2O_2 <17, 22, 28, 29, 31, 32> [35, 45, 47, 51, 53]

Inhibitors
(E)-1,4-diamino-2-butene <22, 28> (<22,28> 0.1 mM, 72% and 87% inhibition [47]) [47]

(Z)-1,4-diamino-2-butene <19, 22, 28> (<22,28> 0.1 mM, 28% inhibition [47]) [47, 48]

1,4-diamino-2-butanone <32, 31> [53]

1,4-diaminocyclohexane <21> (<21> 5 mM, 69% inhibition [42]) [42]

1,4-phenanthroline <11, 12, 17, 19, 22, 27, 28> (<11> 0.0075 mM, 41% inhibition [2]; <12> 0.33 mM, 65% inhibition [4]; <17> 0.1 mM [35]; <27> 1 mM, 71% inhibition [46]) [2, 4, 35, 46, 47, 48]

1,5-diamino-3-pentanone <19, 22, 28, 31, 32> (<22, 28> 1 mM, 50% inhibition [47]) [37, 47, 53]

1-amino-3-phenyl-3-propanone <22, 28> [47]

2,2'-dipyridyl <19, 22, 27, 28> (<27> 1 mM, 74% inhibition [46]) [46, 47, 48]

2,4-dinitrophenylhydrazine <22, 28> (<22, 28> 1 mM, complete inhibition [47]) [47]

2-mercaptoethanol <20> (<20> 0.1 mM, 89% inhibition [39]) [39]

2-methylbenzothiazoline hydrazone <7> (<7> competitive inhibition [44]) [44]

3,5-ethoxy-4-aminomethylpyridine * 2 HCl <23, 24> [32]

4-methoxybenzaldoxime <5> [56]

8-hydroxyquinoline <11, 13> (<11> 0.0075 mM, 27% inhibition [2]; <13> 1 mM, 49.3% inhibition [41]) [2, 41]

Ca^{2+} <8> (<8> 10 mM, 37% inhibition [20]) [20]

Cu^{2+} <20> (<20> 0.5 mM, 92% inhibition [39]) [39]

EDTA <8> (<8> 10 mM, 80% inhibition [20]) [20]

H_2O_2 <12> (<12> uncompetitive vs. p-dimethylaminomethylbenzylamine [11]) [11]

Hg^{2+} <8, 20> (<8> 10 mM, 80% inhibition [20]; <20> 0.5 mM, complete inactivation [39]) [20, 39]

$KMnO_4$ <20> (<20> 0.1 mM, complete inhibition [39]) [39]

L-lobeline <19> [48]

Li^+ <7> (<7> reversible noncompetitive inhibition, irreversible if the enzyme is frozen in the presence of Li^+ [58]) [58]

Mg^{2+} <8> (<8> 10 mM, 37% inhibition [20]) [20]

N-ethylmaleimide <27> (<27> 1 mM, 90% inhibition [46]) [46]

N-isopropyl-α-(2-methyl-hydrazine)p-toluamide-HCl <13> [32]

NH_3 <12> (<12> competitive vs. p-dimethylaminomethylbenzylamine, uncompetitive vs. O_2 [11]) [11]

Na^+ <7> (<7> reversible noncompetitive inhibition, irreversible if the enzyme is frozen in the presence of Li^+ [58]) [58]

NaCl <13> (<13> 200 mM, approx. 50% inhibition, almost complete inhibition of the purified enzyme with 1 M NaCl [17]) [17]

NaN_3 <6, 7, 10, 12, 13, 19> (<12> 3.3 mM, 48% inhibition [4]; <12> uncompetitive inhibition [9]; <12> azide binds to Cu^{2+} ions, competitive inhibition vs. O_2, uncompetitive vs. benzylamine [26]; <12> uncompetitive vs. p-dimethylaminomethylbenzylamine, N_3^- is equitorially coordinated to a tetragonal Cu(II) center [28]) [4, 9, 18, 19, 26, 28, 29, 37, 41]

Ni^{2+} <20> (<20> 0.5 mM, 92% inhibition [39]) [39]

acetone oxime <5, 33> [56]

amiloride <12> (<12> diuretic drug, competitive vs. agmatine and putrescine oxidation [49]) [49]

aminoacetonitrile <32, 31> [53]

aminoguanidine <1-4, 7, 12-14, 19, 20, 21> (<13> 0.00001 mM, 50% inhibition, putrescine or histamine as substrates [16]; <13> 0.02 mM, complete inhibition, oxidation of histamine [17]; <12> model of inhibition mechanism [27]; <20> 0.03 mM, complete inhibition [39]; <13> 1 mM, complete inhibition [41]) [16, 17, 19, 27, 37, 39, 41, 42, 48]

aniline <21> (<21> 10 mM, 43% inhibition [42]) [42]

arcaine sulfate <7, 10> [19, 29]

benzaldoxime <5> [56]

β-aminopropinitrile <32, 31> [53]

β-bromoethylamine <7, 15, 16> [19, 34]

biacetyl monoxime <5> [56]

chinconine <19> [48]

clonidine <12> (<12> antihypertensive drug, competitive vs. agmatine and putrescine oxidation [49]) [49]

cupricin <11> [2]

cuprizone <11, 12, 13, 17> (<11> copper chelating, 0.006 mM, 98% inhibition, competitive vs. benzylamine [2]; <12> competitive binding to enzyme copper is suggested [9]; <17> 0.1 mM [35]) [2, 9, 31, 35, 41]

cyanide <6, 7, 10-12, 13, 19, 21> (<11> 0.1 mM, 76% inhibition [2]; <12> uncompetitive vs. benzylamine, non-competititve vs. O_2 [26]; <13> 1 mM, 76% inhibition [41]) [2, 3, 18, 19, 26, 29, 37, 41, 42]

cyclohexanone oxime <5> [56]

cyclohexylamine <21> (<21> 10 mM, 93% inhibition [42]) [42]

dicyclohexylamine <21> (<21> 10 mM, 79% inhibition [42]) [42]

diethyldithiocarbamate <7, 10, 12, 13, 15-17, 21> (<12> no inhibition [31]; <12> 3.3 mM, 74% inhibition [4]; <13> 1 mM, complete inhibition [41]) [4, 19, 29, 34, 35, 41, 42]

gabexate mesylate <12> (<12> anti coagulant drug, competitive vs. agmatine and putrescine oxidation [49]) [49]

hexyl-2-pyridyl ketoxime <5> [56]

histamine <7, 10, 19> [19, 29, 37]

hydrazines <12> [10]

hydroxylamine <12, 19, 31, 32> (<12> 3.3 mM, 30% inhibition [4]) [4, 37, 53]

iproniazid <13, 20> (<13> 0.1 mM, 68% inhibition [17]; <20> 0.03 mM, 84% inhibition [39]) [17, 39]

isoniazid <7, 10> [19, 29]

isonicotinic acid hydrazide <12> [3]

m-phenylenediamine <12> (<12> competitive vs. cadaverine [12]) [12]

methyl-2-pyridyl ketoxime <5> [56]

methyl-3-pyridyl ketoxime <5> [56]

methyl-4-pyridyl ketoxime <5> [56]

methylglyoxalbis(guanylhydrazone) <5> (<5> without preincubation competitive inhibitor, non-competitive after 1 h preincubation [22]) [22]
neocuproine <12> (<12> 0.033 mM, 61% inhibition [4]) [4]
o-phenylenediamine <12> (<12> competitive vs. cadaverine [12]) [12]
octyl-2-pyridyl ketoxime <5> [56]
p-chloromercuriphenylsulfonate <12, 27> (<12> 0.1 mM, complete inhibition of enzyme from cultured aortic smooth muscle cells [31]; <27> 1 mM, complete inhibition [46]) [31, 46]
p-nitrophenylhydrazine <22, 28> (<22, 28> 1 mM, complete inhibition [47]) [47]
phenelzine <12> (<12> 0.001 mM, complete inhibition of enzyme from cultured aortic smooth muscle cells [31]) [31]
phenylhydrazine <6, 7, 10, 12, 15-17, 19, 22, 28, 29> (<12> irreversible inactivation most likely due to hydrazone formation [9]; <22, 28> 1 mM, complete inhibition [47]; <29> 0.0001 mM, 55% inhibition [51]) [3, 8, 9, 10, 18, 19, 29, 34, 35, 47, 48, 51, 58]
piperidine <21> (<21> 10 mM, 55% inhibition [42]) [42]
propene-1,3-diamine <32> (<32> competitive vs. putrescine [55]) [55]
propyl-2-pyridyl ketoxime <5, 33> [56]
pyridine-2-carbaldoxim <5, 33> [56]
pyridine-3-carbaldoxim <5, 33> [56]
pyridine-4-carbaldoxim <5> [56]
pyridoxine-HCl <8> (<8> 10 mM, 72% inhibition [20]) [20]
quinacrine <13, 22, 28> (<13> 1 mM, complete inhibition [41]; <22>, 675 inhibition [47]; <28> 0.1 mM, 735 inhibition [47]) [41, 47]
salicylaldehyde oxime <5> [56]
sanguinarine <22, 28> (<22, 28> 0.1 mM, 65% inhibition [47]) [47]
semicarbazide <10, 12, 13, 15, 16, 19, 21, 29> (<13> 0.1 mM, 99% inhibition, oxidation of histamine [17]; <12> 0.01 mM, complete inhibition of enzyme from cultured aortic smooth muscle cells [31]; <13> 1 mM, complete inhibition [41]; <29> 0.01 mM, 30% inhibition [51]) [3, 17, 29, 31, 34, 37, 41, 42, 51]
sodium thioglycolate <11> (<11> slight [2]) [2]
thiocyanate <12> (<12> uncompetitive vs. p-dimethylaminomethylbenzylamine, SCN⁻ is equatorially coordinated to a tetragonal Cu(II) center [28]) [28]
tranylcypromine <13> (<13> slight [17]) [17]
triethyltetramine <32, 31> [53]
Additional information <12, 13, 23, 24> (<12> substrate inhibition with diamines [12]; <12> no substrate inhibition with monoamines [12]; <23,24> selective inhibitors [32]; <13> not inhibited by pargyline [41]) [12, 32, 41]

Cofactors/prosthetic groups

2,4,5-trihydroxyphenylalanine quinone <5, 5, 7, 10, 11, 12, 13, 19, 22, 23, 26, 28, 29> (<11> trivial name 6-hydroxydopa or dopaquinone [33]; <19> irreversible inactivation with p-nitrophenylhydrazine gives rise to a coloured enzyme p-nitrophenylhydrazone indicating dopaquinone as cofactor [48]; <5, 6, 7, 10, 11, 12, 13, 23, 26> topa quinone is identified as active site cofactor of eukaryotic amine oxidases, enzyme does not contain pyridoxal 5'-phosphate or pyrrolo quinoline quinone as cofactor as earlier suggested [57]) [33, 43, 44, 45, 47, 48, 51, 57]

FAD <8> (<8> 4 molecules of FAD per dimer [20]) [20]

pyridoxal 5'-phosphate <12> (<12> enzyme may contain pyridoxal phosphate [3,9,10]) [3, 9, 10]

pyrroloquinoline quinone <11, 12, 15, 16, 22> (<11, 12, 22> enzyme may contain covalently-bound pyrroloquinoline quinone [33]; <15, 16> enzyme may contain pyrroloquinoline-quinone as prosthetic group [34]) [33, 34]

Additional information <5, 7, 9, 10, 12, 17> (<12> contains one "active-carbonyl" cofactor per dimer [8]; <5,7> contains one carbonyl-like group per mol [19, 22]; <10> contains 2 carbonyl-like groups per dimer [29]; <9> presence of one carbonyl group per enzyme dimer [21]; <17> 1 mol carbonyl-group per mol enzyme [35]) [8, 19, 21, 22, 29, 33, 35]

Metals, ions

Al^{3+} <20> (<20> activates [39]) [39]

Co^{2+} <12, 20> (<12> can replace Cu^{2+} in the enzyme [4]; <20> activates [39]) [4, 39]

Ni^{2+} <12> (<12> can replace Cu^{2+} in the enzyme [4]) [4]

Zn^{2+} <12, 20> (<12> can replace Cu^{2+} in the enzyme [4]; <20> activates [39]) [4, 39]

copper <5, 6, 7, 9-13, 15-17, 22, 23, 26, 28, 29> (<11, 12> copper protein [2,4,5,9]; <11> copper involved in enzyme activity [2]; <11> 3.7 g atom of copper per mol of enzyme [2]; <17> 2 copper atoms per mol of enzyme [35]; <11> contains cupric copper [2]; <12> contains 8 Cu^{2+} per 1200000 Da, Co^{2+}, Zn^{2+} and Ni^{2+} can replace Cu^{2+}, no effect of Mn^{2+} [4]; <12> 2 mol of Cu^{2+} per dimer [8]; <7> enzyme contains 2 copper atoms [19]; <9> 0.063% i.e. 1.5 copper atoms per dimer [21]; <13> 1.0 g atom of copper per 70000 Da [25]; <7, 15, 16> enzyme contains 0.082% copper [19,34]; <10> enzyme contains 2 Cu^{2+} per dimer [29]; <12> study of cupric ions by magnetic-resonance and kinetic methods, native enzyme contains 2 tightly bound Cu^{2+} ions [26]; <12> copper plays a functional role in enzyme activity [28]; <7> 2 active sites per dimer i. e. one per copper atom, copper free enzyme is stable but showns no activity [40]; <26> 2 mol copper/mol enzyme dimer [43]; <7> removed from enzyme by treatment with diethyldithiocarbamate [44]; <22> 0.5 copper atoms per subunit [47]; <28> 0.09-0.2 copper atoms per subunit [47]; <29> 1.7-2.0 copper atoms per enzyme molecule depending on assay [51]; <26> copper depleted enzyme can be reconstituted with either Cu^{2+}, Zn^{2+}, Co^{2+}, or Ni^{2+}, 79% of activity is restored with Cu^{2+}, 19% is restored with Co^{2+}, 1.7% with Zn^{2+} or Ni^{2+} [54]; <5, 6, 7, 10, 11, 12, 13, 23, 26> 1:1 stoichiometry ratio with organic cofactor dopaquinone, function of copper in the catalytic mechanism is not clear [57]) [2, 4, 5, 8, 9, 19, 21, 25, 26, 28, 29, 34, 35, 40, 43, 44, 47, 51, 54, 57]

manganese <13, 20> (<13> enzyme contains 1.2 gatom of manganese per 70000 Da subunit [25]; <10> no manganese found [29]; <20> Mn^{2+} activates [39]) [25, 39]

Turnover number (min^{-1})

0.014 <7> (p-(dimethylamino)benzylamine) [44]
124.8 <26> (methylamine, <26> Co^{2+} reconstituted enzyme [54]) [54]
127.2 <26> (methylamine, <26> native enzyme [54]) [54]
2220 <22> (bcnzylamine) [45]
3900 <22> (norepinephrine) [45]
4080 <22> (1,4-methylhistamine) [45]
4800 <22> (tyramine) [45]
5700 <22> (histamine) [45]
6000 <22> (tryptamine) [45]
7320 <22> (octopamine) [45]
7920 <22> (aminobutane) [45]
7920 <22> (aminopropane) [45]
8100 <22> (dopamine) [45]
8160 <22> (serotonin) [45]
8220 <22> (2-phenylethylamine) [45]
8520 <22> (aminopentane) [45]
9300 <7> (putrescine) [44]
9540 <22> (aminohexane) [45]
9780 <21> (putrescine) [42]
11300-11900 <12> (benzylamine) [8]
11580 <21> (spermidine) [42]
29880 <21> (cadaverine) [42]

Specific activity (U/mg)

0.0001 <12> (<12> partially purified enzyme from kidney [30]) [30]
0.012 <13> [16]
0.0166 <8> [20]
0.0193 <26> (<26> copper depleted, reconstituted with Zn^{2+} [54]) [54]
0.0195 <26> (<26> copper depleted, reconstituted with Ni^{2+} [54]) [54]
0.023 <27> [46]
0.046 <12> (<12> oxidation of histamine, H$_2$S treated enzyme [13]) [13]
0.054 <12> (<12> oxidation of tyramine, H$_2$S treated enzyme [13]) [13]
0.0571 <26> (<26> copper depleted enzyme [54]) [54]
0.058 <12> (<12> oxidation of serotonine creatine sulfate, H$_2$S treated enzyme [13]) [13]
0.105 <12> [8]
0.11 <12> (<12> oxidation of tryptamine, H$_2$S treated enzyme [13]) [13]
0.13 <12> (<12> oxidation of putrescine, H$_2$S treated enzyme [13]) [13]
0.21 <12> (<12> oxidation of cadeverine, H$_2$S treated enzyme [13]) [13]
0.213 <26> (<26> copper depleted, reconstituted with Co^{2+} [54]) [54]
0.216 <12> (<12> oxidation of N-methyl-β-phenylethylamine, H$_2$S treated enzyme [13]) [13]
0.22 <13> (<13> enzyme from seminal plasma [24]) [24]
0.31 <12> (<12> oxidation of histamine [13]) [13]
0.38 <12> (<12> oxidation of putrescine [13]) [13]
0.412 <19> (<19> enzyme from seedlings [37]) [37]

0.51 <12> (<12> oxidation of cadeverine [13]) [13]
0.58 <18> [36]
0.72 <28> (<28> aminoheptane [47]) [47]
0.73 <13> [41]
0.894 <26> (<26> copper depleted, reconstituted with Cu^{2+} [54]) [54]
1.13 <26> (<26> native enzyme [54]) [54]
1.27 <12> [3]
1.44 <12> [28]
1.6 <12> [5]
1.74 <28> (<28> aminoheptane [47]) [47]
2.08 <12> (<12> enzyme from cultured aortic smooth muscle cells [31]) [31]
2.76 <22> (<22> benzylamine [47]) [47]
3.2 <12> [27]
5.6 <26> (<26> recombinant enzyme, methylamine oxidation [43]) [43]
5.6 <29> (<29> tyramine or phenylethylamine [51]) [51]
6.7 <16> [34]
7 <13> [25]
7.68 <22> (<22> aminoheptane [47]) [47]
21.1 <20> [39]
21.6 <17> [35]
27 <10> [29]
45.9 <5> [55]
48.61 <31> [53]
50.8 <15> [34]
60 <6> (<6> enzyme from shoots and cotyledon [18]) [18]
60.1 <5> (<5> enzyme from embryo [6]) [6]
64.25 <32> [53]
66 <9> [21]
69.62 <5> (<5> enzyme from epicotyl [22]) [22]
70.8 <7> (<7> enzyme from seedlings [19]) [19]
72 <32> [55]
72 <5, 25> (<5,25> enzyme from shoots and cotyledon [18]) [18]
79.6 <5> (<5> enzyme from cotyledon [6]) [6]
100 <7> (<7> approx. value [44]) [44]
135 <19> (<19> cadaverine [48]) [48]
263 <21> [42]
45400 <22> [45]
Additional information <11> (<11> 260.0-380.0 units/mg, 1 unit is defined as the amount of enzyme catalyzing a change of 0.001 absorbance per minute at 25°C [1]) [1]
Additional information <12> (<12> 1100.0-1200.0 units/mg, 1 unit is defined as the amount of enzyme catalyzing a change of 0.001 absorbance per minute at 25°C [4]) [4]
Additional information <21> (<21> 19.6 units/mg, 1 unit is defined as the change in absorbance of 0.1 at 470 nm/min [38]) [38]

K$_m$-Value (mM)

0.00128 <12> (benzylamine, <12> at pH 9.0 [4]) [4]

0.0051 <12> (benzylamine, <12> enzyme from cultured aortic smooth muscle cells, K$_m$ decreases with increasing pH [31]) [31]

0.01 <26> (methylamine, <26> native enzyme [54]) [54]

0.01 <8> (putrescine) [20]

0.013 <12> (histamine, <12> free enzyme [14]) [14]

0.0146 <20> (cadaverine) [39]

0.017 <26> (benzylamine, <26> benzylamine oxidase [43]) [43]

0.0172 <13> (histamine) [41]

0.0174 <12> (benzylamine, <12> at pH 7.2 [4]) [4]

0.019 <9> (cadaverine) [21]

0.019 <13> (histamine) [16]

0.0279 <13> (cadaverine) [41]

0.028 <21> (spermidine) [38]

0.033 <13> (putrescine) [41]

0.038 <9> (putrescine) [21]

0.065 <21> (cadaverine) [42]

0.065 <12> (histamine, <12> at pH 7.4 [12]) [12]

0.065 <21> (spermine) [38]

0.074 <9> (spermidine) [21]

0.075-0.095 <12> (benzylamine) [8]

0.083 <12> (histamine, <12> at pH 6.3 [12]) [12]

0.083 <13> (putrescine) [16]

0.09 <31, 32> (3-hydroxycadaverine) [53]

0.09 <19> (cadaverine) [48]

0.09 <5> (putrescine, <5> embryo enzyme [6]) [6]

0.097 <13> (N^r-methylhistamine) [16]

0.1 <31, 32> ((E)-1,4-diamino-2-butene) [53]

0.1 <31, 32> (cystamine) [53]

0.1 <22> (tryptamine) [45]

0.11 <31, 32> (1,6-diaminohexane) [53]

0.12 <22> (2-phenylethylamine) [45]

0.12 <22> (tyramine) [45]

0.13 <32> ((E)-2-butene-1,4-diamine) [55]

0.13 <31, 32> (cadaverine) [53]

0.13 <1> (dopamine) [15]

0.14 <22> (1,4-methylhistamine) [45]

0.146 <26> (methylamine, <26> recombinant enzyme [43]) [43]

0.15 <22, 28> (aminohexane) [47]

0.16 <5> (putrescine, <5> cotyledon enzyme [6]) [6]

0.17 <12> (histamine, <12> at pH 8.5 [12]) [12]

0.19 <32> (propene-1,3-diamine) [55]

0.19 <21> (putrescine) [42]

0.2 <10> (1,4-diaminobutane) [29]

0.2 <22> (aminohexane) [45]

0.2 <10> (putrescine) [29]

0.21 <19> (cadaverine) [37]
0.24 <22> (benzylamine) [45]
0.24 <6> (putrescine) [18]
0.24 <7> (putrescine) [19]
0.24 <19> (putrescine) [48]
0.25 <27> (1,3-diaminopropane) [46]
0.25 <5> (dopamine) [15]
0.29 <32> (putrescine) [55]
0.29 <22> (serotonin) [45]
0.3 <31, 32> (putrescine) [53]
0.32 <15> (putrescine) [34]
0.35 <22> (aminopentane) [45]
0.36 <19> (1,4-diamino-2-butene) [37]
0.36 <12> (dopamine) [15]
0.36 <22> (octopamine) [45]
0.4 <7> (cadaverine) [19]
0.435 <27> (putrescine) [46]
0.47 <10> (1,5-diaminopentane) [29]
0.47 <10> (cadaverine) [29]
0.5 <31, 32> (agmatine) [53]
0.53 <21> (putrescine) [38]
0.6 <22> (aminobutane) [45]
0.6 <22> (dopamine) [45]
0.6 <22> (histamine) [45]
0.625 <13> (N^1-acetylspermidine) [41]
0.67 <31, 32> (2-hydroxyputrescine) [53]
0.68 <26> (methylamine, <26> Co^{2+} reconstituted enzyme [54]) [54]
0.682 <26> (benzylamine, <26> recombinant enzyme [43]) [43]
0.72 <19> (putrescine) [37]
0.73 <12> (cadaverine, <12> at pH 7.4 [12]) [12]
0.83 <12> (cadaverine, <12> at pH 8.5 [12]) [12]
0.84 <32> (2-butyne-1,4-diamine) [55]
0.87 <12> (putrescine, <12> free enzyme [14]) [14]
0.927 <18> (putrescine) [36]
0.95 <32> ((Z)-2-butene-1,4-diamine) [55]
1.05 <22> (norepinephrine) [45]
1.3 <12> (cadaverine, <12> at pH 6.3 [12]) [12]
1.45 <32> (spermidine) [53]
1.49 <11> (benzylamine) [2]
1.5 <11> (spermidine, <11> free enzyme [14]) [14]
1.6 <22> (aminopropane) [45]
1.67 <31> (spermidine) [53]
1.7 <11> (benzylamine, <11> immobilized enzyme [14]) [14]
1.8 <12> (histamine, <12> immobilized enzyme [14]) [14]
1.8 <12> (putrescine, <12> immobilized enzyme [14]) [14]
2.4 <11> (spermidine, <11> immobilized enzyme [14]) [14]
2.6 <12> (cadaverine, <12> free enzyme [14]) [14]

3 <12> (butylamine, <12> at pH 8.5 [12]) [12]
3.28 <13> (acetylcadaverine) [41]
3.4 <11> (propylamine, <11> immobilized enzyme [14]) [14]
4 <12> (butylamine, <12> at pH 6.3 [12]) [12]
4.1 <11> (benzylamine, <11> free enzyme [14]) [14]
5 <12> (butylamine, <12> at pH 7.4 [12]) [12]
5.5 <11> (propylamine, <11> free enzyme [14]) [14]
5.71 <13> (spermine) [41]
5.9 <21> (spermidine) [42]
6.5 <12> (lysine methylester, <12> at pH 7.4 [12]) [12]
6.6 <12> (propylamine, <12> at pH 7.4 [12]) [12]
7 <16> (putrescine) [34]
8.1 <12> (cadaverine, <12> immobilized enzyme [14]) [14]
9.52 <13> (spermidine) [41]
9.71 <13> (acetylputrescine) [41]
66 <12> (ethylamine, <12> at pH 7.4 [12]) [12]
66 <12> (ethylenediamine, <12> at pH 7.4 [12]) [12]

K_i-Value (mM)
0.0000015 <7> (phenylhydrazine) [19]
0.00004 <16> (phenylhydrazine) [34]
0.00006 <15> (phenylhydrazine) [34]
0.00014 <19> (hydroxylamine) [37]
0.00016 <19> (1,5-diamino-pentanone) [37]
0.00025 <10> (phenylhydrazine) [29]
0.00037 <19> (aminoguanidine) [37]
0.0015 <7> (β-bromoethylamine) [19]
0.0027 <16> (β-bromoethylamine) [34]
0.003 <33> (propyl-2-pyridyl ketoxime) [56]
0.003 <16> (semicarbazide) [34]
0.004 <10> (arcaine sulfate) [29]
0.004 <33> (pyridine-3-carbaldoxime) [56]
0.0055 <7> (semicarbazide) [19]
0.0063 <15> (semicarbazide) [34]
0.007 <15> (β-bromoethylamine) [34]
0.0097 <19> (semicarbazide) [37]
0.01 <12> (putrescine) [49]
0.014 <11> (cuprizone) [2]
0.017 <5> (methylglyoxalbis(guanylhydrazone), <5> after preincubation for 1 h [22]) [22]
0.017 <5> (propyl-2-pyridyl ketoxime) [56]
0.02 <32,31> (1,4-diamino-2-butanone) [53]
0.026 <32,31> (1,5-diamino-3-pentanone) [53]
0.027 <12> (putrescine) [49]
0.037 <5> (methyl-2-pyridyl ketoxime) [56]
0.04 <5> (methyl-4-pyridyl ketoxime) [56]
0.045 <21> (cyclohexylamine) [42]

0.045 <7> (diethyldithiocarbamate) [19]

0.05 <10> (semicarbazide) [29]

0.057 <5> (methyl-3-pyridyl ketoxime) [56]

0.058 <33> (pyridine-2-carbaldoxime) [56]

0.07 <15> (diethyldithiocarbamate) [34]

0.08 <32> (propene-1,3-diamine) [55]

0.09 <16> (diethyldithiocarbamate) [34]

0.1 <5> (pyridine-3-carbaldoxim) [56]

0.12 <10> (NaCN) [29]

0.159 <5> (octyl-2-pyridyl ketoxime) [56]

0.16 <5> (pyridine-4-carbaldoxim) [56]

0.2 <7> (2-methylbenzothiazoline hydrazone) [44]

0.26 <5> (methylglyoxalbis(guanylhydrazone), <5> without preincubation [22]) [22]

0.266 <33> (acetone oxime) [56]

0.3 <12> (H_2O_2, <12> vs. p-dimethylaminomethylbenzylamine [11]) [11]

0.3 <15> (NaCN) [34]

0.33 <5> (hexyl-2-pyridyl ketoxime) [56]

0.391 <5> (pyridine-2-carbaldoxim) [56]

0.43 <10> (diethylthiocarbamate) [29]

0.5 <12> (H_2O_2, <12> vs. O_2 [11]) [11]

0.5 <32,31> (aminoacetonitrile, <32,31> apparent value [53]) [53]

0.5 <32,31> (aminoguanidine, <32,31> apparent value [53]) [53]

0.557 <5> (4-methoxybenzaldoxime) [56]

0.76 <12> (cyanide, <12> vs. O_2, deduced from slope [26]) [26]

0.78 <21> (dicyclohexylamine) [42]

0.83 <16> (NaCN) [34]

0.83 <19> (histamine) [37]

0.9 <12> (putrescine) [49]

1.2 <21> (piperidine) [42]

1.7 <5> (acetone oxime) [56]

2 <12> (N_3^-) [28]

2 <19> (NaN_3) [37]

2 <7> (isoniazid) [19]

2 <32,31> (triethylentetramine) [53]

2.17 <12> (cyanide, <12> vs. benzylamine [26]) [26]

2.3 <5> (cyclohexanone oxime) [56]

2.37 <5> (benzaldoxime) [56]

2.9 <12> (cyanide, <12> vs. O_2, deduced from intercept [26]) [26]

3 <7> (sodium cyanide) [19]

5 <21> (aniline) [42]

6 <10> (histamine) [29]

6 <10> (isoniazid) [29]

6.5 <32> (β-aminopropionitrile) [53]

8.5 <5> (salicylaldehyde oxime) [56]

9 <31> (β-aminopropionitrile) [53]

10 <7> (arcaine sulfate) [19]

10 <12> (m-phenyldiamine) [12]
10 <12> (o-phenyldiamine) [12]
14 <31> (diethylentriamine) [53]
16 <31> (diethylentriamine) [53]
16 <32> (o-phenanthroline) [53]
17 <7> (histamine) [19]
17.6 <5> (diacetyl monoxime) [56]
18 <31> (o-phenanthroline) [53]
20 <32> (hydroxyquinoline) [53]
22.4 <12> (SCN^-) [28]
33.8 <12> (NH_3, <12> vs. p-dimethylaminomethylbenzylamine [11]) [11]
35 <31> (hydroxyquinoline) [53]
36 <12> (NH_3, <12> vs. O_2 [11]) [11]
40 <12> (N_3^-, <12> approx. value [9]) [9]
40 <12> (azide, <12> vs. benzylamine [26]) [26]
50 <31> (2,2'-dipyridiyl) [53]
50 <7> (NaN_3) [19]
54 <10> (NaN_3) [29]
75 <31> (2,2'-dipyridyl) [53]
84 <12> (azide, <12> vs. O_2 [26]) [26]
200 <7> (Li^+) [58]
200 <7> (Na^+) [58]

pH-Optimum

6-6.5 <13> (<13> histane [41]) [41]
6.4-6.6 <13> (<13> histamine [16]) [16]
6.5 <13> (<13> putrescine and cadaverine [41]) [41]
6.6-7 <13> (<13> putrescine [16]) [16]
6.8-7.8 <31, 32> (<31,32> depending on substrate [53]) [53]
7 <11, 15, 17, 18, 20> (<17> cadaverine, histamine, agmatine [35]; <20> histamine [39]; <11> free enzyme [14]) [14, 34, 35, 36, 39]
7-8 <9> (<9> putrescine [21]) [21]
7.1 <7, 10> (<7> putrescine [19]; <10> phosphate buffer [29]) [19, 29]
7.2 <17> (<17> putrescine, spermidine, tryptamine [35]) [35]
7.5 <5, 6, 8, 16, 20, 21, 22, 28> (<6,8> putrescine [18,20]; <20> cadaverine
[39]) [6, 18, 20, 34, 38, 39, 47]
7.6 <10> (<10> in Tris buffer [29]) [29]
7.7 <17, 19> (<17> tyramine [35]) [35, 37]
8 <11, 13> (<11> immobilized enzyme [14]; <13> N^1-acetylspermidine [41])
[14, 41]
8.5 <13, 17> (<17> dopamine, phenylethylamine, benzylamine [35]; <13>
acetylcadaverine [41]) [35, 41]
8.6 <17> (<17> lysine [35]) [35]
9 <12> (<12> 50 mM glycine, 1 mM EDTA [4]) [4]
9.5 <13> (<13> spermidine and spermine [41]) [41]
10-11 <13> (<13> acetylputrescine [41]) [41]

pH-Range

5-10 <21> (<21> pH 5 and pH 10: 50% activity [38]) [38]
6-10 <12> [4]
6.2-7.8 <13> (<13> pH 6.2 and pH 7.8: about 75% activity, putrescine [16])
[16]

Temperature optimum (°C)

30 <8> [20]
40 <12, 17> [4, 35]
50 <20> (<20> 25% activity at 20°C [39]) [39]

Temperature range (°C)

20-40 <12> (<12> relative activity at 20°C: 22.8, at 40°C: 31.2 [4]) [4]
20-63 <17> (<17> about 50% activity at 20°C and 63°C [35]) [35]

4 Enzyme Structure

Molecular weight

80000 <13, 28> (<13> gel filtration, variable PAGE, ultracentrifugation [25];
<28> gel filtration [47]) [25, 47]
110000 <27> (<27> gel filtration [46]) [46]
113000 <21> (<21> gel filtration [42]) [42]
123000 <8> (<8> gel filtration [20]) [20]
126000 <17> (<17> gel filtration [35]) [35]
140000 <10, 20, 31> (<10> gel filtration [29]; <20> gel filtration [39]; <31>
gel filtration [53]) [29, 39, 53]
145000 <19, 21> (<21> gel filtration [38]; <19> a small percentage appears as
290000 Da aggregate, gel filtration [48]) [38, 48]
150000 <9, 15, 16, 22> (<9> gel filtration [21]; <15,16> gel filtration [34];
<22> gel filtration [47]) [21, 34, 47]
172000 <29> (<29> PAGE [51]) [51]
180000 <5> [22]
186000 <12> (<12> sedimentation-equilibrium [26]) [26]
187000 <7> (<7> sedimentation velocity [19]) [19]
196000 <12> (<12> gradient PAGE [8]) [8]
200000 <13> (<13> gel filtration [17]) [17]
235000 <13> (<13> sedimentation equilibrium [25]) [25]
255400 <11> (<11> low speed sedimentation [1]) [1]
1200000 <12> (<12> gel filtration [4]) [4]

Subunits

? <12, 21> (<12> x * 130000, SDS-PAGE [31]; <21> x * 77000, SDS-PAGE
[42]) [17, 31, 42]
dimer <5, 7-10, 12, 13, 15, 16, 17, 19, 20, 21, 22, 27, 29, 31, 32> (<21> 2 *
70000, SDS-PAGE [38]; <20> 1 * 80000 + 1 * 60000, SDS-PAGE [39]; <7> 2 *
78000, SDS-PAGE [19]; <10> 2 * 72000, SDS-PAGE [29]; <8> 2 * 61200 [20];
<9> 2 * 80000, gel electrophoresis under denaturing conditions [21]; <5> 2 *

85000, SDS-PAGE [22]; <17> 2 * 74000, SDS-PAGE [35]; <15,16> 2 * 75000, SDS-PAGE [34]; <12> 2 * 95000, SDS-PAGE [8]; <13> 2 * 90000, SDS-PAGE [17]; <12> 2 * 97000, SDS-PAGE [26]; <13> 2 * 105000, SDS-PAGE [41]; <27> 2 * 58000, SDS-PAGE [46]; <22> 2 * 75000, SDS-PAGE, deduced from nucleotide sequence [47]; <19> 2 * 77000, SDS-PAGE [48]; <29> 2 * 86000, SDS-PAGE [51]; <31,32> 2 * 72000, SDS-PAGE [53]) [8, 17, 19-22, 26, 29, 34, 35, 38, 39, 41, 46, 47, 48, 51, 53]
monomer <13, 28> (<13> 1 * 70000-80000, SDS-PAGE [25]; <28> 1 * 80000, SDS-PAGE [47]) [25, 47]
octamer <12> (<12> 8 * 146000 [4]) [4]

Posttranslational modification

glycoprotein <7, 10, 12, 15, 16, 21, 22, 28, 31, 32> (<12> heterogenity of pig plasma amine oxidase may be due to variable carbohydrate content [8]; <7,15,16> 14% neutral sugar [19,34]; <10> 12% neutral sugar [29]; <21> heterogenity may be due to carbohydrate moiety [42]; <22> less than 0.3% [47]; <28> less than 0.8% [47]; <31> 10.6% neutral sugars [53]; <32> 13% neutral sugars [53]) [8, 19, 29, 34, 42, 47, 53]

5   Isolation/Preparation/Mutation/Application

Source/tissue

axis <21> [38]
blood plasma <11-13> [1, 2, 8-10, 17, 23, 26]
cecum <20> [39]
colonic mucosa <2, 14> [16]
cotyledon <5, 6, 25> [18]
embryo <5, 18> [6, 36]
kidney <12, 13> (<12> cortex [13]) [4, 5, 12-15, 27, 28, 30, 41, 58]
leaf <9, 17, 30> (<9> young folded leaves of clover [21]; <30> expression in developing leaves [52]) [21, 35, 52]
mycelium <28> [47]
placenta <13> [25]
root <19, 30> (<30> expression in root cap cells [52]) [37, 52]
seed <15, 16, 21> [34, 42]
seedling <5, 7, 8, 19, 31, 32> [6, 7, 15, 19, 20, 48, 53]
seminal plasma <13> [24]
shoot <5, 6, 19, 25, 27> [18, 37, 46]
xylem <30> [52]

Localization

cytoplasm <1-3, 12-14> (<1-3> chiefly in cytoplasm [16]; <1,3,13> mainly in cytoplasm but to a considerable extent also in mitochondrial fraction [16]) [16]
mitochondrion <1, 3, 13> (<1,3,13> mainly in cytoplasm but to a considerable extent also in mitochondrial fraction [16]) [16]
particle-bound <4> [16]

periplasm <29> (<29> 83% of activity, 17% is localized in the cytosol [51])
[51]
peroxisome <26> [43, 50]

Purification

<5> (one-step immunoaffinity purification [18]; phosphocellulose, methyl-
glyoxalbis(guanylhydrazone)-Sepharose [22]) [6, 7, 18, 22, 55]
<6> (one-step immunoaffinity purification [18]) [18]
<7> [19]
<8> [20]
<9> (3 isoenzymes [21]) [21]
<10> [29]
<11> (partial purification [30]) [1, 2, 30]
<12> (chromatofocusing, affinity chromatography, partially purified [30]; af-
finity chromatography on AH-Sepharose [5]) [3, 5, 8, 30, 31]
<13> (affinity chromatography on cadaverine-Sepharose [17]; affinity chro-
matography on AH-Sepharose [41]) [16, 17, 24, 25, 30, 41]
<15> [34]
<16> [34]
<17> [35]
<18> (partial [36]) [36]
<19> (CM-23-cellulose, hydroxyapatite, Sephacryl S 300-HR [48]) [37, 48]
<20> [39]
<21> [38, 42]
<22> [45, 47]
<25> (one-step immunoaffinity purification [18]) [18]
<26> (native and recombinant enzyme [43]) [43, 54]
<27> [46]
<28> [47]
<29> (ammonium sulfate, Mono-Q, phenyl-Superose [51]) [51]
<31> (ammonium sulfate, heat denaturation, DEAE-cellulose, hydroxyapatite,
Sephacryl S-300 HR [53]) [53]
<32> (ammonium sulfate, heat denaturation, DEAE-cellulose, hydroxyapatite,
Sephacryl S-300 HR [53]) [53, 55]

Crystallization

<11> [1]
<12> [4]
<26> (sitting drop, orthorhombic crystals, X-ray structure, 2.4 A [50]) [50]

Cloning

<26> (expression Saccharomyces cerevisiae [43]) [43, 54]
<30> (expression in Sf9 insect cells [52]) [52]

6 Stability

pH-Stability

3 <17> (<17> 4 h, complete loss of activity [35]) [35]

3.5 <17> (<17> 4 h, 50% loss of activity [35]) [35]

5-7 <21> (<21> slight decrease in activity, complete loss at pH 4 and pH 8 [42]) [42]

5-9 <17> (<17> 30°C, stable [35]) [35]

6 <20> (<20> unstable below [39]) [39]

6-8 <11> (<11> highest stability [1]) [1]

7-8 <20> (<20> extremely unstable below pH 6.0 and above pH 9.0 [39]) [39]

11 <17> (<17> 4 h, 50% loss of activity [35]) [35]

Temperature stability

0-60 <17> (<17> 10 min, pH 7.0, stable [35]) [35]

40 <20> (<20> stable below, unstable above 50°C [39]) [39]

60 <18, 19, 20> (<18> 20% activity is lost after 10 min [36]; <20> activity is completely lost after 30 min [39]; <19> 5 min at pH 7.0, no loss of activity [48]) [36, 39, 48]

70 <5, 17, 18, 21> (<5> 15 min, embryo enzyme loses 10% of activity, coty-ledon enzyme loses 80% of activity [6]; <17> 10 min, pH 7, 100% loss of activity [35]; <21> 15 min, stable [38]; <18> complete inactivation [36]) [6, 35, 36, 38]

80 <21> (<21> 15 min, complete loss of activity [38]) [38]

100 <8> (<8> 10 min, complete loss of activity [20]) [20]

General stability information

<10>, purified enzyme does not withstand lyophilization [29]

<11>, lyophilization: initial loss of 15-20% of activity [1]

<13>, unstable in dilute solution or in frozen state [41]

<21>, unstable at low ionic strenght, half-life of 8 d in 5 mM potassium phosphate, pH 7.0, at 4°C, 33 d if 100 mM NaCl is added, unstable if 0.01% n-octyl-β-D-glucopyranoside or 0.1% Triton X-100 are added, loss of 25% activity after 5 d [42]

<15, 16>, enzyme does not withstand lyophilization [34]

Storage stability

<8>, 4°C, stable for 15 days [20]

<10>, -20°C, in water or potassium phosphate buffer, indefinitely stable [29]

<10>, 4°C, 50% loss of activity after 15 days [29]

<11>, -5°C to +2°C, crystalline enzyme as a suspension in 0.03 M phosphate buffer, pH 7.0, containing 55% saturated ammonium sulfate, best storage conditions [1]

<11>, 5°C, 0.03 M phosphate buffer, pH 7.0, little loss of activity [1]

<12>, -20°C, 100 mM sodium-potassium phosphate, pH 7.4, stable for at least one year [27]

<13>, -20°C, 24 h, 90% loss of activity [16]

<13>, 500 mM ammonium sulfate, 500 mM potassium phosphate, pH 7.4, stable for at least 3 weeks [41]

<21>, -80°C, 5 mM potassium phosphate, pH 7.0, 100 mM NaCl, 2 months, no loss of activity [42]

<15, 16>, -20°C, pH 7, indefinitely stable [34]

References

[1] Yamada, H.; Yasunobu, K.T.: Monoamine oxidase I. Purification, crystallization and properties of plasma monoamine oxidase. J. Biol. Chem., **237**, 1511-1516 (1962)

[2] Yamada, H.; Yasunobu, K.T.: Monoamine oxidase II. Copper, one of the prosthetic groups of plasma monamine oxidase. J. Biol. Chem., **237**, 3077-3082 (1962)

[3] Mondovi, B.; Costa, M.T.; Agro, A.F.; Rotilio, G.: Pyridoxal phosphate as a prosthetic group of pig kidney diamine oxidase. Arch. Biochem. Biophys., **119**, 373-381 (1967)

[4] Carper, W.R.; Stoddard, D.D.; Martin, D.F.: Pig liver monoamine oxidase I: isolation and characterization. Biochim. Biophys. Acta, **334**, 287-296 (1974)

[5] Floris, G.; Fadda, M.B.; Pellegrini, M.; Corda, M.; Agro, A.F.: Purification of pig kidney diamine oxidase by gel-exclusion chromatography. FEBS Lett., **72**, 179-181 (1976)

[6] Srivastava, S.K.; Prakash, V.: Purification and properties of pea cotyledon and embryo diamine oxidase. Phytochemistry, **16**, 189-190 (1977)

[7] Nylen, U.; Szybek, P.: Kinetic and other characteristics of diamine oxidase of pea seedlings. Acta Chem. Scand. Ser.B, **28**, 1153-1160 (1974)

[8] Falk, M.C.; Staton, A.J.; Williams, T.J.: Heterogeneity of pig plasma amine oxidase: molecular and catalytic properties of chromatographically isolated forms. Biochemistry, **22**, 3746-3751 (1983)

[9] Lindström, A.; Olsson, B.; Petterson, G.: Effect of azide on some spectral and kinetic properties of pig-plasma benzylamine oxidase. Eur. J. Biochem., **48**, 237-243 (1974)

[10] Lindström, A.; Olsson, B.; Petterson, G.: Kinetics of the interaction between pig-plasma benzylamine oxidase and hydrazine derivatives. Eur. J. Biochem., **42**, 177-182 (1974)

[11] Bardsley, W.G.; Crabbe, M.J.C.; Shindler, J.S.: Kinetics of the diamine oxidase reaction. Biochem. J., **131**, 459-469 (1973)

[12] Costa, M.T.; Rotilio, G.; Agro, A.F.; Vallogini, M.P.; Mondovi, B.: On the active site of diamine oxidase: kinetic studies. Arch. Biochem. Biophys., **147**, 8-13 (1971)

[13] Stesina, L.N.; Akopyan, Z.I.; Gorkin, V.Z.: Modification of catalytic properties of amine oxidases. FEBS Lett., **16**, 349-351 (1971)

[14] Stevanato, R.; Porchia, M.; Befani, O.; Mondovi, B.; Rigo, A.: Characterization of free and immobilized amine oxidases. Biotechnol. Appl. Biochem., **11**, 266-272 (1989)

[15] Yu, P.H.: Three types of stereospecificity and the kinetic deuterium isotope effect in the oxidative deamination of dopamine as catalyzed by different amine oxidases. Biochem. Cell Biol., **66**, 853-861 (1988)

[16] Bieganski, T.; Kusche, J.; Lorenz, W.; Hesterberg, R.; Stahlknecht, C.D.; Feussner, K.D.: Distribution and properties of human intestinal diamine oxidase and its relevance for the histamine catabolism. Biochim. Biophys. Acta, **756**, 196-203 (1983)

[17] Baylin, S.B.; Margolis, S.: Purification of histaminase (diamine oxidase) from human pregnancy plasma by affinity chromatography. Biochim. Biophys. Acta, **397**, 294-306 (1975)

[18] Angelini, R.; Di Lisi, F.; Federico, R.: Immunoaffinity purification and characterization of diamine oxidase from Cicer. Phytochemistry, **24**, 2511-2513 (1985)

[19] Floris, G.; Giartosio, A.; Rinaldi, A.: Diamine oxidase from Lens esculenta seedlings: purification and properties. Phytochemistry, **22**, 1871-1874 (1983)

[20] Chaudhuri, M.M.; Ghosh, B.: Purification and characterization of diamine oxidase from rice embryos. Phytochemistry, **23**, 241-243 (1984)

[21] Delhaize, E.; Webb, J.: Purification and characterization of diamine oxidase from clover leaves. Phytochemistry, **26**, 641-643 (1987)

[22] Yanagisawa, H.; Hirasawa, E.; Suzuki, Y.: Purification and properties of diamine oxidase from pea epicotyls. Phytochemistry, **20**, 2105-2108 (1981)

[23] Summers, M.C.; Markovic, R.; Klinman, J.P.: Stereochemistry and kinetic isotope effects in the bovine plasma amine oxidase catalyzed oxidation of dopamine. Biochemistry, **18**, 1969-1979 (1979)

[24] Crabbe, M.J.; Kavanagh, J.P.: The purification and preliminary investigation of fumarase, peroxidase, diamine oxidase and adenosine deaminase from human seminal plasma. Biochem. Soc. Trans., **5**, 735-737 (1977)

[25] Crabbe, M.J.C.; Waight, R.D.; Bardsley, W.G.; Barker, R.W.; Kelly, I.D.; Knowles, P.F.: Human placental diamine oxidase. Improved purification and characterization of a copper- and manganese-containing amine oxidase with novel substrate specificity. Biochem. J., **155**, 679-687 (1976)

[26] Barker, R.; Boden, N.; Cayley, G.; Charlton, S.C.; Henson, R.; Holmes, M.C.; Kelly, I.D.; Knowles, P.F.: Properties of cupric ions in benzylamine oxidase from pig plasma as studied by magnetic-resonance and kinetic methods. Biochem. J., **177**, 289-302 (1979)

[27] Tamura, H.; Horiike, K.; Fukuda, H.; Watanabe, T.: Kinetic studies on the inhibition mechanism of diamine oxidase from porcine kidney by aminoguanidine. J. Biochem., **105**, 299-306 (1989)

[28] Dooley, D.M.; Golnik, K.C.: Spectroscopic and kinetics studies of the inhibition of pig kidney diamine oxidase by anions. J. Biol. Chem., **258**, 4245-4248 (1983)

[29] Rinaldi, A.; Floris, G.; Finazzi-Agro, A.: Purification and properties of diamine oxidase from Euphorbia latex. Eur. J. Biochem., **127**, 417-422 (1982)

[30] Amicosante, G.; Oratore, A.; Crifo, C.; Agro, A.F.: Rapid characterization and partial purification of various animal amine oxidases. Experientia, **40**, 1140-1142 (1984)

[31] Hysmith, R.M.; Boor, P.J.: Purification of benzylamine oxidase from cultured porcine aortic smooth muscle cells. Biochem. Cell Biol., **66**, 821-829 (1988)

[32] Calligham, B.A.; Holt, A.; Elliott, J.: Properties and functions of the semicarbazide-sensitive amine oxidases. Biochem. Soc. Trans., **19**, 228-233 (1991)

[33] Ameyama, M.; Hayashi, M.; Matsushita, K.; Shinagawa, E.; Adachi, O.: Microbial production of pyrroloquinoline quinone. Agric. Biol. Chem., **48**, 561-565 (1984)

[34] Cogoni, A.; Farci, R.; Medda, R.; Rinaldi, A.; Floris, G.: Amine oxidase from Lathyrus cicera and Phaseolus vulgaris: purification and properties. Prep. Biochem., **19**, 95-112 (1989)

[35] Matsuda, H.; Suzuki, Y.: Purification and properties of the diamine oxidase from Vicia faba leaves. Plant Cell Physiol., **22**, 737-746 (1981)

[36] Sindhu, R.K.; Desai, H.V.: Partial purification and characterization of diamine oxidase from groundnut embryo. Indian J. Biochem. Biophys., **17**, 194-197 (1980)

[37] Pec, P.; Zajoncova, L.; Jilek, M.: Diamine oxidase from sainfon (Onobrychis viciifolia). Purification and some properties. Biologia (Bratisl.), **44**, 1177-1184 (1989)

[38] Kang, J.H.; Cho, Y.D.: Purification and properties of diamine oxidase from soybean (Glycine max). Hanguk Saenghwahakhoe Chi, **22**, 361-366 (1989)

[39] Matsumiya, M.; Otake, S.: Lentil seedlings amine oxidase: preparation and properties of the copper-free enzyme. Nippon Suisan Gakkaishi, **52**, 1617-1623 (1986)

[40] Rinald, A.; Giartosio, A.; Floris, G.; Medda, R.; Finazzi Agro, a.: Lentil seedlings amine oxidase: preparation and properties of the copper-free enzyme. Biochem. Biophys. Res. Commun., **120**, 242-24 (1984)

[41] Suzuki, O.; Matsumoto, T.: Purification and properties of diamine oxidase from human kidney. Biogenic Amines, **4**, 237-245 (1987)

[42] Vianello, F.; Di Paolo, M.L.; Stevanato, R.; Gasparini, R.; Rigo, A.: Purification and characterization of amine oxidase from soybean seedlings. Arch. Biochem. Biophys., **307**, 35-39 (1993)

[43] Cai, D.; Klinman, J.P.: Copper amine oxidase: heterologous expression, purification, and characterization of an active enzyme in Saccharomyces cerevisiae. Biochemistry, **33**, 7647-7653 (1994)

[44] Medda, R.; Padiglia, A.; Pedersen, J.Z.; Rotilio, G.; Agro, A.F.; Floris, G.: The reaction mechanism of copper amine oxidase: detection of intermediates by the use of substrates and inhibitors. Biochemistry, **34**, 16375-16381 (1995)

[45] Schilling, B.; Lerch, K.: Amine oxidases from Aspergillus niger: identification of a novel flavin-dependent enzyme. Biochim. Biophys. Acta, **1243**, 529-537 (1995)

[46] Suzuki, Y.: Purification and characterization of diamine oxidase from Triticum aestivum shoots. Phytochemistry, **42**, 291-293 (1996)

[47] Frebort, I.; Tamaki, H.; Ishida, H.; Pec, P.; Luhova, L.; Tsuno, H.; Halata, M.; Asano, Y.; Kato, Y.; et al.: Two distinct quinoprotein amine oxidases are induced by n-butylamine in the mycelia of Aspergillus niger AKU 3302.

Purification, characterization, cDNA cloning and sequencing. Eur. J. Biochem., 237, 255-265 (1996)

[48] Zajoncova, L.; Sebela, M.; Frebort, I.; Faulhammer, H.G.; Navratil, M.; Pec, P.: Quinoprotein amine oxidase from sainfoin seedlings. Phytochemistry, 45, 239-242 (1997)

[49] Federico, R.; Angelini, R.; Ercolini, L.; Venturini, G.; Mattevi, A.; Ascenzi, P.: Competitive inhibition of swine kidney copper amine oxidase by drugs: amiloride, clonidine, and gabexate mesylate. Biochem. Biophys. Res. Commun., 240, 150-152 (1997)

[50] Li, R.; Klinman, J.P.; Mathews, F.S.: Copper amine oxidase from Hansenula polymorpha: the crystal structure determined at 2.4 A resolution reveals the active conformation. Structure, 6, 293-307 (1998)

[51] Hacisalihoglu, A.; Jongejan, J.A.; Duine, J.A.: Distribution of amine oxidases and amine dehydrogenases in bacteria grown on primary amines and characterization of the amine oxidase from Klebsiella oxytoca. Microbiology, 143, 505-512 (1997)

[52] Moller, S.G.; McPherson, M.J.: Developmental expression and biochemical analysis of the Arabidopsis atao1 gene encoding an H_2O_2-generating diamine oxidase. Plant J., 13, 781-791 (1998)

[53] Sebela, M.; Luhova, L.; Frebort, I.; Faulhammer, H.G.; Hirota, S.; Zajoncova, L.; Stuzka, V.; Pec, P.: Analysis of the active sites of copper/topa quinonecontaining amine oxidases from Lathyrus odoratus and L. sativus seedlings. Phytochem. Anal., 9, 211-222 (1998)

[54] Mills, S.A.; Klinman, J.P.: Evidence against reduction of Cu^{2+} to Cu^+ during dioxygen activation in a copper amine oxidase from Yeast. J. Am. Chem. Soc., 122, 9897-9904 (2000)

[55] Sebela, M.; Frebort, I.; Lemr, K.; Brauner, F.; Pec, P.: A study on the reactions of plant copper amine oxidase with C_3 and C_4 aliphatic diamines. Arch. Biochem. Biophys., 384, 88-99 (2000)

[56] Mlickova, K.; Sebela, M.; Cibulka, R.; Frebort, I.; Pec, P.; Liska, F.; Tanizawa, K.: Inhibition of copper amine oxidases by pyridine-derived aldoximes and ketoximes. Biochimie, 83, 995-1002 (2001)

[57] Padiglia, A.; Medda, R.; Bellelli, A.; Agostinelli, E.; Morpurgo, L.; Mondovi, B.; Agro, A.F.; Floris, G.: The reductive and oxidative half-reactions and the role of copper ions in plant and mammalian copper-amine oxidases. Eur. J. Inorg. Chem., 2001, 35-42 (2001)

[58] Padiglia, A.; Medda, R.; Lorrai, A.; Paci, M.; Pedersen, J.Z.; Boffi, A.; Bellelli, A.; Agro, A.F.; Floris, G.: Irreversible inhibition of pig kidney copper-containing amine oxidase by sodium and lithium ions. Eur. J. Biochem., 268, 4686-4697 (2001)

D-Glutamate oxidase

1 Nomenclature

EC number
1.4.3.7

Systematic name
D-glutamate:oxygen oxidoreductase (deaminating)

Recommended name
D-glutamate oxidase

Synonyms
D-glutamic acid oxidase
D-glutamic oxidase

CAS registry number
37255-41-7

2 Source Organism

<1> *Candida boidinii* [1]
<2> *Orconectes limosus* (crayfish) [2]
<3> *Octopus vulgaris* [3]

3 Reaction and Specificity

Catalyzed reaction
D-glutamate + H_2O + O_2 = 2-oxoglutarate + NH_3 + H_2O_2

Reaction type
oxidation
oxidative deamination
redox reaction
reduction

Natural substrates and products
  S D-aspartate + H_2O + O_2 <1-3> (<1, 3> poorer substrate than D-glutamate
         [1, 3]; <2> 20% higher activity with pure oxygen than in air [2]; <3> 70%
         higher activity with pure oxygen than in air [3]) (Reversibility: ? <1-3>
         [1-3]) [1-3]
  P oxaloacetate + NH_3 + H_2O_2 <1-3> [1-3]

S   D-glutamate + H_2O + O_2 <1-3> (<1, 3> better substrate than D-aspartate [1, 3]; <2> highly specific, no deamination of other D-amino acids found, 30% higher activity with pure oxygen than in air [2]; <3> highly specific, 70% higher activity with pure oxygen than in air [3]) (Reversibility: ? <1-3> [1-3]) [1-3]

P   2-oxoglutarate + NH_3 + H_2O_2 <1-3> [1-3]

Substrates and products

S   D-aspartate + H_2O + O_2 <1-3> (<1, 3> poorer substrate than D-glutamate [1, 3]; <2> 20% higher activity with pure oxygen than in air [2]; <3> 70% higher activity with pure oxygen than in air [3]) (Reversibility: ? <1-3> [1-3]) [1-3]

P   oxaloacetate + NH_3 + H_2O_2 <1-3> [1-3]

S   D-glutamate + H_2O + O_2 <1-3> (<1, 3> better substrate than D-aspartate [1,3]; <2> highly specific, no deamination of other D-amino acids found, 30% higher activity with pure oxygen than in air [2]; <3> highly specific, 70% higher activity with pure oxygen than in air [3]) (Reversibility: ? <1-3> [1-3]) [1-3]

P   2-oxoglutarate + NH_3 + H_2O_2 <1-3> [1-3]

S   N-methyl-D-aspartate + H_2O + O_2 <1> (<1> much poorer substrate than D-glutamate or D-aspartate [1]) (Reversibility: ? <1> [1]) [1]

P   oxaloacetate + methylamine + H_2O_2 <1> [1]

Inhibitors

D-malate <1> (<1> competitive inhibitor [1]) [1]

L-aspartate <2> (<2> slight inhibition at 100 mM [2]) [2]

L-glutamate <2> (<2> slight inhibition at 100 mM [2]) [2]

glutarate <1> (<1> competitive inhibitor [1]) [1]

iodoacetamide <2> (<2> 64% inhibition at 1 mM [2]) [2]

iodoacetate <3> (<3> 68% inhibition at 0.1 mM [3]) [3]

m-tartrate <1> (<1> competitive inhibitor [1]) [1]

o-iodosobenzoate <3> (<3> 32% inhibition at 0.1 mM [3]) [3]

p-chloromercuribenzoate <2, 3> (<2> 95% inhibition at 0.1 mM [2]; <3> 74% inhibition at 0.01 mM [3]) [2, 3]

succinate <2> (<2> slight inhibition at 1 mM [2]) [2]

urethane <3> (<3> 31% inhibition at 10 mM [3]) [3]

veronal <2, 3> (<2> strong inhibition [2]; <3> 60% inhibition at 8 mM [3]) [2, 3]

Cofactors/prosthetic groups

FAD <1, 3> (<1,3> cannot be replaced by FMN [1,3]) [1, 3]

Activating compounds

ethanol <3> (<3> 2fold activation in the presence of ethanol [3]) [3]

Specific activity (U/mg)

0.23 <2> [2]

Kₘ-Value (mM)

K_m**-Value (mM)**
   4.5 <3> (D-aspartate) [3]
   5.4 <2> (D-aspartate) [2]
   8 <3> (D-glutamate) [3]
   15.2 <2> (D-glutamate) [2]

pH-Optimum
   7 <1> [1]
   8.1-8.3 <3> [3]
   9 <2> (<2> with D-glutamate [2]) [2]

Temperature optimum (°C)
   37 <1> [1]

4 Enzyme Structure

Molecular weight
   43000 <1> (<1> gel filtration [1]) [1]

Subunits
   monomer <1> (<1> 1 * 45000, SDS-PAGE [1]) [1]

5 Isolation/Preparation/Mutation/Application

Source/tissue
   antennal gland <2> [2]
   cell culture <1> [1]
   hepatopancreas <3> [3]

Purification
   <1> (to homogeneity [1]) [1]
   <2> (partial [2]) [2]
   <3> (to homogeneity, precipitation techniques, calcium phosphate gel treatment [3]) [3]

6 Stability

Temperature stability
   21 <2> (<2> stable for more than 5 h [2]) [2]

General stability information
   <2>, very stable [2]

Storage stability
   <2>, -30°C, no loss of activity for several months [2]

References

[1] Fukunaga, S.; Yno, S.; Takahashi, M.; Taguchi, S.; Kera, Y.; Odani, S.; Yamada, R.H.: Purification and properties of D-glutamate oxidase from Candida boidinii 2201. J. Ferment. Bioeng., **85**, 579-583 (1998)
[2] Urich, K.: D-Glutamtoxydase aus der Antennendrüse des Flusskrebses Orconectes limosus: Reinigung und Charakterisierung. Z. Naturforsch. B, **23**, 1508-1511 (1968)
[3] Rocca, E.; Ghiretti, F.: Purification and properties of D-glutamic acid oxidase from Octopus vulgaris Lam.. Arch. Biochem. Biophys., **77**, 336-349 (1958)

Ethanolamine oxidase 1.4.3.8

1 Nomenclature

EC number
1.4.3.8

Systematic name
ethanolamine:oxygen oxidoreductase (deaminating)

Recommended name
ethanolamine oxidase

Synonyms
oxidase, ethanolamine

CAS registry number
9013-00-7

2 Source Organism

<1> *Pseudomonas sp.* (Bacterium ATCC 13796, Arthrobacter sp. [1,2]) [1, 2]
<2> *Phormia regina* (blowfly [3]) [3]

3 Reaction and Specificity

Catalyzed reaction
ethanolamine + H_2O + O_2 = glycolaldehyde + NH_3 + H_2O_2

Reaction type
oxidation
oxidative deamination
redox reaction
reduction

Natural substrates and products
S ethanolamine + H_2O + O_2 <1> (<1> aerobic catabolism of ethanolamine [2]) (Reversibility: ir <1> [1, 2]) [2]
P glycolaldehyde + NH_3 + H_2O_2

Substrates and products
S 1-amino-2-propanol + H_2O + O_2 <1> (<1> 18% of the activity with ethanolamine [2]) (Reversibility: ? <1> [1, 2]) [1, 2]
P 2-hydroxy-propionic aldehyde + NH_3 + H_2O_2

S 3-amino-1-propanol + H_2O + O_2 <1> (<1> 56% of the activity with etha-
nolamine [2]) (Reversibility: ? <1> [1, 2]) [1, 2]
P 3-hydroxy-propionic aldehyde + NH_3 + H_2O_2
S ethanolamine + H_2O + O_2 <1, 2> (<1,2> under aerobic conditions [1-3])
(Reversibility: ir <1, 2> [1, 2]) [1-3]
P glycolaldehyde + NH_3 + H_2O_2
S Additional information <1> (<1> no activity with: ethylamine, n-propyl-
amine, putrescine, choline and phosphoethanolamine [1,2]) [1, 2]
P ?

Inhibitors

2-mercaptoethanol <1, 2> [1-3]
Co^{2+} <2> [3]
KCN <2> [3]
cysteine <1, 2> [1-3]
dibenzylamine <2> [3]
glutathione <1, 2> (<2> no effect [3]) [1-3]
hydrazine <1, 2> [1-3]
hydroxylamine <1, 2> [1-3]
iproniazide <1> [1, 2]
isoniazide <1> (<1> low inhibition [2]) [1, 2]
quinacrine <1, 2> [1-3]
semicarbazide <1, 2> [1-3]
sodium azide <2> [3]
sodium bisulfite <1, 2> [1-3]
sodium-diethyldithiocarbamate <2> [3]
Additional information <2> (<2> not: 8-hydroxyquinoline [3]) [3]

Activating compounds

Additional information <1> (<1> no stimulation of activity by FAD, pyridox-
alphosphate, NAD^+, $NADP^+$, FMN or tetrazolium dyes [2]) [2]
Additional information <2> (<2> no stimulation of activity by FAD, pyridox-
alphosphate, NAD^+, $NADP^+$ [3]) [3]

Metals, ions

Ca^{2+} <2> (<2> activates the dialyzed enzyme, regeneration of 50% of the
initial activity [3]) [3]
Ni^{2+} <2> (<2> slight stimulation [3]) [3]
Zn^{2+} <2> (<2> activates the dialyzed enzyme, regeneration of 50% of initial
activity [3]) [3]

Specific activity (U/mg)

0.00026 <2> [3]
0.0621 <1> [1, 2]

K_m-Value (mM)

0.16 <2> (ethanolamine) [3]
5 <1> (ethanolamine) [1, 2]

321

pH-Optimum
7.8 <2> [3]
8 <1> [1, 2]

pH-Range
7-8.2 <2> (<2> pH 7.0: about 50% of activity maximum, pH 8.2: 37% of activity maximum [3]) [3]
7-9.8 <1> (<1> half-maximal activities at pH 7 and pH 9.8 [2]) [2]

Temperature optimum (°C)
37 <2> [3]

Temperature range (°C)
26-48 <2> (<2> half-maximal activities at 26°C and 48°C [3]) [3]

5 Isolation/Preparation/Mutation/Application

Source/tissue
egg <2> (<2> low activity [3]) [3]
larva <2> [3]
pupa <2> (<2> low activity [3]) [3]
Additional information <1> (<1> enzyme more active in larvae than in adults [2]) [2]

Localization
cytoplasm <2> [3]
Additional information <2> (<2> no activity in mitochondria [3]) [3]

Purification
<1> [1, 2]
<2> [3]

6 Stability

General stability information
<2>, dialysis causes inactivation, activity cannot be restored by addition of pyridoxal phosphate, FAD, NAD^+, $NADP^+$ alone or in combination with divalent cations [3]

Storage stability
<1>, -5°C, 6 months, stable [1]
<2>, more unstable, cannot be stored [3]

References

[1] Narrod, S.A.; Jakoby, W.B.: Ethanolamine oxidase. Methods Enzymol., **8**, 354-357 (1966)

[2] Narrod, S.A.; Jakoby, W.B.: Metabolism of ethanolamine, an ethanoloxidase. J. Biol. Chem., **239**, 2189-2193 (1964)

[3] Kulkarni, A.P.; Hodgson, E.: Ethanolamine oxidase from the blowfly, Phormia regina (Diptera: Insecta). Comp. Biochem. Physiol. B, **44**, 407-422 (1973)

Tyramine oxidase

1 Nomenclature

EC number
1.4.3.9 (deleted, included in EC 1.4.3.4)

Recommended name
tyramine oxidase

Putrescine oxidase 1.4.3.10

1 Nomenclature

EC number
1.4.3.10

Systematic name
putrescine:oxygen oxidoreductase (deaminating)

Recommended name
putrescine oxidase

Synonyms
PO
oxidase, putrescine

CAS registry number
9076-87-3

2 Source Organism

<1> *Micrococcus rubens* [1, 2, 3, 8, 11, 12, 13, 15]
<2> *Pseudomonas putida* [4]
<3> *Micrococcus luteus* [5]
<4> *Aspergillus nidulans* [6]
<5> *Bos taurus* [7, 14]
<6> *Candida guilliermondii* [9]
<7> *Candida albicans* [9]
<8> *Candida krusei* [9]
<9> *Candida tropicalis* [9]
<10> *Candida parapsilosis* [9]
<11> *Candida stellatoidea* [9]
<12> *Pisum sativum* [10]

3 Reaction and Specificity

Catalyzed reaction
putrescine + O_2 + H_2O = 4-aminobutanal + NH_3 + H_2O_2 (<1>, mechanism [12])

Reaction type
oxidation
oxidative deamination

redox reaction
reduction

Substrates and products

S 1,3-diaminopropane + O_2 + H_2O <5> (Reversibility: ? <5> [7]) [7]
P 3-aminopropanal + NH_3 + H_2O_2
S 1,6-diaminohexane + O_2 + H_2O <1, 5> (<1>, 1% of the activity with putrescine [12]) (Reversibility: ? <1, 5> [7, 12]) [7, 12]
P ?
S 1,7-diaminoheptane + O_2 + H_2O <5> (Reversibility: ? <5> [7]) [7]
P ?
S N-acetylputrescine + O_2 + H_2O <5> (Reversibility: ? <5> [7]) [7]
P N^4-acetylaminobutanal + NH_3 + H_2O_2
S N^1-acetylspermidine + O_2 + H_2O <5> (Reversibility: ? <5> [7]) [7]
P ?
S N^8-acetylspermidine + O_2 + H_2O <5> (Reversibility: ? <5> [7]) [7]
P ?
S benzylamine + O_2 + H_2O <5> (Reversibility: ? <5> [7]) [7]
P benzaldehyde + NH_3 + H_2O_2
S cadaverine + O_2 + H_2O <1, 2, 3, 5, 6> (<2>, at 11% of the activity with putrescine [4]; <6>, 47% of the activity with putrescine [9]; <1>, 7% of the activity with putrescine [12]) (Reversibility: ? <1, 2, 3, 5, 6> [2, 4, 5, 7, 9, 12]) [2, 4, 5, 7, 9, 12]
P 5-aminopentanal + NH_3 + H_2O_2
S ethylamine + O_2 + H_2O <5> (Reversibility: ? <5> [7]) [7]
P ?
S histamine + O_2 + H_2O <5> (<6>, no activity [9]) (Reversibility: ? <5> [7]) [7]
P ?
S putrescine + O_2 + H_2O <1, 2, 3, 4, 5, 6, 7, 8, 9, 10, 11, 12> (Reversibility: ? <1-12> [1-15]) [1, 2, 3, 4, 5, 6, 7, 8, 9, 10, 11, 12, 13, 14, 15]
P 4-aminobutanal + NH_3 + H_2O_2
S spermidine + O_2 + H_2O <1, 2, 5, 6> (<1>, 10% of the activity with putrescine [12]; <2>, at 9% of the activity with putrescine [4]; <5>, no activity at spermidine concentrations of 0.0175 mM, activity at 1 mM [7]; <6>, 51% of the activity with putrescine [9]; <5>, no activity [14]) (Reversibility: ? <1, 2, 5, 6> [2, 4, 7, 9, 12, 13, 15]) [2, 4, 7, 9, 12, 13, 15]
P 4-aminobutyraldehyde + 1,3-diaminopropane + H_2O_2 <1> (<1>, no liberation of ammonia [2,15]) [2, 15]
S spermine + O_2 + H_2O <5, 6> (<6>, 1.7% of the activity with putrescine [9]) (Reversibility: ? <5, 6> [7, 9]) [7, 9]
P N,N'-bis(3-aminopropyl)-4-aminobutanal + NH_3 + H_2O_2
S Additional information <1> (<1>, the enzyme is a weak 2-mercaptoethanol oxidase [3]; <1>, the enzyme modified with 1-ethyl-3-(3-dimethylaminopropyl)carbodiimide shows activity towards monoamines such as n-butylamine, n-hexylamine and n-octylamine, which are not substrates of the native enzyme [13]) [3, 13]
P ?

Inhibitors

1,1,4,4-tetramethyl-1,4-diaminobutane <1> [12]
1,10-diaminodecane <1> [12]
1,12-diaminododecane <1> [12]
1,3-diaminopropane <1> [12]
1,6-diaminohexane <1> [12]
1,7-diaminoheptane <1> [12]
1,8-diaminooctane <1> [12]
1-aminoethanol <1> [12]
1-ethyl-3-(3-dimethylaminopropyl)carbodiimide <1> (<1>, the activity of the modified enzyme towards putrescine is 5.6% of that of the native enzyme. The modified enzyme shows activity towards monoamines such as n-butylamine, n-hexylamine and n-octylamine, which are not substrates of the native enzyme [13]) [13]
3-amino-1-propanol <1> [12]
3-amino-1-propanol <1> [12]
4-amino-1-butanoic acid <1> [12]
5-amino-1-pentanoic acid <1> [12]
5-amino-1-pentanol <1> [12]
6-amino-1-hexanoic acid <1> [12]
6-amino-1-hexanol <1> [12]
8-amino-1-octanoic acid <1> [12]
Cd^{2+} <1> (<1>, inactivation due to dissociation of FAD from the enzyme molecule and denaturation of the apoenzyme [13]) [13]
Co^{2+} <1> (<1>, inactivation due to dissociation of FAD from the enzyme molecule and denaturation of the apoenzyme [13]) [13]
Cu^{2+} <1> (<1>, inactivation due to dissociation of FAD from the enzyme molecule and denaturation of the apoenzyme [13]) [13]
Hg^{2+} <1> (<1>, inactivation due to dissociation of FAD from the enzyme molecule and denaturation of the apoenzyme [13]) [13]
N,N,N',N'-tetramethyl-1,4-1,4-diaminobutane <1> [12]
Ni^{2+} <1> (<1>, inactivation due to dissociation of FAD from the enzyme molecule and denaturation of the apoenzyme [13]) [13]
PCMB <1> [2, 15]
Zn^{2+} <1> (<1>, inactivation due to dissociation of FAD from the enzyme molecule and denaturation of the apoenzyme [13]) [13]
Zn^{2+} <1> (<1>, inactivation due to dissociation of FAD from the enzyme molecule and denaturation of the apoenzyme [13]) [13]
allylamine <1> [12]
aminoguanidine hydrogen carbonate <5> [7]
ammonium <1> [12]
benzylamine <1> [12]
benzylamine <1> [12]
butylamine <1> [12]
butylamine <1> [12]
cadaverine <1> (<1>, noncompetitive against putrescine [3]) [3]
cyanide <1> [2]

dodecylamine <1> [12]
ethylamine <1> [12]
ethylenediamine <1> [12]
heptylamine <1> [12]
hexylamine <1> [12]
hydroxylamine <1> [2]
iproniazid <1> [2]
methylamine <1> [12]
methylglyoxal bis(guanylhydrazine)dihydrochloride monohydrate <5> [7]
pentylamine <1> [12]
phenylethylamine <1> [12]
propylamine <1> [12]
putrescine <2> (<2>, substrate inhibition at 10 mM or above [4]) [4]
spermidine <1> (<1>, competitive inhibition of putrescine oxidation [11]) [11]
spermine <1> (<1>, competitive inhibition of putrescine oxidation [11]) [11]
tyramine <1> [12]

Cofactors/prosthetic groups
FAD <1> (<1>, contains 1 mol of FAD per mol of enzyme [2]; <1>, contains 1 FAD per 82000 Da [3]; <1>, contains FAD [8,13,15]) [2, 3, 8, 13, 15]

Metals, ions
Additional information <6> (<6>, no activation by divalent cations [9]) [9]

Specific activity (U/mg)
9.1 <1> [15]
35 <1> [1]
59.16 <1> [3]
Additional information <2> [4]

K_m-Value (mM)
0.0026 <5> (putrescine) [7]
0.02 <6> (putrescine) [9]
0.025 <5> (spermidine) [7]
0.03 <2> (putrescine) [4]
0.038 <1> (putrescine) [3]
0.124 <1> (O_2) [3]
0.125 <1> (putrescine) [11]
0.2 <6> (spermidine) [9]
0.23 <1> (putrescine) [2]
0.23 <1> (spermidine) [2]
1.1 <6> (cadaverine) [9]
3.3 <4> (putrescine) [6]

K_i-Value (mM)
0.000022 <5> (aminoguanidine, <5>, with putrescine as substrate [7]) [7]
0.000024 <5> (methylglyoxal bis(guanylhydrazine)dihydrochloride monohydrate) [7]

0.002 <1> (1,12-diaminododecane) [12]
0.003 <1> (1,10-diaminodecane) [12]
0.0032 <1> (1,8-diaminooctane) [12]
0.0082 <1> (1,7-diaminoheptane) [12]
0.0082 <1> (dodecylamine) [12]
0.013 <1> (allylamine) [12]
0.014 <1> (heptylamine) [12]
0.023 <1> (1,6-diaminohexane) [12]
0.036 <1> (agmatine) [12]
0.039 <1> (hexylamine) [12]
0.039 <1> (tyramine) [12]
0.047 <1> (benzylamine) [12]
0.0581 <1> (spermine) [11]
0.087 <1> (phenylethylamine) [12]
0.088 <1> (5-amino-1-pentanol) [12]
0.12 <1> (1,3-diaminopropane) [12]
0.14 <1> (N,N,N',N'-tetramethyl-1,4-diaminobutane) [12]
0.181 <1> (spermidine) [11]
0.19 <1> (1,1,4,4-tetramethyl-1,4-diaminobutane) [12]
0.24 <1> (pentylamine) [12]
0.28 <1> (cadaverine) [3]
0.31 <1> (6-amino-1-hexanol) [12]
0.46 <1> (butylamine) [12]
0.5 <1> (3-amino-1-propanol) [12]
0.62 <1> (propylamine) [12]
0.67 <1> (3-amino-1-propanol) [12]
0.95 <1> (ethylenediamine) [12]
2.3 <1> (ethylamine) [12]
3 <1> (2-aminoethanol) [12]
3.4 <1> (methylamine) [12]
20 <1> (ammonium) [12]
59 <1> (5-amino-1-pentanoic acid) [12]
160 <1> (8-amino-1-octanoic acid) [12]
300 <1> (6-amino-1-hexanoic acid) [12]
440 <1> (4-amino-1-butanoic acid) [12]

pH-Optimum
8 <1, 5, 6> (<1>, putrescine [2]) [2, 9, 14]
8-11.7 <2> [4]
8.5 <1> (<1>, spermidine [2,15]) [2, 15]

pH-Range
6.8-9 <6> (<6>, pH 6.8: about 50% of maximal activity, pH 9.0: about 60% of maximal activity [9]) [9]

Temperature optimum (°C)
37 <6> [9]

Temperature range (°C)
20-50 <6> (<6>, 20°C: about 45% of maximal activity, 50°C: about 50% of maximal activity [9]) [9]

4 Enzyme Structure

Molecular weight
83400-87000 <1> (<1>, calculation from sedimentation data [3]) [3]
90000 <1> (<1>, gel filtration [3]) [3]

Subunits
? <1> (<1>, x * 50000, SDS-PAGE [8]) [8]
dimer <1> (<1>, 2 * 46000, SDS-PAGE [3]) [3]

5 Isolation/Preparation/Mutation/Application

Source/tissue
fruit <12> [10]
ovary <12> (<12>, ovary senescence is characterized by increased level of putrescine oxidase, constant level of putrescine oxidase during fruit development [10]) [10]
serum <5> (<5>, foetal [7,14]; <5>, absent from bovine serum after birth [7]) [7, 14]
Additional information <3> (<3>, high level of enzyme is produced in cells from the late exponential growth phase [5]) [5]

Purification
<1> [1, 2, 3, 15]
<5> [14]
<6> (partial [9]) [9]

Cloning
<1> (expression in Escherichia coli [8]) [8]

Application
analysis <1, 2> (<2>, specific measurement of putrescine with putrescine oxidase and aminobutyraldehyde dehydrogenase [4]; <1>, differential determination procedure for putrescine, spermidine and spermine [11]) [4, 11]

6 Stability

pH-Stability
6-10.5 <2> (<2>, stable [4]) [4]

Temperature stability

40 <1, 2> (<1>, stable up to [2]; <2>, pH 7.5, 20 min, 17% loss of activity [4]) [2, 4]

50 <1, 2> (<1>, rapid inactivation at [2]; <2>, pH 7.5, 20 min, 62% loss of activity [4]) [2, 4]

60 <2> (<2>, pH 7.5, 20 min, 93% loss of activity [4]) [4]

Additional information <1> (<1>, the enzyme is resistant to heat denaturation as long as it is maintained in a fully reduced state [3]) [3]

Oxidation stability

<1>, photoreduction of the enzyme in presence of EDTA [3]

Storage stability

<1>, -22°C, stable for months [2]

<1>, 5°C, 0.01 M phosphate buffer, pH 6.0-8.0, little loss of activity [2]

<2>, -20°C, in presence of 25% glycerol, 1 year, stable [4]

References

[1] Okada, M.; Kawashima, S.; Imahori, K.: Purification of putrescine oxidase from Micrococcus rubens by affinity chromatography. Methods Enzymol., **94**, 301-303 (1983)

[2] Yamada, H.: Putrescine oxidase (Micrococcus rubens). Methods Enzymol., **17B**, 726-730 (1971)

[3] Desa, R.J.: Putrescine oxidase from Micrococcus rubens. Purification and properties of the enzyme. J. Biol. Chem., **247**, 5527-5534 (1972)

[4] Shimizu, E.; Tabata, Y.; Hayakawa, R.; Yorifuji, T.: Specific measurement of putrescine with putrescine oxidase and aminobutyraldehyde dehydrogenase. Agric. Biol. Chem., **52**, 2865-2871 (1988)

[5] Rokka, R.; Futamura, N.; Ishida, A.: Role of putrescine oxidase in the generation of hydrogen peroxide by Micrococcus luteus. J. Gen. Appl. Microbiol., **31**, 435-440 (1985)

[6] Spathas, D.H.; Clutterbuck, A.J.; Pateman, J.A.: Putrescine as a nitrogen source for wild type and mutants of Aspergillus nidulans. FEMS Microbiol. Lett., **17**, 345-348 (1983)

[7] Gahl, W.A.; Pitot, H.C.: Polyamine degradation in foetal and adult bovine serum. Biochem. J., **202**, 603-611 (1982)

[8] Ishizuka, H.; Horinouchi, S.; Beppu, T.: Putrescine oxidase of Micrococcus rubens: primary structure and Escherichia coli. J. Gen. Microbiol., **139**, 425-432. (1993)

[9] Gunasekaran, M.; Gunasekaran, U.: Partial purification and properties of putrescine oxidase from Candida guilliermondii. Appl. Biochem. Biotechnol., **76**, 229-236 (1999)

[10] Perez-Amador, M.A.; Carbonell, J.: Arginine decarboxylase and putrescine oxidase in ovaries of Pisum sativum L.: changes during ovary senescence and early stages of fruit development. Plant Physiol., **107**, 865-872 (1995)

[11] Isobe, K.; Tani, Y.; Yamada, H.: Differential determination procedure for putrescine, spermidine and spermine with polymaine oxidase from fungi and putrescine oxidase. Agric. Biol. Chem., **45**, 727-733 (1981)

[12] Swain, W.F.; Desa, R.J.: Mechanism of action of putrescine oxidase. Binding characteristics of the active site of putrescine oxidase from Micrococcus rubens. Biochim. Biophys. Acta, **429**, 331-341 (1976)

[13] Okada, M.; Kawashima, S.; Imahori, K.: Mode of inactivation of putrescine oxidase by 1-ethyl-3-(3-dimethylaminopropyl)carbodiimide or metal ions. J. Biochem., **88**, 481-488 (1980)

[14] Gahl, W.A.; Vale, A.M.; Pitot, H.C.: Separation of putrescine oxidase and spermidine oxidase in foetal bovine serum with the aid of a specific radio-active assay of spermidine oxidase. Biochem. J., **187**, 197-204 (1980)

[15] Adachi, O.; Yamada, H.; Ogata, K.: Purification and properties of putrescine oxidase of Micrococcus rubens. Agric. Biol. Chem., **30**, 1202-1210 (1966)

L-Glutamate oxidase

1 Nomenclature

EC number
1.4.3.11

Systematic name
L-glutamate:oxygen oxidoreductase (deaminating)

Recommended name
L-glutamate oxidase

Synonyms
GLOD
L-glutamic acid oxidase
dehydrogenase, glutamate (acceptor)
glutamate oxidase
glutamic acid oxidase
glutamic dehydrogenase (acceptor)

CAS registry number
39346-34-4

2 Source Organism

<1> *Streptomyces endus* [1]
<2> *Streptomyces violascens* (H82-N-SY7 [2]) [2]
<3> *Streptomyces sp.* (X-119-6 [3]; activity of the mutant strain Z-11-6 is 40fold higher than that of the original natural isolate [7]) [3, 4, 5, 6, 7]

3 Reaction and Specificity

Catalyzed reaction
L-glutamate + O_2 + H_2O = 2-oxoglutarate + NH_3 + H_2O_2

Reaction type
oxidation
oxidative deamination
redox reaction
reduction

Substrates and products

S   4-benzyl-L-glutamate + O_2 + H_2O <1> (<1>, 13% of the activity with L-glutamate [1]) (Reversibility: ? <1> [1]) [1]

P   4-benzyl-2-oxoglutarate + NH_3 + H_2O_2

S   L-Asp + O_2 + H_2O <3> (<3>, at 0.6% of the activity with L-glutamate, at pH 7.4, no measurable activity at pH 6.0 [3,4]) (Reversibility: ? <3> [3, 4]) [3, 4]

P   2-oxosuccinamic acid + NH_3 + H_2O_2

S   L-Gln + O_2 + H_2O <2> (<2>, 2.5% of the activity with L-glutamate at pH 5.0, 32.1% of the activity with L-glutamate at pH 6.8 [2]) (Reversibility: ? <2> [2]) [2]

P   2-oxoglutaric acid 5-amide + NH_3 + H_2O_2

S   L-His + O_2 + H_2O <2> (<2>, 2.4% of the activity with L-glutamate, 13.1% of the activity with L-glutamate at pH 6.8 [2]) (Reversibility: ? <2> [2]) [2]

P   2-oxo-3-(1H-imidazol-4yl)-propionic acid + NH_3 + H_2O_2

S   L-Tyr + O_2 + H_2O <2> (<2>, 2.7% of the activity with L-glutamate [2]) (Reversibility: ? <2> [2]) [2]

P   3-(4-hydroxyphenyl)-2-oxopropionic acid + NH_3 + H_2O_2

S   L-glutamate + O_2 + H_2O <1, 2, 3> (<1>, high specificity for L-glutamate [1,7]; <2>, no activity with $NADP^+$, methylene blue, thionine and ferricyanide as electron acceptor [2]) (Reversibility: ? <1-3> [1-7]) [1, 2, 3, 4, 5, 6, 7]

P   2-oxoglutarate + NH_3 + H_2O_2 <1, 2> [1, 2]

Inhibitors

4-chloro-7-nitrobenzo-2-oxa-1,3-diazol <1> (<1>, 1 mM, 18% loss of activity [1]) [1]

Ag^+ <1, 2> (<1>, 0.1 mM $AgNO_3$, 94% loss of activity [1]; <2>, 1 mM $AgNO_3$, complete inhibition [2]) [1, 2]

Cu^{2+} <2> (<2>, 1 mM $CuCl_2$, 22.4% inhibition [2]; <3>, no inhibition by $CuCl_2$ [4]) [2]

Hg^{2+} <1, 2> (<1>, 0.1 mM $HgCl_2$, complete inactivation [1]; <2>, 1 mM $HgCl_2$, complete inhibition [2]) [1, 2]

N-bromosuccinimide <2> (<2>, 1 mM, 77.6% inhibition [2]) [2]

PCMB <1, 2, 3> (<1>, 1 mM, 16% loss of activity [1]; <2>, 1 mM, 42.8% inhibition [2]) [1, 2, 4]

Cofactors/prosthetic groups

FAD <1, 2, 3> (<1,2>, one molecule of noncovalently bound FAD per subunit [1,2]; <3>, contains 2 mol of FAD per mol of enzyme [4]) [1, 2, 4]

Specific activity (U/mg)

6 <1> [1]

55.1 <3> [4]

K_m-Value (mM)

1.1 <1, 2> (L-glutamate, <1>, pH 7.0 [1]; <2>, pH 5.0 [2]) [1, 2]

1.86 <1> (O_2) [1]

3.3 <2> (L-glutamate, <2>, pH 6.8 [2]) [2]
5 <2> (L-His, <2>, pH 6.8 [2]) [2]
6.7 <2> (L-Gln, <2>, pH 6.8 [2]) [2]
10 <2> (L-Gln, <2>, pH 5.0 [2]) [2]

pH-Optimum
5 <2> [2]
6.5-8 <1> [1]
7-8 <3> [4]

pH-Range
4-7 <2> (<2>, pH 4.0: about 75% of maximal activity, pH 7.0: about 80% of maximal activity [2]) [2]
4-10 <3> (<3>, pH 4.0: about 50% of maximal activity pH 10.0: about 40% of maximal activity [4]) [4]
5.5-8.5 <1> (<1>, pH 5.5: about 30% of maximal activity, pH 8.5: about 80% of maximal activity [1]) [1]

Temperature optimum (°C)
30-45 <1> [1]

Temperature range (°C)
25-60 <1> (<1>, 25°C: about 75% of maximal activity, 60°C: about 55% of maximal activity [1]) [1]

4 Enzyme Structure

Molecular weight
90000 <1> (<1>, gel filtration [1]) [1]
140000 <3> (<3>, gel filtration [4]) [4]

Subunits
dimer <1> (<1>, 2 * 50000, SDS-PAGE [1]) [1]
hexamer <3> (<3>, $\alpha_2\beta_2\gamma_2$, 2 * 44000 + 2 * 19000 + 2 * 9000, SDS-PAGE [4]) [4]

5 Isolation/Preparation/Mutation/Application

Source/tissue
culture medium <1> [1]
Additional information <3> (<3>, wheat bran culture [3,4]) [3, 4]

Localization
extracellular <1, 3> [1, 7]

Purification
<1> [1]
<3> [4, 7]

Application

analysis <2, 3> (<2>, specific and sensitive determination of L-glutamate [2]; <3>, hydrogel-immobilization of L-glutamate oxidases for a novel thick-film biosensor and application in food samples [5]; <3>, immobilization of L-glutamate oxidase and peroxidase for glutamate determination in flow injection analysis system [6]) [2, 5, 6]

6 Stability

pH-Stability

3-7 <2> (<2>, 37°C, 1 h, stable [2]) [2]
5.5-7.5 <1> (<1>, 37°C, 15 min, stable [1]) [1]
5.5-10.5 <3> (<3>, 45°C, 15 min, stable [3]) [3]

Temperature stability

37 <3> (<3>, pH 5.0-10.5, 1 h, stable [4]) [4]
60 <1, 3> (<1>, pH 7.0, 1 h, stable [1]; <3>, pH 5.5-7.0, 15 min, stable [4]) [1, 4]
75 <3> (<3>, pH 5.5, stable up to [3]) [3]
85 <3> (<3>, pH 5.5, 15 min, 50% loss of activity [4]) [4]

References

[1] Boehmer, A.; Mueller, A.; Passarge, M.; Liebs, P.; Honeck, H.; Mueller, H.G.: A novel L-glutamate oxidase from Streptomyces endus. Purification and properties. Eur. J. Biochem., **182**, 327-332 (1989)

[2] Kamei, T.; Asano, K.; Kondo, H.; Matsuzaki, M.; Nakamura, S.: L-Glutamate oxidase from Streptomyces violascens. II. Properties. Chem. Pharm. Bull., **31**, 3609-3616 (1983)

[3] Kusakabe, H.; Midorikawa, Y.; Kuninaka, A.; Yoshino, H.: Occurrence of a new enzyme, L-glutamate oxidase in a wheat bran culture extract of Streptomyces sp. X-119-6. Agric. Biol. Chem., **47**, 179-182 (1983)

[4] Kusakabe, H.; Midorikawa, Y.; Fujishima, T.; Kuninaka, A.; Yoshino, H.: Purification and properties of a new enzyme, L-glutamate oxidase, from Streptomyces sp. X-119-6 grown on wheat bran. Agric. Biol. Chem., **47**, 1323-1328 (1983)

[5] Kwong, A.W.K.; Grundig, B.; Hu, J.; Renneberg, R.: Comparative study of hydrogel-immobilized L-glutamate oxidases for a novel thick-film biosensor and its application in food samples. Biotechnol. Lett., **22**, 267-272 (2000)

[6] Li, G.; Zhang, S.; Yu, J.: Immobilization of L-glutamate oxidase and peroxidase for glutamate determination in flow injection analysis system. Appl. Biochem. Biotechnol., **59**, 53-61 (1996)

[7] Sukhacheva, M.V.; Netrusov, A.I.: Extracellular L-glutamate oxidase of Streptomyces sp. Z-11-6: obtainment and properties. Microbiology, **69**, 17-20 (2000)

Cyclohexylamine oxidase

1 Nomenclature

EC number
1.4.3.12

Systematic name
cyclohexylamine:oxygen oxidoreductase (deaminating)

Recommended name
cyclohexylamine oxidase

CAS registry number
63116-97-2

2 Source Organism

<1> *Pseudomonas sp.* [1]
<2> *Brevibacterium oxydans* (strain IH-35A) [2, 3]

3 Reaction and Specificity

Catalyzed reaction
cyclohexylamine + O_2 + H_2O = cyclohexanone + NH_3 + H_2O_2 (A flavoprotein, FAD. Some other cyclic amines can act instead of cyclohexylamine, but not simple aliphatic and aromatic amides)

Reaction type
oxidation
oxidative deamination
redox reaction
reduction

Natural substrates and products
S   N-methylcyclohexylamine + O_2 + H_2O <2> (<2> highly specific for alicyclic monoamines and aliphatic monoamines [3]) (Reversibility: ? <2> [3]) [3]
P   N-methylcyclohexanone + NH_3 + H_2O_2 <2> [3]
S   cyclohexylamine + O_2 + H_2O <1, 2> (<1> highly specific for alicyclic primary amines, oxygen is the only electron acceptor [1]; <2> highly specific for alicyclic monoamines and aliphatic monoamines [3]) (Reversibility: ? <1, 2> [1-3]) [1-3]

P cyclohexanone + NH_3 + H_2O <1, 2> [1-3]
S cyclopentylamine + O_2 + H_2O <2> (<2> highly specific for alicyclic monoamines and aliphatic monoamines [3]) (Reversibility: ? <2> [3]) [3]
P cyclopentanone + NH_3 + H_2O_2 <2> [3]
S secondary butylamine + O_2 + NH_3 <2> (<2> highly specific for alicyclic monoamines and aliphatic monoamines [3]) (Reversibility: ? <2> [3]) [3]
P butanone + NH_3 + H_2O_2 <2> [3]

Substrates and products

S 1,2,3,4-tetrahydro-2-naphthylamine + O_2 + H_2O <1> (<1> 23% of the activity compared to cyclohexylamine [1]) (Reversibility: ? <1> [1]) [1]
P ? + NH_3 + H_2O_2 <1> [1]
S 1,2-cyclohexanediamine + O_2 + H_2O <1> (<1> 18% of the activity compared to cyclohexylamine [1]) (Reversibility: ? <1> [1]) [1]
P cyclohexaneamine + NH_3 + H_2O_2 <1> [1]
S N-methylcyclohexylamine + O_2 + H_2O <2> (<2> highly specific for alicyclic monoamines and aliphatic monoamines [3]) (Reversibility: ? <2> [3]) [3]
P N-methylcyclohexanone + NH_3 + H_2O_2 <2> [3]
S cycloheptylamine + O_2 + H_2O <1> (<1> 42% of the activity compared to cyclohexylamine [1]) (Reversibility: ? <1> [1]) [1]
P cycloheptanone + NH_3 + H_2O_2 <1> [1]
S cyclohexylamine + O_2 + H_2O <1, 2> (<1> highly specific for alicyclic primary amines, oxygen is the only electron acceptor [1]; <2> highly specific for alicyclic monoamines and aliphatic monoamines [3]) (Reversibility: ? <1, 2> [1-3]) [1-3]
P cyclohexanone + NH_3 + H_2O <1, 2> [1-3]
S cyclopentylamine + O_2 + H_2O <2> (<2> highly specific for alicyclic monoamines and aliphatic monoamines [3]) (Reversibility: ? <2> [3]) [3]
P cyclopentanone + NH_3 + H_2O_2 <2> [3]
S secondary butylamine + O_2 + NH_3 <2> (<2> highly specific for alicyclic monoamines and aliphatic monoamines [3]) (Reversibility: ? <2> [3]) [3]
P butanone + NH_3 + H_2O_2 <2> [3]

Inhibitors

Cd^{2+} <1> (<1> 38% inhibition at 10 mM [1]) [1]
Cu^{2+} <1> (<1> complete inhibition at 10 mM [1]) [1]
Hg^{2+} <1> (<1> complete inhibition at 10 mM [1]) [1]
N-ethylmaleimide <1> (<1> 27% inhibition at 0.05 mM [1]) [1]
Zn^{2+} <1> (<1> 48% inhibition at 10 mM [1]) [1]
p-chloromercuribenzoate <1> (<1> 91% inhibition at 0.0005 mM, partially reversed by glutathione [1]) [1]
quinacrine <1, 2> (<1> 45% inhibition at 2 mM, partially released by FAD [1]) [1, 3]
quinine <2> [3]

Cofactors/prosthetic groups

FAD <1> [1]

Specific activity (U/mg)
  29.6 <1> [1]

K$_m$-Value (mM)
  0.25 <1> (cyclohexylamine) [1]
  1.23 <2> (cyclohexylamine) [3]

pH-Optimum
  6.8 <1> [1]
  7-7.5 <2> [2]
  7.4 <2> [3]

Temperature optimum (°C)
  35 <1> [1]

4 Enzyme Structure

Molecular weight
  50000 <2> (<2> gel filtration [3]) [3]
  80000 <1> (<1> gel filtration [1]) [1]

Subunits
  monomer <2> (<2> 1 * 50000, SDS-PAGE [3]) [3]

Posttranslational modification
  flavoprotein <2> [3]

5 Isolation/Preparation/Mutation/Application

Source/tissue
  cell culture <1, 2> (<2> inducible by growth on cyclohexylamine [2,3]) [1-3]

Purification
  <1> (to homogeneity, fractionation and chromatography steps [1]) [1]
  <2> (to homogeneity, chromatography steps [3]) [3]

6 Stability

pH-Stability
  6-7 <2> (<2> stable between [3]) [3]

Temperature stability
  30 <2> (<2> stable up to [3]) [3]

References

[1] Tokieda, T.; Niimura, T.; Takamura,F.; Yamaha, T.: Purification and some properties of cyclohexylamine oxidase from a Pseudomonas sp.. J. Biochem., **81**, 851-858 (1977)

[2] Iwaki, H.; Shimizu, M.; Tokuyama, T.; Hasegawa, Y.: Biodegradation of cyclohexylamine by Brevibacterium oxydans IH-35A. Appl. Environ. Microbiol., **65**, 2232-2234 (1999)

[3] Iwaki, H.; Shimizu, M.; Tokuyama, T.; Hasegawa, Y.: Purification and characterization of a novel cyclohexylamine oxidase from the cyclohexylamine-degrading Brevibacterium oxydans IH-35A. J. Biosci. Bioeng., **88**, 264-268 (1999)

Protein-lysine 6-oxidase

1 Nomenclature

EC number
1.4.3.13

Systematic name
protein-L-lysine:oxygen 6-oxidoreductase (deaminating)

Recommended name
protein-lysine 6-oxidase

Synonyms
RAS excision protein
lysyl oxidase <1-3> (<1,2> amine oxidase, oxidative deamination [2-7]) [2-7]

CAS registry number
99676-44-5

2 Source Organism

<1> *Gallus gallus* (embryo [1,3-5]) [1-5, 7]
<2> *Homo sapiens* [6, 7]
<3> *mammalia* (expression is sensitively modulated by specific cytokines, growth factors and related intercellular molecular messengers [7]) [7]
<4> *Rattus norvegicus* [7]
<5> *Bos taurus* [7]
<6> *Ovis aries* [7]
<7> *yeast* [7]
<8> *Mus musculus* [7]

3 Reaction and Specificity

Catalyzed reaction
peptidyl-L-lysyl-peptide + O_2 + H_2O = peptidyl-allysyl-peptide + NH_3 + H_2O_2 (Also acts on protein 5-hydroxylysine; <1> information about the activation mechanism of copper [2]; <3> preprotein, lysyl oxidase gene family, amine oxidation through ping pong bi ter kinetic mechanism [7])

Reaction type
oxidation
oxidative deamination

redox reaction
reduction

Natural substrates and products

S   collagen + H_2O + O_2 <1> (Reversibility: ? <1> [1]) [1]
P   allysyl-collagen + NH_3 + H_2O_2 <1> [1]
S   elastin + H_2O + O_2 <1> (Reversibility: ? <1> [1]) [1]
P   allysyl-elastin + NH_3 + H_2O_2 <1> [1]
S   Additional information <2> (<2> lysyl oxidase gene family. Enzyme seems to be involved in binding and crosslinking to other cell-surface and extracellular-matrix proteins [6]) [6]
P   ?

Substrates and products

S   collagen + H_2O + O_2 <1> (Reversibility: ? <1> [1]) [1]
P   allysyl-collagen + NH_3 + H_2O_2 <1> [1]
S   elastin + H_2O + O_2 <1> (Reversibility: ? <1> [1]) [1]
P   allysyl-elastin + NH_3 + H_2O_2 <1> [1]
S   peptidyl-L-hydroxylysyl-peptide + O_2 + H_2O <1> (<1> part of tropocollagen and tropoelastin [5]) (Reversibility: ? <1> [1, 3-5]) [1, 3-5]
P   peptidyl-L-2-amino-5-hydroxy-6-oxohexanoyl-peptide + NH_3 + H_2O_2 <1> [1, 3-5]
S   peptidyl-L-lysyl-peptide + O_2 + H_2O <1, 3> (<1> part of elastin and collagen [3,4]; <1> part of tropocollagen and tropoelastin [5]; <3> crosslinkage in fibrillar collagen and elastin [7]) (Reversibility: ? <1, 3> [1-5, 7]) [1-5, 7]
P   peptidyl-L-allysyl-peptide + NH_3 + H_2O_2 <1> [1-5, 7]
S   Additional information <2> (<2> lysyl oxidase gene family. Enzyme seems to be involved in binding and crosslinking to other cell-surface and extracellular-matrix proteins [6]) [6]
P   ?

Inhibitors

2-mercaptoethanol <1> (<1> weakly inhibitory [4]) [4]
4-deoxypyridoxine <1> (<1> pyridoxal analogue, 50% inhibition at 3 mM [5]) [5]
Zn^{2+} <1> (<1> slightly inhibitory [4]) [4]
basic fibroblast growth factor <3> (<3> reduces enzyme expression in osteogenic cells [7]) [7]
β-aminopropionitrile <1, 4> (<1> lathyrogen, organic nitrile group, blocks active site irreversibly and binds covalently [1]; <1> irreversible [2]; <1> total inhibition at 1 mM [4]; <4> irreversible inhibitor [7]) [1, 2, 4, 5, 7]
cycloheximide <1, 3> (<1> inhibits incorporation of lysine and Cu^{2+} into enzyme, blocks activating process [2]; <3> prevents stimulation by transforming growth factor-β_1 [7]) [2, 7]
diethyldithiocarbamate <1> [4]
dithiothreitol <1> [4]
hydroxylamine <1> [4]

isonicotinic acid hydrazide <1> (<1> lathyrogen, hydrazide group [1]; <1> reversible, 61.8% of full activity restored by pyridoxal or 30.1% by pyrrolo-quinoline-quinone [5]) [1, 5]
phenylhydrazine <1> [4]
semicarbazide <1> (<1> lathyrogen, ureide group [1]; <1> reversible, 21.2% of full activity restored by pyridoxal or 10% by pyrroloquinoline-quinone [5]) [1, 4, 5]

Cofactors/prosthetic groups

lysyltyrosyl quinone <2, 3> (<3> prosthetic group [7]) [6, 7]
trihydroxyphenylalanine quinone <7> [7]

Activating compounds

actinomycin D <1> [2]
pyridoxal <1> (<1> derivative [1,5]; <1> pyridoxal phosphate or closely re-lated compound [4]; <1> needed for biosynthesis of the enzyme [5]) [1, 4, 5]
transforming growth factor-β_1 <2, 4> (<2,4> fibrogenic cytokine, strongly promotes enzyme expression in fibroblasts of neonatal rat lung, human em-bryos and rat vascular smooth muscle cells [7]) [7]

Metals, ions

Cu^{2+} <1-3> (<1> copper metalloenzyme, activation and induction of enzyme synthesis, copper-induced activation maximal at 0.005 mg of copper/ml [2]; <1> copper metalloenzyme contains 0.14% copper, important for enzymatic function [4]; <3> tightly bound in a tetragonally distorted, octahedrally co-ordinated ligand field, one per mol [7]; <2> copper-binding and -dependent [6,7]) [1, 2, 4-7]

Specific activity (U/mg)

Additional information <1> (<1> 0.0214 cpm of titrated water formed per h per g wet weight of tissue [2]; <1> 4.533 cpm/mg protein for elastin and 1.633 cpm/mg for collagen as substrate [3]; <1> 46390 cpm/mg [4]) [2-4]

pH-Optimum

7.7 <1> (<1> assay at [4]) [4]

Temperature optimum (°C)

37 <1> (<1> assay at [2-4]) [2-4]

4 Enzyme Structure

Subunits

? <1-5> (<1> x * 60000, SDS-PAGE [2]; <1> x * 28000, SDS-PAGE [3]; <1> x * 59000, x * 61000, SDS-PAGE, may be isoenzymes or heterologous subunits [4]; <2> x * 97000, including V5-epitope and histidine tag of 2.3 kDa, Wes-tern-Blot, x * 83600, calculated [6]; <3> x * 50000, preprotein [7]; <3,4> x * 32000, catalytically active enzyme after proteolytic cleavage [7]; <5> x * 46000, predicted, x * 32000, SDS-PAGE [7]) [2-4, 6, 7]

Additional information <1> (<1> different molecular forms of enzyme exist [3]) [3]

Posttranslational modification

glycoprotein <2, 3> (<3> N-glycosylated as 50 kDa preproprotein [7]) [6, 7] proteolytic modification <3> (<3> proteolytically cleaved to the 32 kDa catalytically active enzyme [7]) [7]

5 Isolation/Preparation/Mutation/Application

Source/tissue

aorta <1, 4, 5> [1, 2, 4, 7]
basal cell <3> [7]
biliary epithelium <3> [7]
bone <1> (<1> leg [1,5]) [1, 5]
cartilage <1> [1, 3, 5]
chondrocyte <3> [7]
endothelium <3> [7]
fibroblast <2, 3> [7]
glomerular epithelium <3> [7]
heart <2> [6]
lung <4> [7]
ovary <2> [6]
placenta <2> [6]
skin <2, 4> [7]
small intestine <2> [6]
smooth muscle <3, 4> (<4> nucleus [7]) [7]
spleen <2> [6]
testis <2> [6]

Purification

<1> (affinity, gel-filtration, ion-exchange [2]; affinity, ion-exchange [3,4]) [2-4]

Cloning

<1> [7]
<2> (LOXL3 cDNA is expressed in human HT-1080 fibrosarcoma cells [6]) [6, 7]
<3> (expressed in Escherichia coli [7]) [7]
<8> [7]

Application

medicine <3> (<3> cutis laxa, indicated by hyperextensible skin with marked deficiency of skin elastic fibers, is caused by deficiency of lysyl oxidase expression. Fibrosis is a symptom of scleroderma, where lysyl oxidase is markedly elevated in affected tissues. Menkes syndrome may involve abnormalities in the expression of several genes coding for connective tissue proteins [7]; <4> increase in the expression of lysyl oxidase during wound healing in skin and in aorta, lung and skin of developing and aging rats. Implication of lysyl oxidase as a target gene for IRF-1 in tumor suppression. Lysly oxidase has unusual and important roles in cellular homeostasis [7]) [7]

6 Stability

General stability information
<1>, remarkably stable in concentrated solutions of urea [3]

Storage stability
<1>, -15°C, powdered tissue, full enzyme activity for at least 4 months [4]

References

[1] Levene, C.I.; Carrington, M.J.: The inhibition of protein-lysine 6-oxidase by various lathyrogens. Evidence for two different mechanisms. Biochem. J., **232**, 293-296 (1985)

[2] Rayton, J.K.; Harris, E.D.: Induction of lysyl oxidase with copper. Properties of an in vitro system. J. Biol. Chem., **254**, 621-626 (1979)

[3] Stassen, F.L.H.: Properties of highly purified lysyl oxidase from embryonic chick cartilage. Biochim. Biophys. Acta, **438**, 49-60 (1976)

[4] Harris, E.D.; Gonnerman, W.A.; Savage, J.E.; O'Dell, B.L.: Connective tissue amine oxidase. II. Purification and partial characterization of lysyl oxidase from chick aorta. Biochim. Biophys. Acta, **341**, 332-344 (1974)

[5] Levene, C.I.; O'Shea, M.P.; Carrington, M.J.: Protein lysine 6-oxidase (lysyl oxidase) cofactor: methoxatin (PQQ) or pyridoxal?. Int. J. Biochem., **20**, 1451-1456 (1988)

[6] Mäki, J.M.; Kivirikko, K.I.: Cloning and characterization of a fourth human lysyl oxidase isoenzyme. Biochem. J., **355**, 381-387 (2001)

[7] Smith-Mungo, L.I.; Kagan, H.M.: Lysyl oxidase: properties, regulation and multiple functions in biology. Matrix Biol., **16**, 387-398 (1998)

1 Nomenclature

EC number
1.4.3.14

Systematic name
L-lysine:oxygen 2-oxidoreductase (deaminating)

Recommended name
L-lysine oxidase

Synonyms
L-lysine α-oxidase
L-lysine-a-oxidase <2> [11]
L-lysyl-α-oxidase
lysine-oxidase

CAS registry number
70132-14-8

2 Source Organism

<-5> no activity in *Fusarium* [3]
<-4> no activity in *Penicillium* [3]
<-3> no activity in *Mucor* [3]
<-2> no activity in *Rhizopus* [3]
<-1> no activity in *Aspergillus* [3]
<1> *Trichoderma viride* (Y244-2, ATCC 20536 [3-6]; mold, grown in wheat bran culture [3-6]; i4 [9]) [1-6, 9, 10]
<2> *Trichoderma harzianum* (Rifai [8,11]) [2, 8, 11]
<3> *Trichoderma longibrachiatum* [2]
<4> *Trichoderma aureoviride* [2]
<5> *Trichoderma pseudokonigii* (wheat bran culture of Ts75-2 [7]) [7]
<6> *Mus musculus* (fasted [10]) [10]

3 Reaction and Specificity

Catalyzed reaction
L-lysine + O_2 + H_2O = 6-amino-2-oxohexanoate + NH_3 + H_2O_2 (also acts, more slowly, on L-ornithine, L-phenylalanine, L-arginine and L-histidine; <1-

4> the substrate specificity of Trichoderma strains changes as a function of the age of the culture and its morphological condition [2]; <6> catabolic pathway of L-lysine through the pipecolic acid pathway [10])

Reaction type
oxidation
oxidative deamination
redox reaction
reduction

Natural substrates and products

S L-lysine + O_2 + H_2O <1, 2, 5, 6> (<1> also other α,ω-diaminomonocarboxylates and L-lysine hydroxamate are oxidized [4]; <1> preferred substrate [6]; <1> lysine metabolism [5,6]; <5> most favored substrate [7]; <1> preferred substrate [9]) (Reversibility: ? <1, 2, 5, 6> [1, 3-10]) [1, 3-10]

P 6-amino-2-oxohexanoate + NH_3 + H_2O_2 <1, 2, 5, 6> [1, 3-10]

Substrates and products

S L-arginine + O_2 + H_2O <1, 5> (Reversibility: ? <1, 5> [4, 6, 7, 9]) [4, 6, 7, 9]

P 5-guanidino-2-oxopentanoate + NH_3 + H_2O_2 <1, 5> [4, 6, 7, 9]

S L-histidine + O_2 + H_2O <1> (<1> only 1% in comparison with L-lysine [9]) (Reversibility: ? <1> [4, 6, 9]) [4, 6, 9]

P 3-(1H-imidazol-4-yl)-2-oxopropanoate + NH_3 + H_2O_2 <1> [4, 6, 9]

S L-leucine + O_2 + H_2O <1, 6> (Reversibility: ? <1, 6> [1, 10]) [1, 10]

P 4-methyl-2-oxopentanoate + NH_3 + H_2O_2 <1, 6> [1, 10]

S L-lysine + O_2 + H_2O <1, 2, 5, 6> (<1> also other α,ω-diaminomonocarboxylates and L-lysine hydroxamate are oxidized [4]; <1> preferred substrate [6]; <1> lysine metabolism [5,6]; <5> most favored substrate [7]; <1> preferred substrate [9]) (Reversibility: ? <1, 2, 5, 6> [1, 3-10]) [1, 3-10]

P 6-amino-2-oxohexanoate + NH_3 + H_2O_2 <1, 2, 5, 6> [1, 3-10]

S L-lysine + semicarbazide + ? <1> (Reversibility: ? <1> [1]) [1]

P 4-methyl-2-oxopentanoate + ? <1> [1]

S L-ornithine + O_2 + H_2O <1, 5> (Reversibility: ? <1, 5> [4, 6, 7, 9]) [4, 6, 7, 9]

P 5-amino-2-oxopentanoic acid + NH_3 + H_2O_2 <1, 5> [4, 6, 7, 9]

S L-phenylalanine + O_2 + H_2O <1, 5> (<1> 16% in comparison to L-lysine [9]) (Reversibility: ? <1, 5> [4, 6, 7, 9]) [4, 6, 7, 9]

P phenylpyruvate + NH_3 + H_2O_2 <1, 5> [4, 6, 7, 9]

S L-tyrosine + O_2 + H_2O <1> (Reversibility: ? <1> [4]) [4]

P 3-(4-hydroxyphenyl)pyruvate + NH_3 + H_2O_2 <1> [4]

S Additional information <1-4> (<1-4> no activity with L-cysteine, L-asparagine-NH_2, DL-serine, DL-asparagine and DL-glutamic acid [2]; <1> very low activity with several other amino acids [2]; <2> very low activity with DL phenylalanine and DL-methionine [2]; <4> very low activity with DL-phenylalanine, L-histidine, L-arginine, DL-methionine and L-glycine

[2]; <1> several amino acids, amino acid derivatives and other substances that are assumed to be possible substrates are tested [9]) [2, 9]

P ?

Inhibitors

1,4-diaminobutane <1> (<1> 20% inhibition [9]) [9]
1,5-diaminopentane <1> (<1> 24% inhibition [9]) [9]
1,6-diaminohexane <1> (<1> 43% inhibition [9]) [9]
6-aminoccaproic acid <1> (<1> 30% inhibition [9]) [9]
$CoSO_4$ <1> [9]
$CuCl_2$ <1> [9]
$HgCl_2$ <1> [4, 9]
$ZnSO_4$ <1> [9]
p-chloromercuribenzoate <1> [4, 9]
semicarbazide <1> (<1> weak competitive, reversible inhibition [1]) [1]

Cofactors/prosthetic groups

FAD <1, 2, 6> (<1> 2 mol per mol of enzyme, prosthetic group [4]; <2> firmly attached [8]; <1> 1 mol per subunit, non-covalently bound [9]; <6> tightly bound [10]) [1, 4, 5, 8-10]

Activating compounds

L-glutamine <1> (<1> increases enzyme production [5]) [5]
L-histidine <1> (<1> increases enzyme production [5]) [5]
NH_4NO_3 <1> (<1> stimulation of enzyme production [5]) [5]
$NaNO_3$ <1> (<1> stimulation of enzyme production [5]) [5]
adenine <1> (<1> increases enzyme production [5]) [5]
glycine <1> (<1> increases enzyme production [5]) [5]
purine <1> (<1> increases enzyme production [5]) [5]
Additional information <1> (<1> no significant effect of sugars and vitamines [5]) [5]

Specific activity (U/mg)

61.3 <5> [7]
66.15 <1> [4]
90 <1> [9]
Additional information <1, 6> (<2> the magnitude of the L-lysine-α-oxidase activity is demonstrated as a function of the sources of carbon and nitrogen, the number of conidia forming during the growth is correlated completely with the magnitude of the L-lysine-α-oxidase activity [2]; <1> 0.68 U/ml [3]; <1> 1.2 U/ml [4]; <6> activity is higher in the brain of mice which are fasted [10]) [2-4, 10]

K_m-Value (mM)

0.026 <1> (L-lysine) [9]
0.027 <5> (L-lysine) [7]
0.04 <1> (L-lysine) [4]
0.04 <1> (L-lysine, <1> 70 mM potassium phosphate buffer pH 8.0 [1]) [1]
0.044 <1> (L-ornithine) [4]

0.4 <1> (L-lysine, <1> 100 mM potassium phosphate buffer pH 7.4 and 500 mM semicarbazide [1]) [1]
0.46 <5> (L-arginine) [7]
0.625 <1> (L-ornithine) [9]
0.68 <1> (L-arginine) [9]
0.75 <5> (L-ornithine) [7]
1.6 <1> (O_2) [4]
14 <1> (L-phenylalanine) [4]
25.5 <5> (L-phenylalanine) [7]

K_i-Value (mM)

56 <1> (semicarbazide, <1> 500 mM, 10 mM L-lysine, 3% inhibition [1]) [1]

pH-Optimum

7 <5> [7]
7-8 <1> (<1> Tris- and phosphate buffer [1]) [1]
7.4 <1, 6> (<1,6> assay at [1,6,10]) [1, 6, 10]
7.4-9.2 <1> (<1> for L-lysine, reaction rate declines markedly below pH 7.0 [4]) [4]
8 <1> (<1> assay at [4,5]) [4, 5]
8-9 <1> [9]

pH-Range

5-11 <1> (<1> enzyme activity does not fall below 60%, at pH 3.0 the activity is still 30% [9]) [9]
Additional information <1-4> (<1-4> the Trichoderma strains show different amino acid oxidation activity over a varied pH range [2]) [2]

Temperature optimum (°C)

37 <1-4, 6> (<1-4,6> assay at [1,2,4-6,10]) [1, 2, 4-6, 10]
50 <1, 5> (<1> for L-lysine [4]) [4, 7, 9]

Temperature range (°C)

30-70 <1> (<1> 82% relative activity at 30°C, 103% at 50°C and 58°C at 70°C [4]) [4]

4 Enzyme Structure

Molecular weight

110000 <1> (<1> gel filtration [9]) [9]
112000-119000 <1> (<1> gel filtration, sedimentation equilibrium centrifugation [4]) [4]

Subunits

dimer <1, 5> (<5> 2 * 58000 [7]; <1> 2 * 56000, SDS-PAGE [4]; 2 * 55000, SDS-PAGE [9]) [4, 7, 9]

5 Isolation/Preparation/Mutation/Application

Source/tissue

brain <6> [10]

Localization

extracellular <1> [4-6]

Purification

<1> (ammonium sulfate precipitation [3]; ammonium sulfate precipitation, ion-exchange, gel-filtration [4]; ammonium sulfate precipitation, ion-exchange [6]; acetone precipitation, ion-exchange [9]) [3, 4, 6, 9]

<6> (gel-filtration [10]) [10]

<1-4> (rapid loss of activity during isolation and purification [2]) [2]

Application

medicine <1-4> (<1-4> antitumor activity [2-4,8,10,11]; <2> suppresses efficiently genital herpes simplex virus type II (HSV-2) infection in guinea pig, showing the best results after combined treatment with enzyme gel and intramuscular injection on day 5 postinfection [11]) [2-6, 8, 10, 11]

6 Stability

pH-Stability

5-9 <5> [7]

Additional information <1> (<1> stable at extreme pH values [9]) [9]

Temperature stability

30 <1> (<1> 30 min, 0.1 M potassium phosphate buffer, reaction rate is between 85% at pH 3.0 and 90% at pH 11.0 [9]) [9]

45 <1> (<1> stable for 30 min in the pH range from 5.0 or 7.0 to 10.0 [4,6]) [4, 6]

50 <5> (<5> up to, 90 min, pH 7.0 [7]) [7]

55 <1> (<1> 0.1 M potassium phosphate buffer, pH 7.4, 30 min, 100% of original activity [4]) [4]

60 <1> (<1> 0.1 M potassium phosphate buffer, pH 7.4, 30 min, 92% of original activity [4]) [4]

65 <1> (<1> 0.1 M potassium phosphate buffer, pH 7.4, 30 min, 42% of original activity [4]) [4]

70 <1> (<1> 0.1 M potassium phosphate buffer, pH 7.4, 30 min, 4% of original activity [4]; <1> total loss of activity if preincubated at 70°C [9]) [4, 9]

75 <1> (<1> 30 min, complete loss of activity [6]) [6]

Additional information <1> (<1> heat labile [4-6]; <1> stable at relatively high temperatures [9]) [4-6, 9]

Organic solvent stability

acetone <1> (<1> 60%, complete inactivation after 60 min [9]) [9]

ethanol <1> (<1> 70%, immediately inactivated [9]) [9]

Storage stability

<1>, -20°C, 0.05 M potassium phosphate buffer of Tris-HCl buffer pH 8.0, at least 14 months stable without any loss of activity [9]

<1>, -20°C, 0.1 M potassium phosphate buffer, pH 7.4, at least 3 months, with little loss of activity [4]

<1>, 4°C-37°C, 0.05 M potassium phosphate buffer of Tris-HCl buffer pH 8.0, stable for 4-5 days [9]

<1-4>, -4°C, practically no loss of activity over several months [2]

<1-4>, 4°C, activity preserved [2]

References

[1] Danson, J.W.; Trawick, M.L.; Cooper, A.J.: Spectrophotometric assays for L-lysine α-oxidase and γ-glutamylamine cyclotransferase. Anal. Biochem., **303**, 120-130 (2002)

[2] Smirnova, I.P.; Berezov, T.T.: Substrate specifity of L-amino acid oxidases of certain strains of the genus Trichoderma. Mikrobiologiya, **58**, 49-53 (1989)

[3] Kusakabe, H.; Midorikawa, Y.; Kuninaka, A.; Yoshino, H.: Distribution of extracellular oxygen related enzymes in molds. Agric. Biol. Chem., **47**, 1385-1387 (1983)

[4] Kusakabe, H.; Kodama, K.; Kuninaka, A.; Yoshino, H.; Misono, H.; Soda, K.: A new antitumor enzyme, L-lysine α-oxidase from Trichoderma viride. Purification and enzymological properties. J. Biol. Chem., **255**, 976-981 (1980)

[5] Kusakabe, H.; Kodama, K.; Kuninaka, A.; Yoshino, H.; Soda, K.: Extracellular production of L-lysine α-oxidase in wheat bran culture of a strain of Trichoderma viride. Agric. Biol. Chem., **43**, 2531-2535 (1979)

[6] Kusakabe, H.; Kodama, K.; Machida, H.; Midorikawa, Y.; Kuninaka, A.; Misono, H.; Soda, K.: Occurence of a novel enzyme, L-lysine oxidase with antitumor activity in culture of Trichoderma viride. Agric. Biol. Chem., **43**, 337-343 (1979)

[7] Hu, H.M.; Cheng, S.W.; Huang, M.C.; Tang, S.J.; Chang, M.C.; Tsai, Y.C.: Purification and characterization of L-lysine oxidase from Trichoderma pseudokonigii and its effect on growth of mouse erythroleukemia cells. Zhongguo Nongye Huaxue Huizhi, **32**, 361-371 (1994)

[8] Treshalina, H.M.; Lukasheva, E.V.; Sedakova, L.A.; Firsova, G.A.; Guerassimova, G.K.; Gogichaeva, N.V.; Berezov, T.T.: Anticancer enzyme L-lysine α-oxidase: properties and application perspectives. Appl. Biochem. Biotechnol., **88**, 267-273 (2000)

[9] Weber, E.; Tonder, K.; Reinbothe, C.; Unverhau, K.; Weide, H.; Aurich, H.: L-Lysine α-oxidase from Trichoderma viride i4. Purification and characterization. J. Basic Microbiol., **34**, 265-276 (1994)

[10] Murthy, S.N.; Janardanasarma, M.K.: Identification of L-amino acid/L-lysine α-amino oxidase in mouse brain. Mol. Cell. Biochem., **197**, 13-23 (1999)

[11] Smirnova, I.P.; Alekseev, S.B.; Diorditsa, S.V.; Vesa, V.S.; Zaitsev, I.Z.: Effect of L-lysine-α-oxidase on the development of genital herpes infection in guinea pigs. Bull. Exp. Biol. Med., **128**, 1226-1228 (2000)

D-Glutamate(D-aspartate) oxidase   1.4.3.15

1 Nomenclature

EC number
1.4.3.15

Systematic name
D-glutamate(D-aspartate):oxygen oxidoreductase (deaminating)

Recommended name
D-glutamate(D-aspartate) oxidase

Synonyms
D-glutamic-aspartic oxidase
D-monoaminodicarboxylic acid oxidase

2 Source Organism

<1> *Aspergillus ustus* (strain f [1]) [1]
<2> *Neurospora crassa* [2]

3 Reaction and Specificity

Catalyzed reaction
D-glutamate + H_2O + O_2 = 2-oxoglutarate + NH_3 + H_2O_2

Reaction type
oxidation
oxidative deamination
redox reaction
reduction

Natural substrates and products
S D-aspartic acid + H_2O + O_2 <1> (Reversibility: ? <1> [1]) [1]
P oxaloacetate + NH_3 + H_2O_2
S D-glutamic acid + H_2O + O_2 <1> (Reversibility: ? <1> [1]) [1]
P 2-oxoglutarate + NH_3 + H_2O_2

Substrates and products
S D-aspartic acid + H_2O + O_2 <1, 2> (<1> D-glutamate and D-aspartate are
oxidized at the same rate [1]) (Reversibility: ? <1, 2> [1, 2]) [1, 2]
P oxaloacetate + NH_3 + H_2O_2 <1> [1]

S   D-glutamic acid + H_2O + O_2 <1, 2> (<1> D-glutamate and D-aspartate are oxidized at the same rate [1]) (Reversibility: ? <1, 2> [1, 2]) [1, 2]

P   2-oxoglutarate + NH_3 + H_2O_2 <1> [1]

S   Additional information <1> (<1> DL-α-aminoadipic acid not deaminated, α-aminomalonic acid not deaminated, possible acceptors: oxygen, ferricyanide, 2,6-dichlorophenol-indophenol and methylene blue [1]) [1]

P   ?

Inhibitors

FMN <1> (<1> competes with FAD [1]) [1]

KCN <1> (<1> remarkable inhibition at 1 mM, inhibition is only slight with 2,6-dichlorophenol-indophenol electron acceptor but higher when indophenol is used [1]) [1]

Cofactors/prosthetic groups

FAD <1> (<1> flavoprotein [1]) [1]

Turnover number (min^{-1})

5 <1> (D-glutamate) [1]

Temperature optimum (°C)

30 <1> (<1> assay at [1]) [1]

5 Isolation/Preparation/Mutation/Application

Source/tissue

mycelium <1> [1]

Purification

<1> (90% pure [1]) [1]

6 Stability

General stability information

<1>, very labile when FAD is removed from the enzyme [1]

Storage stability

<1>, -20°C, several months [1]

References

[1] Mitzushima, S: Studies on the metabolism of D-amino acid in microorganisms. Part 4. On the nature of the purified D-glutamic-aspartic oxidase of Aspergillus ustus. J. Gen. Appl. Microbiol., **3**, 233-239 (1957)

[2] Onishi, E.; Macleod, H.; Horowitz, N.H.: Mutants of Neurospora deficient in D-amio-acid oxidase. J. Biol. Chem., **237**, 138-142 (1962)

L-Aspartate oxidase

1 Nomenclature

EC number
1.4.3.16

Systematic name
L-aspartate:oxygen oxidoreductase (deaminating)

Recommended name
L-aspartate oxidase

Synonyms
LASPO
oxidase, L-aspartate

CAS registry number
69106-47-4

2 Source Organism

<1> *Escherichia coli* (fumarate reductase in anaerobic conditions [6,7]) [1, 2, 5-9]
<2> *Gossypium hirsutum* (actually glutamic oxaloacetic transaminase [4]) [3, 4]
<3> *Pyrococcus horikoshii* (OT-3 [10]) [10]

3 Reaction and Specificity

Catalyzed reaction
L-aspartate + H_2O + O_2 = oxaloacetate + NH_3 + H_2O_2

Reaction type
oxidation
oxidative deamination
redox reaction
reduction

Natural substrates and products
S L-aspartate + H_2O + O_2 <1> (<1> first enzyme of quinolinate synthetase system [1,2,5]) (Reversibility: ? <1> [1, 2, 5]) [1, 2, 5]
P oxaloacetate + NH_3 + H_2O_2

Substrates and products

S L-aspartate + H_2O + O_2 <1-3> (<1> specific for L-aspartate [2]; <1> fumarate or succinate can be electron acceptor instead of O_2 [6,7]; <3> 2,6-dichlorophenol-indophenol, ferricyanide and fumarate can be electron acceptor [10]) (Reversibility: ? <1-3> [1-10]) [1-10]

P oxaloacetate + NH_3 + H_2O_2

Inhibitors

L-aspartate <1> (<1> substrate inactivation [1]) [1]
NAD^+ <1, 2> (<1> competitive to FAD [1,5]; <2> 50% at 1 mM [3]) [1, 3, 5]
iodoacetate <1> [1]
meso-tartrate <1> [1]
tetranitromethane <1> [1]
urea <1> [1]

Cofactors/prosthetic groups

FAD <1> (<1> flavoprotein [1, 5]; <1> 1 mol of FAD covalently bound per mol of enzyme [1,6,9]) [1, 5-9]

Activating compounds

L-aspartate <1> (<1> substrate activation above 1.0 mM [5]) [5]
α-ketoglutatate <2> [4]

Turnover number (min^{-1})

2 <1> (L-aspartate) [1]
156 <1> (L-aspartate, <1> oxygen electron acceptor [6]) [6]
333 <1> (L-aspartate, <1> fumarate electron acceptor [6]) [6]

Specific activity (U/mg)

3.2 <3> [10]
Additional information <1> [1, 5]

K$_m$-Value (mM)

0.5 <1> (L-aspartate, <1> at aspartate concentration 1-20 mM [1]) [1]
0.66 <2> (L-aspartate) [3]
4.1 <1> (L-aspartate, <1> at L-asparate concentration 0.2-1 mM [1]) [1]

pH-Optimum

8 <1> (<1> assay at [2]) [2]
8.6 <3> [10]

Temperature optimum (°C)

40 <1> (<1> at pH 8.0 [1]) [1]
90 <3> [10]

4 Enzyme Structure

Molecular weight

60280 <1> (<1> sequence analysis [2]) [2]
83000 <2> (<2> gel filtration [3]) [3]
151000 <3> (<3> gel filtration [10]) [10]

Subunits

monomer <1> (<1> 1 * 60000, SDS-PAGE [1]) [1]
trimer <3> (<3> 3 * 51000, SDS-PAGE [10]) [10]

5 Isolation/Preparation/Mutation/Application

Source/tissue

callus <2> [3]

Purification

<1> (overexpression in Escherichia coli [6,7,9]) [1, 2, 5-7, 9]
<2> [3]
<3> (overexpression in Escherichia coli [10]) [10]

Crystallization

<1> (Laspo R386L [8]) [8]

Cloning

<1> [2]

Engineering

H244A <1> (<1> binds substrate analogues with higher dissociation constants and presents lower k_{cat}/K_m values in the reduction of fumarate [7]) [7]
H244S <1> (<1> binds substrate analogues with higher dissociation constants and presents lower k_{cat}/K_m values in the reduction of fumarate [7]) [7]
H351A <1> (<1> binds substrate analogues with higher dissociation constants and presents lower k_{cat}/K_m values in the reduction of fumarate [7]) [7]
H351S <1> (<1> binds substrate analogues with higher dissociation constants and presents lower k_{cat}/K_m values in the reduction of fumarate [7]) [7]
R386L <1> (<1> binds substrate analogues with higher dissociation constants and presents lower k_{cat}/K_m values in the reduction of fumarate [7]) [7]
Additional information <1> (<1> mutant reduces affinity to FAD [9]) [9]

6 Stability

General stability information

<1>, urea, 7 M, denaturation [1]
<3>, stable at 4°C for at least 2 months without loss of activity [10]

References

[1] Seifert, J.; Kunz, N.; Flachmann, R.; Laufer, A.; Jany, K.D.; Gassen, H.G.:
Expression of the E. coli nadB gene and characterization of the gene product L-aspartate oxidase. Biol. Chem. Hoppe-Seyler, **371**, 239-248 (1990)
[2] Flachmann, R.; Kunz, N.; Seifert, J.; Gutlich, M.; Wientjes, F.J.; Läufer, A.;
Gassen, H.G.: Molecular biology of pyridine nucleotide biosynthesis in

Escherichia coli. Cloning and characterization of quinolinate synthesis genes nadA and nadB. Eur. J. Biochem., **175**, 221-229 (1988)

[3] Hosokawa, Y.; Mitchell, E.; Gholson, R.K.: Higher plants contain L-asparate oxidase, the first enzyme of the Escherichia coli quinolinate synthetase system. Biochem. Biophys. Res. Commun., **111**, 188-193 (1983)

[4] Wilder, J.P.; Sae-Lee, J.A.; Mitchell, E.D.; Gholson, R.K.: The L-aspartate oxidase reported to be present in higher plants is actually glutamic oxaloacetic transaminase. Biochem. Biophys. Res. Commun., **123**, 836-841 (1984)

[5] Nasu, S.; Wicks, F.D.; Gholson, R.K.: L-Aspartate oxidase, a newly discovered enzyme of Escherichia coli, is the B protein of quinolinate synthetase. J. Biol. Chem., **257**, 626-632 (1982)

[6] Tedeschi, G.; Negri, A.; Mortarino, M.; Ceciliani, F.; Simonic, T.; Faotto, L.; Ronchi, S.: L-Aspartate oxidase from Escherichia coli. II. Interaction with C_4 dicarboxylic acids and identification of a novel L-aspartate:fumarate oxidoreductase activity. Eur. J. Biochem., **239**, 427-433 (1996)

[7] Tedeschi, G.; Ronchi, S.; Simonic, T.; Treu, C.; Mattevi, A.; Negri, A.: Probing the active site of L-aspartate oxidase by site-directed mutagenesis: Role of basic residues in fumarate reduction. Biochemistry, **40**, 4738-4744 (2001)

[8] Bossi, R.T.; Negri, A.; Tedeschi, G.; Mettevi, A.: Structure of FAD-bound L-aspartate oxidase: Insight into substrate specificity and catalysis. Biochemistry, **41**, 3018-3024 (2002)

[9] Mortarino, M.; Negri, A.; Tedeschi, G.; Simonic, T.; Duga, S.; Gassen, H.G.; Ronchi, S.: L-Aspartate oxidase from Escherichia coli. I. Characterization of coenzyme binding and product inhibition. Eur. J. Biochem., **239**, 418-426 (1996)

[10] Sakuraba, H.; Satomura, T.; Kawakami, R.; Yamamoto, S.; Kawarabayasi, Y.; Kikuchi, H.; Ohshima, T.: L-Aspartate oxidase is present in the anaerobic hyperthermophilic archaeon Pyrococcus horikoshii OT-3: characteristics and role in the de novo biosynthesis of nicotinamide adenine dinucleotide proposed by genome sequencing. Extremophiles, **6**, 275-281 (2002)

Tryptophan α,β-oxidase

1 Nomenclature

EC number
1.4.3.17

Systematic name
L-tryptophan:oxygen α,β-oxidoreductase

Recommended name
tryptophan α,β-oxidase

Synonyms
L-tryptophan 2',3'-oxidase
L-tryptophan α,β-dehydrogenase
tryptophan side chain oxidase II <2> [3, 4]

CAS registry number
156859-19-7

2 Source Organism

<1> *Chromobacterium violaceum* (strain ATCC12472 [1,2]) [1, 2]
<2> *Pseudomonas sp.* [3, 4]

3 Reaction and Specificity

Catalyzed reaction
L-tryptophan + O_2 = α,β-didehydrotryptophan + H_2O_2

Reaction type
oxidation
redox reaction
reduction

Substrates and products
S   L-tryptophan + O_2 <1> (Reversibility: ? <1> [1]) [1]
P α,β-didehydro-L-tryptophan + H_2O_2 <1> [1]
S   L-tryptophanamide + O_2 <1> (Reversibility: ? <1> [1]) [1]
P α,β-didehydro-L-tryptophanamide + H_2O_2 <1> [1]
S   N-acetyl-L-tryptophan + O_2 <1> (Reversibility: ? <1> [1]) [1]
P   N-acetyl-α,β-didehydro-L-tryptophan + H_2O_2 <1> [1]
S   N-acetyl-L-tryptophanamide + O_2 <1> (Reversibility: ? <1> [1, 2]) [1, 2]

P N-acetyl-α,β-didehydrotryptophanamide + H_2O_2 <1> [1, 2]
S adrenocorticotropic hormone + O_2 <1> (Reversibility: ? <1> [1]) [1]
P ? + H_2O_2 <1> [1]
S indole-3-propionate + O_2 <1> (Reversibility: ? <1> [1]) [1]
P ? + H_2O_2 <1> [1]
S luteinizing hormone-releasing hormone + O_2 <1> (Reversibility: ? <1> [1]) [1]
P ? + H_2O_2 <1> [1]
S pentagastrin + O_2 <1> (Reversibility: ? <1> [1]) [1]
P ? + H_2O_2 <1> [1]
S tryptophan No.14 of α-globin + O_2 <2> (Reversibility: ? <2> [3]) [3]
P α,β-dehydrotryptophan + H_2O_2 <2> [3]
S tryptophan No.15 of β-globin + O_2 <2> (Reversibility: ? <2> [3]) [3]
P α,β-dehydrotryptophan + H_2O_2 <2> [3]

Inhibitors
5-hydroxy-L-tryptophan <1> (<1> competitive inhibition, K_i: 0.037 mmol [1]) [1, 2]
D-tryptophan <1> (<1> competitive inhibition, K_i: 0.0066 mmol [1]) [1, 2]
methyl-3-indole <1> (<1> competitive inhibition, K_i: 0.0013 mmol [1]) [1, 2]
sodium dodecyl sulfate <2> (<2> no modification with 0.4% SDS [3]) [3]
tryptamine <1> (<1> competitive inhibition, K_i: 0.0054 mmol [1]) [1, 2]
urea <2> (<2> no modification with 8 M urea [3]) [3]
Additional information <1> (<1> no inhibition with L-histidine, N-acetyl-L-phenylalaninamide and N-acetyl-L-tyrosinamide [1,2]) [1, 2]

Cofactors/prosthetic groups
Additional information <1> (<1> NAD^+, $NADP^+$, FAD, or FMN show no effect on the activity [2]) [2]

Metals, ions
$MgCl_2$ <1> (<1> 10 mM, residual activity 103% [1]) [1]

Turnover number (min^{-1})
17 <1> (adrenocorticotropic hormone, <1> dehydrogenation of the tryptophanyl side chain [1]) [1]
20 <1> (luteinizing hormone-releasing hormone, <1> dehydrogenation of the tryptophanyl side chain [1]) [1]
32 <1> (pentagastrin, <1> dehydrogenation of the tryptophanyl side chain [1]) [1]
35.6 <1> (L-tryptophanamide) [1]
39.6 <1> (L-tryptophan) [1]
45.2 <1> (N-acetyl-L-tryptophanamide) [1, 2]
47.1 <1> (indole-3-propionate) [1]
47.4 <1> (N-acetyl-L-tryptophan) [1]

Specific activity (U/mg)
0.02 <2> (<2> substrate tryptophanamide [4]) [4]
0.07 <2> (<2> substrate tryptophan-leucine [4]) [4]

0.31 <2> (<2> substrate tryptophan [4]) [4]
0.74 <2> (<2> substrate Leu-Trp-Leu [4]) [4]
1.08 <2> (<2> substrate indolepropionic acid [4]) [4]
1.42 <2> (<2> substrate N-acetyltryptophanamide [4]) [4]
1.82 <2> (<2> substrate N-acetyltryptophan [4]) [4]
2.32 <2> (<2> substrate leucin-tryptophan [4]) [4]
30.3 <1> (<1> after purification [1]) [1]

K$_m$-Value (mM)
0.0058 <1> (L-tryptophan) [1]
0.0091 <1> (L-tryptophanamide) [1]
0.0195 <1> (N-acetyl-L-tryptophanamide) [1, 2]
0.02 <1> (indole-3-propionate) [1]
0.026 <1> (pentagastrin, <1> dehydrogenation of the tryptophanyl side chain [1]) [1]
0.071 <1> (luteinizing hormone-releasing hormone, <1> dehydrogenation of the tryptophanyl side chain [1]) [1]
0.0791 <1> (N-acetyl-L-tryptophan) [1]
0.93 <1> (adrenocorticotropic hormone, <1> dehydrogenation of the tryptophanyl side chain [1]) [1]

pH-Optimum
5 <1> [1]

pH-Range
3-9.5 <1> [2]

Temperature optimum (°C)
80 <1> [2]

4 Enzyme Structure

Molecular weight
150000 <2> [4]
680000 <1> (<1> gel filtration [1,2]) [1, 2]

Subunits
octamer <1> (<1> α, 4 * 14000 + β 4 * 74000, organized in an α_4,β_4 manner, heterooligomeric structure, one heme molecule per α,β protomer, SDS-PAGE [1,2]) [1, 2]

5 Isolation/Preparation/Mutation/Application

Localization
cytoplasm <1> [2]

Purification

<1> (cell disruption with Eaton press, ultracentrifugation, ammonium sulfate precipitation,DEAE-column, concentration using centrifugal microcentrators, gel filtration, purified 108fold, yield 33.8% [2]) [2]

<2> (gel filtration, phosphocellulose column chromatography, SDS-PAGE [4]) [4]

6 Stability

pH-Stability

7 <1> (<1> after dilution in 0.1 M bis-Tris buffer [1]) [1]

Temperature stability

80 <1> (<1> active up to 80°C [1]) [1]

General stability information

<1>, active with 0.2 M dithiothreitol, 1% SDS, or 4.5 M urea [1]

Storage stability

<1>, -80°C, for several months and for 24-48 h at room temperature in the absence of any additives [2]

References

[1] Genet, R.; Benetti, P.H.; Hammadi, A.; Menez, A.: L-Tryptophan 2',3'-oxidase from Chromobacterium violaceum. J. Biol. Chem., **270**, 23540-23545 (1995)

[2] Genet, R.; Donoyelle, C.; Menez, A.: Purification and partial characterization of an amino acid α,β-dehydrogenase, L-tryptophan 2',3'-oxidase from Chromobacterium violaceum. J. Biol. Chem., **269**, 18177-18184 (1994)

[3] Takai, K.; Sasai, Y.; Morimoto, H.; Yamazaki, H.; Yoshii, H.; Inoue, S.: Enzymatic dehydrogenation of tryptophan residues of human globins by tryptophan side chain oxidase II. J. Biol. Chem., **259**, 4452-4457 (1984)

[4] Ito, S.; Takai, K.; Tokuyama, T.; Hayaishi, O.: Enzymatic modification of tryptophan residues by tryptophan side chain oxidase I and II from Pseudomonas. J. Biol. Chem., **256**, 7834-7843 (1981)

Cytokinin oxidase

1 Nomenclature

EC number
1.4.3.18

Systematic name
cytokinin:oxygen oxidoreductase

Recommended name
cytokinin oxidase

Synonyms
CKO
CKX
N^6-(D2-isopentenyl)adenosine oxidase
cytokinin dehydrogenase
isopentenyladenosine oxidase

CAS registry number
55326-39-1

2 Source Organism

<1> *Zea mays* [1, 3, 7, 8]
<2> *Vinca rosea* [7, 8]
<3> *Nicotiana tabacum* [4, 7, 8]
<4> *Phaseolus vulgaris* (cv Great Northern [2,5-7]) [2, 5-7, 8]
<5> *Phaseolus lunatus* (cv Kingston [5]) [5, 7, 8]
<6> *Triticum aestivum* [7-9]
<7> *Populus x eruoamericana* [7]
<8> *Glycine max* [7]

3 Reaction and Specificity

Catalyzed reaction
N^6-(3-methylbut-2-enyl)adenine + H_2O + O_2 = adenine + 3-methylbut-2-enal + H_2O_2 (a flavoprotein (FAD). Catalyses the oxidation of cytokinins, a family of N^6-substituted adenine derivatives that are plant hormones, where the substituent is an isopentenyl group)

Reaction type

oxidation

reduction

Substrates and products

S N-benzyladenine + H_2O + O_2 <4> (Reversibility: ? <4> [2, 6]) [2, 6]

P ?

S N^6-(3-methylbut-2-enyl)adenine + H_2O + O_2 <1, 6> (Reversibility: ? <1, 6> [1-4, 9]) [1-4, 9]

P adenine + 3-methylbut-2-enal + H_2O_2

S N^6-hexyladenine + H_2O + O_2 <4> (Reversibility: ? <4> [2]) [2]

P ?

S N^6-isopentenyladenine + H_2O + O_2 <1-6> (Reversibility: ? <1-6> [2, 3, 8, 9]) [2, 3, 8, 9]

P ?

S cis-zeatin + H_2O + O_2 <4> (Reversibility: ? <4> [6]) [6]

P ?

S kinetin + H_2O + O_2 <4> (Reversibility: ? <4> [2]) [2]

P ?

S thidiazuron + H_2O + O_2 <4> (Reversibility: ? <4> [2]) [2]

P ?

S zeatin + H_2O + O_2 <1-6> (Reversibility: ? <1-6> [6, 8, 9]) [6, 8, 9]

P ?

Inhibitors

N-(2-chloro-4-pyridyl)-N'-phenylurea <1> [7]

cordycepin <4> [2]

rifampicin <4> [2]

Metals, ions

Cu^{2+} <1-6> (<1-6> enhances with imidazole complex [4-6,8]) [1, 4-6, 8]

Specific activity (U/mg)

24 <6> [9]

Additional information <1> [1]

K_m-Value (mM)

0.0003 <6> (N^6-isopentenyladenine) [9]

pH-Optimum

6.5 <4> [5]

7.5 <6> [9]

8.4 <5> [5]

Temperature optimum (°C)

30 <6> [9]

4 Enzyme Structure

Molecular weight
25100 <2> (<2>, gel filtration [7]) [7]
40000 <6> (<6>, gel filtration [9]) [9]
40000-60000 <6> (<6>, gel filtration [7]) [7]
55000-94400 <1> (<1>, gel filtration [7]) [7]

5 Isolation/Preparation/Mutation/Application

Source/tissue
callus <3-5> [2, 4, 5]
embryo <6> [9]
kernel <1> [1, 3]
leaf <3> [4]
root <3> [4]
stem <3> [4]

Purification
<1> [1]
<6> (partially [9]) [9]

References

[1] Burch, L.R.; Horgan, R.: The purification of cytokinin oxidase from Zea mays kernels. Phytochemistry, **28**, 1313-1319 (1989)
[2] Chatfield, J.M.; Armstrong, D.J.: Regulation of cytokinin oxidase activity in callus tissues of Phaseolus vulgaris L. cv Great Northern. Plant Physiol., **80**, 493-499 (1986)
[3] McGaw, B.A.; Horgan, R.: Cytokinin catabolism and cytokinin oxidase. Phytochemistry, **22**, 1103-1105 (1983)
[4] Motyka, V.; Faiss, M.; Strnad, M.; Kaminek, M.; Schmuelling, T.: Changes in cytokinin content and cytokinin oxidase activity in response to derepression of ipt gene transcription in transgenic tobacco calli and plants. Plant Physiol., **112**, 1035-1043 (1996)
[5] Kaminek, M.; Armstrong, D.J.: Genotypic variation in cytokinin oxidase from Phaseolus callus cultures. Plant Physiol., **93**, 1530-1538 (1990)
[6] Chatfield, J.M.; Armstrong, D.J.: Cytokinin oxidase from Phaseolus vulgaris callus tissues. Enhanced in vitro activity of the enzyme in the presence of copper-imidazole complexes. Plant Physiol., **84**, 726-731 (1987)
[7] Hare, P.D.; van Staden, J.: Cytokinin oxidase: biochemical features and physiological significance. Physiol. Plant., **91**, 128-136 (1994)
[8] Jones, R.J.; Schreiber, B.M.N.: Role and function of cytokinin oxidase in plants. Plant Growth Regul., **23**, 123-134 (1997)
[9] Laloue, M.; Fox, J.E.: Cytokinin oxidase from wheat. Plant Physiol., **90**, 899-906 (1989)

Glycine oxidase

1 Nomenclature

EC number
1.4.3.19

Systematic name
glycine:oxygen oxidoreductase (deaminating)

Recommended name
glycine oxidase

Synonyms
GO

CAS registry number
39307-16-9

2 Source Organism

<1> *Bacillus subtilis* [1, 2, 3, 4, 5]

3 Reaction and Specificity

Catalyzed reaction
D-alanine + H_2O + O_2 = pyruvate + NH_3 + H_2O_2
N-ethylglycine + H_2O + O_2 = glyoxylate + ethylamine + H_2O_2
glycine + H_2O + O_2 = glyoxylate + NH_3 + H_2O_2 (<1>, ternary complex sequential mechanism [3])
sarcosine + H_2O + O_2 = glyoxylate + methylamine + H_2O_2

Reaction type
oxidation
redox reaction
reduction

Natural substrates and products
S cycopropylglycine + H_2O + O_2 <1> (<1>, enzyme is required for the biosynthesis of the thiazole moiety of thiamine diphosphate [5]) (Reversibility: ? <1> [5]) [5]
P ?

Substrates and products

S D-2-aminobutyrate + H_2O + O_2 <1> (<1>, strictly stereospecific in only oxidation of the D-isomer [1]; <1>, 2.2% of the activity with sarcosine [2]) (Reversibility: ? <1> [1, 2]) [1, 2]

P ? + NH_3 + H_2O_2

S D-Ala + H_2O + O_2 <1> (<1>, 7.4% of the activity with sarcosine [2]; <1>, no activity with D-Ala [5]) (Reversibility: ? <1> [1, 2, 4, 5]) [1, 2, 4, 5]

P pyruvate + NH_3 + H_2O_2 <1> [1, 4]

S D-Arg + H_2O + O_2 <1> (<1>, low activity, strictly stereospecific in only oxidation of the D-isomer [1]) (Reversibility: ? <1> [1]) [1]

P ? + NH_3 + H_2O_2

S D-His + H_2O + O_2 <1> (<1>, low activity, strictly stereospecific in only oxidation of the D-isomer [1]) (Reversibility: ? <1> [1]) [1]

P ? + NH_3 + H_2O_2

S D-Ile + H_2O + O_2 <1> (<1>, low activity, strictly stereospecific in only oxidation of the D-isomer [1]) (Reversibility: ? <1> [1]) [1]

P ? + NH_3 + H_2O_2

S D-Leu + H_2O + O_2 <1> (<1>, low activity, strictly stereospecific in only oxidation of the D-isomer [1]) (Reversibility: ? <1> [1]) [1]

P ? + NH_3 + H_2O_2

S D-Pro + H_2O + O_2 <1> (<1>, 15.1% of the activity with sarcosine [2]; <1>, strictly stereospecific in only oxidation of the D-isomer [1]) (Reversibility: ? <1> [1, 2, 3, 4]) [1, 2, 3, 4]

P ? + NH_3 + H_2O_2

S D-Val + H_2O + O_2 <1> (<1>, low activity [1]; <1>, strictly stereospecific in only oxidation of the D-isomer [1]; <1>, 4.8% of the activity with sarcosine [2]) (Reversibility: ? <1> [1, 2]) [1, 2]

P ? + NH_3 + H_2O_2

S D-pipecolate + H_2O + O_2 <1> (<1>, strictly stereospecific in only oxidation of the D-isomer [1]) (Reversibility: ? <1> [1]) [1]

P ? + NH_3 + H_2O_2

S N-ethylglycine + H_2O + O_2 <1> (<1>, 85.3% of the activity with sarcosine [2]) (Reversibility: ? <1> [1, 2, 4]) [1, 2, 4]

P glyoxylate + NH_3 + H_2O_2 <1> [1, 4]

S N-methyl-D-Ala + H_2O + O_2 <1> (<1>, strictly stereospecific in only oxidation of the D-isomer [1]; <1>, 16.9% of the activity with sarcosine [2]) (Reversibility: ? <1> [1]) [1, 2]

P ? + NH_3 + H_2O_2

S cycopropylglycine + H_2O + O_2 <1> (Reversibility: ? <1> [5]) [5]

P ?

S ethyl glycine ester + H_2O + O_2 <1> (Reversibility: ? <1> [1]) [1]

P glyoxylic acid ethyl ester + NH_3 + H_2O_2

S glycine + H_2O + O_2 <1> (<1>, 77.4% of the activity with sarcosine [2]) (Reversibility: ? <1> [1, 2, 3, 4]) [1, 2, 3, 4]

P glyoxylate + NH_3 + H_2O_2 <1> [1, 4]

S sarcosine + H_2O + O_2 <1> (<1>, strictly stereospecific in only oxidation of the D-isomer [1]) (Reversibility: ? <1> [1, 2, 3, 4]) [1, 2, 3, 4]

P glyoxylate + methylamine + H_2O_2 <1> [1, 4]
S Additional information <1> (<1>, glycine oxidase is converted to a two-electron reduced form upon anaerobic reduction with the individual substrates and its reductive half-reaction is reversible [3]) [3]
P ?

Inhibitors

Ag^{2+} <1> (<1>, 2.0 mM [2]) [2]
Cd^{2+} <1> (<1>, 2.0 mM [2]) [2]
Cu^{2+} <1> (<1>, 2.0 mM [2]) [2]
Hg^{2+} <1> (<1>, 2.0 mM [2]) [2]
PCMB <1> (<1>, 2.0 mM [2]) [2]
glycolate <1> (<1>, competitive inhibitor with respect to sarcosine [4]) [4]

Cofactors/prosthetic groups

FAD <1> (<1>, FAD-dependent enzyme [5]) [5]
flavin <1> (<1>, flavoprotein [1,2]; <1>, addition of exogenous flavin does not increase activity [1,4]; <1>, enzyme contains one molecule of non-covalently bound flavin adenine dinucleotide per subunit [4]) [1, 2, 4]

Turnover number (min^{-1})

24 <1> (D-Ala) [5]
66 <1> (D-Ala) [2]
68.4 <1> (cyclopropylglycine) [5]
72 <1> (Gly) [5]
78 <1> (D-Pro) [2]
78 <1> (Gly) [2]
84 <1> (N-ethylglycine) [2]
96 <1> (sarcosine) [2]
210 <1> (D-Pro) [3]
240 <1> (Gly) [3]
252 <1> (sarcosine) [3]

K$_m$-Value (mM)

0.22 <1> (sarcosine) [2]
0.38 <1> (O_2, <1>, reaction with Gly [3]) [3]
0.42 <1> (O_2, <1>, reaction with sarcosine [3]) [3]
0.44 <1> (O_2, <1>, reaction with D-Pro [3]) [3]
0.6-0.8 <1> (sarcosine, <1>, value depends on assay procedure [4]) [4]
0.66 <1> (N-ethylglycine) [2]
0.7 <1> (sarcosine) [4]
0.99 <1> (Gly) [2]
1.12 <1> (Gly) [5]
1.17 <1> (cyclopropylglycine) [5]
2.1 <1> (N-ethylglycine) [4]
2.5 <1> (ethylglycinate) [4]
2.6 <1> (sarcosine) [3]
3.8 <1> (Gly) [3]
16.9 <1> (D-Pro) [4]

46 <1> (D-Pro) [2]
76.5 <1> (D-Pro) [3]
81 <1> (D-Ala) [2]
162 <1> (D-Ala) [4]
490 <1> (D-Ala) [4]

pH-Optimum
8 <1> [2]
10 <1> (<1>, His-tagged enzyme [1]) [1]

Temperature optimum (°C)
45 <1> [2]

4 Enzyme Structure

Molecular weight
159000 <1> (<1>, gel filtration [2]) [2]
166400 <1> (<1>, His-tagged chimeric enzyme, gel filtration [1]) [1]
180000 <1> (<1>, native enzyme, gel filtration [1]) [1]

Subunits
? <1> (<1>, x * 49400, SDS-PAGE [4]) [4]
tetramer <1> (<1>, 4 * 42000, SDS-PAGE [2]; <1>, 4 * 46600, native enzyme, SDS-PAGE [1]; <1>, 4 * 49400, His-tagged chimeric enzyme, SDS-PAGE [1]) [1, 2]

5 Isolation/Preparation/Mutation/Application

Localization
soluble <1> [1]

Purification
<1> (single-step purification of His-tagged enzyme by nickel-chelate chromatography [1]; recombinant enzyme expressed in Escherichia coli [2,4]) [1, 2, 4, 5]

Cloning
<1> (overexpression of wild-type enzyme and His-tagged enzyme in Escherichia coli [1]) [1]

6 Stability

pH-Stability
7-8.5 <1> (<1>, maximal stability of the His-tagged enzyme [1]) [1]
7.5-8.5 <1> (<1>, 25°C, 5 h, most stable [2]) [2]

Temperature stability

4-35 <1> (<1>, pH 7.0, 30 min, His-tagged enzyme is stable [1]) [1]
25 <1> (<1>, 5 h, most stable between pH 7.5 and 8.5 [2]) [2]
45 <1> (<1>, 10 min, stable up to [2]) [2]

Storage stability

<1>, -80°C, pH 7.5, native enzyme and His-tagged enzyme, stable for months [1]
<1>, -80°C, stable for several months [4]

References

[1] Job, V.; Marcone, G.L.; Pilone, M.S.; Pollegioni, L.: Glycine oxidase from Bacillus subtilis. Characterization of a new flavoprotein. J. Biol. Chem., **277**, 6985-6993 (2002)

[2] Nishiya, Y.; Imanaka, T.: Purification and characterization of a novel glycine oxidase from Bacillus subtilis. FEBS Lett., **438**, 263-266 (1998)

[3] Molla, G.; Motteran, L.; Job, V.; Pilone, M.S.; Pollegioni, L.: Kinetic mechanisms of glycine oxidase from Bacillus subtilis. Eur. J. Biochem., **270**, 1474-1482 (2003)

[4] Job, V.; Marcone, G.L.; Pilone, M.S.; Pollegioni, L.: Glycine oxidase from Bacillus subtilis. Characterization of a new flavoprotein. J. Biol. Chem., **277**, 6985-6993 (2002)

[5] Settembre, E.C.; Dorrestein, P.C.; Park, J.H.; Augustine, A.M.; Begley, T.P.; Ealick, S.E.: Structural and mechanistic studies on ThiO, a glycine oxidase essential for thiamin biosynthesis in Bacillus subtilis. Biochemistry, **42**, 2971-2981 (2003)

D-Proline reductase (dithiol) 1.4.4.1

1 Nomenclature

EC number
1.4.4.1 (transferred to EC 1.21.4.1)

Recommended name
D-proline reductase (dithiol)

1 Nomenclature

EC number
1.4.4.2

Systematic name
glycine:lipoylprotein oxidoreductase (decarboxylating and acceptor-amino-methylating)

Recommended name
glycine dehydrogenase (decarboxylating)

Synonyms
glycine cleavage system P-protein
P-protein
protein P1
decarboxylase, glycine
glycine decarboxylase
glycine decarboxylase P-protein
glycine-cleavage complex

CAS registry number
37259-67-9

2 Source Organism

<1> *Bos taurus* [8]
<2> *Gallus gallus* [1, 4, 8-12]
<3> *Pisum sativum* (pea) [2, 5, 7, 19-21, 23]
<4> *Eubacterium acidaminophilum* [3, 6]
<5> *Triticum aestivum* (wheat) [5]
<6> *Rattus norvegicus* [13, 16]
<7> *Peptococcus glycinophilus* [14, 15]
<8> *Arthrobacter globiformis* [17]
<9> *Saccharomyces cerevisiae* [18]
<10> *C3-plants* [21, 22]

3 Reaction and Specificity

Catalyzed reaction

glycine + lipoylprotein = S-aminomethyldihydrolipoylprotein + CO_2 (<2> H-protein ping-pong mechanism [1]; <1, 2> mechanism [8, 9]; <2> sequential random bi bi mechanism in which no abortive dead end complex is formed [9])

Reaction type

decarboxylation
oxidation
redox reaction
reduction

Natural substrates and products

S glycine + lipoylprotein <10, 1-9> (<1-8> lipoyl protein: H-protein, lipoamide can also act as acceptor [1-17]; <2> glycine decarboxylation catalyzed by P-protein alone is extremely low [1]; <2> reaction is stimulated by lipoic acid which is a functional group of the H-protein [1,12]) (Reversibility: ? <10, 1-9> [1-23]) [1-23]

P S-aminomethyldihydrolipoylprotein + CO_2

Substrates and products

S glycine + lipoylprotein <10, 1-9> (<1-8> lipoyl protein: H-protein, lipoamide can also act as acceptor [1-17]; <2> glycine decarboxylation catalyzed by P-protein alone is extremely low [1]; <2> reaction is stimulated by lipoic acid which is a functional group of the H-protein [1,12]) (Reversibility: ? <10, 1-9> [1-23]) [1-23]

P S-aminomethyldihydrolipoylprotein + CO_2

S Additional information <2, 3, 6, 8> (<6> the enzyme is a component of the reversible glycine cleavage system, previously known as glycine synthase [16,17]; <2> the glycine cleavage system consists of 4 protein components: 1. P-protein is a pyridoxal containing protein: a Schiff base is formed between the hydroxyl group of the pyridoxal phosphate and the α-NH_2 of glycine, the amino group and the α-carbon of the glycine are transferred to the lipoamide cofactor of the second enzyme of the complex the H-protein, the α-carbonyl group of glycine is lost as CO_2, 2. H-protein, 3. T-protein: catalyzes the passage of α-carbon from lipoamide of H protein to tetrahydrofolate, α-NH_2 from glycine is lost as NH_4^+, 4. L-protein: catalyzes oxidation of reduced lipoamide back to its original form with concomitant reduction of NAD^+ to NADH [1]; <4> glycine decarboxylase does not occur as a complex of all four proteins, but L-protein takes part in both glycine decarboxylase as well as in glycine reductase reaction [6]; <2,3> P-protein also catalyzes exchange of carbonyl carbon of glycine with CO_2, reaction greatly stimulated by addition of H protein [1,2,12]) [1, 2, 12, 16, 17]

P ?

Inhibitors

CO_2 <2> (<2> product inhibition [9]) [9]
Co^{2+} <2> (<2> inhibition of glycine-CO_2 exchange by binding of metal with H-protein-bound intermediate of glycine decarboxylation [11]) [11]
Cu^{2+} <2> (<2> inhibition of glycine-CO_2 exchange by binding of metal with H-protein-bound intermediate of glycine decarboxylation [11]) [11]
Fe^{2+} <2> (<2> slight inhibition of glycine-CO_2 exchange by binding of metal with H-protein-bound intermediate of glycine decarboxylation [11]) [11]
K_2HPO_4 <2> (<2> inhibition of glycine-CO_2 exchange reaction [12]) [12]
KCl <2> (<2> inhibition of glycine-CO_2 exchange reaction [12]) [12]
N-ethylmaleimide <6> [16]
N^4-methylglutamine <2> (<2> inhibition of glycine-CO_2 exchange reaction [12]) [12]
Ni^{2+} <2> (<2> inhibition of glycine-CO_2 exchange by binding of metal with H-protein-bound intermediate of glycine decarboxylation [11]) [11]
Zn^{2+} <2> (<2> inhibition of glycine-CO_2 exchange by binding of metal with H-protein-bound intermediate of glycine decarboxylation [11]) [11]
glycine methyl ester <2> [9]
modified H-protein <2> (<2> lipoic acid prosthetic group and cysteinyl residues modified with N-ethylmaleimide [9]) [9]
serine <10> (<10> competitive inhibitor [21]) [21]
Additional information <2, 4, 7> (<4,7> inhibitors of the multienzyme complex [3,14]; <2> inactivation after incubation with glycine in presence of aminomethyl carrier protein (H-protein), it is a suicide reaction of the P-protein as a side reaction of the glycine decarboxylation [10]) [3, 10, 14]

Cofactors/prosthetic groups

pyridoxal 5'-phosphate <10, 2-4, 6, 7> (<2> bound [1]; <4> loosly bound [3]; <2> pyridoxal phosphate binding site [4]; <7> 2 mol of pyridoxal phosphate bound per mol of enzyme [15]) [1-4, 14-16, 22]

Activating compounds

2-mercaptoethanol <2> (<2> thiol compound required for maximal activity on glycine-CO_2 exchange reaction [12]) [12]
GSH <2> (<2> thiol compound required for maximal activity on glycine-CO_2 exchange reaction [12]) [12]
dithiothreitol <2, 3> (<3> P and H protein alone jointly catalyze the glycine-CO_2 exchange reaction in presence of pyridoxal phosphate and dithiothreitol [2]; <2> thiol compound required for maximal activity on glycine-CO_2 exchange reaction [12]) [2, 12]

Specific activity (U/mg)

0.053 <6> [16]
0.25 <3> [23]
1.045 <6> [13]
1.9 <8> [17]
4.13 <2> [1]

K$_m$-Value (mM)

0.0034 <2> (glycine) [9]
0.0046 <7> (pyridoxal phosphate) [15]
0.0074 <6> (glycine) [13]
3.4 <2> (CO$_2$, <2> glycine-CO$_2$ exchange [12]) [12]
5.8 <2> (glycine) [9]
6 <10> (glycine) [21]
6.6 <6> (glycine) [13]
20 <2> (glycine, <2> glycine-CO$_2$ exchange [12]) [12]
31 <7> (bicarbonate) [15]
32 <7> (glycine) [15]
40 <2> (glycine) [1]

pH-Optimum

6.6 <2> (<2> glycine-CO$_2$ exchange [12]) [12]
6.7 <6> (<6> glycine-CO$_2$ exchange [13]) [13]
7 <7> (<7> glycine-CO$_2$ exchange [14]) [14]
7.1 <2> (<2> glycine + lipoylprotein [9]) [9]

Temperature optimum (°C)

25 <3> (<3> assay at [19]) [19]
37 <2> (<2> assay at [1]) [1]

4 Enzyme Structure

Molecular weight

105000 <3> (<3> predicted from amino acids [21]) [21]
114400 <9> (<9> calculation from cDNA [18]) [18]
152000 <6> (<6> gel filtration [13]) [13]
208000 <2> (<2> sucrose density gradient centrifugation [1]) [1]
210000 <3, 6> (<3> gel filtration [2,7]; <6> sucrose density gradient centrifugation [13]) [2, 7, 13]
225000 <4> (<4> gel filtration [3]) [3]
270000 <8> (<8> gel filtration [17]) [17]

Subunits

dimer <10, 2, 3, 6> (<2> α_2, 2 * 100000, SDS-PAGE [1]; <3> α_2, 2 * 98000, SDS-PAGE [2]; <3> α_2, 2 * 97000, lithium dodecyl sulfate PAGE [7]; <6> α_2, 2 * 105000, SDS-PAGE [13]; <10> α_2, 2 * 97000 [22]) [1, 2, 7, 13, 22]
tetramer <4> (<4> $\alpha_2\beta_2$, 2 * 59500 + 2 * 54100, SDS-PAGE [3]) [3]

Posttranslational modification

proteolytic modification <10> [21]

5 Isolation/Preparation/Mutation/Application

Source/tissue

leaf <3, 5> (<3,5> spatial and temporal influences on cell-specific distribution in leaves [5]; <3> matrix [7,19]) [2, 5, 7, 19, 20, 23]
liver <2, 6> [1, 4, 9, 11, 13, 16]

Localization

cytoplasm <4> (<4> protein P1/P2 complex is located predominantly in the cytoplasm [6]) [6]
mitochondrion <2, 3, 5, 6> [1, 2, 5, 7, 11, 13, 16, 19, 20, 23]

Purification

<2> [1]
<3> [2, 7, 19]
<4> [3]
<6> [13, 16]
<8> [17]

Cloning

<9> (expression of P-protein in yeast strains DS273U and DS275U [18]) [18]

6 Stability

Temperature stability

70 <6> (<6> 1 min, complete loss of activity [16]) [16]

Storage stability

<2>, -20°C, several weeks [1]
<3>, -80°C [7]
<6>, 0-20°C, 24 h, P-protein in purified P,L-protein fraction, 50-60% loss of activity [16]

References

[1] Hiraga, K.; Kikuchi, G.: The mitochondrial glycine cleavage system. Purification and properties of glycine decarboxylase from chicken liver mitochondria. J. Biol. Chem., **255**, 11664-11670 (1980)

[2] Walker, J.L.; Oliver, D.J.: Glycine decarboxylase multienzyme complex. Purification and partial characterization from pea leaf mitochondria. J. Biol. Chem., **261**, 2214-2221 (1986)

[3] Freudenberg, W.; Andreesen, J.R.: Purification and partial characterization of the glycine decarboxylase multienzyme complex from Eubacterium acidaminophilum. J. Bacteriol., **171**, 2209-2215 (1989)

[4] Fujiwara, K.; Okamura-Ikeda, K.; Motokawa, Y.: Amino acid sequence of the phosphopyridoxyl peptide from P-protein of the chicken liver glycine cleavage system. Biochem. Biophys. Res. Commun., **149**, 621-627 (1987)

[5] Tobin, A.K.; Thorpe, J.R.; Hylton, C.M.; Rawsthorne, S.: Spatial and temporal influences on the cell-specific distribution of glycine decarboxylase in leaves of wheat (Triticum aestivum L.) and pea (Pisum sativum L.). Plant Physiol., **91**, 1219-1225 (1989)

[6] Freudenberg, W.; Mayer, F.; Andreesen, J.R.: Immunocytochemical localization of proteins P1, P2, P3 of glycine decarboxylase and of the selenoprotein PA of glycine reductase, all involved in anaerobic glycine metabolism of Eubacterium acidaminophilum. Arch. Microbiol., **152**, 182-188 (1989)

[7] Bourguignon, J.; Neuburger, M.; Douce, R.: Resolution and characterization of the glycine-cleavage reaction in pea leaf mitochondria. Properties of the forward reaction catalysed by glycine decarboxylase and serine hydroxymethyltransferase. Biochem. J., **255**, 169-178 (1988)

[8] Fujiwara, K.; Okamura-Ikeda, K.; Ohmura, Y.; Motokawa, Y.: Mechanism of the glycine cleavage reaction: retention of C-2 hydrogens of glycine on the intermediate attached to H-protein and evidence for the inability of serine hydroxymethyltransferase to catalyze the glycine decarboxylation. Arch. Biochem. Biophys., **251**, 121-127 (1986)

[9] Fujiwara, K.; Motokawa, Y.: Mechanism of the glycine cleavage reaction. Steady state kinetic studies of the P-protein-catalyzed reaction. J. Biol. Chem., **258**, 8156-8162 (1983)

[10] Hiraga, K.; Kikuchi, G.: The mitochondrial glycine cleavage system: inactivation of glycine decarboxylase as a side reaction of the glycine decarboxylation in the presence of aminomethyl carrier protein. J. Biochem., **92**, 1489-1498 (1982)

[11] Hiraga, K.; Kikuchi, G.: The mitochondrial glycine cleavage system: differential inhibition by divalent cations of glycine synthesis and glycine decarboxylation in the glycine-CO_2 exchange. J. Biochem., **92**, 937-944 (1982)

[12] Hiraga, K.; Kikuchi, G.: The mitochondrial glycine cleavage system. Functional association of glycine decarboxylase and aminomethyl carrier protein. J. Biol. Chem., **255**, 11671-11676 (1980)

[13] Hayasaka, K.; Kochi, H.; Hiraga, K.; Kikuchi, G.: Purification and properties of glycine decarboxylase, a component of the glycine cleavage system, from rat liver mitochondria and immunochemical comparison of this enzyme from various sources. J. Biochem., **88**, 1193-1199 (1980)

[14] Klein, S.M.; Sagers, R.D.: Glycine metabolism. I. Properties of the system catalyzing the exchange of bicarbonate with the carboxyl group of glycine in Peptococcus glycinophilus. J. Biol. Chem., **241**, 197-205 (1966)

[15] Klein, S.M.; Sagers, R.: Glycine metabolism. II. Kinetic and optical studies on the glycine decarboxylase system from Peptococcus glycinophilus. J. Biol. Chem., **241**, 206-209 (1966)

[16] Motokawa, Y.; Kikuchi, G.: Glycine metabolism by rat liver mitochondria. Reconstruction of the reversible glycine cleavage system with partially purified protein components. Arch. Biochem. Biophys., **164**, 624-633 (1974)

[17] Kochi, H.; Kikuchi, G.: Mechanism of the reversible glycine cleavage reaction in Arthrobacter globiformis. I. Purification and function of protein components required for the reaction. J. Biochem., 75, 1113-1127 (1974)

[18] Sinclair, D.A.; Hong, S.P.; Dawes, I.W.: Specific induction by glycine of the gene for the P-subunit of glycine decarboxylase from Saccharomyces cerevisiae. Mol. Microbiol., 19, 611-623 (1996)

[19] Besson, V.; Rebeille, F.; Neuburger, M.; Douce, R.; Cossins, E.A.: Effects of tetrahydrofolate polyglutamates on the kinetic parameters of serine hydroxymethyltransferase and glycine decarboxylase from pea leaf mitochondria. Biochem. J., 292, 425-430 (1993)

[20] Rebeille, F.; Neuburger, M.; Douce, R.: Interaction between glycine decarboxylase, serine hydroxymethyltransferase and tetrahydrofolate polyglutamates in pea leaf mitochondria. Biochem. J., 302, 223-228 (1994)

[21] Oliver, D.J.; Raman, R.: Glycine decarboxylase: protein chemistry and molecular biology of the major protein in leaf mitochondria. J. Bioenerg. Biomembr., 27, 407-414 (1995)

[22] Douce, R.; Bourguignon, J.; Macherel, D.; Neuburger, M.: The glycine decarboxylase system in higher plant mitochondria: structure, function and biogenesis. Biochem. Soc. Trans., 22, 184-188 (1994)

[23] Zhang, Q.; Wiskich, J.T.: Activation of glycine decarboxylase in pea leaf mitochondria by ATP. Arch. Biochem. Biophys., 320, 250-256 (1995)

Glutamate synthase (ferredoxin)   1.4.7.1

1 Nomenclature

EC number
1.4.7.1

Systematic name
L-glutamate:ferredoxin oxidoreductase (transaminating)

Recommended name
glutamate synthase (ferredoxin)

Synonyms
Fd-GOGAT
ferredoxin-dependent glutamate synthase
ferredoxin-glutamate synthase
glutamate synthase (ferredoxin-dependent)

CAS registry number
62213-56-3

2 Source Organism

<1> *Medicago sativa* (induced biosynthesis during nodule development [3])
[3]
<2> *Bouvardia ternifolia* (Schlecht) [9]
<3> *Glycine max* [12]
<4> *Chlamydomonas reinhardtii* [1, 4, 10, 19]
<5> *Zea mays* (maize [7,8]) [7, 8]
<6> *Lycopersicon esculentum* (tomato [5,11]) [5, 11]
<7> *Spinacia oleracea* (spinach [13,20,21]) [2, 6, 13, 20, 21, 25, 27]
<8> *Oryza sativa* (rice [15,18]) [15, 18, 23, 29]
<9> *Vicia faba* (bean [22]) [22]
<10> *Nicotiana tabacum* [14]
<11> *Nostoc muscorum* [16]
<12> *Pisum sativum* [17]
<13> *Pinus pinaster* [24]
<14> *Pinus halepensis* [24]
<15> *Pinus sylvestris* [24]
<16> *Pinus pinea* [24]
<17> *Larix decidua* [24]
<18> *Abies pinsapo* [24]

<19> *Hordeum vulgare* (L. cv. maris mink, barley [26]) [26]
<20> *Synechocystis sp.* (PCC 6803, cyanobacterium [28]) [28, 31]
<21> *Vitis vinifera* (grapevine [30]) [30]

3 Reaction and Specificity

Catalyzed reaction

2 L-glutamate + 2 oxidized ferredoxin = L-glutamine + 2-oxoglutarate + 2 reduced ferredoxin (<20> proposed electron-transfer pathway [31])

Reaction type

oxidation
redox reaction
reduction
transamination

Natural substrates and products

S L-glutamine + 2-oxoglutarate + reduced ferredoxin <1-12, 19, 20, 21> (<4> highly specific for glutamine and 2-oxoglutarate [10]; <6> enzyme may be linked to light driven electron transport [11]) (Reversibility: ? <1-12, 19, 20, 21> [1-22, 26, 28, 30]) [1-22, 26, 28, 30]
P L-glutamate + oxidized ferredoxin <1-12, 19, 20, 21> [1-22, 26, 28, 30]

Substrates and products

S L-glutamate + oxidized iodonitrotetrazolium <20> (<20> enzyme exhibits L-glutamate:iodonitrotetrazolium oxidoreductase activity at pH 7.5-9.5 [31]) (Reversibility: ? <20> [31]) [31]
P 2-oxoglutarate + NH$_3$ + reduced iodonitrotetrazolium <20> [31]
S L-glutamine + 2-oxoglutarate + reduced ferredoxin <1-12, 19, 20, 21> (<4> about 35% of activity with methylviologen [4]; <4> highly specific for glutamine and 2-oxoglutarate [10]) (Reversibility: ? <1-12, 19, 20, 21> [1-22, 26, 28, 30]) [1-22, 26, 28, 30]
P L-glutamate + oxidized ferredoxin <1-12, 19, 20> [1-22, 26, 28, 30]
S L-glutamine + 2-oxoglutarate + reduced methylviologen <1-10, 12, 19, 20> (<1-6, 8, 10, 12> artificial elctron donor [1-12, 14, 15, 17-22]; <7> only in the presence of ferredoxin [13]; <10> 60% of ferredoxin activity [14]; <19> similar activity as with ferredoxin [26]; <20> significantly higher activity with ferredoxin than with methyl viologen [31]) (Reversibility: ? <1-10, 12, 19, 20> [1-15, 17-22, 26, 28]) [1-10, 12-15, 17-22, 26, 28, 31]
P L-glutamate + oxidized methylviologen <1-6, 8, 10, 12, 19, 20> [1-10, 12-15, 17-22, 26, 28, 31]

Inhibitors

2-oxoglutarate <20> (<20> at high concentrations [28]) [28]
3-(3,4-dichlorophenyl)-1,1-dimethylurea <12> [17]
3-phosphoglycerate <6> [5]
3-phosphoserine <6> (<6> competitive inhibition [5]) [5]

6-diazo-5-oxo-norleucine <4, 20> (<4> 5 mM, complete inhibition [10]; <20> 0.05 mM, 99% inhibition [28]; <20> reversible inactivator [31]) [10, 28, 31]

Cd^{2+} <8> (<8> 5 mM, 69% inhibition [18]) [18]

Co^{2+} <8> (<8> 5 mM, 65% inhibition [18]) [18]

L-alanine <6> [5]

L-arginine <6> [5]

L-leucine <6> [5]

L-lysine <6> [5]

L-methionine <6> [5]

L-ornithine <6> [5]

L-serine <4, 6> (<4> 20 mM, 70% inhibition [1]) [1, 5]

L-threonine <6> [5]

Mn^{2+} <8> (<8> 5 mM, 45% inhibition [18]) [18]

N-ethylmaleimide <4> (<4> 15 mM, 93% inactivation after 15 h, L-glutamine protects from inactivation [1]; <4> 0.05 mM, 94% inhibition [10]) [1, 10]

Ni^{2+} <8> (<8> 5 mM, 73% inhibition [18]) [18]

aspartate <6> (<6> competitive inhibition [5]) [5]

atebrin <7> (<7> 1 mM, 45% inhibition [13]) [13]

azaserine <4, 7, 8, 11, 12, 20> (<4> 0.5 mM, complete inhibition [10]]; <7> 1 mM, complete inhibition [13]; <20> 0.05 mM, 62% inhibition [28]) [10, 13, 16-19, 28]

bromocresol green <4> (<4> 1 mM, 89% inhibition [10]) [10]

flavianate <4> [10]

glycine <4, 6> (<4> 20 mM, 70% inhibition [1]) [1, 5]

glycolate <4> (<4> 20 mM, more than 60% inhibition [1]) [1]

glyoxylate <4> (<4> 20 mM, more than 60% inhibition [1]) [1]

iodoacetamide <4> (<4> 90% inactivation after 15 h, L-glutamine protects from inactivation [1]) [1]

p-chloromercuribenzoate <7> (<7> 0.1 mM, complete inhibtion [13]) [13]

p-chloromercuriphenylsulfonate <7> (<7> 0.1 mM, complete inhibition [13]) [13]

p-hydroxymercuribenzoate <4> (<4> 1 mM, complete inactivation after 1 h [1]) [1, 10]

Cofactors/prosthetic groups

FMN <7, 20> (<7> in contrast to earlier conclusions the enzyme contains 1 mol of FMN per mol enzyme but no FAD [25]; <20> one FMN per enzyme molecule [28]) [2, 13, 25, 28, 31]

ferredoxin <7> (<7> enzyme forms an electrostatically stabilized complex with ferredoxin [2]) [2, 6]

flavin <4> (<4> enzyme may contain a flavin prosthetic group [10]) [10]

Metals, ions

Ca^{2+} <8> (<8> 1 mM, 6% activation [18]) [18]

iron <7, 20> (<7> enzyme contains an iron-sulfur-center [13]; <7> enzyme may contain two [2Fe-2S] clusters rather than a single [4Fe-4S] cluster [2];

<7> enzyme contains a [3Fe-4S] cluster [25]; <20> enzyme contains a single [3Fe-4S] cluster [28]) [2, 13, 25, 27, 28, 31]

Turnover number (min^{-1})

220 <20> (L-glutamate, <20> L-glutamate:iodonitrotetrazolium oxidoreductase activity at pH 9.5 [31]) [31]

8214 <8> (L-glutamine) [18]

Specific activity (U/mg)

0.0045 <2> (<2> activity in young leaves [9]) [9]

0.018 <19> (<19> activity in root crude extracts [26]) [26]

0.077 <5> (<5> enzyme separated from NADH-dependent activity on Sephadex G-200 [8]) [8]

0.084 <19> (<19> activity in leaf crude extracts [26]) [26]

0.091 <8> (<8> activity in roots under airobic conditiones, approx. value [29]) [29]

0.125 <8> (<8> activity in roots after anaerobic treatment for 24 h, approx. value [29]) [29]

3.1 <19> (<19> activity in first leaf after 5 d development [26]) [26]

4.2 <19> (<19> activity in 30 d old plants in the fourth leaf, activity declines in older leaves [26]) [26]

7 <9> (<9> ferredoxin affinity chromatographyie purified enzyme [22]) [22]

8.16 <13> [24]

10.4 <4> [10]

10.4 <4> [10]

13.67 <7> [21]

35.88 <8> [18]

35.9 <7, 8> [18, 20]

36.6 <20> (<20> recombinant enzyme [28]) [28]

101.7 <7> [13]

Additional information <10> (<10> 16.3 nmol glutamate/min/g fresh weight, activity in green cells, assay with methyl viologen [14]) [14]

Additional information <10> (<10> 31.3 nmol glutamate/min/g fresh weight, activity in green cells, assay with ferredoxin [14]) [14]

Additional information <12> (<12> 0.0009 mmol/min/shoot, activity in 17 d old shoots, activity increases during seedling development [17]) [17]

K$_m$-Value (mM)

0.0002 <6> (ferredoxin) [11]

0.0018 <1> (ferredoxin) [3]

0.0019 <3> (ferredoxin) [12]

0.002 <7> (ferredoxin) [6]

0.0035 <20> (ferredoxin, <20> native enzyme [28]) [28]

0.0045 <20> (ferredoxin, <20> recombinant enzyme [28]) [28]

0.0048 <8> (ferredoxin, <8> in etiolated leaf tissue [15]) [15]

0.005 <11> (ferredoxin, <11> approx. value [16]) [16]

0.0055 <8> (ferredoxin, <8> in green leaf tissue [15]) [15, 18]

0.015 <4> (ferredoxin) [10]

0.02 <8> (ferredoxin, <8> in root tissue [15]) [15]
0.044 <20> (L-glutamate, <20> L-glutamate:iodonitrotetrazolium oxidoreductase activity at pH 9.5 [31]) [31]
0.062 <7> (2-oxoglutarate) [13]
0.089 <7> (2-oxoglutarate) [21]
0.1 <20> (iodonitrotetrazolium, <20> L-glutamate:iodonitrotetrazolium oxidoreductase activity at pH 9.5 [31]) [31]
0.133 <3> (methyl viologen) [12]
0.17 <4> (2-oxoglutarate) [19]
0.182 <1> (L-glutamine) [3]
0.19 <4> (L-glutamine) [19]
0.2 <6> (2-oxoglutarate) [11]
0.33 <8> (2-oxoglutarate) [18]
0.5 <20> (2-oxoglutarate, <20> native enzyme [28]) [28]
0.5 <6> (L-glutamine) [11]
0.53 <7> (L-glutamine) [13]
0.556 <7> (L-glutamine) [21]
0.6 <20> (2-oxoglutarate, <20> recombinant enzyme [28]) [28]
0.606 <1> (2-oxoglutarate) [3]
0.7 <4> (L-glutamine) [10]
1 <4> (2-oxoglutarate) [10]
2.2 <20> (L-glutamine, <20> recombinant enzyme [28]) [28]
2.5 <20> (L-glutamine, <20> native enzyme [28]) [28]
4.7 <4> (methyl viologen) [10]

K_i-Value (mM)
0.1 <6> (aspartate) [5]
4.9 <6> (phosphoserine) [5]

pH-Optimum
7.3 <8> [18]
7.5 <4> [10, 19]
7.5-8.5 <7> [21]
7.8 <6> [5]

Temperature optimum (°C)
35 <4> [10]

Temperature range (°C)
35 <4> [10]

4 Enzyme Structure

Molecular weight
140000 <7> (<7> gel filtration [13]) [13]
141000 <6> (<6> gel filtration [5]) [5]
144000 <4> (<4> sucrose gradient density centrifugation [10]) [10]
145900 <4> (<4> amino acid composition [1]) [1]

163000 <13> (gel filtration [24]) [24]
165000 <4> (<4> gel filtration [19]) [19]
171000 <5> (<5> gel filtration [8]) [8]
175000 <20> (<20> recombinant enzyme, gel filtration [28]) [28]
178000 <20> (<20> native enzyme, gel filtration [28]) [28]
180000 <7> (<7> gel filtration [6]) [6]
224000 <8> (<8> gel filtration [18]) [18]

Subunits

? <1, 3, 6-7, 21> (<7> x * 180000, SDS-PAGE [20]; <3> x * 165000, SDS-PAGE
[12]; <6> x * 153000, SDS-PAGE [5]; <1> x * 68200, SDS-PAGE [3]; <21> x +
160000, SDS-PAGE, immunodetection [30]) [3, 5, 12, 20, 30]
dimer <8> (<8> 2 * 115000, SDS-PAGE [18]) [18]
monomer <4-8, 13> (<6> 1 * 153000, SDS-PAGE [5]; <5> 1 * 145000, immu-
noprecipitation, SDS-PAGE [7]; <4> 1 * 151000, SDS-PAGE [10]; <7> 1 *
170000, SDS-PAGE [13]; <8> SDS-PAGE [18]; <3> SDS-PAGE [12]; <13> 1 *
168000, SDS-PAGE [24]; <5> 1 * 145000, SDS-PAGE [7]) [3, 5, 7, 10, 12, 13,
18, 20, 24]

5 Isolation/Preparation/Mutation/Application

Source/tissue

callus <2, 10> [9, 14]
cotyledon <13-18> (<14> more abundant than in stems, barely detectable in
roots [24]) [24]
endosperm <5> [8]
internode <21> [30]
leaf <1-9, 12, 19, 21> [3-15, 17, 18, 20-22, 23, 26, 30]
root <12, 4-6, 8, 10, 19, 21> (<8> expression is induced by NO_3^- [23]) [7, 8,
10, 11, 14, 15, 17, 23, 26, 30]
root nodule <1, 3> [3, 12]
stem <13-18> [24]

Localization

chloroplast <2, 4-6, 7, 8, 10, 12, 13, 21> (<13> stroma [24]) [4, 6, 7, 9-11, 14,
17, 18, 24, 30]

Purification

<4> (protamine sulfate, DEAE-Sephacel, hydroxylapatite, ferredoxin-Sephar-
ose [10]; partially purified [19]) [1, 10, 19]
<5> [8]
<6> (ammonium sulfate, DEAE cellulose, glutamate agarose [11]) [11]
<7> (affinity chromatography on ferredoxin Sepharose [13]) [13, 20, 21]
<8> (affinity chromatography on ferredoxin Sepharose [18]) [18]
<9> (affinity chromatography on ferredoxin Sepharose [22]) [22]
<13> (ammonium sulfate, DEAE-cellulose, DEAE-Sephacel, Phenyl-Sephar-
ose [24]) [24]

<20> (recombinant enzyme, affinity chromatography on ferredoxin-Sepharose, native enzyme, partially purified [28]) [28]
<21> (recombinant enzyme [30]) [30]

Cloning
<5> (in vitro translation in reticulocyte lysate [7]) [7]
<15> (cDNA insert coding for 550 C-terminal amino acids, expression in Escherichia coli [24]) [24]
<20> (expression in Escherichia coli [28]) [28, 31]
<21> (expression in Escherichia coli [30]) [30]

6 Stability

Temperature stability
50 <4> (<4> full activity after 10 min, 50% activity remains after heating at 55°C for 10 min, complete loss of activity at 60°C [10]) [10]
55 <20> (<20> 50% activity remains after 20 min, recombinant enzyme [28]) [28]

General stability information
<7>, unstable [20]

Storage stability
<8>, -20°C, 5 mM dithiothreitol, 20 days, 54% loss of activity [18]

References

[1] Gotor, C.; Martinez-Rivas, J.M.; Marquez, A.J.; Vega, J.M.: Functional properties of purified ferredoxin-glutamate synthase from Chlamydomonas reinhardtii. Phytochemistry, **29**, 711-717 (1990)
[2] Hirasawa, M.; Chang, K.T.; Morrow jr.; K.J.; Knaff, D.B.: Circular dichroism, binding and immunological studies on the interaction between spinach ferredoxin and glutamate synthase. Biochim. Biophys. Acta, **977**, 150-156 (1989)
[3] Suzuki, A.; Carrayol, E.; Zehnacker, C.; Deroche, M.E.: Glutamate synthase in Medicago sativa L. Occurrence and properties of FD-dependent enzyme in plant cell fraction during root nodule development. Biochem. Biophys. Res. Commun., **156**, 1130-1138 (1988)
[4] Fischer, P.; Klein, U.: Localization of nitrogen-assimilation enzymes in the chloroplast of Chlamydomonas reinhardtii. Plant Physiol., **88**, 947-952 (1988)
[5] Avila, C.; Botella, J.R.; Canovas, F.M.; Nunez de Castro, I.; Valpuesta, V.: Different characteristics of the two glutamate synthases in the green leaves of Lycopersicon esculentum. Plant Physiol., **85**, 1036-1039 (1987)

[6] Hirasawa, M.; Boyer, J.M.; Gray, K.A.; Davis, D.J.; Knaff, D.B.: The interaction of ferredoxin with chloroplast ferredoxin-linked enzymes. Biochim. Biophys. Acta, **851**, 23-28 (1986)

[7] Commere, B.; Vidal, J.; Suzuki, A.; Gadal, P.; Caboche, M.: Detection of the messenger RNA encoding for the ferredoxin-dependent glutamate synthase in maize leaf. Plant Physiol., **80**, 859-862 (1986)

[8] Misra, S.; Oaks, A.: Ferredoxin and pyridine nucleotide-dependent glutamate synthase activities in maize endosperm tissue. Plant Sci., **39**, 1-5 (1985)

[9] Murillo, E.; S·nchez de Jimenez, E.: Glutamate synthase in greening callus of Bouvardia ternifolia Schlecht. Planta, **163**, 448-452 (1985)

[10] Galvan, F.; Marquez, A.J.; Vega, J.M.: Purification and molecular properties of ferredoxin-glutamate synthase from Chlamydomonas reinhardtii. Planta, **162**, 180-187 (1984)

[11] Avila, C.; Canovas, F.; Nunez de Castro, I.; Valpuesta, V.: Separation of two forms of glutamate synthase in leaves of tomato (Lycopersicon esculentum). Biochem. Biophys. Res. Commun., **122**, 1125-1130 (1984)

[12] Suzuki, A.; Vidal, J.; Nguyen, J.; Gadal, P.: Occurence of ferredoxin-dependent glutamate synthase in plant cell fraction of soybean root nodules (Glycine max). FEBS Lett., **173**, 204-208 (1984)

[13] Hirasawa, M.; Tamura, G.: Flavin and iron-sulfur containing ferredoxin-linked glutamate synthase from spinach leaves. J. Biochem., **95**, 983-994 (1984)

[14] Suzuki, A.; Nato, A.; Gadal, P.: Glutamate synthase isoforms in Tobacco cultured cells. Plant Sci. Lett., **33**, 93-101 (1984)

[15] Suzuki, A.; Jacquot, J.P.; Gadal, P.: Glutamate synthase in rice roots. Studies on the electron donor specificity. Phytochemistry, **22**, 1543-1546 (1983)

[16] Haeger, K.P.; Danneberg, G.; Bothe, H.: The glutamate synthase in heterocysts of Nostoc muscorum. FEMS Microbiol. Lett., **17**, 179-183 (1983)

[17] Matoh, T.; Takahashi, E.: Changes in the activities of ferredoxin- and NADH-glutamate synthase during seedling development of peas. Planta, **154**, 289-294 (1982)

[18] Suzuki, A.; Gadal, P.: Glutamate synthase from rice leaves. Plant Physiol., **69**, 848-852 (1982)

[19] Cullimore, J.V.; Sims, A.P.: Occurrence of two forms of glutamate synthase in Chlamydomonas reinhardii. Phytochemistry, **20**, 597-600 (1981)

[20] Tamura, G.; Oto, M.; Hirasawa, M.; Aketagawa, J.: Isolation and partial characterization of homogeneous glutamate synthase from Spinacia oleracea. Plant Sci. Lett., **19**, 209-215 (1980)

[21] Tamura, G.; Kanki, M.; Hirasawa, M.; Oto, M.: The purification and properties of glutamate synthase from spinach leaves, and its dependence on ferredoxin. Agric. Biol. Chem., **44**, 925-927 (1980)

[22] Wallsgrove, R.M.; Miflin, B.J.: Ferredoxin-Sepharose as an affinity absorbent for the purification of glutamate synthase and other ferredoxin-dependent enzymes. Biochem. Soc. Trans., **5**, 269-271 (1977)

[23] Redinbaugh, M.G.; Campbell, W.H.: Glutamine synthetase and ferredoxin-dependent glutamate synthase expression in the maize (Zea mays) root pri-

mary response to nitrate (evidence for an organ-specific response). Plant Physiol., **101**, 1249-1255 (1993)

[24] Garcia-Gutierrez, A.; Canton, F.R.; Gallardo, F.; Sanchez-Jimenez, F.; Canovas, F.M.: Expression of ferredoxin-dependent glutamate synthase in dark-grown pine seedlings. Plant Mol. Biol., **27**, 115-128 (1995)

[25] Hirasawa, M.; Hurley, J.K.; Salamon, Z.; Tollin, G.; Knaff, D.B.: Oxidation-reduction and transient kinetic studies of spinach ferredoxin-dependent glutamate synthase. Arch. Biochem. Biophys., **330**, 209-215 (1996)

[26] Pajuelo, P.; Pajuelo, E.; Forde, B.G.; Marquez, A.J.: Regulation of the expression of ferredoxin-glutamate synthase in barley. Planta, **203**, 517-525 (1997)

[27] Vanoni, M.A.; Curti, B.: Glutamate synthase: a complex iron-sulfur flavoprotein. Cell. Mol. Life Sci., **55**, 617-638 (1999)

[28] Navarro, F.; Martin-Figueroa, E.; Candau, P.; Florencio, F.J.: Ferredoxin-dependent iron-sulfur flavoprotein glutamate synthase (GlsF) from the Cyanobacterium synechocystis sp. PCC 6803: expression and assembly in Escherichia coli. Arch. Biochem. Biophys., **379**, 267-276 (2000)

[29] Reggiani, R.; Nebuloni, M.; Mattana, M.; Brambilla, I.: Anaerobic accumulation of amino acids in rice roots: role of the glutamine synthetase/glutamate synthase cycle. Amino Acids, **18**, 207-217 (2000)

[30] Loulakakis, K.A.; Primikirios, N.I.; Nikolantonakis, M.A.; Roubelakis-Angelakis, K.A.: Immunocharacterization of Vitis vinifera L. ferredoxin-dependent glutamate synthase, and its spatial and temporal changes during leaf development. Planta, **215**, 630-638 (2002)

[31] Ravasio, S.; Dossena, L.; Martin-Figueroa, E.; Florencio, F.J.; Mattevi, A.; Morandi, P.; Curti, B.; Vanoni, M.A.: Properties of the recombinant ferredoxin-dependent glutamate synthase of Synechocystis PCC6803. Comparison with the Azospirillum brasilense NADPH-dependent enzyme and its isolated α-subunit. Biochemistry, **41**, 8120-8133 (2002)

D-Amino-acid dehydrogenase

1.4.99.1

1 Nomenclature

EC number
1.4.99.1

Systematic name
D-amino-acid:(acceptor) oxidoreductase (deaminating)

Recommended name
D-amino-acid dehydrogenase

CAS registry number
37205-44-0

2 Source Organism

<1> *Pyrobaculum islandicum* [1]
<2> *Pyrobaculum islandicum* [2]
<3> *Methanosarcina barkeri* [2]
<4> *Halobacterium salinarium* [2]
<5> *Escherichia coli* [3-6, 12]
<6> *Salmonella typhimurium* [7, 8]
<7> *Pseudomonas fluorescens* [9, 10]
<8> *Pseudomonas aeruginosa* [11]

3 Reaction and Specificity

Catalyzed reaction
a D-amino acid + H_2O + acceptor = a 2-oxo acid + NH_3 + reduced acceptor
(A flavoprotein, FAD. Acts to some extent on all D-amino acids, except D-aspartate and D-glutamate)

Reaction type
oxidation
oxidative deamination
redox reaction
reduction

Natural substrates and products
S D-alanine + H_2O + FAD <2> (<2> no activity for Methanosarcina barkeri and Halobacterium salinarium enzyme [2]) (Reversibility: ? <2> [2]) [2]

387

P pyruvate + NH_3 + $FADH_2$ <2> [2]
S D-alanine + H_2O + acceptor <6> (<6> best substrate tested [8]) (Reversibility: ? <6> [8]) [8]
P pyruvate + NH_3 + reduced acceptor <6> [8]
S D-alanine + H_2O + coenzyme Q <5> (<5> physiological electron acceptor [6]) (Reversibility: ? <5> [6]) [6]
P pyruvate + NH_3 + reduced coenzyme Q <5> [6]
S D-alanine + H_2O + cytochrome c <5> (Reversibility: ? <5> [6]) [6]
P pyruvate + NH_3 + reduced cytochrome c <5> [6]
S D-asparagine + H_2O + acceptor <6> (Reversibility: ? <6> [8]) [8]
P 2-oxosuccinamic acid + NH_3 + reduced acceptor <6> [8]
S D-aspartate + H_2O + FAD <2> (<2> no activity for Methanosarcina barkeri and Halobacterium salinarium enzyme [2]) (Reversibility: ? <2> [2]) [2]
P oxaloacetate + NH_3 + $FADH_2$ <2> [2]
S D-cysteine + H_2O + acceptor <6> (<6> low activity [8]) (Reversibility: ? <6> [8]) [8]
P 3-mercapto-2-oxopropanoate + NH_3 + reduced acceptor <6> [8]
S D-glutamate + H_2O + FAD <2-4> (Reversibility: ? <2-4> [2]) [2]
P 2-oxo-glutarate + NH_3 + $FADH_2$ <2-4> [2]
S D-histidine + H_2O + acceptor <6> (<6> low activity [8]) (Reversibility: ? <6> [8]) [8]
P 3-(1H-imidazol-4-yl)-2-oxopropanoate + NH_3 + reduced acceptor <6> [8]
S D-leucine + H_2O + acceptor <6> (<6> low activity [8]) (Reversibility: ? <6> [8]) [8]
P 4-methyl-2-oxopentanoate + NH_3 + reduced acceptor <6> [8]
S D-methionine + H_2O + acceptor <6> (Reversibility: ? <6> [8]) [8]
P 4-methylsulfanyl-2-oxobutanoate + NH_3 + reduced acceptor <6> [8]
S D-ornithine + H_2O + acceptor <6> (<6> low activity [8]) (Reversibility: ? <6> [8]) [8]
P ? + NH_3 + reduced acceptor <6> [8]
S D-phenylalanine + H_2O + acceptor <6> (Reversibility: ? <6> [8]) [8]
P phenylpyruvate + NH_3 + reduced acceptor <6> [8]
S D-proline + H_2O + FAD <2-4> (Reversibility: ? <2-4> [2]) [2]
P Δ^1-pyrroline-2-carboxylat + NH_3 + $FADH_2$ <2-4> [2]
S D-proline + H_2O + acceptor <1> (<1> mostly preferred electron acceptor: 2,6-dichlorophenolindophenol, lower activity with phenazine methosulfate [1]) (Reversibility: ? <1> [1]) [1]
P Δ^1-pyrroline-2-carboxylate + NH_3 + reduced acceptor <1> [1]
S D-serine + H_2O + FAD <2-4> (Reversibility: ? <2-4> [2]) [2]
P 3-hydroxy-2-oxopropanoate + NH_3 + $FADH_2$ <2-4> [2]
S D-serine + H_2O + acceptor <6> (<6> low activity [8]) (Reversibility: ? <6> [8]) [8]
P 3-hydroxy-2-oxopropanoate + NH_3 + reduced acceptor <6> [8]
S D-threonine + H_2O + acceptor <6> (<6> low activity [8]) (Reversibility: ? <6> [8]) [8]
P 3-hydroxy-2-oxobutanoate + NH_3 + reduced acceptor <6> [8]

S D-tyrosine + H_2O + acceptor <6> (<6> low activity [8]) (Reversibility: ? <6> [8]) [8]

P 3-(4-hydroxyphenyl)-2-oxopropanoate + NH_3 + reduced acceptor <6> [8]

Substrates and products

S (DL)-α-amino-n-butyrate + H_2O + 2,6-dichlorophenolindophenol <8> (Reversibility: ? <8> [11]) [11]

P 2-oxobutyrate + NH_3 + reduced 2,6-dichlorophenolindophenol <8> [11]

S (DL)-arginine + H_2O + 2,6-dichlorophenolindophenol <5> (<5> low activity [3]) (Reversibility: ? <5> [5]) [3]

P 5-guanidino-2-oxopentanoate + NH_3 + reduced 2,6-dichlorophenolindophenol <5> [3]

S D-alanine + H_2O + 2,6-dichlorophenolindophenol <1, 5, 7, 8> (<1> 26% of the activity compared to D-proline [1]; <5,8> best substrate tested [3,11]; <5> reduces a common proton translocating step in the membrane-bound respiratory chain [4]; <5> preferred substrate [6]; <7> low activity [9]; <5> coupled action of alanine racemase and D-alanine dehydrogenase to active transport of amino acids in membrane vesicles [12]) (Reversibility: ? <1, 5, 7, 8> [1, 3-6, 9, 11, 12]) [1, 3-6, 9, 11, 12]

P pyruvate + NH_3 + reduced 2,6-dichlorophenolindophenol <1, 5, 7, 8> [1, 3-6, 9, 11, 12]

S D-alanine + H_2O + FAD <2> (<2> no activity for Methanosarcina barkeri and Halobacterium salinarium enzyme [2]) (Reversibility: ? <2> [2]) [2]

P pyruvate + NH_3 + $FADH_2$ <2> [2]

S D-alanine + H_2O + acceptor <6> (<6> best substrate tested [8]) (Reversibility: ? <6> [8]) [8]

P pyruvate + NH_3 + reduced acceptor <6> [8]

S D-alanine + H_2O + coenzyme Q <5> (<5> physiological electron acceptor [6]) (Reversibility: ? <5> [6]) [6]

P pyruvate + NH_3 + reduced coenzyme Q <5> [6]

S D-alanine + H_2O + cytochrome c <5> (Reversibility: ? <5> [6]) [6]

P pyruvate + NH_3 + reduced cytochrome c <5> [6]

S D-alanine + H_2O + ferricyanide <5, 6> (<5> best electron acceptor tested [6]) (Reversibility: ? <5, 6> [6-8]) [6-8]

P pyruvate + NH_3 + ferrocyanide <5, 6> [6-8]

S D-alanine + H_2O + methylene blue <7> (<7> low activity [9]) (Reversibility: ? <7> [9]) [9]

P pyruvate + NH_3 + reduced methylene blue <7> [9]

S D-α-alanine methyl ester + H_2O + 2,6-dichlorophenolindophenol <5> (<5> 35% of the activity compared to D-alanine [3]) (Reversibility: ? <5> [3]) [3]

P 2-oxopropanoic acid methyl ester + NH_3 + reduced 2,6-dichlorophenolindophenol <5> [3]

S D-α-amino-n-butyrate + H_2O + 2,6-dichlorophenolindophenol <5> (<5> 38% of the activity compared to D-alanine [3]) (Reversibility: ? <5> [3, 6]) [3, 6]

P 2-oxobutyrate + NH_3 + reduced 2,6-dichlorophenolindophenol <5> [3, 6]

S   D-arginine + H_2O + 2,6-dichlorophenolindophenol <1> (<1> 10% of the
activity compared to D-proline [1]) (Reversibility: ? <1> [1]) [1]

P   2-oxo-5-guanidinopentanoate + NH_3 + reduced 2,6-dichlorophenolindo-
phenol <1> [1]

S   D-asparagine + H_2O + 2,6-dichlorophenolindophenol <5, 7, 8> (<5> 26%
of the activity compared to D-alanine [3]) (Reversibility: ? <5, 7, 8> [3, 9,
11]) [3, 9, 11]

P   2-oxosuccinamic acid + NH_3 + reduced 2,6-dichlorophenolindophenol
<5, 7, 8> [3, 9, 11]

S   D-asparagine + H_2O + acceptor <6> (Reversibility: ? <6> [8]) [8]

P   2-oxosuccinamic acid + NH_3 + reduced acceptor <6> [8]

S   D-asparagine + H_2O + ferricyanide <6> (Reversibility: ? <6> [7]) [7]

P   2-oxosuccinamic acid + NH_3 + ferrocyanide <6> [7]

S   D-asparagine + H_2O + methylene blue <7> (Reversibility: ? <7> [9]) [9]

P   2-oxosuccinamic acid + NH_3 + reduced methylene blue <7> [9]

S   D-aspartate + H_2O + 2,6-dichlorophenolindophenol <1> (<1> 21% of the
activity compared to D-proline [1]) (Reversibility: ? <1> [1]) [1]

P   oxaloacetate + NH_3 + reduced 2,6-dichlorophenolindophenol <1> [1]

S   D-aspartate + H_2O + FAD <2> (<2> no activity for Methanosarcina barkeri and Halobacterium salinarium enzyme [2]) (Reversibility: ? <2> [2])
[2]

P   oxaloacetate + NH_3 + $FADH_2$ <2> [2]

S   D-cysteine + H_2O + acceptor <6> (<6> low activity [8]) (Reversibility: ?
<6> [8]) [8]

P   3-mercapto-2-oxopropanoate + NH_3 + reduced acceptor <6> [8]

S   D-cysteine + H_2O + methylene blue <7> (Reversibility: ? <7> [9]) [9]

P   3-mercapto-2-oxopropanoate + NH_3 + reduced methylene blue <7> [9]

S   D-cystine + H_2O + methylene blue <7> (Reversibility: ? <7> [9]) [9]

P   ? + NH_3 + reduced methylene blue <7> [9]

S   D-glutamate + H_2O + 2,6-dichlorophenolindophenol <1> (<1> 23% of
the activity compared to D-proline [1]) (Reversibility: ? <1> [1]) [1]

P   2-oxo-glutarate + NH_3 + reduced 2,6-dichlorophenolindophenol <1> [1]

S   D-glutamate + H_2O + FAD <2-4> (Reversibility: ? <2-4> [2]) [2]

P   2-oxo-glutarate + NH_3 + $FADH_2$ <2-4> [2]

S   D-glutamine + H_2O + 2,6-dichlorophenolindophenol <5> (Reversibility: ?
<5> [6]) [6]

P   4-carbamoyl-2-oxobutanoate + NH_3 + reduced 2,6-dichlorophenolindo-
phenol <5> [6]

S   D-histidine + H_2O + 2,6-dichlorophenolindophenol <1, 5, 7, 8> (<1> 30%
of the activity compared to D-proline [1]; <5,8> low activity [3,11]) (Reversibility: ? <1, 5, 7, 8> [1, 3, 9, 11]) [1, 3, 9, 11]

P   3-(1H-imidazol-4-yl)-2-oxopropanoate + NH_3 + reduced 2,6-dichlorophe-
nolindophenol <1, 5, 7, 8> [1, 3, 9, 11]

S   D-histidine + H_2O + acceptor <6> (<6> low activity [8]) (Reversibility: ?
<6> [8]) [8]

P   3-(1H-imidazol-4-yl)-2-oxopropanoate + NH_3 + reduced acceptor <6> [8]

S   D-histidine + H_2O + ferricyanide <6> (Reversibility: ? <6> [7, 8]) [7, 8]

P 3-(1H-imidazol-4-yl)-2-oxopropanoate + NH_3 + ferrocyanide <6> [7, 8]

S D-histidine + H_2O + methylene blue <7> (Reversibility: ? <7> [9]) [9]

P 3-(1H-imidazol-4-yl)-2-oxopropanoate + NH_3 + reduced methylene blue <7> [9]

S D-isoleucine + H_2O + 2,6-dichlorophenolindophenol <1, 7> (<1> 49% of the activity compared to D-proline [1]) (Reversibility: ? <1, 7> [1, 9]) [1, 9]

P 3-methyl-2-oxopropanoate + NH_3 + reduced 2,6-dichlorophenolindophenol <1, 7> [1, 9]

S D-isoleucine + H_2O + methylene blue <7> (Reversibility: ? <7> [9]) [9]

P 3-methyl-2-oxopentanoate + NH_3 + reduced methylene blue <7> [9]

S D-kynurenine + H_2O + 2,6-dichlorophenolindophenol <7> (Reversibility: ? <7> [9]) [9]

P 4-(2-aminophenyl)-2,4-dioxobutanoate + NH_3 + reduced 2,6-dichlorophenolindophenol <7> [9]

S D-kynurenine + H_2O + methylene blue <7> (Reversibility: ? <7> [9]) [9]

P 4-(2-aminophenyl)-2,4-dioxobutanoate + NH_3 + reduced methylene blue <7> [9]

S D-leucine + H_2O + 2,6-dichlorophenolindophenol <1, 7> (<1> 39% of the activity compared to D-proline [1]) (Reversibility: ? <1, 7> [1, 9]) [1, 9]

P 4-methyl-2-oxopentanoate + NH_3 + reduced 2,6-dichlorophenolindophenol <1, 7> [1, 9]

S D-leucine + H_2O + acceptor <6> (<6> low activity [8]) (Reversibility: ? <6> [8]) [8]

P 4-methyl-2-oxopentanoate + NH_3 + reduced acceptor <6> [8]

S D-leucine + H_2O + ferricyanide <6> (Reversibility: ? <6> [7]) [7]

P 4-methyl-2-oxopentanoate + NH_3 + ferrocyanide <6> [7]

S D-leucine + H_2O + methylene blue <7> (Reversibility: ? <7> [9]) [9]

P 4-methyl-2-oxopentanoate + NH_3 + reduced methylene blue <7> [9]

S D-methionine + H_2O + 2,6-dichlorophenolindophenol <5, 7, 8> (<5> 35% of the activity compared to D-alanine [3]) (Reversibility: ? <5, 7, 8> [3, 6, 9, 11]) [3, 6, 9, 11]

P 4-methylsulfanyl-2-oxobutanoate + NH_3 + reduced 2,6-dichlorophenolindophenol <5, 7, 8> [3, 6, 9, 11]

S D-methionine + H_2O + acceptor <6> (Reversibility: ? <6> [8]) [8]

P 4-methylsulfanyl-2-oxobutanoate + NH_3 + reduced acceptor <6> [8]

S D-methionine + H_2O + ferricyanide <6> (Reversibility: ? <6> [7, 8]) [7, 8]

P 4-methylsulfanyl-2-oxobutanoate + NH_3 + ferrocyanide <6> [7, 8]

S D-norleucine + H_2O + 2,6-dichlorophenolindophenol <7> (Reversibility: ? <7> [9]) [9]

P 2-oxohexanoate + NH_3 + reduced 2,6-dichlorophenolindophenol <7> [9]

S D-norleucine + H_2O + methylene blue <7> (Reversibility: ? <7> [9]) [9]

P 2-oxohexanoate + NH_3 + reduced methylene blue <7> [9]

S D-norvaline + H_2O + methylene blue <7> (Reversibility: ? <7> [9]) [9]

P 2-oxopentanoate + NH_3 + reduced methylene blue <7> [9]

S D-ornithine + H_2O + acceptor <6> (<6> low activity [8]) (Reversibility: ? <6> [8]) [8]

P ? + NH_3 + reduced acceptor <6> [8]

S D-phenylalanine + H_2O + 2,6-dichlorophenolindophenol <1, 5, 7, 8> (<1> 28% of the activity compared to D-proline [1]; <5> preferred substrate [6]; <8> low activity [11]) (Reversibility: ? <1, 5, 7, 8> [1, 6, 9, 11]) [1, 6, 9, 11]

P phenylpyruvate + NH_3 + reduced 2,6-dichlorophenolindophenol <1, 5, 7, 8> [1, 6, 9, 11]

S D-phenylalanine + H_2O + acceptor <6> (Reversibility: ? <6> [8]) [8]

P phenylpyruvate + NH_3 + reduced acceptor <6> [8]

S D-phenylalanine + H_2O + ferricyanide <6> (Reversibility: ? <6> [7]) [7]

P phenylpyruvate + NH_3 + ferrocyanide <6> [7]

S D-phenylalanine + H_2O + methylene blue <7> (Reversibility: ? <7> [9]) [9]

P phenylpyruvate + NH_3 + reduced methylene blue <7> [9]

S D-proline + H_2O + 2,6-dichlorophenolindophenol <1, 5> (<1> preferred substrate [1]) (Reversibility: ? <1, 5> [1, 6]) [1, 6]

P Δ^1-pyrroline-2-carboxylate + NH_3 + reduced 2,6-dichlorophenolindophenol <1, 5> [1, 6]

S D-proline + H_2O + FAD <2-4> (Reversibility: ? <2-4> [2]) [2]

P Δ^1-pyrroline-2-carboxylat + NH_3 + $FADH_2$ <2-4> [2]

S D-proline + H_2O + acceptor <1> (<1> mostly preferred electron acceptor: 2,6-dichlorophenolindophenol, lower activity with phenazine methosulfate [1]) (Reversibility: ? <1> [1]) [1]

P Δ^1-pyrroline-2-carboxylate + NH_3 + reduced acceptor <1> [1]

S D-serine + H_2O + 2,6-dichlorophenolindophenol <1, 5, 7> (<1> 13% of the activity compared to D-proline [1]; <5> 20% of the activity compared to D-alanine [3]; <7> low activity [9]) (Reversibility: ? <1, 5, 7> [1, 3, 6, 9]) [1, 3, 6, 9]

P 3-hydroxy-2-oxopropanoate + NH_3 + reduced 2,6-dichlorophenolindophenol <1, 5, 7> [1, 3, 6, 9]

S D-serine + H_2O + FAD <2-4> (Reversibility: ? <2-4> [2]) [2]

P 3-hydroxy-2-oxopropanoate + NH_3 + $FADH_2$ <2-4> [2]

S D-serine + H_2O + acceptor <6> (<6> low activity [8]) (Reversibility: ? <6> [8]) [8]

P 3-hydroxy-2-oxopropanoate + NH_3 + reduced acceptor <6> [8]

S D-serine + H_2O + methylene blue <7> (<7> low activity [9]) (Reversibility: ? <7> [9]) [9]

P 3-hydroxy-2-oxopropanoate + NH_3 + reduced methylene blue <7> [9]

S D-threonine + H_2O + 2,6-dichlorophenolindophenol <1, 7> (<1> 15% of the activity compared to D-proline [1]; <7> low activity [9]) (Reversibility: ? <1, 7> [1, 9]) [1, 9]

P 3-hydroxy-2-oxobutanoate + NH_3 + reduced 2,6-dichlorophenolindophenol <1, 7> [1, 9]

S D-threonine + H_2O + acceptor <6> (<6> low activity [8]) (Reversibility: ? <6> [8]) [8]

P 3-hydroxy-2-oxobutanoate + NH_3 + reduced acceptor <6> [8]
S D-threonine + H_2O + methylene blue <7> (<7> low activity [9]) (Reversibility: ? <7> [9]) [9]
P 3-hydroxy-2-oxobutanoate + NH_3 + reduced methylene blue <7> [9]
S D-tryptophan + H_2O + 2,6-dichlorophenolindophenol <1, 7> (<1> 14% of the activity compared to D-proline [1]) (Reversibility: ? <1, 7> [1, 9]) [1, 9]
P 3-indole-2-oxopropanoate + NH_3 + reduced 2,6-dichlorophenolindophenol <1, 7> [1, 9]
S D-tryptophan + H_2O + ferricyanide <6> (Reversibility: ? <6> [7]) [7]
P 3-indole-2-oxopropanoate + NH_3 + ferrocyanide <6> [7]
S D-tryptophan + H_2O + methylene blue <7> (Reversibility: ? <7> [9]) [9]
P 3-indole-2-oxopropanoate + NH_3 + reduced methylene blue <7> [9]
S D-tyrosine + H_2O + 2,6-dichlorophenolindophenol <7> (Reversibility: ? <7> [9]) [9]
P 3-(4-hydroxyphenyl)-2-oxopropanoate + NH_3 + reduced 2,6-dichlorophenolindophenol <7> [9]
S D-tyrosine + H_2O + acceptor <6> (<6> low activity [8]) (Reversibility: ? <6> [8]) [8]
P 3-(4-hydroxyphenyl)-2-oxopropanoate + NH_3 + reduced acceptor <6> [8]
S D-tyrosine + H_2O + methylene blue <7> (Reversibility: ? <7> [9]) [9]
P 3-(4-hydroxyphenyl)-2-oxopropanoate + NH_3 + reduced methylene blue <7> [9]
S D-valine + H_2O + 2,6-dichlorophenolindophenol <1, 5, 7> (<1> 41% of the activity compared to D-proline [1]; <5> low activity [3]) (Reversibility: ? <1, 5, 7> [1, 3, 6, 9]) [1, 3, 6, 9]
P 2-oxoisopentanoate + NH_3 + reduced 2,6-dichlorophenolindophenol <1, 5, 7> [1, 3, 6, 9]
S D-valine + H_2O + methylene blue <7> (Reversibility: ? <7> [9]) [9]
P 2-oxoisopentanoate + NH_3 + reduced methylene blue <7> [9]
S allo-4-hydroxy-D-proline + H_2O + 2,6-dichlorophenolindophenol <1, 5, 8> (<1> 89% of the activity compared to D-proline [1]; <5,8> low activity [3,11]) (Reversibility: ? <1, 5, 8> [1, 3, 11]) [1, 3, 11]
P ? + NH_3 + reduced 2,6-dichlorophenolindophenol <1, 5, 8> [1, 3, 11]

Inhibitors

2,6-dichlorophenolindophenol <7> (<7> inhibition of the dichlorophenolindophenol-specific enzyme at high concentrations [9]) [9]
D-amino acids <7> (<7> various D-amino acids competitively inhibit D-kynurenine deamination [9]) [9]
D-tryptophan <6> (<6> 84% inhibition of D-methionine deamination at 20 mM [7]) [7]
NaCN <6> (<6> 34% inhibition of D-histidine deamination at 0.2 mM [8]) [8]
NaN_3 <6> (<6> 8% inhibition of D-histidine deamination [8]) [8]
nicotinic acid <6> (<6> 16% inhibition of D-histidine deamination at 12 mM [8]) [8]

p-aminobenzoic acid <6> (<6> 20% inhibition of D-histidine deamination at 12 mM [8]) [8]

p-chloromercuribenzoic acid <7> (<7> 90% inhibition of the methylene blue specific, 70% inhibition of the dichlorophenolindophenol specific enzyme at 0.5 mM [9]) [9]

Cofactors/prosthetic groups
FAD <1, 2, 5, 7> [1, 2, 6, 9]

Activating compounds
cardiolipin <5> (<5> 1.9fold activation of lipid-free enzyme preparation [3]) [3]

diphosphatidyl glycerol <5> (<5> 1.5fold activation of lipid-free enzyme preparation [3]) [3]

phosphatidyl choline <5> (<5> 1.8fold activation of lipid-free enzyme preparation [3]) [3]

phosphatidyl glycerol <5> (<5> 1.4fold activation of lipid-free enzyme preparation [3]) [3]

phosphoglyceric acid <5> (<5> 1.3fold activation of lipid-free enzyme preparation [3]) [3]

Metals, ions
Mg^{2+} <5> (<5> reduced phospholipid activation of lipid-free enzyme preparation if Mg^{2+} is omitted, but no increase of activity with Mg^{2+} alone [3]) [3]

iron <5> (<5> iron-sulfur protein [6]) [6]

Turnover number (min^{-1})
190 <5> (D-glutamine) [6]
320 <5> (D-serine) [6]
500 <5> (D-proline) [6]
880 <5> (D-phenylalanine) [6]
930 <5> (D-alanine) [6]
940 <5> (D-α-amino butyrate) [6]
960 <5> (D-methionine) [6]

Specific activity (U/mg)
0.113 <7> (<7> methylene blue specific enzyme [9]) [9]
0.213 <7> (<7> 2,6-dichlorophenolindophenol specific enzyme [9]) [9]
6.8 <5> [6]
11.2 <1> [1]
13.2 <1> (<1> recombinant enzyme [1]) [1]

K$_m$-Value (mM)
0.008 <7> (methylene blue, <7> specific acceptor, cosubstrate: D-kynurenine [9]) [9]

0.01 <5> (2,6-dichlorophenolindophenol, <5> coenzyme Q mediated [6]) [6]

0.013 <5> (ferricyanide, <5> with D-alanine [6]) [6]

0.07 <7> (2,6-dichlorophenolindophenol, <7> specific acceptor, cosubstrate: D-kynurenine [9]) [9]

0.14 <1> (2,6-dichlorophenolindophenol, <1> with D-proline [1]) [1]

0.15 <5> (cytochrome c, <5> with D-alanine [6]) [6]

0.29 <1> (2,6-dichlorophenolindophenol, <1> with allo-4-hydroxy-D-proline [1]) [1]

0.45 <7> (D-kynurenine, <7> both methylene blue and 2,6-dichlorophenolindophenol specific enzymes [9]) [9]

0.47 <7> (D-tryptophan, <7> both methylene blue and 2,6-dichlorophenolindophenol specific enzymes [9]) [9]

0.63 <5> (2,6-dichlorophenolindophenol, <5> with D-alanine [6]) [6]

0.86 <5> (D-α-amino-n-butyrate, <5> lipid-free enzyme preparation [3]) [3]

2 <5> (D-alanine, <5> with ferricyanide [6]) [6]

3 <5> (D-alanine, <5> with 2,6-dichlorophenolindophenol [6]) [6]

4.2 <1> (D-proline, <1> electron acceptor: 2,6-dichlorophenolindophenol [1]) [1]

5.3 <5> (D-alanine, <5> with coenzyme Q [6]) [6]

6 <5> (D-phenylalanine, <5> with 2,6-dichlorophenolindophenol [6]) [6]

6.25 <5> (D-alanine, <5> lipid-free enzyme preparation [3]) [3]

6.4 <5> (D-alanine, <5> with 2,6-dichlorophenolindophenol, coenzyme Q mediated [6]) [6]

8 <6> (D-alanine) [8]

9.5 <1> (allo-4-hydroxy-D-proline, <1> electron acceptor: 2,6-dichlorophenolindophenol [1]) [1]

11 <5> (D-methionine, <5> with 2,6-dichlorophenolindophenol [6]) [6]

12 <5> (D-α-amino butyrate, <5> with 2,6-dichlorophenolindophenol [6]) [6]

12.2 <5> (D-asparagine, <5> lipid-free enzyme preparation [3]) [3]

14 <5> (D-glutamine, <5> with 2,6-dichlorophenolindophenol [6]) [6]

19 <5> (D-serine, <5> with 2,6-dichlorophenolindophenol [6]) [6]

24 <5> (D-asparagine, <5> native, membrane-bound enzyme preparation [3]) [3]

25 <5> (D-α-amino-n-butyrate, <5> native, membrane-bound enzyme [3]) [3]

30 <5> (D-alanine, <5> native, membrane-bound enzyme [3]) [3]

38 <6> (D-phenylalanine) [8]

40 <6> (D-methionine) [8]

40 <5> (D-proline, <5> with 2,6-dichlorophenolindophenol [6]) [6]

48 <6> (D-histidine) [8]

50 <5> (D-methionine, <5> native, membrane-bound and lipid-free enzyme preparation [3]) [3]

pH-Optimum

7-8 <7> (<7> both methylene blue and 2,6-dichlorophenolindophenol specific enzymes [9]) [9]

7.5 <1> (<1> D-proline dehydrogenation [1]) [1]

7.9 <5> (<5> D-alanine dehydrogenation, lipid-free enzyme preparation [3]) [3]

8.9 <5> (<5> D-alanine dehydrogenation, native, membrane-bound enzyme [3]) [3]

9 <6> [8]

pH-Range

6.5 <1> (<1> half-maximal activity [1]) [1]

8.5 <1> (<1> half-maximal activity [1]) [1]

Temperature optimum (°C)

70 <1> (<1> maximal activity at and above [1]) [1]

4 Enzyme Structure

Molecular weight

117000 <5> (<5> gel filtration [6]) [6]

145000 <1> (<1> gel filtration [1]) [1]

Subunits

dimer <5> (<5> α_1, 45000, β_1, 55000, SDS-PAGE [6]) [6]

tetramer <1> (<1> 4 * 42000, SDS-PAGE [1]) [1]

5 Isolation/Preparation/Mutation/Application

Source/tissue

cell culture <1-8> (<6> inducible by D- or L-alanine [8]; <7> one methylene blue specific, one 2,6-dichlorophenolindophenol specific enzyme, the methylene blue specific enzyme is only detectable in D-tryptophan grown cells [9,10]) [1-12]

Localization

cytoplasmic membrane <1, 5> (<5> inner surface [5]) [1, 3-6, 12]

Purification

<1> (to homogeneity, chromatography steps [1]; to homogeneity, recombinant enzyme [1]) [1]

<5> (partial [6]) [6]

<7> (near homogeneity, two enzymes with different acceptor specificities, fractionation and chromatography techniques [9]) [9]

Cloning

<1> (expression in Escherichia coli [1]) [1]

6 Stability

pH-Stability

4 <1> (<1> completely stable for 10 min [1]) [1]

10 <1> (<1> completely stable for 10 min [1]) [1]

Temperature stability

30-80 <1> (<1> full activity within 10 min [1]) [1]

37 <5> (<5> 20% loss of activity in 10 min of native enzyme, 90% loss of activity in 10 min of lipid-free enzyme preparation [3]) [3]
100 <1> (<1> loss of activity in 10 min [1]) [1]

General stability information

<7>, very unstable [9]

Storage stability

<5>, -20°C, 0.1 M potassium phosphate, pH 7.5, 0.02% Triton X-100, more than 1 mg protein/ml, 3 months [6]

References

[1] Satomura, T.; Kawakami, R.; Sakuraba, H.; Ohshima, T.: Dye-linked D-proline dehydrogenase from hyperthermophilic archaeon Pyrobaculum islandicum is a novel FAD-dependent amino acid dehydrogenase. J. Biol. Chem., **277**, 12861-12867 (2002)

[2] Nagata, Y.; Tanaka, K.; Iida, T.; Kera, Y.; Yamada, R.; Nakajima, Y.; Fujiwara, T.; Fukumori, Y.; Yamanaka, T.; Koga, Y.; Tsuji, S.; Kawaguchi-Nagata, K.: Occurrence of D-amino acids in a few archaea and dehydrogenase activities in hyperthermophile Pyrobaculum islandicum. Biochim. Biophys. Acta, **1435**, 160-166 (1999)

[3] Jones, H.; Venables, W.A.: Effects of solubilization on some properties of the membrane-bound respiratory enzyme D-amino acid dehydrogenase of Escherichia coli. FEBS Lett., **151**, 189-192 (1983)

[4] Haldar, K.; Olsiewski, P.J.; Walsh, C.; Kaczorowski, G.J.; Bhaduri, A.; Kaback, H.R.: Simultaneous reconstitution of Escherichia coli membrane vesicles with D-lactate and D-amino acid dehydrogenases. Biochemistry, **21**, 4590-4596 (1982)

[5] Olsiewski, P.J.; Kaczorowski, G.J.; Walsh, C.T.; Kaback, H.R.: Reconstitution of Escherichia coli membrane vesicles with D-amino acid dehydrogenase. Biochemistry, **20**, 6272-6279 (1981)

[6] Olsiewski, P.J.; Kaczorowski, G.J.; Walsh, C.: Purification and properties of D-amino acid dehydrogenase, an inducible membrane-bound iron-sulfur flavoenzyme from Escherichia coli B. J. Biol. Chem., **255**, 4487-4494 (1980)

[7] Wild, J.; Filutowicz, M.; Klopotowski, T.: Utilization of D-amino acids by dadR mutants of Salmonella typhimurium. Arch. Microbiol., **118**, 71-77 (1978)

[8] Wild, J.; Walczak, W.; Krajewska-Grynkiewicz, K.; Klopotowski, T.: D-Amino acid dehydrogenase: The enzyme of the first step of D-histidine and D-methionine racemization in Salmonella typhimurium. Mol. Gen. Genet., **128**, 131-146 (1974)

[9] Tsukada, K.: D-Amino acid dehydrogenase (Pseudomonas fluorescens). Methods Enzymol., **17B**, 623-624 (1971)

[10] Tsukada, K.: D-Amino acid dehydrogenases of Pseudomonas fluorescens. J. Biol. Chem., **241**, 4522-4528 (1966)

[11] Pioli, D.; Venables, W.A.; Franklin, F.C.H.: D-Alanine dehydrogenase. Its role on the utilization of alanine isomers as growth substrates by Pseudomonas aeruginosa PA01. Arch. Microbiol., **110**, 287-293 (1976)
[12] Kaczorowski, G.; Shaw, L.; Fuentes, M.; Walsh, C.: Coupling of alanine racemase and D-alanine dehydrogenase to active transport of amino acids in Escherichia coli B membrane vesicles. J. Biol. Chem., **250**, 2855-2865 (1975)

Taurine dehydrogenase

1.4.99.2

1 Nomenclature

EC number
1.4.99.2

Systematic name
taurine:(acceptor) oxidoreductase (deaminating)

Recommended name
taurine dehydrogenase

CAS registry number
50812-14-1

2 Source Organism

<1> *bacterium* (gram-negative) [1, 2]

3 Reaction and Specificity

Catalyzed reaction
taurine + H_2O + acceptor = sulfoacetaldehyde + NH_3 + reduced acceptor

Reaction type
oxidation
oxidative deamination
redox reaction
reduction

Natural substrates and products
S taurine + H_2O + acceptor <1> (<1> acts only with phenazine methosulfate as electron acceptor [1,2]) (Reversibility: ? <1> [1, 2]) [1, 2]
P sulfoacetaldehyde + NH_3 + reduced acceptor <1> [1, 2]

Substrates and products
S hypotaurine + H_2O + phenazine methosulfate <1> (<1> poor substrate [1,2]) (Reversibility: ? <1> [1, 2]) [1, 2]
P ? + NH_3 + reduced phenazine methosulfate <1> [1, 2]
S taurine + H_2O + acceptor <1> (<1> acts only with phenazine methosulfate as electron acceptor [1,2]) (Reversibility: ? <1> [1, 2]) [1, 2]
P sulfoacetaldehyde + NH_3 + reduced acceptor <1> [1, 2]

S   taurine + H_2O + phenazine methosulfate <1> (<1> only acts with phenazine methosulfate as electron acceptor [1,2]) (Reversibility: ? <1> [1, 2]) [1, 2]

P   sulfoacetaldehyde + NH_3 + reduced phenazine methosulfate <1> [1, 2]

Inhibitors

2-heptyl-4-hydroxyquinoline N-oxide <1> (<1> some inhibition [1,2]) [1, 2]
atebrin <1> (<1> some inhibition [1,2]) [1, 2]
isonicotinic acid 2-isopropylhydrazine <1> (<1> strong inhibition [1,2]) [1, 2]
methylene blue <1> (<1> some inhibition [1,2]) [1, 2]
neocuproine <1> (<1> some inhibition [1,2]) [1, 2]
p-chloromercuribenzoate <1> (<1> strong inhibition [1,2]) [1, 2]
phenelzine <1> (<1> strong inhibition [1,2]) [1, 2]

Specific activity (U/mg)

0.48 <1> [1, 2]

K_m-Value (mM)

0.056 <1> (phenazine methosulfate) [1, 2]
20 <1> (taurine) [1, 2]

pH-Optimum

8.4 <1> [1, 2]

5 Isolation/Preparation/Mutation/Application

Source/tissue

cell culture <1> [1, 2]

Localization

membrane <1> [1, 2]

Purification

<1> (homogenous, precipitation, chromatography steps [1,2]) [1, 2]

6 Stability

Temperature stability

40 <1> (<1> 40% loss of activity in 20 min [1,2]) [1, 2]

General stability information

<1>, stable to freeze-thawing [1, 2]

Storage stability

<1>, -20°C, 1.5 mg protein per ml, 1 month [1, 2]

References

[1] Kondo, H.; Ishimoto, M.: Taurine dehydrogenase. Methods Enzymol., **143**, 496-499 (1987)
[2] Kondo, H.; Kagotani, K.; Oshima, M.; Ishimoto, M.: Purification and some properties of taurine dehydrogenase from a bacterium. J. Biochem., **73**, 1269-1278 (1973)

Amine dehydrogenase 1.4.99.3

1 Nomenclature

EC number
1.4.99.3

Systematic name
primary-amine:(acceptor) oxidoreductase (deaminating)

Recommended name
amine dehydrogenase

Synonyms
MADH
QH-AmDH <4> (<4> different from MADH [15]) [15]
amine: oxidoreductase (acceptor deaminating)
dehydrogenase, amine
methylamine dehydrogenase
primary-amine dehydrogenase
quinohemoprotein amine dehydrogenase <4> (<4> different from MADH
[15]) [15]

CAS registry number
55476-92-1
60496-14-2

2 Source Organism

<1> *Pseudomonas putida* (inducible enzyme [7]) [5-7, 18]
<2> *Pseudomonas sp.* (K95 [2]; AM1 (cells grown in methylamine, inducible
enzyme [4]) [3, 4, 9, 10]) [1-4, 9, 10]
<3> *Mycobacterium convolutum* (inducible enzyme [8]) [8]
<4> *Paracoccus denitrificans* (strain IFO 12442 [15]) [11, 15, 16, 17]
<5> *Thiobacillus versutus* [12, 13]
<6> *Methylobacterium extorquens* (AM1 [14]) [14]

3 Reaction and Specificity

Catalyzed reaction
$RCH_2NH_2 + H_2O$ + acceptor = $RCHO + NH_3$ + reduced acceptor

Reaction type

oxidation
oxidative deamination
redox reaction
reduction

Natural substrates and products

S methylamine + amicyanin + H_2O <5> (<5> amicyanin is the in vivo electron acceptor [12]) (Reversibility: ? <5> [12]) [12]
P formaldehyde + reduced amicyanin + NH_3

Substrates and products

S 1,12-diaminododecane + acceptor + H_2O <2> (Reversibility: ? <2> [2]) [2]
P ? + NH_3 + reduced acceptor
S 1,8-diaminooctane + H_2O + acceptor <2> (Reversibility: ? <2> [2]) [2]
P ?
S 12-aminododecane + acceptor + H_2O <2> (Reversibility: ? <2> [2]) [2]
P dodecanal + NH_3 + reduced acceptor
S RCH_2NH_2 + acceptor + H_2O <1, 2, 4, 5> (<2,4> acceptor 2,6-dichlorophenolindophenol [1,11]; <1,2> acceptor phenazine methosulfate [1,2,4,7]; <4> acceptor phenazine ethosulfate or amicyanin [11]; <2> acceptor 3-(4,5-dimethyl-2-thiazolyl)-2,5-diphenyl-2H-tetrazolium bromide and potassium ferricyanide [1]; <1> cytochrome c or artificial electron acceptor [6]; <1> acceptor 2,6-dichlorophenol-indophenol, ferricyanide or cytochrome c [7]; <5> acceptor cytochrome c-550 [13]; <4> acceptor potassium ferricyanide, phenazine ethosulfate, 2,6-dichloroindophenol [15]) (Reversibility: ? <1, 2, 4, 5> [1, 2, 4, 6, 7, 11, 12, 13, 15]) [1, 2, 4, 6, 7, 11, 12, 13, 15]
P RCHO + NH_3 + reduced acceptor
S benzylamine + acceptor + H_2O < <2, 4> (Reversibility: ? >2, 4# [1, 15]) [1, 15]
P benzaldehyde + NH_3 + reduced acceptor
S butylamine + acceptor + H_2O <2, 4> (Reversibility: ? <2, 4> [1, 15]) [1, 15]
P butanal + NH_3 + reduced acceptor
S histamine + acceptor + H_2O <4> (Reversibility: ? <4> [17]) [17]
P ? + NH_3 + reduced acceptor
S methylamine + acceptor + H_2O <3, 4> (Reversibility: ? <3, 4> [8, 11-13, 15]) [8, 11-13, 15]
P methanal + NH_3 + reduced acceptor
S n-hexylamine + acceptor + H_2O <2> (Reversibility: ? <2> [2]) [2]
P hexanal + NH_3 + reduced acceptor
S n-nonylamine + acceptor + H_2O <2> (Reversibility: ? <2> [2]) [2]
P nonanal + NH_3 + reduced acceptor
S n-pentylamine + acceptor + H_2O <2> (Reversibility: ? <2> [1]) [1]
P pentanal + NH_3 + reduced acceptor

S phenylethylamine + acceptor + H_2O <2, 4> (Reversibility: ? <2, 4> [1, 15]) [1, 15]

P phenylacetaldehyde + NH_3 + reduced acceptor

S propylamine + acceptor + H_2O <2, 4> (Reversibility: ? <2, 4> [1, 15]) [1, 15]

P propionaldehyde + NH_3 + reduced acceptor

S Additional information <1-3> (<1> not: FMN, NAD^+, $NADP^+$ [7]; <1-3> broad specificity [2, 4-8]; <2> nonspecific oxidizing both short and long primary monoamines and diamines, polyamines, L-noradrenaline, histamine, benzylamine and di-n-hexylamine [2]; <1> little or no activity with isoamines, L-ornithine, L-lysine and certain diamines or polyamines [6]) [1, 2, 4-8]

P ?

Inhibitors

8-hydroxyquinoline <1> (<1> slight [6]) [6]

KCN <1> [5, 6]

N-ethylmaleimide <2> [2]

borohydride <2> [10]

cuprizone <1, 2> [4-6]

hydrazine <4> [15]

hydroxylamine <1, 2, 4> [4, 6, 15]

iodoacetate <1> [6]

isoniazid <1, 2> [2, 4, 5, 6]

n-butyraldehyde <2> [1]

neocuproine <2> (<2> slight [2]) [2]

p-chloromercuribenzoate <1, 2> [4, 6]

phenylhydrazine <4> [15]

quinacrine <1, 2> [4, 6]

quinine <2> (<2> slight [2]) [2]

semicarbazide <1, 2, 4> [2, 4, 5, 6, 15]

Additional information <2> (<2> immobilized enzyme: little change in sensitivity to inhibition [3]) [3]

Cofactors/prosthetic groups

cysteine tryptophylquinone <1> (<1> in γ-subunit [18]) [18]

heme c <1, 4> (<1> hemoprotein which contains 2.01 mol per mol of enzyme [5,6]; <1> heme present only in the heavier of 2 subunits [5]; <1> contains a di-heme cytochrome c in the α-subunit [18]; <4> contains 1 mol heme c per mol enzyme in the α-subunit [15]) [5, 6, 15]

pyrroloquinoline quinone <2> (<2> covalently bound [1]) [1]

quinoid cofactor <4> (<4> α-subunit contains unknown quinoid cofactor [15]) [15]

quinone <2> (<2> contains a quinone similar to but not identical with the prosthetic group of EC 1.1.99.8 [9]) [9]

tryptophan tryptophylquinone <6> [14]

tryptophan tryptophylquinone <4> (<4> contains a tryptophan tryptophylquinone prosthetic group [11]) [11]

Metals, ions

Ca^{2+} <1, 2> (<2> slight [4]; <1> no effect [6]) [4, 5]
Co^{2+} <1, 2> (<1,2> slight [4,6]) [4, 6]
Mn^{2+} <1, 2> (<2> slight [4]) [4, 6]
Na^+ <1, 2> (<2> slight [4]; <1> no effect [6]) [4, 6]
Zn^{2+} <1> [6]
Additional information <4, 5> (<5> cations inhibit reduction of oxidized MADH at pH 8, monovalent cations increase oxidation rate, divalent cations decrease oxidation rate [12]; <4> monovalent cations affect spectral properties and rate of gated electron transfer from reduced enzyme to electron acceptor [16]) [12, 16]

Turnover number (min^{-1})

90 <4> (2,6-dichloroindophenol) [15]
138 <5> (cytochrome c-550) [13]
426 <4> (phenazine ethosulfate) [15]
780 <4> (ethylamine) [15]
816 <4> (methylamine, <4> with phenylazine as electron acceptor [11]) [11]
840 <4> (methylamine) [15]
900 <4> ($K_3Fe(CN)_6$) [15]
1320 <4> (benzylamine) [15]
1320 <4> (butylamine) [15]
1560 <4> (phenethylamine) [15]
1620 <4> (propylamine) [15]
1800 <5> (methylamine, <5> with amicyanin as electron acceptor at pH 7 [12]) [12]
2910 <4> (methylamine, <4> with amicyanin as electron acceptor [11]) [11]
3300 <5> (methylamine, <5> with amicyanin as electron acceptor at pH 10 [12]) [12]

Specific activity (U/mg)

3.5 <2> [3, 4]
3.86 <2> [10]
3.9 <2> [2]
4.67 <1> [6]
13.5 <2> [1]

K$_m$-Value (mM)

0.003 <2> (1,12-diaminododecane) [2]
0.0059 <4> (methylamine, <4> with amicyanin as electron acceptor [11]) [11]
0.0064 <4> (methylamine, <4> with phenylazine as electron acceptor [11]) [11]
0.014 <4> (phenazine ethosulfate) [15]
0.033 <2> (12-aminododecanoic acid) [2]
0.042 <5> (cytochrome c-550) [13]
0.1 <4> (ethylamine) [15]
0.12 <4> (2,6-dichloroindophenol) [15]

0.78 <4> ($K_3Fe(CN)_6$) [15]

1.3 <4> (methylamine) [15]

Additional information <4> (<4> less than 0.001 for propylamine, butylamine, benzylamine, phenethylamine [15]; α F55A mutated enzyme has 400fold lower K_m for histamine than native enzyme in solution [16]) [15, 16]

pH-Optimum

7 <2> (<2> 1,12-diaminododecane [2]) [2]

7.4 <2> (<2> methylamine [3]) [3]

7.5 <2, 5> (<2> n-butylamine [1]; <5> methylamine [12]) [1, 12]

8 <2> (<2> 12-aminododecanoic acid [2]) [2]

9 <2> (<2> putrescine [1]) [1]

pH-Range

Additional information <2> [4]

Temperature optimum (°C)

40 <2> (<2> 1,12-diaminododecane [2]) [2]

4 Enzyme Structure

Molecular weight

13960 <6> (<6> L subunit, MALDI [14]) [14]

13970 <6> (<6> L subunit, calculated from sequence [14]) [14]

38500 <3> (<3> gel filtration [8]) [8]

41160 <6> (<6> H subunit, calculated from sequence [14]) [14]

41170 <6> (<6> H subunit, MALDI [14]) [14]

56000 <2> (<2> gel filtration [2]) [2]

95300 <1> (<1> sedimentation equilibrium [6]) [6]

100000 <2, 4> (<2,4> gel filtration [1,15]) [1, 15]

112000 <1> (<1> gel filtration [6]) [6]

133000 <2> (<2> ultracentrifugation [10]) [10]

Subunits

dimer <1, 2, 4> (<2> 1 * 60000 + 1 * 39000, SDS-PAGE [1]; <1> 1 * 58000 + 1 * 42000, SDS-PAGE [6]; <4> α,β 1 * 59500 + 1 * 36500, SDS-PAGE [15]) [1, 6, 15]

monomer <2> (<2> 1 * 47000, SDS-PAGE [2]) [2]

tetramer <6> (<6> 2 * 14000 + 2 * 41000, X-ray crystallography [14]) [14]

trimer <1> (<1> α,β,γ, X-ray crystallography [18]) [18]

5 Isolation/Preparation/Mutation/Application

Localization

periplasm <4> [15]

Purification

    <1> [6]

    <2> [1, 2, 3]

    <4> [15]

Crystallization

    <2> [1]

    <6> (<6> one week in 19% PEG4000, 70 mM acetate buffer, pH 4.1 [14]) [14]

Cloning

    <4> (<4> heterologous expression of mutated α-subunit in Rhodobacter sphaeroides [16]) [16]

    <5> (<5> heterologous expression of cytochrome c-550 in Escherichia coli [13]) [13]

Engineering

    F55A <4> (<4> mutation of the α-subunit [16,17]) [16, 17]

    F55E <4> (<4> mutation of the α-subunit [16]) [16]

    K14E <5> (<5> mutation of cytochrome c-550 [13]) [13]

    K14Q <5> (<5> mutation of cytochrome c-550 [13]) [13]

Application

    Additional information <4> (<4> immobilized enzyme used as histamine biosensor [16]) [16]

6 Stability

pH-Stability

    2.6-10.6 <2> (<2> 75 min [4]) [4]

    6-11 <2> (<2> 2 weeks, room temperature [1]) [1]

Temperature stability

    50 <2> (<2> 10 min [1]) [1]

    65 <1> (<1> 3 min, 97% loss of activity [7]) [7]

    80 <2> (<2> 15 min, stable, inactivated above [4]) [4]

    84 <2> (<2> 30 min, 40% loss of activity (immobilized enzyme), about 85% loss of activity (soluble enzyme) [3]) [3]

General stability information

    <2>, dialysis against diethyldithiocarbamate, stable [10]

    <2>, freezing and thawing, stable [4]

    <2>, immobilized enzyme [3]

    <2>, no difference in temperature-stability and pH-stability at 1 mg protein/ml and 0.0005 mg/ml [1]

Storage stability

    <2>, -15°C, 1 mg/ml protein concentration [4]

References

[1] Shinagawa, E.; Matsushita, K.; Nakashima, K.; Adachi, O.; Ameyama, M.: Crystallization and properties of amine dehydrogenase from Pseudomonas sp.. Agric. Biol. Chem., **52**, 2255-2263 (1988)

[2] Niimura, Y.; Omori, T.; Minoda, Y.: Purification and properties of an amine dehydrogenase from Pseudomonas K95 grown on 1,12-diaminododecane (DAD). Agric. Biol. Chem., **50**, 1445-1451 (1986)

[3] Boulton, C.A.; Large, P.J.: Properties of Pseudomonas AM1 primary-amine dehydrogenase immobilized on agarose. Biochim. Biophys. Acta, **570**, 22-30 (1979)

[4] Eady, R.R.; Large, P.J.: Purification and properties of an amine dehydrogenase from Pseudomonas AM1 and its role in growth on methylamine. Biochem. J., **106**, 245-255 (1968)

[5] Durham, D.R.; Perry, J.J.: Amine dehydrogenase of Pseudomonas putida: properties of the heme-prosthetic group. J. Bacteriol., **135**, 981-986 (1978)

[6] Durham, D.R.; Perry, J.J.: Purification and characterization of a heme-containing amine dehydrogenase from Pseudomonas putida. J. Bacteriol., **134**, 837-843 (1978)

[7] Durham, D.R.; Perry, J.J.: The inducible amine dehydrogenase in Pseudomonas putida NP and its role in the metabolism of benzylamine. J. Gen. Microbiol., **105**, 39-44 (1978)

[8] Cerniglia, C.E.; Perry, J.J.: Metabolism of n-propylamine, isopropylamine, and 1,3-propane diamine by Mycobacterium convolutum. J. Bacteriol., **124**, 285-289 (1975)

[9] De Beer, R.; Duine, J.A.; Frank, J.; Large, P.J.: The prosthetic group of methylamine dehydrogenase from Pseudomonas AM1: evidence for a quinone structure. Biochim. Biophys. Acta, **622**, 370-374 (1980)

[10] Eady, R.R.; Large, P.J.: Microbial oxidation of amines. Spectral and kinetic properties of the primary amine dehydrogenase of Pseudomonas AM1. Biochem. J., **123**, 757-771 (1971)

[11] Brooks, H.B.; Jones, L.H.; Davidson, V.L.: Deuterium kinetic isotope effect and stopped-flow kinetic studies of the quinoprotein methylamine dehydrogenase. Biochemistry, **32**, 2725-2729 (1993)

[12] Gorren, A.C.F.; Duine, J.A.: The effects of pH and cations on the spectral and kinetic properties of methylamine dehydrogenase from Thiobacillus versutus. Biochemistry, **33**, 12202-12209 (1994)

[13] Ubbink, M.; Hunt, N.I.; Hill, H.A.O.; Canters, G.W.: Kinetics of the reduction of wild-type and mutant cytochrome c-550 by methylamine dehydrogenase and amicyanin from Thiobacillus versutus. Eur. J. Biochem., **222**, 561-571 (1994)

[14] Labesse, G.; Ferrari, D.; Chen, Z.W.; Rossi, G.L.; Kuusk, V.; McIntire, W.S.; Mathews, F.S.: Crystallographic and spectroscopic studies of native, aminoquinol, and monovalent cation-bound forms of methylamine dehydrogenase from Methylobacterium extorquens AM1. J. Biol. Chem., **273**, 25703-25712 (1998)

[15] Takagi, K.; Torimura, M.; Kawaguchi, K.; Kano, K.; Ikeda, T.: Biochemical and electrochemical characterization of quinohemoprotein amine dehydrogenase from Paracoccus denitrificans. Biochemistry, **38**, 6935-6942 (1999)

[16] Sun, D.; Davidson, V.L.: Re-engineering monovalent cation binding sites of methylamine dehydrogenase: effects on spectral properties and gated electron transfer. Biochemistry, **40**, 12285-12291 (2001)

[17] Bao, L.; Sun, D.; Tachikawa, H.; Davidson, V.L.: Improved sensitivity of a histamine sensor using an engineered methylamine dehydrogenase. Anal. Chem., **74**, 1144-1148 (2002)

[18] Satoh, A.; Kim, J.K.; Miyahara, I.; Devreese, B.; Vandenberghe, I.; Hacisalihoglu, A.; Okajima, T.; Kuroda, S.i.; Adachi, O.; Duine, J.A.; Van Beeumen, J.; Tanizawa, K.; Hirotsu, K.: Crystal structure of quinohemoprotein amine dehydrogenase from Pseudomonas putida: identification of a novel quinone cofactor encaged by multiple thioether cross-bridges. J. Biol. Chem., **277**, 2830-2834 (2002)

Aralkylamine dehydrogenase   1.4.99.4

1 Nomenclature

EC number
1.4.99.4

Systematic name
aralkylamine:(acceptor) oxidoreductase (deaminating)

Recommended name
aralkylamine dehydrogenase

Synonyms
AADH
AauA <4> (<4> small subunit polypeptide [8]) [8]
AauB <4> (<4> large subunit polypeptide [8]) [8]
aromatic amine dehydrogenase
dehydrogenase, arylamine
tyramine dehydrogenase

CAS registry number
85030-73-5

2 Source Organism

<1> *Alcaligenes faecalis* (strain IFO 14479) [1, 4-7]
<2> *Pseudomonas aeruginosa* (PAO 1, inducible enzyme) [2]
<3> *Pseudomonas sp.* [3]
<4> *Alcaligenes faecalis* [8]
<5> *Alcaligenes xylosoxydans* [8]

3 Reaction and Specificity

Catalyzed reaction
$RCH_2NH_2 + H_2O$ + acceptor = $RCHO + NH_3$ + reduced acceptor

Reaction type
oxidation
oxidative deamination <3> [3]
redox reaction
reduction

Natural substrates and products

S aromatic amines + acceptor + H_2O <1, 2> (<2> catabolism [2]; <1> physiological acceptor: azurin, not amicyanin [5,7]) (Reversibility: ? <1, 2> [2, 5, 7]) [2, 5, 7]

P ? + reduced acceptor + NH_3

Substrates and products

S β-phenylethylamine + acceptor + H_2O <4> (Reversibility: ? <4> [8]) [8]

P ? + NH_3 + reduced acceptor

S dopamine + acceptor + H_2O <1, 3> (Reversibility: ? <1, 3> [1, 3, 6]) [1, 3, 6]

P 4-(2-oxoethyl)-1,2-benzenediol + NH_3 + reduced acceptor

S n-hexylamine + acceptor + H_2O <3> (Reversibility: ? <3> [3]) [3]

P hexanal + NH_3 + reduced acceptor

S n-octylamine + acceptor + H_2O <3> (Reversibility: ? <3> [3]) [3]

P octanal + NH_3 + reduced acceptor

S serotonin + acceptor + H_2O <1, 3> (Reversibility: ? <1, 3> [1, 3]) [1, 3]

P 3-(2-oxoethyl)-1H-indol-5-ol + NH_3 + reduced acceptor

S tryptamine + acceptor + H_2O <1, 3> (Reversibility: ? <1, 3> [1, 3]) [1, 3]

P 1H-indole-3-acetaldehyde + NH_3 + reduced acceptor

S tyramine + acceptor + H_2O <1-3> (<1> electron acceptor: phenazine methosulfate or phenazine ethosulfate [4]; <1> electron acceptor: azurin [5,7]) (Reversibility: ? <1-3> [1-5, 7]) [1-5, 7]

P 4-hydroxyphenylacetaldehyde + NH_3 + reduced acceptor

S Additional information <1, 3> (<1-3> slow reaction with aromatic amines [1-3]; <3> long-chain aliphatic amines, phenazine methosulfate is the only effective electron acceptor [3]; <3> not: methylamine, ethylamine, polyamines [3]) [1, 3]

P ?

Inhibitors

aminoguanidine <1> [4]

β-phenylethylamine <3> (<3> substrate inhibition [3]) [3]

hydrazine <1> [4]

hydroxylamine <1, 3> [1, 3, 4]

iproniazid <1> [4]

isoniazid <1> [4]

phenylhydrazine <1, 3> [1, 3, 4]

semicarbazide <1, 3> [1, 3, 4]

tryptamine <3> (<3> substrate inhibition [3]) [3]

tyramine <1, 3> (<1,3> substrate inhibition [3,4]) [3, 4]

Cofactors/prosthetic groups

tryptophan tryptophylquinone <1> [4, 5, 7]

Additional information <1> (<1> probably chromophore with quinone structure [1]) [1]

Metals, ions

copper <3> (<3> metal content: 0.1 mol per mol of enzyme [3]) [3]
iron <3> (<3> metal content: 0.6 mol per mol of enzyme [3]) [3]

Turnover number (min^{-1})

546 <1> (dopamine) [6]
1626 <1> (dopamine) [6]
3840 <1> (azurin, <1> in 0.01 M potassium phosphate buffer [5]) [5]
5460 <1> (azurin, <1> in 0.25 M potassium phosphate buffer [5]) [5]

Specific activity (U/mg)

0.39 <1> (<1> cell extract from cells grown in β-phenylethylamine [4]) [4]
9.11 <1, 3> (<1,3> after purification [1,3]) [1, 3]
14.3 <1> (<1> after purification [4]) [4]
87 <5> [8]
95 <4> [8]

K$_m$-Value (mM)

0.0012 <3> (tyramine) [3]
0.0045 <3> (β-phenylethylamine) [3]
0.0054 <1> (tyramine) [4]
0.0092 <3> (tryptamine) [3]
0.044 <1> (azurin, <1> in 0.25 M potassium phosphate buffer [5]) [5]
0.11 <3> (phenazine methosulfate) [3]
0.13 <1> (dopamine) [6]
0.137 <1> (azurin, <1> in 0.01 M potassium phosphate buffer [5]) [5]
0.25 <1> (phenazine methosulfate) [4]
0.292 <1> ((1,1-2H2)dopamine) [6]
0.45 <1> (phenazine ethosulfate) [4]

K$_i$-Value (mM)

0.0084 <1> (isoniazid) [4]
0.186 <1> (iproniazid) [4]
1.08 <1> (tyramine, <1> substrate inhibition [4]) [4]

pH-Optimum

7.5 <3> (<3> tyramine, β-phenylethylamine [3]) [3]
8 <3> (<3> serotonin, tryptamine [3]) [3]

pH-Range

5.5-9.5 <3> (<3> 5.5: about 25% of activity maximum, 9.5: about 10% of activity maximum [3]) [3]

Temperature optimum (°C)

25 <1> (<1> assay at [1]) [1]

4 Enzyme Structure

Molecular weight
103000 <1, 3> (<1,3> sedimentation equilibrium [1,3]) [1, 3]
108900 <1> (<1> sedimentation equilibrium [4]) [4]
114000 <1> (<1> estimated from subunit molecular weights [4]) [4]

Subunits
tetramer <1, 3, 4> (<1> $\alpha_2\beta_2$, 2 * 46000 + 2 * 8000, high speed gel filtration
in 6 M guanidine hydrochloride [1]; <3> $\alpha_2\beta_2$, 2 * 46000 + 2 * 8000, SDS-
PAGE [3]; <1> $\alpha_2\beta_2$, 2 * 39000 + 2 * 18000, SDS-PAGE [4]; <4> $\alpha_2\beta_2$, 2 *
38222 + 2 * 19652, calculated from DNA sequence [8]) [1, 3, 4, 8]

5 Isolation/Preparation/Mutation/Application

Localization
membrane <2> [2]
periplasm <4> [8]

Purification
<1> [1, 4]
<3> [3]

Crystallization
<1, 3> [1, 3]

Cloning
<2> (<2> cloning of portions of PAO 1 chromosome with genes coding for
enzymes of catabolic pathway of aromatic biogenic amines [2]) [2]

6 Stability

Temperature stability
20-60 <1> (<1> full activity after incubation for 20 min [4]) [4]

General stability information
<1>, stable against guanidine hydrochloride up to 2 M [4]

References

[1] Nozaki, M.: Aromatic amine dehydrogenase from Alcaligenes faecalis. Methods Enzymol., **142**, 650-655 (1987)
[2] Cuskey, S.M.; Peccoraro, V.; Olsen, R.H.: Initial catabolism of aromatic biogenic amines by Pseudomonas aeruginosa PAO: pathway description, mapping of mutations, and cloning of essential genes. J. Bacteriol., **169**, 2398-
2404 (1987)

[3] Iwaki, M.; Yagi, T.; Horiike, K.; Saeki, Y.; Ushijima, T.; Nozaki, M.: Crystallization and properties of aromatic amine dehydrogenase from Pseudomonas sp. Arch. Biochem. Biophys., **220**, 253-262 (1983)

[4] Govindaraj, S.; Eisenstein, E.; Jones, L.H.; Sanders-Loehr, J.; Chistoserdov, A.Y.; Davidson, V.L.; Edwards, S.L.: Aromatic amine dehydrogenase, a second tryptophan tryptophylquinone enzyme. J. Bacteriol., **176**, 2922-2929 (1994)

[5] Hyun, Y.L.; Davidson, V.L.: Electron transfer reactions between aromatic amine dehydrogenase and azurin. Biochemistry, **34**, 12249-12254 (1995)

[6] Hyun, Y.L.; Davidson, V.L.: Unusually large isotope effect for the reaction of aromatic amine dehydrogenase. A common feature of quinoproteins?. Biochim. Biophys. Acta, **1251**, 198-200 (1995)

[7] Hyun, Y.L.; Zhu, Z.; Davidson, V.L.: Gated and ungated electron transfer reactions from aromatic amine dehydrogenase to azurin. J. Biol. Chem., **274**, 29081-29086 (1999)

[8] Chistoserdov, A.Y.: Cloning, sequencing and mutagenesis of the genes for aromatic amine dehydrogenase from Alcaligenes faecalis and evolution of amine dehydrogenases. Microbiology, **147**, 2195-2202 (2001)

Glycine dehydrogenase (cyanide-forming) 1.4.99.5

1 Nomenclature

EC number
1.4.99.5

Systematic name
glycine:acceptor oxidoreductase (hydrogen-cyanide-forming)

Recommended name
glycine dehydrogenase (cyanide-forming)

Synonyms
HCA <5> [2]
HCN synthase
HCS <5> [2]
cyanogenic enzyme system <5> [2]
hydrogen cyanide synthase
synthase, hydrogen cyanide (9Cl)

CAS registry number
210488-48-5

2 Source Organism

<1> *Anacystis nidulans* [5]
<2> *Chromobacterium violaceum* [1-5]
<3> *Nostoc muscorum* [5]
<4> *Plectonema boryanum* [5]
<5> *Pseudomonas aeruginosa* (strain 9-D2 [1, 2]; wild-type PAO1 [3, 4]; strain ADD1976 [3]; strain PAO1A4 [4]) [1-5]
<6> *Pseudomonas aureofaciens* [5]
<7> *Pseudomonas chlororaphis* [5]
<8> *Pseudomonas fluorescens* (strain CHA0 [3-5]; mutant CHA5, CHA21, CHA77 [3]; wild-type P3 [3]) [3-5]
<9> *Pseudomonas sp.* [3]
<10> *Rhizobium leguminosarum* (produces HCN as free-living bacterium [5]) [5]

3 Reaction and Specificity

Catalyzed reaction

glycine + 2 A + H^+ = HCN + CO_2 + 2 AH_2 (The enzyme from Pseudomonas sp. contains FAD. The enzyme is membrane-bound, and the 2-electron acceptor is a component of the respiratory chain. The enzyme can act with various artificial electron acceptors, including phenazine methosulfate)

Natural substrates and products

S   glycine + O_2 <1-10> (Reversibility: ? <1-10> [1-5]) [1-5]
P   HCN + CO_2 + 2 H_2O

Substrates and products

S   glycine + 2,2'-di-(p-nitrophenyl-5,5')-(diphenyl)-3,3'-(3,3'-dimethoxy-4,4'-diphenylene) ditetrazolium chloride <5> (<5> NBT [2]) (Reversibility: ? <5> [2]) [2]
P   HCN + ?
S   glycine + 2,6-dichlorophenol-indophenol <5> (<5> DCIP [2]) (Reversibility: ? <5> [2]) [2]
P   HCN + CO_2 + ?
S   glycine + 2-(p-iodophenyl)-3-(p-nitrophenyl)-5-phenyl-tetrazolium chloride <5> (<5> INT [2]) (Reversibility: ? <5> [2]) [2]
P   HCN + CO_2 +?
S   glycine + O_2 <1-10> (<5> dissolved oxygen regulates the synthesis of the cyanogenic enzyme system [1]) (Reversibility: ? <1-10> [1-5]) [1-5]
P   HCN + CO_2 + H_2O
S   glycine + phenazine methosulfate <5, 9> (<5> PMS provided greatest activity, oxygen allows only a limited response [2]) (Reversibility: ? <5, 9> [2, 3]) [2, 3]
P   HCN + CO_2 + ?
S   glycine + potassium ferricyanide <5> (Reversibility: ? <5> [2]) [2]
P   HCN + CO_2 + ?
S   glycine + pyocyanin <5> (Reversibility: ? <5> [2]) [2]
P   HCN + ?
S   Additional information <5> (<5> sodium nitrate, sodium nitrite, NAD^+, $NADP^+$, FAD and FMN fail as electron acceptors to support cyanide biosynthesis [2]) [2]
P   ?

Inhibitors

O_2 <5, 8, 9> (<8,9> very sensitive to molecular oxygen, becomes rapidly inactivated in presence [3,4]) [1-4]
chloramphenicol <5> [1]
pyrrolnitrin <9> (<9> strongly inhibits cyanide formation in vitro [3]) [3]

Activating compounds

ANR <8> (<8> anaerobic regulator required for maximal expression of the HCN biosynthetic genes [4,5]) [4, 5]

FAD <9> (<9> stimulates the reaction [3]) [3]

GacA <8> (<8> global activator required for maximal expression of the HCN biosynthetic genes [5]) [5]

pyocyanin <5> (<5> secondary metabolite, supports HCS mediated cyanide production [2]) [2]

Metals, ions

Fe^{2+} <8> (<8> iron regulation of the hcnABC genes encoding hydrogen cyanide synthase, iron limitation has negative effect on hcn expression [4,5]) [5]

5 Isolation/Preparation/Mutation/Application

Localization

membrane <5, 8> (<5,8> membrane-bound flavoenzyme [3,5]) [3, 5]

Purification

<5> (partially [2,3]) [2, 3]

<9> (partially [3]) [3]

Cloning

<8> (hcnABC gene cluster encoding hydrogen cyanide synthase, heterologous expressed in Escherichia coli [3-5]) [3-5]

6 Stability

Oxidation stability

<5>, extremely unstable, oxygen inactivation, protected from inactivation by anaerobic conditions [1]

<5>, synthesis of HCN is repressed under highly aerobic conditions, cyanogenic enzyme system is rapidly inactivated, both in vivo and in vitro, in the presence of oxygen [2]

General stability information

<5>, extremely unstable [1]

References

[1] Castric, K.F.; McDevitt, D.A.; Castric, P.A.: Influence of aeration on hydrogen cyanide biosynthesis b[Pseudomonas aeruginosa. Curr. Microbiol., 5, 223-226 (1981)

[2] Castric, P.: Influence of oxygen on the Pseudomonas aeruginosa hydrogen cyanide synthase. Curr. Microbiol., 29, 19-21 (1994)

[3] Laville, J.; Blumer, C.; von Schroetter, C.; Gaia, V.; Defago, G.; Keel, C.; Haas, D.: Characterization of the HCNABC gene cluster encoding hydrogen cyanide synthase and anaerobic regulation by ANR in the strictly aerobic bio-

control agent Pseudomonas fluorescens CHA0. J. Bacteriol., **180**, 3187-3196 (1998)

[4] Blumer, C.; Haas, D.: Iron regulation of the hcnABC genes encoding hydrogen cyanide synthase depends on the anaerobic regulator ANR rather than on the global activator GacA in Pseudomonas fluorescens CHA0. Microbiology, **146**, 2417-2424 (2000)

[5] Blumer, C.; Haas, D.: Mechanism, regulation, and ecological role of bacterial cyanide biosynthesis. Arch. Microbiol., **173**, 170-177 (2000)